MATHEMATICS FOR THE NONMATHEMATICIAN

MATHEMATICS FOR THE NONMATHEMATICIAN

MORRIS KLINE
Professor of Mathematics, Emeritus
Courant Institute of Mathematical Sciences
New York University

DOVER PUBLICATIONS, INC.
NEW YORK

Published in Canada by General Publishing Company, Ltd., 30 Lesmill Road, Don Mills, Toronto, Ontario.
Published in the United Kingdom by Constable and Company, Ltd., 10 Orange Street, London WC2H 7EG.

This Dover edition, first published in 1985, is an unabridged republication of the work first published by Addison-Wesley Publishing Company, Inc., Reading, Massachusetts, in 1967 under the title *Mathematics for Liberal Arts*. The Instructor's Manual published with the original edition, containing additional answers and solutions to the problems in the text, has been added to this edition.

Manufactured in the United States of America
Dover Publications, Inc., 31 East 2nd Street, Mineola, N.Y. 11501

Library of Congress Cataloging in Publication Data

Kline, Morris, 1908–
Mathematics for the nonmathematician.

Reprint. Originally published: Mathematics for liberal arts. Reading, Mass.: Addison-Wesley, © 1967. (Addison-Wesley series in introductory mathematics)
Includes bibliographies and index.
1. Mathematics—1961– I. Title.
QA37.2.K6 1985 510 84-25923
ISBN 0-486-24823-2

PREFACE

"*. . . I consider that without understanding as much of the abstruser part of geometry, as Archimedes or Apollonius, one may understand enough to be assisted by it in the contemplation of nature; and that one needs not know the profoundest mysteries of it to be able to discern its usefulness. . . . I have often wished that I had employed about the speculative part of geometry, and the cultivation of the specious [symbolic] algebra I had been taught very young, a good part of that time and industry that I spent about surveying and fortification. . . .*"

ROBERT BOYLE

I believe as firmly as I have in the past that a mathematics course addressed to liberal arts students must present the scientific and humanistic import of the subject. Whereas mathematics proper makes little appeal and seems even less pointed to most of these students, the subject becomes highly significant to them when it is presented in a cultural context. In fact, the branches of elementary mathematics were created primarily to serve extra-mathematical needs and interests. In the very act of meeting such needs each of these creations has proved to have inestimable importance for man's understanding of the nature of his world and himself.

That so many professors have chosen to teach mathematics as an integral part of Western culture, as evidenced by their reception of my earlier book, *Mathematics: A Cultural Approach*, has been extremely gratifying. That book will continue to be available. In the present revision and abridgment, which has been designed to meet the needs of particular groups of students, the spirit of the original text has been preserved. The historical approach has been retained because it is intrinsically interesting, provides motivation for the introduction of various topics, and gives coherence to the body of material. Each topic or branch of mathematics dealt with is shown to be a response to human interests, and the cultural import of the technical development is presented. I adhered to the principle that the level of rigor should be suited to the mathematical age of the student rather than to the age of mathematics.

As in the earlier text, several of the topics are treated quite differently from what is now fashionable. These are the real number system, logic, and set theory. I tried to present these topics in a context and with a level of emphasis which I believe to be appropriate for an elementary course in mathematics. In this book, the axiomatic approach to the real numbers is formulated after the various types of numbers and their properties are derived from physical situations and uses. The treatment of logic is confined to the fundamentals of Aristotelian logic. And set theory serves as an illustration of a different kind of algebra.

The changes made in this revision are intended to suit special groups. Some students need more review and drill on elementary concepts and techniques than the earlier book provides. Others, chiefly those preparing for teaching on the elementary level, need to learn more about elementary mathematics than their high school courses covered. Teachers of twelfth-year high school courses and one-semester college courses often found the extensive amount of material in *Mathematics: A Cultural Approach* rather disconcerting because it offered so much more than could be covered.

To meet the needs of these groups I have made the following changes:

1. Four of the chapters devoted entirely to cultural influences have been dropped. The size of the original book has thereby been reduced considerably.
2. A few applications of mathematics to science have been omitted, primarily to reduce the size of the text.
3. Some of the chapters on technical topics, Chapter 3 on logic and mathematics, Chapter 4 on number, Chapter 5 on elementary algebra, and Chapter 21 on arithmetics and their algebras have been expanded.
4. Additional drill exercises have been added within a few chapters, and a set of review exercises providing practice in technique has been added to each of a number of chapters.
5. Improvements in presentation have been made in a number of places.

With respect to use in courses, it is probably true of the present text, as it is of the earlier one, that it contains more material than can be covered in some courses. However, many of the chapters as well as sections in chapters are not essential to the logical continuity. These chapters and sections have been starred (✱). Thus Chapter 10 on painting shows historically how mathematicians were led to projective geometry (Chapter 11), but from a logical standpoint, Chapter 10 is not needed in order to understand the succeeding chapter. Chapter 19 on musical sounds is an application of the material on the trigonometric functions in Chapter 18 but is not essential to the continuity. The two chapters on the calculus are not used in the succeeding chapters. Desirable as it may be to give students some idea of what the calculus is about, it may still be necessary in some classes to omit these chapters. The same can be said of the chapters on statistics (Chapter 22) and probability (Chapter 23).

As for sections within chapters, Chapter 6 on Euclidean geometry may well serve as an illustration. The mathematical material of this chapter is intended as a review of some basic ideas and theorems of Euclidean geometry and as an introduction to the conic sections. Some of the familiar applications are given in Section 6–3 (see the Table of Contents) and probably should be taken up. However the applications to light in Sections 6–4 and 6–6 and the discussion of cultural influences in Section 6–7 can be omitted.

Some of the material, whether or not included in the following recommendations for particular groups, can be left to student reading. In fact, the first two chapters were deliberately fashioned so that they could be *read* by students. The objective here, in addition to presenting intrinsically important ideas, was to induce students to read a mathematics book, to give them the confidence to do so, and to get them into the habit of doing so. It seems necessary to counter the students' impression, resulting no doubt from elementary and high school instruction in mathematics, that whereas history texts are to be read, mathematics texts are essentially reference books for formulas and homework exercises.

For courses emphasizing the number concept and its extension to algebra, it is possible to take advantage of the logical independence of numerous chapters and use Chapters 3 through 5 on reasoning, arithmetic, and algebra and Chapter 21 on different algebras. To pursue the development of this theme into the area of functions one can include Chapters 13 and 15.

Courses emphasizing geometry can utilize Chapters 6, 7, 11, 12, and 20 on Euclidean geometry, trigonometry, projective geometry, coordinate geometry, and non-Euclidean geometry respectively. Some algebra, that reviewed in Chapter 5, is involved in Chapters 7 and 12. If knowledge of the material of Chapter 5 cannot be presupposed, this chapter must precede the treatment of geometry.

The essence of the two preceding suggestions may be diagrammed thus:

Chapters 3, 4, 5

↓ Algebra — Chapters 21, 13, 15

↓ Geometry — Chapters 6, 7, 11, 12, 20

Of course, starred sections in these chapters are optional.

For a one-semester liberal arts course, the basic content can be as follows:

Chapter 2	on a historical orientation,
Chapter 3	on logic and mathematics,
Chapters 4 and 5	on the number system and elementary algebra,
Chapter 6	through Section 6–5, on Euclidean geometry,
Chapter 7	through Section 7–3, on trigonometry,
Chapter 12	on coordinate geometry,
Chapter 13	on functions and their uses,
Chapter 14	through Section 14–4, on parametric equations,
Chapter 15	through Section 15–10, on the further use of functions in science,
Chapter 20	on non-Euclidean geometry,
Chapter 21	on different algebras.

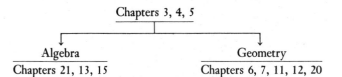

Any additional material would enrich the course but would not be needed for continuity.

Though the teacher's problem of presenting material outside the domain of mathematics proper is far simpler with this text than with the earlier one, it may still be necessary to assure him that he need not hesitate to undertake this task. The feeling that one must be an authority in a subject to say anything about it is unfounded. We are all laymen outside the field of our own specialty, and we should not be ashamed to point this out to students. In contiguous areas we are merely giving indications of ideas that students may pursue further in other courses or in independent reading.

I hope that this text will serve the needs of the groups of students to which it is addressed and that, despite the somewhat greater emphasis on technical matters, it will convey the rich significance of mathematics.

I wish to thank my wife Helen for her critical scrutiny of the contents and her conscientious reading of the proofs. I wish to express, also, my thanks to members of the Addison-Wesley staff for very helpful suggestions and for their continuing support of a culturally oriented approach to mathematics.

New York, 1967 M.K.

CONTENTS

MATHEMATICS FOR THE NONMATHEMATICIAN

WHY MATHEMATICS?

In mathematics I can report no deficience, except it be that men do not sufficiently understand the excellent use of the Pure Mathematics. . . .

FRANCIS BACON

One can wisely doubt whether the study of mathematics is worth while and can find good authority to support him. As far back as about the year 400 A.D., St. Augustine, Bishop of Hippo in Africa and one of the great fathers of Christianity, had this to say:

> *The good Christian should beware of mathematicians and all those who make empty prophecies. The danger already exists that the mathematicians have made a covenant with the devil to darken the spirit and to confine man in the bonds of Hell.*

Perhaps St. Augustine, with prophetic insight into the conflicts which were to arise later between the mathematically minded scientists of recent centuries and religious leaders, was seeking to discourage the further development of the subject. At any rate there is no question as to his attitude.

At about the same time that St. Augustine lived, the Roman jurists ruled, under the Code of Mathematicians and Evil-Doers, that "to learn the art of geometry and to take part in public exercises, an art as damnable as mathematics, are forbidden."

Even the distinguished seventeenth-century contributor to mathematics, Blaise Pascal, decided after studying mankind that the pure sciences were not suited to it. In a letter to Fermat written on August 10, 1660, Pascal says: "To speak freely of mathematics, I find it the highest exercise of the spirit; but at the same time I know that it is so useless that I make little distinction between a man who is only a mathematician and a common artisan. Also, I call it the most beautiful profession in the world; but it is only a profession; and I have often said that it is good to make the attempt [to study mathematics], but not to use our forces: so that I would not take two steps for mathematics, and I am confident that you are strongly of my opinion." Pascal's famous injunction was, "Humble thyself, impotent reason."

The philosopher Arthur Schopenhauer, who despised mathematics, said many nasty things about the subject, among others that the lowest activity of the spirit is arithmetic, as is shown by the fact that it can be performed by a machine. Many other great men, for example, the poet Johann Wolfgang Goethe and the historian Edward Gibbon, have felt likewise and have not hesitated to express themselves. And so the student who dislikes the subject can claim to be in good, if not living, company.

In view of the support he can muster from authorities, the student may well inquire why he is asked to learn mathematics. Is it because Plato, some 2300 years ago, advocated mathematics to train the mind for philosophy? Is it because the Church in medieval times taught mathematics as a preparation for theological reasoning? Or is it because the commercial, industrial, and scientific life of the Western world needs mathematics so much? Perhaps the subject got into the curriculum by mistake, and no one has taken the trouble to throw it out. Certainly the student is justified in asking his teacher the very question which Mephistopheles put to Faust:

Is it right, I ask, is it even prudence,
To bore thyself and bore the students?

Perhaps we should begin our answers to these questions by pointing out that the men we cited as disliking or disapproving of mathematics were really exceptional. In the great periods of culture which preceded the present one, almost all educated people valued mathematics. The Greeks, who created the modern concept of mathematics, spoke unequivocally for its importance. During the Middle Ages and in the Renaissance, mathematics was never challenged as one of the most important studies. The seventeenth century was aglow not only with mathematical activity but with popular interest in the subject. We have the instance of Samuel Pepys, so much attracted by the rapidly expanding influence of mathematics that at the age of thirty he could no longer tolerate his own ignorance and begged to learn the subject. He began, incidentally, with the multiplication table, which he subsequently taught to his wife. In 1681 Pepys was elected president of the Royal Society, a post later held by Isaac Newton.

In perusing eighteenth-century literature, one is struck by the fact that the journals which were on the level of our *Harper's* and the *Atlantic Monthly* contained mathematical articles side by side with literary articles. The educated man and woman of the eighteenth century knew the mathematics of their day, felt obliged to be *au courant* with all important scientific developments, and read articles on them much as modern man reads articles on politics. These people were as much at home with Newton's mathematics and physics as with Pope's poetry.

The vastly increased importance of mathematics in our time makes it all the more imperative that the modern person know something of the nature and role of mathematics. It is true that the role of mathematics in our civilization is not always obvious, and the deeper and more complex modern applications are not readily comprehended even by specialists. But the essential nature and accomplishments of the subject can still be understood.

Perhaps we can see more easily why one should study mathematics if we take a moment to consider what mathematics is. Unfortunately the answer cannot be given in a single sentence or a single chapter. The subject has many facets or, some might say, is Hydra-headed. One can look at mathematics as a language, as a particular kind of logical structure, as a body of knowledge about number and space, as a series of methods for deriving conclusions, as the essence of our knowledge of the physical world, or merely as an amusing intellectual activity. Each of these features would in itself be difficult to describe accurately in a brief space.

Because it is impossible to give a concise and readily understandable definition of mathematics, some writers have suggested, rather evasively, that mathematics is what mathematicians do. But mathematicians are human beings, and most of the things they do are uninteresting and some, embarrassing to relate. The only merit in this proposed definition of mathematics is that it points up the fact that mathematics is a human creation.

A variation on the above definition which promises more help in understanding the nature, content, and values of mathematics, is that mathematics is what *mathematics does*. If we examine mathematics from the standpoint of what it is intended to and does accomplish, we shall undoubtedly gain a truer and clearer picture of the subject.

Mathematics is concerned primarily with what can be accomplished by reasoning. And here we face the first hurdle. Why should one reason? It is not a natural activity for the human animal. It is clear that one does not need reasoning to learn how to eat or to discover what foods maintain life. Man knew how to feed, clothe, and house himself millenniums before mathematics existed. Getting along with the opposite sex is an art rather than a science mastered by reasoning. One can engage in a multitude of occupations and even climb high in the business and industrial world without much use of reasoning and certainly without mathematics. One's social position is hardly elevated by a display of his knowledge of trigonometry. In fact, civilizations in which reasoning and mathematics played no role have endured and even flourished. If one were willing to reason, he could readily supply evidence to prove that reasoning is a dispensable activity.

Those who are opposed to reasoning will readily point out other methods of obtaining knowledge. Most people are in fact convinced that their senses are really more than adequate. The very common assertion "seeing is believing" expresses the common reliance upon the senses. But everyone should recognize that the senses are limited and often fallible and, even where accurate, must be interpreted. Let us consider, as an example, the sense of sight.

How big is the sun? Our eyes tell us that it is about as large as a rubber ball. This then is what we should believe. On the other hand, we do not see the air around us, nor for that matter can we feel, touch, smell, or taste it. Hence we should not believe in the existence of air.

To consider a somewhat more complicated situation, suppose a teacher should hold up a fountain pen and ask, What is it? A student coming from some primitive society might call it a shiny stick, and indeed this is what the eyes see. Those who call it a fountain pen are really calling upon education and experience stored in their minds. Likewise, when we look at a tall building from a distance, it is experience which tells us that the building is tall. Hence the old saying that "we are prone to see what lies behind our eyes, rather than what appears before them."

Every day we see the sun where it is not. For about five minutes before what we call sunset, the sun is actually below the geometrical horizon and should therefore be invisible. But the rays of light from the sun curve toward us as they travel in the *earth's* atmosphere, and the observer at P (Fig. 1–1) not only "sees" the sun but thinks the light is coming from the direction $O'P$. Hence he believes the sun is in that direction.

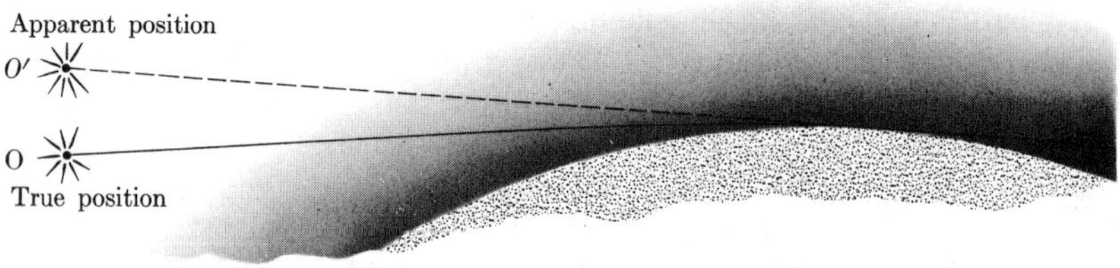

Fig. 1–1. Deviation of a ray by the earth's atmosphere.

The senses are obviously helpless in obtaining some kinds of knowledge, such as the distance to the sun, the size of the earth, the speed of a bullet (unless one wishes to feel its velocity), the temperature of the sun, the prediction of eclipses, and dozens of other facts.

If the senses are inadequate, what about experimentation or, in simple cases, measurement? One can and in fact does learn a great deal by such means. But suppose one wants to find a very simple quantity, the area of a rectangle. To obtain it by measurement, one could lay off unit squares to cover the area and then count the number of squares. It is at least a little simpler to measure the lengths of the sides and then use a formula obtained by reasoning, namely, that the area is the product of length and width. In the only slightly more complicated problem of determining how high a projectile will go, we should certainly not consider traveling with the projectile.

As to experimentation, let us consider a relatively simple problem of modern technology. One wishes to build a bridge across a river. How long and how thick should the many beams be? What shape should the bridge take? If it is to be supported by cables, how long and how thick should these be? Of course one could arbitrarily choose a number of lengths and thicknesses for the beams and cables and build the bridge. In this event, it would only be fair that the experimenter be the first to cross this bridge.

It may be clear from this brief discussion that the senses, measurement, and experimentation, to consider three alternative ways of acquiring knowledge, are by no means adequate in a variety of situations. Reasoning is essential. The lawyer, the doctor, the scientist, and the engineer employ reasoning daily to derive knowledge that would otherwise not be obtainable or perhaps obtainable only at great expense and effort. Mathematics more than any other human endeavor relies upon reasoning to produce knowledge.

One may be willing to accept the fact that mathematical reasoning is an effective procedure. But just what does mathematics seek to accomplish with its reasoning? The primary objective of all mathematical work is to help man study nature, and in this endeavor mathematics cooperates with science. It may seem, then, that mathematics is merely a useful tool and that the real pursuit is science. We shall not attempt at this stage to separate the roles of mathematics and science and to evaluate the relative merits of their contributions. We shall simply state that their methods are different and that mathematics is at least an equal partner with science.

We shall see later how observations of nature are framed in statements called axioms. Mathematics then discloses by reasoning secrets which nature may never have intended to reveal. The determination of the pattern of motion of celestial bodies, the discovery and control of radio waves, the understanding of molecular, atomic, and nuclear structures, and the creation of artificial satellites are a few basically mathematical achievements. Mathematical formulation of physical data and mathematical methods of deriving new conclusions are today the substratum in all investigations of nature.

The fact that mathematics is of central importance in the study of nature reveals almost immediately several values of this subject. The first is the practical value. The construction of bridges and skyscrapers, the harnessing of the power of water, coal, electricity, and the atom, the effective employment of light, sound, and radio in illumination, communication, navigation, and even entertainment, and the advantageous employment of chemical knowledge in the design of materials, in the production of useful forms of oil, and in medicine are but a few of the practical achievements already attained. And the future promises to dwarf the past.

However, material progress is not the most compelling reason for the study of nature, nor have practical results usually come about from investigations so directed. In fact, to overemphasize practical values is to lose sight of the greater significance of human thought. The deeper reason for the study of nature is to try to understand the ways of nature, that is, to satisfy sheer intellectual curiosity. Indeed, to ask disinterested questions about nature is one of the distinguishing marks of mankind. In all civilizations some people at least have tried to answer such questions as: How did the universe come about? How old is the universe and the earth in particular? How large are the sun and the earth? Is man an accident or part of a larger design? Will the solar system continue to function or will the earth some day fall into the sun? What is light? Of course, not all people are interested in such questions. Food, shelter, sex, and television are enough to keep many happy. But others, aware of the pervasive natural mysteries, are more strongly obsessed to resolve them than any business man is to acquire wealth and power.

Beyond improvement in the material life of man and beyond satisfaction of intellectual curiosity, the study of nature offers intangible values of another sort, especially the abolition of fear and terror and their replacement by a deep, quiet satisfaction in the ways of nature. To the uneducated and to those uninitiated in the world of science, many manifestations of nature have appeared to be agents of destruction sent by angry gods. Some of the beliefs in ancient and even medieval Europe may be of special interest in view of what happened later. The sun was the center of all life. As winter neared and the days became shorter, the people believed that a battle between the gods of light and darkness was taking place. Thus the god Wodan was supposed to be riding through heaven on a white horse followed by demons, all of whom sought every opportunity to harm people. When, however, the days began to lengthen and the sun began to show itself higher in the sky each day, the people believed that the gods of light had won. They ceased all work and celebrated this victory. Sacrifices were offered to the benign gods. Symbols of fertility such as fruit and nuts, whose growth is, of course, aided by the sun, were placed on the altars. To symbolize further the desire for light and the joy in light, a huge log was placed in the fire to burn for twelve days, and candles were lit to heighten the brightness.

The beliefs and superstitions which have been attached to events we take in stride are incredible to modern man. An eclipse of the sun, a threat to the continuance of the light and heat which causes crops to grow, meant that the heavenly body was being swallowed up by a dragon. Many Hindu people believe today that a demon residing in the sky attacks the sun once in a while and that this is what causes the eclipse. Of course, when prayers, sacrifices, and ceremonies were followed by the victory of the sun or moon, it was clear that these rituals were the effective agent and so had to be pursued on every such occasion. In addition, special magic potions drunk during eclipses insured health, happiness, and wisdom.

To primitive peoples of the past, thunder, lightning, and storms were punishments visited by the gods on people who had apparently sinned in some way. The stories in the Old Testament of the flood and of the destruction of Sodom and Gomorrah by fire and brimstone are examples of such acts of wrath by the God of the Hebrews. Hence there was continual concern and even dread about what the gods might have in mind for helpless humans. The only recourse was to propitiate the divine powers, so that they would bring good fortune instead of evil.

Fears, dread, and superstitions have been eliminated, at least in our Western civilization, by just those intellectually curious people who have studied nature's mighty displays. Those "seemingly unprofitable amusements of speculative brains" have freed us from serfdom, given us undreamed of powers, and, in fact, have replaced negative doctrines by positive mathematical laws which reveal a remarkable order and uniformity in nature. Man has emerged as the proud possessor of knowledge which has enabled him to view nature calmly and objectively. An eclipse of the sun occurring on schedule is no longer an occasion for trembling but for quiet satisfaction that we know nature's ways. We breathe freely, knowing that nature will not be willful or capricious.

Indeed, man has been remarkably successful in his study of nature. History is said to repeat itself, but, in general, the circumstances of the supposed repetition are not the same as those of the earlier occurrence. As a consequence, the history of man has not been too effective a guide for the future. Nature is kinder. When nature repeats herself, and she does so constantly, the repetitions are exact facsimiles of previous events, and therefore man can anticipate nature's behavior and be prepared for what will take place. We have learned to recognize the patterns of nature and we can speak today of the uniformity of nature and delight in the regularity of her behavior.

The successes of mathematics in the study of inanimate nature have inspired in recent times the mathematical study of human nature. Mathematics has not only contributed to the very practical institutions such as banking, insurance, pension systems, and the like, but it has also supplied some substance, spirit, and methodology to the infant sciences of economics, politics, and sociology. Number, quantitative studies, and precise reasoning have replaced vague, subjective, and ineffectual speculations and have already given evidence of greater values to come.

As man turns to thoughts about himself and his fellow man, other questions occur to him which are as fundamental as any he can ask. Why is man born? What purposes does he serve or should he serve? What future awaits him? The knowledge acquired about our physical universe has profound implications for the origin and role of man. Moreover, as mathematics and science have amassed increasing knowledge and power, they have gradually encompassed the biological and psychological sciences, which in turn have shed further light on man's physical and mental life. Thus it has come about that mathematics and science have profoundly affected philosophy and religion.

Perhaps the most profound questions in the realm of philosophy are, What is truth and how does man acquire it? Though we have no final answer to these questions, the contribution of mathematics toward this end is paramount.

For two millenniums mathematics was the prime example of truths man had unearthed. Hence all investigations of the problem of acquiring truths necessarily reckoned with mathematics. Though some startling developments in the nineteenth century altered completely our understanding of the nature of mathematics, the effectiveness of the subject, especially in representing and analyzing natural phenomena, has still kept mathematics the focal point of all investigations into the nature of knowledge. Not the least significant aspect of this value of mathematics has been the insight it has given us into the ways and powers of the human mind. Mathematics is the supreme and most remarkable example of the mind's power to cope with problems, and as such it is worthy of study.

Among the values which mathematics offers are its services to the arts. Most people are inclined to believe that the arts are independent of mathematics, but we shall see that mathematics has fashioned major styles of painting and architecture, and the service mathematics renders to music has not only enabled man to understand it, but has spread its enjoyment to all corners of our globe.

Practical, scientific, philosophical, and artistic problems have caused men to investigate mathematics. But there is one other motive which is as strong as any of these — the search for beauty. Mathematics is an art, and as such affords the pleasures which all the arts afford. This last statement may come as a shock to people who are used to the conventional concept of the true arts and mentally contrast these with mathematics to the detriment of the latter. But the average person has not thought through what the arts really are and what they offer. All that many people actually see in painting, for example, are familiar scenes and perhaps bright colors. These qualities, however, are not the ones which make painting an art. The real values must be learned, and a genuine appreciation of art calls for much study.

Nevertheless, we shall not insist on the aesthetic values of mathematics. It may be fairer to rest on the position that just as there are tone-deaf and color-blind people, so may there be some who temperamentally are intolerant of cold argumentation and the seemingly overfine distinctions of mathematics.

To many people, mathematics offers intellectual challenges, and it is well known that such challenges do engross humans. Games such as bridge, crossword puzzles, and magic squares are popular. Perhaps the best evidence is the attraction of puzzles such as the following: A wolf, a goat, and cabbage are to be transported across a river by a man in a boat which can hold only one of these in addition to the man. How can he take them across so that the wolf does not eat the goat or the goat the cabbage? Two husbands and two wives have to cross a river in a boat which can hold only two people. How can they cross so that no woman is in the company of a man unless her husband is also present? Such puzzles go back to Greek and Roman times. The mathematician Tartaglia, who lived in the sixteenth century, tells us that they were after-dinner amusements.

People do respond to intellectual challenges, and once one gets a slight start in mathematics, he encounters these in abundance. In view of the additional values to be derived from the subject, one would expect people to spend time on mathematical problems as opposed to the more superficial, and in some instances cheap, games which lack depth, beauty, and importance. The tantalizing and compelling pursuit of mathematical problems offers mental absorption, peace of mind amid endless challenges, repose in activity, battle without conflict, and the beauty which the ageless mountains present to senses tried by the kaleidoscopic rush of events. The appeal offered by the detachment and objectivity of mathematical reasoning is superbly described by Bertrand Russell.

> *Remote from human passions, remote even from the pitiful facts of nature, the generations have gradually created an ordered cosmos, where pure thought can dwell as in its natural home and where one, at least, of our nobler impulses can escape from the dreary exile of the actual world.*

The creation and contemplation of mathematics offer such values.

Despite all these arguments for the study of mathematics, the reader may have justifiable doubts. The idea that thinking about numbers and figures leads to deep and powerful conclusions which influence almost all other branches of thought may seem incredible. The study of numbers and geometrical figures may not seem a sufficiently attractive and promising enterprise. Not even the founders of mathematics envisioned the potentialities of the subject.

So we start with some doubts about the worth of our enterprise. We could encourage the reader with the hackneyed maxim, nothing ventured, nothing gained. We could call to his attention the daily testimony to the power of mathematics offered by almost every newspaper and journal. But such appeals are hardly inspiring. Let us proceed on the very weak basis that perhaps those more experienced in what the world has to offer may also have the wisdom to recommend worth-while studies.

Hence, despite St. Augustine, the reader is invited to tempt hell and damnation by engaging in a study of the subject. Certainly he can be assured that the subject is within his grasp and that no special gifts or qualities of mind are needed to learn mathematics. It is even debatable whether the creation of mathematics requires special talents as does the creation of music or great paintings, but certainly the appreciation of what others have done does not demand a "mathematical mind" any more than the appreciation of art requires an "artistic mind." Moreover, since we shall not draw upon any previously acquired knowledge, even this potential source of trouble will not arise.

Let us review our objectives. We should like to understand what mathematics is, how it functions, what it accomplishes for the world, and what it has to offer in itself. We hope to see that mathematics has content which serves the physical and social scientist, the philosopher, logician, and the artist; content which influences the doctrines of the statesman and the theologian;

content which satisfies the curiosity of the man who surveys the heavens and the man who muses on the sweetness of musical sounds; and content which has undeniably, if sometimes imperceptibly, shaped the course of modern history. In brief, we shall try to see that mathematics is an integral part of the modern world, one of the strongest forces shaping its thoughts and actions, and a body of living though inseparably connected with, dependent upon, and in turn valuable to all other branches of our culture. Perhaps we shall also see how by suffusing and influencing all thought it has set the intellectual temper of our times.

EXERCISES

1. A wolf, a goat, and a cabbage are to be rowed across a river in a boat holding only one of these three objects besides the oarsman. How should he carry them across so that the goat should not eat the cabbage or the wolf devour the goat?
2. Another hoary teaser is the following: A man goes to a tub of water with two jars, one holding 3 pt and the other 5 pt. How can he bring back exactly 4 pt?
3. Two husbands and two wives have to cross a river in a boat which can hold only two people. How can they cross so that no woman is in the company of a man unless her husband is also present?

Recommended Reading

RUSSELL, BERTRAND: "The Study of Mathematics," an essay in the collection entitled *Mysticism and Logic*, Longmans, Green and Co., New York, 1925.

WHITEHEAD, ALFRED NORTH: "The Mathematical Curriculum," an essay in the collection entitled *The Aims of Education*, The New American Library, New York, 1949.

WHITEHEAD, ALFRED NORTH: *Science and the Modern World*, Chaps. 2 and 3, Cambridge University Press, Cambridge, 1926.

CHAPTER 2

A HISTORICAL ORIENTATION

An educated mind is, as it were, composed of all the minds of preceding ages.

LE BOVIER DE FONTENELLE

2-1 INTRODUCTION

Our first objective will be to gain some historical perspective on the subject of mathematics. Although the logical development of mathematics is not markedly different from the historical, there are nevertheless many features of mathematics which are revealed by a glimpse of its history rather than by an examination of concepts, theorems, and proofs. Thus we may learn what the subject now comprises, how the various branches arose, and how the character of the mathematical contributions made by various civilizations was conditioned by these civilizations. This historical survey may also help us to gain some provisional understanding of the nature, extent, and uses of mathematics. Finally, a preview may help us to keep our bearings. In studying a vast subject, one is always faced with the danger of getting lost in details. This is especially true in mathematics, where one must often spend hours and even days in seeking to understand some new concepts or proofs.

2-2 MATHEMATICS IN EARLY CIVILIZATIONS

Aside possibly from astronomy, mathematics is the oldest and most continuously pursued branch of human thought. Moreover, unlike science, philosophy, and social thought, very little of the mathematics that has ever been created has been discarded. Mathematics is also a cumulative development, that is, newer creations are built logically upon older ones, so that one must usually understand older results to master newer ones. These facts recommend that we go back to the very origins of mathematics.

As we examine the early civilizations, one remarkable fact emerges immediately. Though there have been hundreds of civilizations, many with great art, literature, philosophy, religion, and social institutions, very few possessed any mathematics worth talking about. Most of these civilizations hardly got past the stage of being able to count to five or ten.

In some of these early civilizations a few steps in mathematics were taken. In prehistoric times, which means roughly before 4000 B.C., several civilizations at least learned to think about numbers as abstract concepts. That is, they recognized that three sheep and three arrows have something in common, a quantity called three, which can be thought about independently of any physical objects. Each of us in his own schooling goes through this same process of divorcing numbers from physical objects. The appreciation of "number" as an abstract idea is a great, and perhaps the first, step in the founding of mathematics.

Another step was the introduction of arithmetical operations. It is quite an idea to add the numbers representing two collections of objects in order to arrive at the total instead of counting the objects in the combined collections. Similar remarks apply to subtraction, multiplication, and division. The early methods of carrying out these operations were crude and complicated compared with ours, but the ideas and the applications were there.

Only a few ancient civilizations, Egypt, Babylonia, India, and China, possessed what may be called the rudiments of mathematics. The history of mathematics, and indeed the history of Western civilization, begins with what occurred in the first two of these civilizations. The role of India will emerge later, whereas that of China may be ignored because it was not extensive and moreover had no influence on the subsequent development of mathematics.

Our knowledge of the Egyptian and Babylonian civilizations goes back to about 4000 B.C. The Egyptians occupied approximately the same region that now constitutes modern Egypt and had a continuous, stable civilization from ancient times until about 300 B.C. The term "Babylonian" includes a succession of civilizations which occupied the region of modern Iraq. Both of these peoples possessed whole numbers and fractions, a fair amount of arithmetic, some algebra, and a number of simple rules for finding the areas and volumes of geometrical figures. These rules were but the incidental accumulations of experience, much as people learned through experience what foods to eat. Many of the rules were in fact incorrect but good enough for the simple applications made then. For example, the Egyptian rule for finding the area of a circle amounts to using 3.16 times the square of the radius; that is, their value of π was 3.16. This value, though not accurate, was even better than the several values the Babylonians used, one of these being 3, the value found in the Bible.

What did these early civilizations do with their mathematics? If we may judge from problems found in ancient Egyptian papyri and in the clay tablets of the Babylonians, both civilizations used arithmetic and algebra largely in commerce and state administration, to calculate simple and compound interest on loans and mortgages, to apportion profits of business to the owners, to buy and sell merchandise, to fix taxes, and to calculate how many bushels of grain would make a quantity of beer of a specified alcoholic content. Geometrical rules were applied to calculate the areas of fields, the estimated yield of pieces of land, the volumes of structures, and the quantity of bricks or stones needed to erect a temple or pyramid. The ancient Greek historian Herodotus says that because the annual overflow of the Nile wiped out the boundaries of the farmers' lands, geometry was needed to redetermine the boundaries. In fact, Herodotus speaks of geometry as the gift of the Nile. This bit of history is a partial truth. The redetermination of boundaries was undoubtedly an application, but geometry existed in Egypt long before the date of 1400 B.C. mentioned by Herodotus for its origin. Herodotus would have been more accurate to say that Egypt is a gift of the Nile, for it is true today as it was then that the only fertile land in Egypt is that along the Nile; and this because the river deposits good soil on the land as it overflows.

Applications of geometry, simple and crude as they were, did play a large role in Egypt and Babylonia. Both peoples were great builders. The Egyptian temples, such as those at Karnak and Luxor, and the pyramids still appear to be admirable engineering achievements even in this age of skyscrapers. The Babylonian temples, called ziggurats, also were remarkable pyramidal structures. The Babylonians were, moreover, highly skilled irrigation engineers, who built a system of canals to feed their hot dry lands from the Tigris and Euphrates rivers.

Perhaps a word of caution is necessary with respect to the pyramids. Because these are impressive structures, some writers on Egyptian civilization have jumped to the conclusion that the mathematics used in the building of pyramids must also have been impressive. These writers point out that the horizontal dimensions of any one pyramid are exactly of the same length, the sloping sides all make the same angle with the ground, and the right angles are right. However, not mathematics but care and patience were required to obtain such results. A cabinetmaker need not be a mathematician.

Mathematics in Egypt and Babylonia was also applied to astronomy. Of course, astronomy was pursued in these ancient civilizations for calendar reckoning and, to some extent, for navigation. The motions of the heavenly bodies give us our fundamental standard of time, and their positions at given times enable ships to determine their location and caravans to find their bearings in the deserts. Calendar reckoning is not only a common daily and commercial need, but it fixes religious holidays and planting times. In Egypt it was also needed to predict the flood of the Nile, so that farmers could move property and cattle away beforehand.

It is worthy of note that by observing the motion of the sun, the Egyptians managed to ascertain that the year contains 365 days. There is a conjecture that the priests of Egypt knew that $365\frac{1}{4}$ was a more accurate figure but kept the knowledge secret. The Egyptian calendar was taken over much later by the Romans and then passed on to Europe. The Babylonians, by contrast, developed a lunar calendar. Since the duration of the month as measured from new moon to new moon varies from 29 to 30 days, the twelve-month year adopted by the Babylonians did not coincide with the year of the seasons.

Hence the Babylonians added extra months, up to a total of seven, in every 19-year cycle. This scheme was also adopted by the Hebrews.

Astronomy served not only the purposes just described, but from ancient times until recently it also served astrology. In ancient Babylonia and Egypt the belief was widespread that the moon, the planets, and the stars directly influenced and even controlled affairs of the state. This doctrine was gradually extended and later included the belief that the health and welfare of the individual were also subject to the will of the heavenly bodies. Hence it seemed reasonable that by studying the motions and relative positions of these bodies man could determine their influences and even predict his future.

When one compares Egyptian and Babylonian accomplishments in mathematics with those of earlier and contemporary civilizations, one can indeed find reason to praise their achievements. But judged by other standards, Egyptian and Babylonian contributions to mathematics were almost insignificant, although these same civilizations reached relatively high levels in religion, art, architecture, metallurgy, chemistry, and astronomy. Compared with the accomplishments of their immediate successors, the Greeks, the mathematics of the Egyptians and Babylonians is the scrawling of children just learning how to write as opposed to great literature. They barely recognized mathematics as a distinct subject. It was a tool in agriculture, commerce, and engineering, no more important than the other tools they used to build pyramids and ziggurats. Over a period of 4000 years hardly any progress was made in the subject. Moreover, the very essence of mathematics, namely, reasoning to establish the validity of methods and results, was not even envisioned. Experience recommended their procedures and rules, and with this support they were content. Egyptian and Babylonian mathematics is best described as empirical and hardly deserves the appellation mathematics in view of what, since Greek times, we regard as the chief features of the subject. Some flesh and bones of concrete mathematics were there, but the spirit of mathematics was lacking.

The lack of interest in theoretical or systematic knowledge is evident in all activities of these two civilizations. The Egyptians and Babylonians must have noted the paths of the stars, planets, and moon for thousands of years. Their calendars, as well as tables which are extant, testify to the scope and accuracy of these observations. But no Egyptian or Babylonian strove, so far as we know, to encompass all these observations in one major plan or theory of heavenly motions. Nor does one find any other scientific theory or connected body of knowledge.

2-3 THE CLASSICAL GREEK PERIOD

We have seen so far that mathematics, initiated in prehistoric times, struggled for existence for thousands of years. It finally obtained a firm grip on life in the highly congenial atmosphere of Greece. This land was invaded about

1000 B.C. by people whose origins are not known. By about 600 B.C. these people occupied not only Greece proper but many cities in Asia Minor on the Mediterranean coast, islands such as Crete, Rhodes, and Samos, and cities in southern Italy and Sicily. Though all of these areas bred famous men, the chief cultural center during the classical period, which lasted from about 600 B.C. to 300 B.C., was Athens.

Greek culture was not entirely indigenous. The Greeks themselves acknowledge their indebtedness to the Babylonians and especially to the Egyptians. Many Greeks traveled in Egypt and in Asia Minor. Some went there to study. Nevertheless, what the Greeks created differs as much from what they took over from the Egyptians and Babylonians as gold differs from tin. Plato was too modest in his description of the Greek contribution when he said, "Whatever we Greeks receive we improve and perfect." The Greeks not only made finished products out of the raw materials imported from Egypt and Babylonia, but they created totally new branches of culture. Philosophy, pure and applied sciences, political thought and institutions, historical writings, almost all our literary forms (except fictional prose), and new ideals such as the freedom of the individual are wholly Greek contributions.

The supreme contribution of the Greeks was to call attention to, employ, and emphasize the power of human reason. This recognition of the power of reasoning is the greatest single discovery made by man. Moreover, the Greeks recognized that reason was the distinctive faculty which humans possessed. Aristotle says, "Now what is characteristic of any nature is that which is best for it and gives most joy. Such to man is the life according to reason, since it is that which makes him man."

It was by the application of reasoning to mathematics that the Greeks completely altered the nature of the subject. In fact, mathematics as we understand the term today is entirely a Greek gift, though in this case we need not heed Virgil's injunction to fear such benefactions. But how did the Greeks plan to employ reason in mathematics? Whereas the Egyptians and Babylonians were content to pick up scraps of useful information through experience or trial and error, the Greeks abandoned empiricism and undertook a systematic, rational attack on the whole subject. First of all, the Greeks saw clearly that numbers and geometric forms occur everywhere in the heavens and on earth. Hence they decided to concentrate on these important concepts. Moreover, they were explicit about their intention to treat general abstract concepts rather than particular physical realizations. Thus they would consider the ideal circle rather than the boundary of a field or the shape of a wheel. They then observed that certain facts about these concepts are both obvious and basic. It was evident that equal numbers added to or subtracted from equal numbers give equal numbers. It was equally evident that two right angles are necessarily equal and that a circle can be drawn when center and radius are given. Hence they selected some of these obvious facts as a starting point and called them *axioms*. Their next idea was to apply reasoning, with these facts as premises, and to use only the most reliable methods of reasoning man possesses. If the reasoning were successful, it would produce new knowledge. Also, since they were to reason about general concepts, their conclusions would apply to all objects of which the concepts were representative. Thus if they could prove that the area of a circle is π times the square of the radius, this fact would apply to the area of a circular field, the floor area of a circular temple, and the cross section of a circular tree trunk. Such reasoning about general concepts might not only produce knowledge of hundreds of physical situations in one proof, but there was always the chance that reasoning would produce knowledge which experience might never suggest. All these advantages the Greeks expected to derive from reasoning about general concepts on the basis of evident reliable facts. A neat plan, indeed!

It is perhaps already clear that the Greeks possessed a mentality totally different from that of the Egyptians and Babylonians. They reveal this also in the plans they had for the use of mathematics. The application of arithmetic and algebra to the computation of interest, taxes, or commercial transactions, and of geometry to the computation of the volumes of granaries was as far from their minds as the most distant star. As a matter of fact, their thoughts were on the distant stars. The Greeks found mathematics valuable in many respects, as we shall learn later, but they saw its main value in the aid it rendered to the study of nature; and of all the phenomena of nature, the heavenly bodies attracted them most. Thus, though the Greeks also studied light, sound, and the motions of bodies on the earth, astronomy was their chief scientific interest.

Just what did the Greeks seek in probing nature? They sought no material gain and no power over nature; they sought merely to satisfy their minds. Because they enjoyed reasoning and because nature presented the most imposing challenge to their understanding, the Greeks undertook the purely intellectual study of nature. Thus the Greeks are the founders of science in the true sense.

The Greek conception of nature was perhaps even bolder than their conception of mathematics. Whereas earlier and later civilizations viewed nature as capricious, arbitrary, and terrifying, and succumbed to the belief that magic and rituals would propitiate mysterious and feared forces, the Greeks dared to look nature in the face. They dared to affirm that nature was rationally and indeed mathematically designed, and that man's reason, chiefly through the aid of mathematics, would fathom that design. The Greek mind rejected traditional doctrines, supernatural causes, superstitions, dogma, authority, and other such trammels on thought and undertook to throw the light of reason on the processes of nature. In seeking to banish the mystery and seeming arbitrariness of nature and in abolishing belief in dreaded forces, the Greeks were pioneers.

For reasons which will become clearer in a later chapter, the Greeks favored geometry. By 300 B.C., Thales, Pythagoras and his followers, Plato's disciples, notably Eudoxus, and hundreds of other famous men had built up an enormous logical structure, most of which Euclid embodied in his *Elements*. This is, of course, the geometry we still study in high school. Though they made some contributions to the study of the properties of numbers and to the solution of equations, almost all of their work was in geometric form, and so there was no improvement over the Egyptians and Babylonians in the representation of, and calculation with, numbers or in the symbolism and techniques of algebra. For these contributions the world had to wait many more centuries. But the vast development in geometry exerted an enormous influence in succeeding civilizations and supplied the inspiration for mathematical activity in civilizations which might otherwise never have acquired even the very concept of mathematics.

The Greek accomplishments in mathematics had, in addition, the broader significance of supplying the first impressive evidence of the power of human reason to deduce new truths. In every culture influenced by the Greeks, this example inspired men to apply reason to philosophy, economics, political theory, art, and religion. Even today Euclid is the prime example of the power and accomplishments of reason. Hundreds of generations since Euclid's days have learned from his geometry what reasoning is and what it can accomplish. Modern man as well as the ancient Greeks learned from the Euclidean document how exact reasoning should proceed, how to acquire facility in it, and how to distinguish correct from false reasoning. Although many people depreciate this value of mathematics, it is interesting nevertheless that when these people seek to offer an excellent example of reasoning, they inevitably turn to mathematics.

This brief discussion of Euclidean geometry may show that the subject is far from being a relic of the dead past. It remains important as a stepping stone in mathematics proper and as a paradigm of reasoning. With their gift of reason and with their explicit example of the power of reason, the Greeks founded Western civilization.

2-4 THE ALEXANDRIAN GREEK PERIOD

The intellectual life of Greece was altered considerably when Alexander the Great conquered Greece, Egypt, and the Near East. Alexander decided to build a new capital for his vast empire and founded the city in Egypt named after him. The center of the new Greek world became Alexandria instead of Athens. Moreover, Alexander made deliberate efforts to fuse Greek and Near Eastern cultures. Consequently, the civilization centered at Alexandria, though predominantly Greek, was strongly influenced by Egyptian and Babylonian contributions. This Alexandrian Greek civilization lasted from about 300 B.C. to 600 A.D.

The mixture of the theoretical interests of the Greeks and the practical outlook of the Babylonians and Egyptians is clearly evident in the mathematical and scientific work of the Alexandrian Greeks. The purely geometric investigations of the classical Greeks were continued, and two of the most famous Greek mathematicians, Apollonius and Archimedes, pursued their studies during the Alexandrian period. In fact, Euclid also lived in Alexandria, but his writings reflect the achievements of the classical period. For practical applications, which usually require quantitative results, the Alexandrians revived the crude arithmetic and algebra of Egypt and Babylonia and used these empirically founded tools and procedures, along with results derived from exact geometrical studies. There was some progress in algebra, but what was newly created by men such as Nichomachus and Diophantus was still short of even the elementary methods we learn in high school.

The attempt to be quantitative, coupled with the classical Greek love for the mathematical study of nature, stimulated two of the most famous astronomers of all time, Hipparchus and Ptolemy, to calculate the sizes and distances of the heavenly bodies and to build a sound and, for those times, accurate astronomical theory, which is still known as Ptolemaic theory. Hipparchus and Ptolemy also created the chief tool they needed for this purpose, the mathematical subject known as trigonometry.

During the centuries in which the Alexandrian civilization flourished, the Romans grew strong, and by the end of the third century B.C. they were a world power. After conquering Italy, the Romans conquered the Greek mainland and a number of Greek cities scattered about the Mediterranean area. Among these was the famous city of Syracuse in Sicily, where Archimedes spent most of his life, and where he was killed at the age of 75 by a Roman soldier. According to the account given by the noted historian Plutarch, the soldier shouted to Archimedes to surrender, but the latter was so absorbed in studying a mathematical problem that he did not hear the order, whereupon the soldier killed him.

The contrast between Greek and Roman cultures is striking. The Romans have also bequeathed gifts to Western civilization, but in the fields of mathematics and science their influence was negative rather than positive. The Romans were a practical people and even boasted of their practicality. They sought wealth and world power and were willing to undertake great engineering enterprises, such as the building of roads and viaducts, which might help them to expand, control, and administer their empire, but they would spend no time or effort on theoretical studies which might further these activities. As the great philosopher Alfred North Whitehead remarked, "No Roman ever lost his life because he was absorbed in the contemplation of a mathematical diagram."

Indirectly as well as directly, the Romans brought about the destruction of the Greek civilization at Alexandria, directly by conquering Egypt and indirectly by seeking to suppress Christianity. The adherents to this new religious movement, though persecuted cruelly by the Romans, increased in number while the Roman Empire grew weaker. In 313 A.D. Rome legalized Christianity and, under the Emperor Theodosius (379–395), adopted it as the official religion of the empire. But even before this time, and certainly after it, the Christians began to attack the cultures and civilizations which had opposed them. By pillage and the burning of books, they destroyed all they could reach of ancient learning. Naturally the Greek culture suffered, and many works wiped out in these holocausts are now lost to us forever.

The final destruction of Alexandria in 640 A.D. was the deed of the Arabs. The books of the Greeks were closed, never to be reopened in this region.

2-5 THE HINDUS AND ARABS

The Arabs, who suddenly appeared on the scene of history in the role of destroyers, had been a nomadic people. They were unified under the leadership of the prophet Mohammed and began an attempt to convert the world to Mohammedanism, using the sword as their most decisive argument. They conquered all the land around the Mediterranean Sea. In the Near East they took over Persia and penetrated as far as India. In southern Europe they occupied Spain, southern France, where they were stopped by Charles Martel, southern Italy and Sicily. Only the Byzantine or Eastern Roman Empire was not subdued and remained an isolated center of Greek and Roman learning. In rather surprisingly quick time as the history of nations goes, the Arabs settled down and built a civilization and culture which maintained a high level from about 800 to 1200 A.D. Their chief centers were Bagdad in what is now Iraq, and Cordova in Spain. Realizing that the Greeks had created wonderful works in many fields, the Arabs proceeded to gather up and study what they could still find in the lands they controlled. They translated the works of Aristotle, Euclid, Apollonius, Archimedes, and Ptolemy into Arabic. In fact, Ptolemy's chief work, whose title in Greek meant "Mathematical Collection," was called the *Almagest* (The Greatest Work) by the Arabs and is still known by this name. Incidentally, other Arabic words which are now common mathematical terms are algebra, taken from the title of a book written by Al-Khowarizmi, a ninth-century Arabian mathematician, and algorithm, now meaning a process of calculation, which is a corruption of the man's name.

Though they showed keen interest in mathematics, optics, astronomy, and medicine, the Arabs contributed little that was original. It is also peculiar that, although they had at least some of the Greek works and could therefore see what mathematics meant, their own contributions, largely in arithmetic and algebra, followed the empirical, concrete approach of the Egyptians and Babylonians. They could on the one hand appreciate and critically review the precise, exact, and abstract mathematics of the Greeks while, on the other, offer methods of solving equations which, though they worked, had no reasoning to support them. During all the centuries in which Greek works were in their possession, the Arabs manfully resisted the lures of exact reasoning in their own contributions.

We are indebted to the Arabs not only for their resuscitation of the Greek works but for picking up some simple but useful ideas from India, their neighbor on the East. The Indians, too, had built up some elementary mathematics comparable in extent and spirit with the Egyptian and Babylonian developments. However, after about 200 A.D., mathematical activity in India became more appreciable, probably as a result of contacts with the Alexandrian Greek civilization. The Hindus made a few contributions of their own, such as the use of special number symbols from 1 to 9, the introduction of 0, and the use of positional notation with base ten, that is, our modern method of writing numbers. They also created negative numbers. These ideas were taken over by the Arabs and incorporated in their mathematical works.

Because of internal dissension the Arab Empire split into two independent parts. The Crusades launched by the Europeans and the inroads made by the Turks further weakened the Arabs, and their empire and culture disintegrated.

2-6 EARLY AND MEDIEVAL EUROPE

Thus far Europe proper has played no role in the history of mathematics. The reason is simple. The Germanic tribes who occupied central Europe and the Gauls of western Europe were barbarians. Among primitive civilizations, theirs were primitive indeed. They had no learning, no art, no science, not even a system of writing.

The barbarians were gradually civilized. While the Romans were still successful in holding the regions now called France, England, southern Germany, and the Balkans, the barbarians were in contact with, and to some extent influenced by, the Romans. When the Roman Empire collapsed, the Church, already a strong organization, took on the task of civilizing and converting the barbarians. Since the Church did not favor Greek learning and since at any rate the illiterate Europeans had first to learn reading and writing, one is not surprised to find that mathematics and science were practically unknown in Europe until about 1100 A.D.

2-7 THE RENAISSANCE

Insofar as the history of mathematics is concerned, the Arabs served as the agents of destiny. Trade with the Arabs and such invasions of the Arab lands as the Crusades acquainted the Europeans, who hitherto possessed only fragments of the Greek works, with the vast stores of Greek learning possessed by the Arabs. The ideas in these works excited the Europeans, and scholars set about acquiring them and translating them into Latin. Through another accident of history another group of Greek works came to Europe. We have already noted that the Eastern Roman or Byzantine Empire had survived the Germanic and the Arab aggrandizements. But in the fifteenth century the Turks captured the Eastern Roman Empire, and Greek scholars carrying precious manuscripts fled the region and went to Europe.

We shall leave for a later chapter a fuller account of how the European world was aroused by the renaissance of the novel and weighty Greek ideas, and of the challenge these ideas posed to the European beliefs and way of life.* From the Greeks the Europeans acquired arithmetic, a crude algebra, the vast development of Euclidean geometry, and the trigonometry of Hipparchus and Ptolemy. Of course, Greek science and philosophy also became known in Europe.

The first major European development in mathematics occurred in the work of the artists. Imbued with the Greek doctrines that man must study himself and the real world, the artists began to paint reality as they actually perceived it instead of interpreting religious themes in symbolic styles. They applied Euclidean geometry to create a new system of perspective which permitted them to paint realistically. Specifically, the artists created a new style of painting which enabled them to present on canvas, scenes making the same impression on the eye as the actual scenes themselves. From the work of the artists, the mathematicians derived ideas and problems that led to a new branch of mathematics, projective geometry.

Stimulated by Greek astronomical ideas, supplied with data and the astronomical theory of Hipparchus and Ptolemy, and steeped in the Greek doctrine that the world is mathematically designed, Nicolaus Copernicus sought to show that God had done a better job than Hipparchus and Ptolemy had described. The result of Copernicus' thinking was a new system of astronomy in which the sun was immobile and the planets revolved around the sun. This heliocentric theory was considerably improved by Kepler. Its effects on religion, philosophy, science, and on man's estimations of his own importance were profound. The heliocentric theory also raised scientific and mathematical problems which were a direct incentive to new mathematical developments.

Just how much mathematical activity the revival of Greek works might have stimulated cannot be determined, for simultaneously with the translation and absorption of these works, a number of other revolutionary developments altered the social, economic, religious, and intellectual life of Europe. The introduction of gunpowder was followed by the use of muskets and later cannons. These inventions revolutionized methods of warfare and gave the newly emerging social class of free common men an important role in that domain. The compass became known to the Europeans and made possible long-range navigation, which the merchants sponsored for the purpose of

* See Chapter 9.

finding new sources of raw materials and better trade routes. One result was the discovery of America and the consequent influx of new ideas into Europe. The invention of printing and of paper made of rags afforded books in large quantities and at cheap prices, so that learning spread far more than it ever had in any earlier civilizations. The Protestant Revolution stirred debate and doubts concerning doctrines that had been unchallenged for 1500 years. The rise of a merchant class and of free men engaged in labor in their own behalf stimulated an interest in materials, methods of production, and new commodities. All of these needs and influences challenged the Europeans to build a new culture.

2-8 DEVELOPMENTS FROM 1550 TO 1800

Since many of the problems raised by the motion of cannon balls, navigation, and industry called for quantitative knowledge, arithmetic and algebra became centers of attention. A remarkable improvement in these mathematical fields followed. This is the period in which algebra was built as a branch of mathematics and in which much of the algebra we learn in high school was created. Almost all the great mathematicians of the sixteenth and seventeenth centuries, Cardan, Tartaglia, Vieta, Descartes, Fermat, and Newton, men we shall get to know better later, contributed to the subject. In particular, the use of letters to represent a class of numbers, a device which gives algebra its generality and power, was introduced by Vieta. In this same period, logarithms were created to facilitate the calculations of astronomers. The history of arithmetic and algebra illustrates one of the striking and curious features of the history of mathematics. Ideas that seem remarkably simple once explained were thousands of years in the making.

The next development of consequence, coordinate geometry, came from two men, both interested in method. One was René Descartes. Descartes is perhaps even more famous as a philosopher than as a mathematician, though he was one of the major contributors to our subject. As a youth Descartes was already troubled by the intellectual turmoil of his age. He found no certainty in any of the knowledge taught him, and he therefore concentrated for years on finding the method by which man can arrive at truths. He found the clue to this method in mathematics, and with it constructed the first great modern philosophical system. Because the scientific problems of his time involved work with curves, the paths of ships at sea, of the planets, of objects in motion near the earth, of light, and of projectiles, Descartes sought a better method of proving theorems about curves. He found the answer in the use of algebra. Pierre de Fermat's interest in method was confined to mathematics proper, but he too appreciated the need for more effective ways of working with curves and also arrived at the idea of applying algebra. In this development of coordinate geometry we have one of the remarkable examples of how the times influence the direction of men's thoughts.

We have already noted that a new society was developing in Europe. Among its features were expanded commerce, manufacturing, mining, large-scale agriculture, and a new social class—free men working as laborers or as independent artisans. These activities and interests created problems of materials, methods of production, quality of the product, and utilization of devices to replace or increase the effectiveness of manpower. The people involved, like the artists, had become aware of Greek mathematics and science and sensed that it could be helpful. And so they too sought to employ this knowledge in their own behalf. Thereby arose a new motive for the study of mathematics and science. Whereas the Greeks had been content to study nature merely to satisfy their own curiosity and to organize their conclusions in patterns pleasing to the mind, the new goal, effectively proclaimed by Descartes and Francis Bacon, was to make nature serve man. Hence mathematicians and scientists turned earnestly to an enlarged program in which both understanding and mastery of nature were to be sought.

However, Bacon had cautioned that nature can be commanded only when one learns to obey her. One must have facts of nature on which to base reasoning about nature. Hence mathematicians and scientists sought to acquire facts from the experience of artists, technicians, artisans, and engineers. The alliance of mathematics and experience was gradually transformed into an alliance of mathematics and experimentation, and a new method for the pursuit of the truths of nature, first clearly perceived and formulated by Galileo Galilei (1564–1642) and Newton, was gradually evolved. The plan, perhaps oversimply stated, was that experience and experiment were to supply basic mathematical principles and mathematics was to be applied to these principles to deduce new truths, just as new truths are deduced from the axioms of geometry.

The most pressing scientific problem of the seventeenth century was the study of motion. On the practical side, investigations of the motion of projectiles, of the motion of the moon and planets to aid navigation, and of the motion of light to improve the design of the newly discovered telescope and microscope, were the primary interests. On the theoretical side, the new heliocentric astronomy invalidated the older, Aristotelian laws of motion and called for totally new principles. It was one thing to explain why a ball fell to earth on the assumption that the earth was immobile and the center of the universe, and another to explain this phenomenon in the light of the fact that the earth was rotating and revolving around the sun. A new science of motion was created by Galileo and Newton, and in the process two brand-new developments were added to mathematics. The first of these was the notion of a function, a relationship between variables best expressed for most purposes as a formula. The second, which rests on the notion of a function but represents the greatest advance in method and content since Euclid's days, was the calculus. The subject matter of mathematics and the power of mathematics expanded so greatly that at the end of the seventeenth century Leibniz could say,

> *Taking mathematics from the beginning of the world to the time when Newton lived, what he had done was much the better half.*

With the aid of the calculus Newton was able to organize all data on earthly and heavenly motions into one system of mathematical mechanics which encompassed the motion of a ball falling to earth and the motion of the planets and stars. This great creation produced universal laws which not only united heaven and earth but revealed a design in the universe far more impressive than man had ever conceived. Galileo's and Newton's plan of applying mathematics to sound physical principles not only succeeded in one major area but gave promise, in a rapidly accelerating scientific movement, of embracing all other physical phenomena.

We learn in history that the end of the seventeenth century and the eighteenth century were marked by a new intellectual attitude briefly described as the Age of Reason. We are rarely told that this age was inspired by the successes which mathematics, to be sure in conjunction with science, had achieved in organizing man's knowledge. Infused with the conviction that reason, personified by mathematics, would not only conquer the physical world but could solve all of man's problems and should therefore be employed in every intellectual and artistic enterprise, the great minds of the age undertook a sweeping reorganization of philosophy, religion, ethics, literature, and aesthetics. The beginnings of new sciences such as psychology, economics, and politics were made during these rational investigations. Our principal intellectual doctrines and outlook were fashioned then, and we still live in the shadow of the Age of Reason.

While these major branches of our culture were being transformed, eighteenth-century scientists continued to win victories over nature. The calculus was soon extended to a new branch of mathematics called differential equations, and this new tool enabled scientists to tackle more complex problems in astronomy, in the study of the action of forces causing motions, in sound, especially musical sounds, in light, in heat (especially as applied to the development of the steam engine), in the strength of materials, and in the flow of liquids and gases. Other branches, which can be merely mentioned, such as infinite series, the calculus of variations, and differential geometry, added to the extent and power of mathematics. The great names of the Bernoullis, Euler, Lagrange, Laplace, d'Alembert, and Legendre, belong to this period.

2-9 DEVELOPMENTS FROM 1800 TO THE PRESENT

During the nineteenth century, developments in mathematics came at an ever increasing rate. Algebra, geometry and analysis, the last comprising those subjects which stem from calculus, all acquired new branches. The great mathematicians of the century were so numerous that it is impractical to list them. We shall encounter some of the greatest of these, Karl Friedrich Gauss and Bernhard Riemann, in our work. We might mention also Henri Poincaré and David Hilbert, whose work extended into the twentieth century.

Undoubtedly the primary cause of this expansion in mathematics was the expansion in science. The progress made in the seventeenth and eighteenth centuries had sufficiently illustrated the effectiveness of science in penetrating the mysteries of the physical world and in giving man control over nature, to cause an all the more vigorous pursuit of science in the nineteenth century. In that century also, science became far more intimately linked with engineering and technology than ever before. Mathematicians, working closely with the scientists as they had since the seventeenth century, were presented with thousands of significant physical problems and responded to these challenges.

Perhaps the major scientific development of the century, which is typical in its stimulation of mathematical activity, was the study of electricity and magnetism. While still in its infancy this science yielded the electric motor, the electric generator, and telegraphy. Basic physical principles were soon expressed mathematically, and it became possible to apply mathematical techniques to these principles, to deduce new information just as Galileo and Newton had done with the principles of motion. In the course of such mathematical investigations, James Clerk Maxwell discovered electromagnetic waves of which the best known representatives are radio waves. A new world of phenomena was thus uncovered, all embraced in one mathematical system. Practical applications, with radio and television as most familiar examples, soon followed.

Remarkable and revolutionary developments of another kind also took place in the nineteenth century, and these resulted from a re-examination of elementary mathematics. The most profound in its intellectual significance was the creation of non-Euclidean geometry by Gauss. His discovery had both tantalizing and disturbing implications: tantalizing in that this new field contained entirely new geometries based on axioms which differ from Euclid's, and disturbing in that it shattered man's firmest conviction, namely that mathematics is a body of truths. With the truth of mathematics undermined, realms of philosophy, science, and even some religious beliefs went up in smoke. So shocking were the implications that even mathematicians refused to take non-Euclidean geometry seriously until the theory of relativity forced them to face the full significance of the creation.

For reasons which we trust will become clearer further on, the devastation caused by non-Euclidean geometry did not shatter mathematics but released it from bondage to the physical world. The lesson learned from the history of non-Euclidean geometry was that though mathematicians may start with axioms that seem to have little to do with the observable behavior of nature, the axioms and theorems may nevertheless prove applicable. Hence mathematicians felt freer to give reign to their imaginations and to consider abstract concepts such as complex numbers, tensors, matrices, and n-dimensional spaces. This development was followed by an even greater advance in mathematics and, surprisingly, an increasing use of mathematics in the sciences.

Even before the nineteenth century, the rationalistic spirit engendered by the success of mathematics in the study of nature penetrated to the social scientists. They began to emulate the physical scientists, that is, to search for the basic truths in their fields and to attempt reorganization of their subjects on the mathematical pattern. But these attempts to deduce the laws of man and society and to erect sciences of biology, economics, and politics did not succeed, although they did have some indirect beneficial effects.

The failure to penetrate social and biological problems by the deductive method, that is, the method of reasoning from axioms, caused social scientists to take over and develop further the mathematical theories of statistics and probability, which had already been initiated by mathematicians for various purposes ranging from problems of gambling to the theory of heat and astronomy. These techniques have been remarkably successful and have given some scientific methodology to what were largely speculative domains.

This brief sketch of the mathematics which will fall within our purview may make it clear that mathematics is not a closed book written in Greek times. It is rather a living plant that has flourished and languished with the rise and fall of civilizations. Since about 1600 it has been a continuing development which has become steadily vaster, richer, and more profound. The character of mathematics has been aptly, if somewhat floridly, described by the nineteenth-century English mathematician James Joseph Sylvester.

> *Mathematics is not a book confined within a cover and bound between brazen clasps, whose contents it needs only patience to ransack; it is not a mine, whose treasures may take long to reduce into possession, but which fill only a limited number of veins and lodes; it is not a soil, whose fertility can be exhausted by the yield of successive harvests; it is not a continent or an ocean, whose area can be mapped out and its contour defined; it is as limitless as the space which it finds too narrow for its aspirations; its possibilities are as infinite as the worlds which are forever crowding in and multiplying upon the astronomer's gaze; it is incapable of being restricted within assigned boundaries or being reduced to definitions of permanent validity as the consciousness, the life, which seems to slumber in each monad, in every atom of matter, in each leaf and bud and cell and is forever ready to burst forth into new forms of vegetable and animal existence.*

Our sketch of the development of mathematics has attempted to indicate the major eras and civilizations in which the subject has flourished, the variety of interests which induced people to pursue mathematics, and the branches of mathematics that have been created. Of course, we intend to investigate more carefully and more fully what these creations are and what values they have furnished to mankind. One fact of history may be noted by way of summary here. Mathematics as a body of reasoning from axioms stems from one source, the classical Greeks. All other civilizations which have pursued or are pursuing mathematics acquired this concept of mathematics from the Greeks. The Arab and Western European were the next civilizations to take over and expand on the Greek foundation. Today countries such as the United States, Russia, China, India, and Japan are also active. Though the last three of these did possess some native mathematics, it was limited and empirical as in Babylonia and Egypt. Modern mathematical activity in these five countries and wherever else it is now taking hold was inspired by Western European thought and actually learned by men who studied in Europe and returned to build centers of teaching in their own countries.

2-10 THE HUMAN ASPECT OF MATHEMATICS

One final point about mathematics is implicit in what we have said. We have spoken of problems which gave rise to mathematics, of cultures which emphasized some directions of thinking as opposed to others, and of branches of mathematics, as though all these forces and activities were as impersonal as the force of gravitation. But ideas and thinking are conveyed by people. Mathematics is a human creation. Although most Greeks did believe that mathematics existed independently of human beings as the planets and mountains seem to, and that all that human beings do is discover more and more of the structure, the prevalent belief today is that mathematics is entirely a human product. The concepts, the axioms, and the theorems established are all created by human beings in man's attempt to understand his environment, to give play to his artistic instincts, and to engage in absorbing intellectual activity.

The lives and activities of the men themselves are also fascinating. While mathematicians produce formulas, no formula produces mathematicians. They have come from all levels of society. The special talent, if there is such, which makes mathematicians has been found in Casanovas and ascetics, among business men and philosophers, among atheists and the profoundly religious, among the retiring and the worldly. Some, like Blaise Pascal and Gauss, were precocious; Évariste Galois was dead at 21, and Niels Hendrik Abel at 27. Others, like Karl Weierstrass and Henri Poincaré, matured more normally and were productive throughout their lives. Many were modest; others extremely egotistical and vain beyond toleration. One finds scoundrels, such as Cardan, and models of rectitude. Some were generous in their recognition of other great minds; others were resentful and jealous and even stole ideas to boost their own reputations. Disputes about priority of discovery abound.

The point in learning about these human variations, aside from satisfying our instinct to pry into other people's lives, is that it explains to a large extent why the progress of the highly rational subject of mathematics has been highly irrational. Of course, the major historical forces, which we sketched above, limit the actions and influence the outlook of individuals, but we also find in the history of mathematics all the vagaries which he have learned to associate with human beings. Leading mathematicians have failed to recognize bright ideas suggested by younger men, and the authors died neglected. Big men and little men made unsuccessful attempts to solve problems which their successors solved with ease. On the other hand, some supposed proofs offered even by masters were later found to be false. Generations and even ages failed to note new ideas, despite the fact that all that was needed was not a technical achievement but merely a point of view. The examples of the blindness of human beings to ideas which later seem simple and obvious furnish fascinating insight into the working of the human mind.

Recognition of the human element in mathematics explains in large measure the differences in the mathematics produced by different civilizations and the sudden spurts made in new directions by virtue of insights supplied by genius. Though no subject has profited as much as mathematics has by the cumulative effect of thousands of workers and results, in no subject is the role of great minds more readily discernible.

EXERCISES

1. Name a few civilizations which contributed to mathematics.
2. What basis did the Egyptians and Babylonians have for believing in their mathematical methods and formulas?
3. Compare Greek and pre-Greek understanding of the concepts of mathematics.
4. What was the Greek plan for establishing mathematical conclusions?
5. What was the chief contribution of the Arabs to the development of mathematics?
6. In what sense is mathematics a creation of the Greeks rather than of the Egyptians and Babylonians?
7. Criticize the statement "Mathematics was created by the Greeks and very little was added since their time."

Topics for Further Investigation

To write on the following topics use the books listed under Recommended Reading.

1. The mathematical contributions of the Egyptians or Babylonians.
2. The mathematical contributions of the Greeks.

Recommended Reading

BALL, W. W. ROUSE: *A Short Account of the History of Mathematics,* Dover Publications, Inc., New York, 1960.

BELL, ERIC T.: *Men of Mathematics,* Simon and Schuster, New York, 1937.

CHILDE, V. GORDON: *Man Makes Himself,* The New American Library, New York, 1951.

EVES, HOWARD: *An Introduction to the History of Mathematics,* Rev. ed., Holt, Rinehart and Winston, Inc., New York, 1964.

NEUGEBAUER, OTTO: *The Exact Sciences in Antiquity,* Princeton University Press, Princeton, 1952.

SCOTT, J. F.: *A History of Mathematics,* Taylor and Francis, Ltd., London, 1958.

SMITH, DAVID EUGENE: *History of Mathematics,* Vol. I, Dover Publications, Inc., New York, 1958.

STRUIK, DIRK J.: *A Concise History of Mathematics,* Dover Publications, Inc., New York, 1948.

CHAPTER 3

LOGIC AND MATHEMATICS

Geometry will draw the soul toward truth and create the spirit of philosophy.

PLATO

3-1 INTRODUCTION

Mathematics has its own ways of establishing knowledge, and the understanding of mathematics is considerably promoted if one learns first just what those ways are. In this chapter we shall study the concepts which mathematics treats; the method, called deductive proof, by which mathematics establishes its conclusions; and the principles or axioms on which mathematics rests. Study of the contents and logical structure of mathematics leaves untouched the subject of how the mathematician knows what conclusions to establish and how to prove them. We shall therefore present a brief and preliminary discussion of the creation of mathematics. This topic will recur as we examine the subject matter itself in subsequent chapters.

Since mathematics, as we conceive the subject today, was fashioned by the Greeks, we shall also attempt to see what features of Greek thought and culture caused these people to remodel what the Egyptians and Babylonians had pursued for several thousand years.

3-2 THE CONCEPTS OF MATHEMATICS

The first major step which the Greeks made was to insist that mathematics must deal with abstract concepts. Let us see just what this means. When we first learn about numbers we are taught to think about collections of particular objects such as two apples, three men, and so on. Gradually and rather subconsciously we begin to think about the numbers 2, 3, and other whole numbers without having to associate them with physical objects. We soon reach the more advanced stage of adding, subtracting, and performing other operations with numbers without having to handle collections of objects in order to understand these operations or to see that the results agree with experience. Thus we soon become convinced that 4 times 5 must be 20, whether these numbers represent quantities of apples, horses, or even purely imaginary objects. By this time we are really dealing with concepts or ideas, for the whole numbers do not exist in nature. Any whole number is rather an abstraction of a property which is common to many different collections or sets of objects.

The whole numbers then are ideas, and the same is true of fractions such as $\frac{2}{3}$, $\frac{5}{7}$, and so on. In the latter case, too, the formulation of the physical relationship of a part of an object to the whole, whether it refers to pies, bushels of wheat, or to a smaller monetary value in relation to a larger one, again leads to an abstraction. Mathematicians formulate operations with fractions, that is, combining parts of an object, taking one part away from the other, or taking a part of a part, in such a way that the result of any operation on abstract fractions agrees with the corresponding physical occurrence. Thus the mathematical process of, say adding $\frac{2}{3}$ and $\frac{4}{5}$, which yields $\frac{22}{15}$, expresses the addition of $\frac{2}{3}$ of a pie and $\frac{4}{5}$ of a pie, and the result tells us how many parts of a pie one would actually have.

Whole numbers, fractions, and the various operations with whole numbers and fractions are abstractions. Although this fact is rather easy to understand, we tend to lose sight of it and cause ourselves unnecessary confusion. Let us consider an example. A man goes into a shoe store and buys 3 pairs of shoes at 10 dollars per pair. The storekeeper reasons that 3 pairs times 10 dollars is 30 dollars and asks for 30 dollars in return for the 3 pairs of shoes. If this reasoning is correct, then it is equally correct for the customer to argue that 3 pairs times 10 dollars is 30 pairs of shoes and to walk out with 30 pairs of shoes without handing the storekeeper one cent. The customer may end up in jail, but he may console himself while he languishes there that his reasoning is as sound as the storekeeper's.

The source of the difficulty is, of course, that one cannot multiply shoes by dollars. One can multiply the number 3 by the number 10 and obtain the number 30. The practical and no doubt obligatory physical interpretation of the answer in the above situation is that one must pay 30 dollars rather than walk out with 30 pairs of shoes. We see, therefore, that one must distinguish between the purely mathematical operation of multiplying 3 by 10 and the physical objects with which these numbers may be associated.

The same point is involved in a slightly different situation. Mathematically $\frac{2}{1}$ is equal to $\frac{4}{2}$. But the corresponding physical fact may not be true. One may be willing to accept 4 half-pies instead of 2 whole pies, but no woman would accept 4 half-dresses in place of 2 dresses or 4 half-shoes in place of 1 pair of whole shoes.

The Egyptians and Babylonians did reach the stage of working with pure numbers dissociated from physical objects. But like young children of our civilization, they hardly recognized that they were dealing with abstract entities. By contrast, the Greeks not only recognized numbers as ideas but emphasized that this is the way we must regard them. The Greek philosopher

Plato, who lived from 428 to 348 B.C. and whose ideas are representative of the classical Greek period, says in his famous work, the *Republic*,

> *We must endeavor that those who are to be the principal men of our State go and learn arithmetic, not as amateurs, but they must carry on the study until they see the nature of numbers with the mind only; . . . arithmetic has a very great and elevating effect, compelling the soul to reason about abstract number, and rebelling against the introduction of visible or tangible objects into the argument.*

The Greeks not only emphasized the distinction between pure numbers and the physical applications of such numbers, but they preferred the former to the latter. The study of the properties of pure numbers, which they called *arithmetica*, was esteemed as a worthy activity of the mind, whereas the use of numbers in practical applications, *logistica*, was deprecated as a mere skill.

Geometrical thinking prior to the classical Greek period was even less advanced than thinking about numbers. To the Egyptians and Babylonians the words "straight line" meant no more than a stretched rope or a line traced in sand, and a rectangle was a piece of land of a particular shape. The Greeks began the practice of treating point, line, triangle, and other geometrical notions as concepts. They did of course appreciate that these mental notions are suggested by physical objects, but they stressed that the concepts differ from the physical examples as sharply as the concept of time differs from the passage of the sun across the sky. The stretched string is a physical object illustrating the concept of line, but the mathematical line has no thickness, no color, no molecular structure, and no tension.

The Greeks were explicit in asserting that geometry deals with abstractions. Speaking of mathematicians, Plato says,

> *And do you not know also that although they make use of the visible forms and reason about them, they are thinking not of these, but of the ideals which they resemble; not of the figures which they draw, but of the absolute square and the absolute diameter . . . they are really seeking to behold the things themselves, which can be seen only with the eye of the mind?*

On the basis of elementary abstractions, mathematics creates others which are even more remote from anything real. Negative numbers, equations involving unknowns, formulas, and other concepts we shall encounter are abstractions built upon abstractions. Fortunately, every abstraction is ultimately derived from, and therefore understandable in terms of, intuitively meaningful objects or phenomena. The mind does play its part in the creation of mathematical concepts, but the mind does not function independently of the outside world. Indeed the mathematician who treats concepts that have no physically real or intuitive origins is almost surely talking nonsense. The intimate connection between mathematics and objects and events in the physical world is reassuring, for it means that we can not only hope to understand the mathematics proper, but also expect physically meaningful and valuable conclusions.

The use of abstractions is not peculiar to mathematics. The concepts of force, mass, and energy, which are studied in physics, are abstractions from real phenomena. The concept of wealth, an abstraction from material possessions such as land, buildings, and jewelry, is studied in economics. The concepts of liberty, justice, and democracy are familiar in political science. Indeed, with respect to the use of abstract concepts, the distinction between mathematics on the one hand and the physical and social sciences on the other is not a sharp one. In fact, the influence of mathematics and mathematical ways of thinking on the physical sciences especially has led to ever increasing use of abstract concepts including some, as we shall see, which may have no direct real counterpart at all, any more than a mathematical formula has a direct real counterpart.

The very fact that other studies also engage in abstractions raises an important question. Mathematics is confined to some abstractions, numbers and geometrical forms, and to concepts built upon these basic ones. Abstractions such as mass, force, and energy belong to physics, and still other abstractions belong to other subjects. Why doesn't mathematics also treat forces, wealth, and justice? Certainly these concepts are also worthy of study. Did the mathematicians make an agreement with physicists, economists, and others to divide the concepts among themselves? The restriction of mathematics to numbers and geometrical forms is partly a historical accident and partly a deliberate decision made by the Greeks. Numbers and geometrical forms had already been introduced by the Egyptians and Babylonians, and their utility in daily life was established. Since the Greeks learned the rudiments of mathematics from these civilizations, the sheer weight of tradition might have caused them to continue the practice of regarding mathematics as the study of numbers and geometrical figures. But people as original and bold in thought as the Greeks would not have been bound merely by tradition, had they not found in numbers and geometrical forms sharp and clear notions which appealed to their delight in the processes of exact thinking. However, an even more compelling reason was their belief that numerical and geometrical properties and relationships were basic, that they underlay the phenomena of the physical world and the design of the entire universe. Hence to understand the world one should seek this mathematical essence. The brilliance and depth of their conception of the universe will be revealed more and more as we proceed.

When one compares the pre-Greek and Greek understanding of the concepts of mathematics and notes the sharp transition from the concrete to the abstract, another question presents itself. The Greeks eliminated the physical substance and retained only the idea. Why did they do it? Surely it is more difficult to think about abstractions than about concrete things. Also it would seem that an attempt to study nature by concentrating on just a few aspects of physical objects rather than on the objects themselves would fall far short of effectiveness.

Insofar as the emphasis on abstractions is concerned, the Greeks saw at once what any thinking people would see sooner or later. One advantage of treating abstractions is the gain in generality. When a child learns that $5 + 5 = 10$, he acquires in one swoop a fact which applies to hundreds of situations. Likewise a theorem proved about the abstract triangle applies to a triangular piece of land, a musical percussion instrument, and a triangle determined by three heavenly bodies at any instant of time. It has been said that the process of abstraction amounts to giving the same name to different things, but this very recognition that different objects possess the common property named in the abstraction carries with it the implication that anything true of the abstraction will apply to the several objects. Part of the secret of the power of mathematics is that it deals with abstractions.

Another advantage of abstraction was also clear to the Greeks. Abstracting from a physical situation just those properties which are to be studied frees the mind from burdensome and irrelevant details and enables one to concentrate on the features of interest. When one wishes to determine the area of a piece of land, only shape and size are relevant, and it is desirable to think only about these and not about the fertility of the soil.

The emphasis on mathematical abstractions by the classical Greeks was part and parcel of their outlook on the entire universe. They were concerned with truths, and leading philosophical schools, notably the Pythagoreans and the Platonists, maintained that truths could be established only about abstractions. Let us follow their argument. The physical world presents various objects to the senses. But the impressions received by the senses are inexact, transitory, and constantly changing; indeed, the senses may be even deceived, as by mirages. However, truth, by its very meaning, must consist of permanent, unchanging, definite entities and relationships. Fortunately, the intelligence of man excited to reflection by the impressions of sensible objects may rise to higher conceptions of the realities faintly exhibited to the senses, and so man may rise to the contemplation of ideas. These are eternal realities and the true goal of thought, whereas mere "things are the shadows of ideas thrown on the screen of experience."

Thus Plato would say that there is nothing real in a horse, a house, or a beautiful woman. The reality is in the universal type or idea of a horse, a home, or a woman. The ideas, among which Plato emphasized Beauty, Justice, Intelligence, Goodness, Perfection, and the State, are independent of the superficial appearances of things, of the flux of life, and of the biases and warped desires of man; they are in fact constant and invariable, and knowledge concerning them is firm and indestructible. Real and eternal knowledge concerns these ideas, rather than sensuous objects. This distinction between the intelligible world and the world revealed by the senses is all-important in Plato.

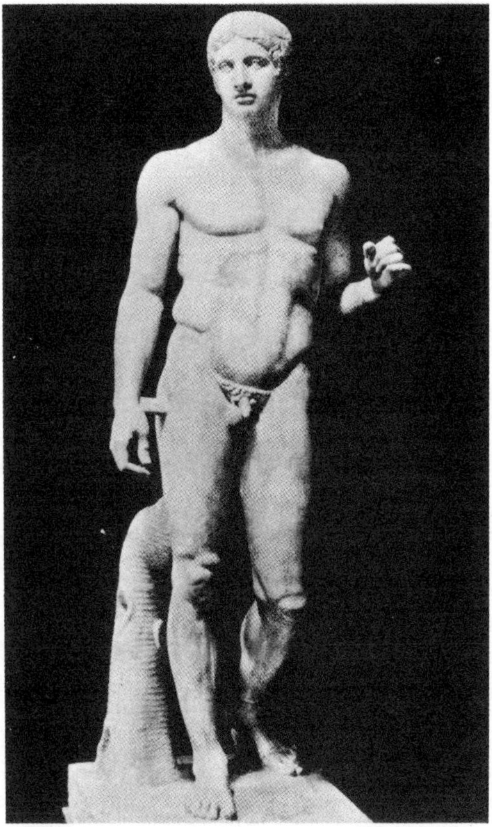

Fig. 3–1.
Polyclitus: *Spear-bearer (Daryphorus).* National Museum, Naples.

To put Plato's doctrine in everyday language, fundamental knowledge does not concern itself with what John ate, Mary heard, or William felt. Knowledge must rise above individuals and particular objects and tell us about broad classes of objects and about man as a whole. True knowledge must therefore of necessity concern abstractions. Plato admits that physical or sensible objects suggest the ideas just as diagrams of geometry suggest abstract geometrical concepts. Hence there is a point to studying physical objects, but one must not lose himself in trivial and confusing minutiae.

The abstractions of mathematics possessed a special importance for the Greeks. The philosophers pointed out that, to pass from a knowledge of the world of matter to the world of ideas, man must train his mind to grasp the

FIG. 3–2.
Bust of Caesar. Vatican.

ideas. These highest realities blind the person who is not prepared to contemplate them. He is, to use Plato's famous simile, like one who lives continuously in the deep shadows of a cave and is suddenly brought out into the sunlight. The study of mathematics helps make the transition from darkness to light. Mathematics is in fact ideally suited to prepare the mind for higher forms of thought because on the one hand it pertains to the world of visible things and on the other hand it deals with abstract concepts. Hence through the study of mathematics man learns to pass from concrete figures to abstract forms; moreover, this study purifies the mind by drawing it away from the contemplation of the sensible and perishable and leading it to the eternal ideas. These latter abstractions are on the same mental level as the concepts of mathematics. Thus, Socrates says, "The understanding of mathematics is necessary for a sound grasp of ethics."

To sum up Plato's position we may say that while a little knowledge of geometry and calculation suffices for practical needs, the higher and more advanced portions tend to lift the mind above mundane considerations and enable it to apprehend the final aim of philosophy, the idea of the Good. Mathematics, then, is the best preparation for philosophy. For this reason Plato recommended that the future rulers, who were to be philosopher-kings, be trained for ten years, from age 20 to 30, in the study of the exact sciences, arithmetic, plane geometry, solid geometry, astronomy, and harmonics (music). The oft-repeated inscription over the doors of Plato's Academy, stating that no one ignorant of mathematics should enter, fully expresses the importance

Fig. 3-3. *Parthenon,* Athens.

he attached to the subject, although modern critics of Plato read into these words his admission that one would not be able to learn it after entering. This value of mathematical training led one historian to remark, "Mathematics considered as a science owes its origins to the idealistic needs of the Greek philosophers, and not as fable has it, to the practical demands of Egyptian economics."

The preference of the Greeks for abstractions is equally evident in the art of the great sculptors, Polyclitus, Praxiteles, and Phidias. One has only to glance at the face in Fig. 3–1 to observe that Greek sculpture of the classical period dwelt not on particular men and women but on types, ideal types. Idealization extended to standardization of the ratios of the parts of the body to each other. Polyclitus believed, in fact, that there were ideal numerical ratios which fix the proportions of the human body. Perfect art must follow these ideal proportions. He wrote a book, *The Canon,* on the subject and constructed the "Spear-bearer" to illustrate the thesis. These abstract types contrast sharply with what is found in numerous busts and statues of private individuals and military and political leaders made by Romans (Fig. 3–2).

Greek architecture also reveals the emphasis on ideal forms. The simple and austere buildings were always rectangular in shape; even the ratios of the dimensions employed were fixed. The Parthenon at Athens (Fig. 3–3) is an example of the style and proportions found in almost all Greek temples.

EXERCISES

1. Suppose 5 trucks pass by with 4 men in each. To answer the question of how many men there are in all the trucks, a person reasons that 4 men times 5 trucks is 20 men. On the other hand, if there are 4 men each owning 5 trucks, the total number of trucks is 20 trucks. Hence 4 men times 5 trucks yields 20 trucks. How do you know that the answer is 20 men in one case and 20 trucks in the other?
2. If the product of 25¢ and 25¢ is obtained by multiplying 0.25 by 0.25 the result is 0.0625 or $6\frac{1}{4}$¢. Does it pay to multiply money?
3. Can you suggest some abstract political or ethical concepts?
4. Suppose 30 books are to be distributed among 5 people. Since 30 books divided by 5 people yields 6 books, each person gets 6 books. Criticize the reasoning.
5. A store advertises that it will give a credit of $1 for each purchase amounting to $1. A man who spends $6 reasons that he should receive a credit of $6 times $1, or $6. But $6 is 600¢ and $1 is 100¢. Hence 600¢ times 100¢ is 60,000¢, or $600. It would seem that it is more profitable to operate with the almost worthless cent than with dollars! What is wrong?
6. What does the statement that mathematics deals with abstractions mean?
7. Why did the Greeks make mathematics abstract?

3-3 IDEALIZATION

The geometrical notions of mathematics are abstract in the sense that shapes are mental concepts which actual physical objects merely approximate. The sides of a rectangular piece of land may not be exactly straight nor would each angle be exactly 90°. Hence, in adopting such abstract concepts, mathematics does idealize. But in studying the physical world, mathematics also idealizes in another sense which is equally important. Very often mathematicians undertake to study an object which is not a sphere and yet choose to regard it as such. For example, the earth is not a sphere but a spheroid, that is, a sphere flattened at the top and bottom. Yet in many physical problems which are treated mathematically the earth is represented as a perfect sphere. In problems of astronomy a large mass such as the earth or the sun is often regarded as concentrated at one point.

In making such idealizations, the mathematician deliberately distorts or approximates at least some features of the physical situation. Why does he do it? The reason usually is that he simplifies the problem and yet is quite sure that he has not introduced any gross errors. If one is to investigate, for example, the motion of a shell which travels ten miles, the difference between the assumed spherical shape of the earth and the true spheroidal shape does not matter. In fact, in the study of any motion which takes place over a limited region, say one mile, it may be sufficient to treat the earth as a flat surface. On the other hand, if one were to draw a very accurate map of the earth, he would take into account that the shape is spheroidal. As another example, to find the distance to the moon, it is good enough to assume that the moon is a point in space. However, to find the size of the moon, it is clearly pointless to regard the moon as a point.

The question does arise, how does the mathematician know when idealization is justified? There is no simple answer to this question. If he has to solve a series of like problems, he may solve one using the correct figure, and another, using a simplified figure, and compare results. If the difference does not matter for his purposes, he may then retain the simpler figure for the remaining problems. Sometimes he can estimate the error introduced by using the simpler figure and may find that this error is too small to matter. Or the mathematician may make the idealization and use the result because it is the best he can do. Then he must accept experience as his guide in deciding whether the result is good enough.

To idealize by deliberately introducing a simplification is to lie a little, but the lie is a white one. Using idealizations to study the physical world does impose a limitation on what mathematics accomplishes, but we shall find that even where idealizations are employed, the knowledge gained is of immense value.

EXERCISES

1. Distinguish between abstraction and idealization.
2. Is it correct to assume that the lines of sight to the sun from two places A and B on the earth's surface are parallel?
3. Suppose you wished to measure the height of a flagpole. Would it be wise to regard the flagpole as a line segment?

3-4 METHODS OF REASONING

There are many ways, more or less reliable, of obtaining knowledge. One can resort to authority as one often does in obtaining historical knowledge. One may accept revelation as many religious people do. And one may rely upon experience. The foods we eat are chosen on the basis of experience. No one determined in advance by careful chemical analysis that bread is a healthful food.

We may pass over with a mere mention such sources of knowledge as authority and revelation, for these sources cannot be helpful in building mathematics or in acquiring knowledge of the physical world. It is true that in the medieval period of Western European culture men did contend that all desirable knowledge of nature was revealed in the Bible. However in no significant period of scientific thought has this view played any role. Experience, on the other hand, is a useful source of knowledge. But there are difficulties in employing this method. We should not wish to build a fifty-story building in order to decide whether a steel beam of specified dimensions is strong enough to be used in the foundation. Moreover, even if one should happen to choose workable dimensions, the choice may be wasteful of materials. Of course, experience is of no use in determining the size of the earth or the distance to the moon.

Closely related to experience is the method of experiment which amounts to setting up and going through a series of purposive, systematic experiences. It is true that experimentation fundamentally is experience, but it is usually accompanied by careful planning which eliminates extraneous factors, and the experience is repeated enough times to yield reliable information. However, experimentation is subject to much the same limitations as experience.

Are authority, revelation, experience, and even experimentation the only methods of obtaining knowledge? The answer is no. The major method is reasoning, and within the domain of reasoning there are several forms. One can reason by analogy. A boy who is considering a college career may note that his friend went to college and handled it successfully. He argues that since he is very much like his friend in physical and mental qualities, he too should succeed in college work. The method of reasoning just illustrated is to find a similar situation or circumstance and to argue that what was true for the similar case should be true of the one in question. Of course, one must be able to find a similar situation and one must take the chance that the differences do not matter.

Another common method of reasoning is induction. People use this method of reasoning every day. Because a person may have had unfortunate experiences in dealing with a few department stores, he concludes that all department stores are bad to deal with. Or, for example, experimentation would show that iron, copper, brass, oil, and other substances expand when heated, and one consequently concludes that *all* substances expand when heated. Inductive reasoning is in fact the common method used in experimentation. An experiment is generally performed many times, and if the same result is obtained each time, the experimenter concludes that the result will always follow. The essence of induction is that one observes repeated occurrences of the *same phenomenon* and concludes that the phenomenon will always occur. Conclusions obtained by induction seem well warranted by the evidence, especially when the number of instances observed is large. Thus the sun is observed so often to rise in the morning that one is sure it has risen even on those mornings when it is hidden by clouds.

There is still a third method of reasoning, called deduction. Let us consider some examples. If we accept as basic facts that honest people return found money and that John is honest, we may conclude unquestionably that John will return money that he finds. Likewise, if we start with the facts that no mathematician is a fool and that John is a mathematician, then we may conclude with certainty that John is not a fool. In deductive reasoning we start with certain statements, called premises, and assert a conclusion which is a *necessary* or *inescapable* consequence of the premises.

All three methods of reasoning, analogy, induction, and deduction, and other methods we could describe, are commonly employed. There is one essential difference, however, between deduction on the one hand and all other methods of reasoning on the other. Whereas the conclusion drawn by analogy or induction has only a probability of being correct, the conclusion drawn by deduction necessarily holds. Thus one might argue that because lions are similar to cows and cows eat grass, lions also eat grass. This argument by analogy leads to a false conclusion. The same is true for induction: although experiment may indeed show that two dozen different substances expand when heated, it does not necessarily follow that all substances do. Thus water, for example, when heated from 0° to 4° centigrade* does not expand; it contracts.

Since deductive reasoning has the outstanding advantage of yielding an indubitable conclusion, it would seem obvious that one should always use this method in preference to the others. But the situation is not that simple. For one thing analogy and induction are often easier to employ. In the case of analogy, a similar situation may be readily available. In the case of induction, experience often supplies the facts with no effort at all. The fact that the sun rises every morning is noticed by all of us almost automatically. Furthermore, deductive reasoning calls for premises which it may be impossible to obtain despite all efforts. Fortunately we can use deductive reasoning in a variety of situations. For example, we can use it to find the distance to the moon. In this instance, both analogy and induction are powerless, whereas, as we shall see later, deduction will obtain the result quickly. It is also apparent that where deduction can replace induction based on expensive experimentation, deduction is preferred.

Because we shall be concerned primarily with deductive reasoning, let us become a little more familiar with it. We have given several examples of deductive reasoning and have asserted that the conclusions are inescapable consequences of the premises. Let us consider, however, the following example. We shall accept as premises that

All good cars are expensive

and

All Locomobiles are expensive.

We might conclude that

All Locomobiles are good cars.

---

* In scientific texts, "celsius" is considered to be the more precise term.

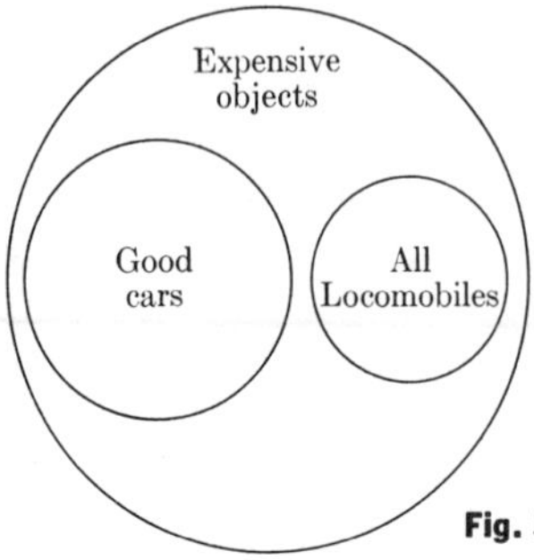

Fig. 3-4

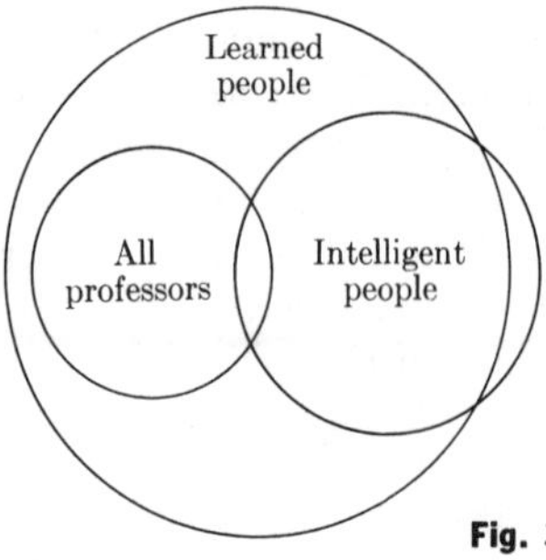

Fig. 3-5

The reasoning here is intended as deductive; that is, the presumption in drawing this conclusion is that it is an inevitable consequence of the premises. Unfortunately, the reasoning is not correct. How can we see that it is not correct? A good way of picturing deductive arguments which enables us to see whether or not they are correct is called the circle test.

We note that the first premise deals with cars and expensive objects. Let us think of all the expensive objects in this world as represented by the points of a circle, the largest circle in Fig. 3–4. The statement that all good cars are expensive means that all good cars are a part of the collection of expensive objects. Hence we draw another circle within the circle of expensive objects, and the points of this smaller circle represent all the good cars. The second premise says that all Locomobiles are expensive. Hence if we represent all Locomobiles by the points of a circle, this circle, too, must be drawn within the circle of expensive objects. However we do not know, on the basis of the two premises, where to place the circle representing all Locomobiles. It can, as far as we know, fall in the position shown in the figure. Then we cannot conclude that all Locomobiles are good cars, because if that conclusion were inevitable, the circle representing Locomobiles must fall inside the circle representing good cars.

Many people do conclude from the above premises that all Locomobiles are good cars and the reason that they err is that they confuse the premise "All good cars are expensive" with the statement that "All expensive cars are good." Were the latter statement our first premise then the deductive argument would be valid or correct.

Let us consider another example. Suppose we take as our premises that

All professors are learned people

and

Some professors are intelligent people.

May we necessarily conclude that

Some intelligent people are learned?

It may or may not be obvious that this conclusion is correct. Let us use the circle test. We draw a circle representing the class of learned people (Fig 3–5). Since the first premise tells us that all professors are learned people, the circle representing the class of professors must fall within the circle representing learned people. The second premise introduces the class of intelligent people, and we now have to determine where to draw that circle. This class must include some professors. Hence the circle must intersect the circle of professors. Since the latter is inside the circle of learned people, some intelligent people must fall within the class of learned people.

These examples of deductive reasoning may make another point clear. In determining whether a given argument is correct or valid, we must rely only upon the facts given in the premises. We may not use information which is not explicitly there. For example, we may believe that learned people are intelligent because to acquire learning they must possess intelligence. But this belief or fact, if it is a fact, cannot enter into the argument. Nothing that one may happen to know or believe about learned or intelligent people is to be used unless explicitly stated in the premises. In fact, as far as the validity of the argument is concerned, we might just as well have considered the premises

All x's are y's,

Some x's are z's,

and the conclusion, then, is

Some z's are y's.

Here we have used x for professor, y for learned person, and z for intelligent person. The use of x, y, and z does make the argument more abstract and more difficult to retain in the mind, but it emphasizes that we must look only at the information in the premises and avoids bringing in extraneous information about professors, learned people, and intelligent people. When we write the argument in this more abstract form, we also see more clearly that what determines the validity of the argument is the *form* of the premises rather than the meaning of x, y, and z.

A great deal of deductive reasoning falls into the patterns we have been illustrating. There are, however, variations that should be noted. It is quite customary, especially in the geometry we learn in high school, to state theorems in what is called the "if . . . then" form. Thus one might say, if a triangle is isosceles, then its base angles are equal. One could as well say, all isosceles triangles have equal base angles; or, the base angles of an isosceles triangle are equal. All three versions say the same thing.

Connected with the "if . . . then" form of a premise is a related statement which is often misunderstood. The statement "if a man is a professor, he is learned" offers no difficulty. As noted in the preceding paragraph, it is equivalent to "all professors are learned." However the statement "only if a man is a professor, is he learned" has quite a different meaning. It means that to be learned one must be a professor or that if a man is learned, he must be a professor. Thus the addition of the word only has the significance of interchanging the "if" clause and the "then" clause.

We shall encounter numerous instances of deductive reasoning in our work. The subject of deductive reasoning is customarily studied in logic, a discipline which treats more thoroughly the valid forms of reasoning. However, we shall not need to depend upon formal training in logic. In most cases, common experience will enable us to ascertain whether the reasoning is or is not valid. When in doubt, we can use the circle test. Moreover, mathematics itself is the superb field from which to learn reasoning and is the best exercise in logic. The laws of logic were in fact formulated by the Greeks on the basis of their experiences with mathematical arguments.

EXERCISES

1. A coin is tossed ten times and each time it falls heads. What conclusion does inductive reasoning warrant?
2. Characterize deductive reasoning.
3. What superior features does deductive reasoning possess compared with induction and analogy?
4. Can you prove deductively that George Washington was the best president of the United States?
5. Can one always apply deductive reasoning to prove a desired statement?
6. Can you prove deductively that monogamy is the best system of marriage?
7. Are the following purportedly deductive arguments valid?
   a) All good cars are expensive. A Daffy is an expensive car. Therefore a Daffy is a good car.
   b) All New Yorkers are good citizens. All good citizens give to charity. Therefore all New Yorkers give to charity.
   c) All college students are clever. All young boys are clever. Therefore all young boys are college students.
   d) The same premises as in (c), but the conclusion: All college students are young boys.
   e) It rains every Monday and it is raining today; hence today must be Monday.
   f) No decent people curse; Americans are decent; therefore Americans do not curse.
   g) No decent people curse; Americans curse; therefore some Americans are not decent.
   h) No decent people curse; some Americans are not decent; therefore some Americans curse.
   i) No undergraduates have a bachelor-of-arts degree; no freshmen have a bachelor-of-arts degree. Therefore all freshmen are undergraduates.

8. If someone gave you a valid deductive argument but the conclusion was not true, where would you look for the difficulty?
9. Distinguish between the validity of a deductive argument and the truth of the conclusion.

3-5 MATHEMATICAL PROOF

We have seen so far in our discussion of reasoning that there are several methods of reasoning and that all are useful. These methods can be applied to mathematical problems. Let us suppose that one wished to determine the sum of the angles of a triangle. He could draw on paper many different triangles or construct some out of wood or metal and measure the angles. In each case he would find that the sum is as close to 180° as the eye and hand can determine. By inductive reasoning he could conclude that the sum of the angles in every triangle is 180°. As a matter of fact, the Babylonians and Egyptians did in effect use inductive reasoning to establish their mathematical results. They must have determined by measurement that the area of a triangle is one-half the base times the altitude and, having used this formula repeatedly and having obtained reliable results, they concluded that the formula is correct.

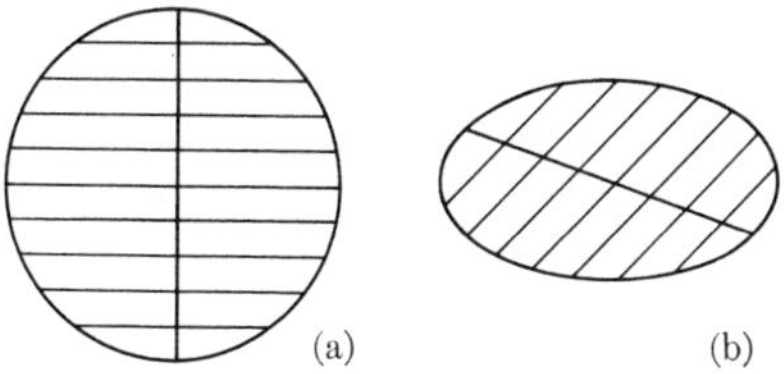

Fig. 3-6. The mid-points of parallel chords lie on a straight line.

To see that reasoning by analogy can be used in mathematics, let us note first that the centers of a set of parallel chords of a circle lie on a straight line (Fig. 3-6a). In fact this line is a diameter of the circle. Now an ellipse (Fig. 3-6b) is very much like a circle. Hence one might conclude that the centers of a set of parallel chords of an ellipse also lie on a straight line.

Deduction is certainly applicable in mathematics. The proofs which one learns in Euclidean geometry are deductive. As another illustration we might consider the following algebraic argument. Suppose one wishes to solve the equation $x - 3 = 7$. One knows that equals added to equals give equals. If we added 3 to both sides of the preceding equation, we would be adding equals to an equality. Hence the addition of 3 to both sides is justified. When this is done, the result is $x = 10$, and the equation is solved.

Thus all three methods are applicable. There is a lot to be said for the use of induction and analogy. The inductive argument for the sum of the angles of a triangle can be carried out in a matter of minutes. The argument by analogy given above is also readily made. On the other hand, finding deductive proofs for these same conclusions might take weeks or might never be accomplished by the average person. As a matter of fact, we shall soon encounter some examples of conjectures for which the inductive evidence is overwhelming but for which no deductive proof has been thus far obtained even by the best mathematicians.

Despite the usefulness and advantages of induction and analogy, mathematics does not rely upon these methods to establish its conclusions. *All mathematical proofs must be deductive.* Each proof is a chain of deductive arguments, each of which has its premises and conclusion.

Before examining the reasons for this restriction to deductive proof, we might contrast the method of mathematics with those of the physical and social sciences. The scientist feels free to draw conclusions by any method of reasoning and, for that matter, on the basis of observation, experimentation, and experience. He may reason by analogy as, for example, when he reasons about sound waves by observing water waves or when he reasons about a possible cure for a disease affecting human beings by testing the cure on animals. In fact reasoning by analogy is a powerful method in science. The scientist may also reason inductively: if he observes many times that hydrogen and oxygen combine to form water, he will conclude that this combination will always form water. At some stages of his work the scientist may also reason deductively and, in fact, even employ the concepts and methods of mathematics proper.

To contrast further the method of mathematics with that of the scientist —and perhaps to illustrate just how stubborn the mathematician can be—we might consider a rather famous example. Mathematicians are concerned with whole numbers, or integers, and among these they distinguish the prime numbers. A prime is a number which has no integral divisors other than itself and 1. Thus 11 is a prime number, whereas 12 is not because it is divisible by 2 for example. Now by actual trial one finds that each of the first few even numbers can be expressed as the sum of two prime numbers. For example, $2 = 1 + 1$; $4 = 2 + 2$; $6 = 3 + 3$; $8 = 3 + 5$; $10 = 3 + 7$; . . . . If one investigates larger and larger even numbers, one finds without exception that every even number can be expressed as the sum of two primes. Hence by *inductive* reasoning one could conclude that *every* even number is the sum of two prime numbers.

But the mathematician does not accept this conclusion as a theorem of mathematics because it has not been proved deductively from acceptable premises. The conjecture that every even number is the sum of two primes, known as Goldbach's hypothesis because it was first suggested by the eighteenth-century mathematician Christian Goldbach, is an unsolved problem of mathematics. The mathematician will insist on a deductive proof even if it takes thousands of years, as it literally has in some instances, to find one. However a scientist would not hesitate to use this inductively well supported conclusion.

Of course, the scientist should not be surprised to find that some of his conclusions are false because, as we have seen, induction and analogy do not lead to sure conclusions. But it does seem as though the scientist's procedure is wiser since he can take advantage of any method of reasoning which will help him advance his knowledge. The mathematician by comparison appears to be narrow-minded or shortsighted. He achieves a reputation for certainty, but at the price of limiting his results to those which can be established deductively. How wise the mathematician may be in his insistence on deductive proof we shall learn as we proceed.

The decision to confine mathematical proof to deductive reasoning was made by the Greeks of the classical period. And they not only rejected all other methods of proof in mathematics, but they also discarded all the knowledge which the Egyptians and Babylonians had acquired over a period of four thousand years because it had only an empirical justification. Why did the Greeks do it?

The intellectuals of the classical Greek period were largely absorbed in philosophy and these same men, because they possessed intellectual interests, were the very ones who developed mathematics as a system of thought. The Ionians, the Pythagoreans, the Sophists, the Platonists, and the Aristotelians were the leading philosophers who gave mathematics its definitive form. The credit for initiating this step probably belongs to one school of Greek philosopher-mathematicians, known as the Ionian school. However, if credit can be assigned to any one person, it belongs to Thales, who lived about 600 B.C. Though a native of Miletus, a Greek city in Asia Minor, Thales spent many years in Egypt as a merchant. There he learned what the Egyptians had to offer in mathematics and science, but apparently he was not satisfied, for he would accept no results that could not be established by deductive reasoning from clearly acceptable axioms. In his wisdom Thales perceived what we shall perceive as we follow the story of mathematics, that the obvious is far more suspect than the abstruse.

Thales probably supplied the proof of many geometrical theorems. He acquired great fame as an astronomer and is believed to have predicted an eclipse of the sun in 585 B.C. A philosopher-astronomer-mathematician might readily be accused of being an impractical stargazer, but Aristotle tells us otherwise. In a year when olives promised to be plentiful, Thales shrewdly cornered all the oil presses to be found in Miletus and in Chios. When the olives were ripe for pressing, Thales was in a position to rent out the presses at his own price. Thales might perhaps have lived in history as a leading businessman, but he is far better known as the father of Greek philosophy and mathematics. From his time onward, deductive proof became the standard in mathematics.

It is to be expected that philosophers would favor deductive reasoning. Whereas scientists select particular phenomena for observation and experimentation and then draw conclusions by induction or analogy, philosophers are concerned with broad knowledge about man and the physical world. To establish universal truths, such as that man is basically good, that the world is designed, or that man's life has purpose, deductive reasoning from acceptable principles is far more feasible than induction or analogy. As Plato put it in his *Republic*, "If persons cannot give or receive a reason, they cannot attain that knowledge which, as we have said, man ought to have."

There is another reason that philosophers favor deductive reasoning. These men seek truths, the eternal verities. We have seen that of all the methods of reasoning only deductive reasoning grants sure and exact conclusions. Hence this is the method which philosophers would almost necessarily adopt. Not only do induction and analogy fail to yield absolutely unquestionable conclusions, but many Greek philosophers would not have accepted as facts the data with which these methods operate, because these are acquired by the senses. Plato stressed the unreliability of sensory perceptions. Empirical knowledge, as Plato put it, yields opinion only.

The Greek preference for deduction had a sociological basis. Contrary to our own society wherein bankers and industrialists are respected most, in classical Greek society, the philosophers, mathematicians, and artists were the leading citizens. The upper class regarded earning a living as an unfortunate necessity. Work robbed man of time and energy for intellectual activities, the duties of citizenship and discussion. These Greeks did not hesitate to express their disdain for work and business. The Pythagoreans, who, as we shall see, delighted in the properties of numbers and applied numbers to the study of nature, derided the use of numbers in commerce. They boasted that they sought knowledge rather than wealth. Plato, too, maintained that knowledge rather than trade was the goal in studying arithmetic. Freemen, he declared, who allowed themselves to become preoccupied with business should be punished, and a civilization which is concerned mainly with the material wants of man is no more than a "city of happy pigs." Xenophon, the famous Greek general and historian, says, "What are called the mechanical arts carry a social stigma and are rightly dishonored in our cities." Aristotle wanted an ideal society in which citizens would not have to practice any mechanical arts. Among the Boeotians, one of the independent tribes of Ancient Greece, those who defiled themselves with commerce were by law excluded from state positions for ten years.

Who did the daily work of providing food, shelter, clothing, and the other necessities of life? Slaves and free men ineligible for citizenship ran the businesses and the households, did unskilled and technical work, managed the industries, and carried on the professions such as medicine. They produced even the articles of refinement and luxury.

In view of this attitude of the Greek upper class towards commerce and trade, it is not hard to understand the classical Greek's preference for deduction. People who do not "live" in the workaday world can learn little from experience, and people who will not observe and use their hands to experiment will not have the facts on which to base reasoning by analogy or induction. In fact the institution of slavery in classical Greek society fostered a divorce of theory from practice and favored the development of speculative and deductive science and mathematics at the expense of experimentation and practical applications.

Over and above the various cultural forces which inclined the Greeks toward deduction were a farsightedness and a wisdom which mark true genius. The Greeks were the first to recognize the power of reason. The mind was a faculty not only additional to the senses but more powerful than the senses. The mind can survey all the whole numbers, but the senses are limited to perceiving only a few at a time. The mind can encompass the earth and the heavens; the sense of sight is confined to a small angle of vision. Indeed the mind can predict future events which the senses of contemporaries will not live to perceive. This mental faculty could be exploited. The Greeks saw clearly that if man could obtain some truths, he could establish others entirely by reasoning, and these new truths, together with the original ones, enabled man to establish still other truths. Indeed the possibilities would multiply at an enormous rate. Here was a means of acquiring knowledge which had been either overlooked or neglected.

This was indeed the plan which the Greeks projected for mathematics. By starting with some truths about numbers and geometrical figures they could deduce others. A chain of deductions might lead to a significant new fact which would be labeled a theorem to call attention to its importance. Each theorem added to the stock of truths that could serve as premises for new deductive arguments, and so one could build an immense body of knowledge about the basic concepts.

Although the Greeks may have been guilty of overemphasizing the power of the mind unaided by experience and observation to obtain truths, there is no doubt that in insisting on deductive proof as the sole method, they rose above the practical level of carpenters, surveyors, farmers, and navigators. At the same time they elevated the subject of mathematics to a system of thought. Moreover the preference for reason which they exhibited gave this faculty the high prestige which it now enjoys and permitted it to exercise its true powers. When we have surveyed some of the creations of the mind that succeeding civilizations building on the Greek plan contributed, we shall appreciate the true depth of the Greek vision.

EXERCISES

1. Compare Greek and pre-Greek standards of proof in mathematics. Reread the relevant parts of Chapter 2.
2. Distinguish science and mathematics with respect to ways of establishing conclusions.

3. Explain the statement that the Greeks converted mathematics from an empirical science to a deductive system.
4. Are the following deductive arguments valid?
   a) All even numbers are divisible by 4. Ten is an even number. Hence 10 is divisible by 4.
   b) Equals divided by equals give equals. Dividing both sides of $3x=6$ by 3 is dividing equals by equals. Hence $x=2$.
5. Does it follow from the fact that the square of any odd number is odd that the square of any even number is even?
6. Criticize the argument:
   The square of every even number is even because $2^2=4$, $4^2=16$, $6^2=36$, and it is obvious that the square of any larger even number also is even.
7. If we accept the premises that the square of any odd number is odd and the square of any even number is even, does it follow deductively that if the square of a number is even, the number must be even?
8. Why did the Greeks insist on deductive proof in mathematics?
9. Let us take for granted that if a triangle has two equal sides, the opposite angles are equal and that we have a triangle in which all three sides are equal. Prove deductively that all three angles are equal in the triangle under consideration. You may also use the premise that things equal to the same thing are equal to one another.
10. How did the Greeks propose to obtain new truths from known ones?

3-6 AXIOMS AND DEFINITIONS

From our discussion of deductive reasoning we know that to apply such reasoning we must have premises. Hence the question arises, what premises does the mathematician use? Since the mathematician reasons about numbers and geometrical figures, he must of course have facts about these concepts. These cannot be obtained deductively because then there would have to be prior premises, and if one continued this process backward, there would be no starting point. The Greeks readily found premises. It seemed indisputable, for example, that two points determine one and only one line and that equals added to equals give equals.

To the Greeks the premises on which mathematics was to be built were self-evident truths, and they called these premises axioms. Socrates and Plato believed, as did many later philosophers, that these truths were already in our minds at birth and that we had but to recall them. And since the Greeks believed that axioms were truths and since deductive reasoning yielded unquestionable conclusions, they also believed that theorems were truths. This view is no longer held, and we shall see later in this book why mathematicians abandoned it. We now know that axioms are suggested by experience and observation. Naturally, to be as certain as we can of these axioms we select those facts which seem clearest and most reliable in our experience. But we must recognize that there is no guarantee that we have selected truths about the world. Some mathematicians prefer to use the word assumptions instead of axioms to emphasize this point.

The mathematician also takes care to state his axioms at the outset and to be sure as he performs his reasoning that no assumptions or facts are used which were not so stated. There is an interesting story told by former President Charles W. Eliot of Harvard which illustrates the likelihood of introducing unwarranted premises. He entered a crowded restaurant and handed his hat to the doorman. When he came out, the doorman at once picked Eliot's hat out of hundreds on the racks and gave it to him. He was amazed that the doorman could remember so well and asked him, "How did you know that was my hat?" "I didn't," replied the doorman. "Why, then, did you hand it to me?" The doorman's reply was, "Because you handed it to me, sir."

Undoubtedly no harm would have been done if the doorman had assumed that the hat he returned to President Eliot belonged to the man. But the mathematician interested in obtaining conclusions about the physical world might be wasting his time if he unwittingly introduced an assumption that he had no right to make

There is one other element in the logical structure of mathematics about which we shall say a few words now and return to in a later chapter (Chapter 20). Like other studies mathematics uses definitions. Whenever we have occasion to use a concept whose description requires a lengthy statement, we introduce a single word or phrase to replace that lengthy statement. For example, we may wish to talk about the figure which consists of three distinct points which do not lie on the same straight line and of the line segments joining these points. It is convenient to introduce the word triangle to represent this long description. Likewise the word circle represents the set of all points which are at a fixed distance from a definite point. The definite point is called the center, and the fixed distance is called the radius. Definitions promote brevity.

EXERCISES

1. What belief did the Greeks hold about the axioms of mathematics?
2. Summarize the changes which the Greeks made in the nature of mathematics.
3. Is it fair to say that mathematics is the child of philosophy?

3-7 THE CREATION OF MATHEMATICS

Because mathematical proof is strictly deductive and merely reasonable or appealing arguments may not be used to establish a conclusion, mathematics has been described as a deductive science, or as the science which derives necessary conclusions, that is, conclusions which necessarily or inevitably follow from the axioms. This description of mathematics is incomplete. Mathematicians must also discover what to prove and how to go about establishing proofs. These processes are also part of mathematics and they are *not* deductive.

How does the mathematician discover what to prove and the deductive arguments that lead to the conclusions? The most fertile source of mathematical ideas is nature herself. Mathematics is devoted to the study of the physical world, and simple experience or the more careful scrutiny of nature suggests idea after idea. Let us consider here a few simple examples. Once mathematicians had decided to devote themselves to geometric forms, it was only natural that such questions should arise as, what are the area, perimeter, and sum of the angles of common figures? Moreover, it is even possible to see how the precise statement of the theorem to be proved would follow from direct experience with physical objects. The mathematician might measure the sum of the angles of various triangles and find that these measurements all yield results close to 180°. Hence the suggestion that the sum of the angles in every triangle is 180° occurs as a possible theorem. To decide the question, which has more area, a polygon or a circle having the same perimeter, one might cut out cardboard figures and weigh them. The relative weights would suggest the statement of the theorem to be proved.

After some theorems have been suggested by direct physical problems, others are readily conceived by generalizing or varying the conditions. Thus knowing the problem of determining the sum of the angles of a triangle, one might ask, What is the sum of the angles of a quadrilateral, a pentagon, and so forth? That is, once the mathematician begins an investigation which is suggested by a physical problem, he can easily find new problems which go beyond the original one.

In the domains of arithmetic and algebra direct calculation with numbers, which is analogous to measurement in geometry, will suggest possible theorems. Anyone who has played with integers, for example, has doubtless observed the following facts:

$$1 = 1,$$

$$1 + 3 = 4 = 2^2,$$

$$1 + 3 + 5 = 9 = 3^2,$$

$$1 + 3 + 5 + 7 = 16 = 4^2,$$

We note that each number on the right is the square of the number of odd numbers appearing on the left; thus in the fourth line, there are four numbers on the left side, and the right side is 4^2. The general result which these calculations suggest is that if the first n odd numbers were on the left side, then the sum would be n^2. Of course, this possible theorem is not proved by the above calculations. Nor could it ever be proved by such calculations, for no mortal man could make the infinite set of computations required to establish the conclusion for *every n*. The calculations do, however, give the mathematician something to work on.

These simple illustrations of how observation, measurement, and calculation suggest possible theorems are not too striking or very profound. We shall see in the course of later work how physical problems suggest more significant mathematical theorems. However, experience, measurement, calculation, and generalization do not include the most fertile source of possible theorems. And it is especially true in seeking methods of proof that more than routine techniques must be utilized. In both endeavors the most important source is the creative act of the human mind.

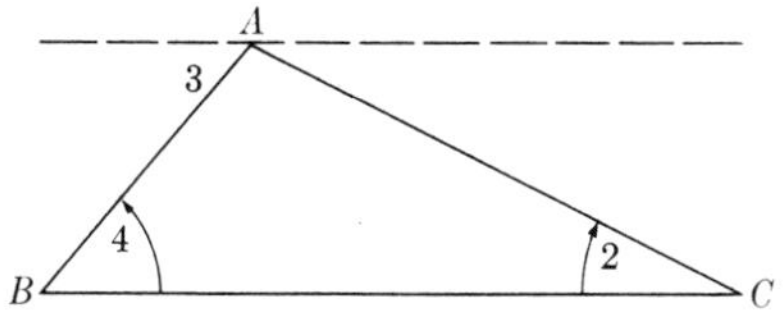

Fig. 3–7

Let us consider the matter of proof. Suppose one has discovered by measurements that the sum of the angles of various triangles is 180°. One must now prove this result deductively. No obvious method will do the job. Some new idea is required, and the reader who remembers his elementary geometry will recall that the proof is usually made by drawing a line through one vertex (A in Fig. 3–7) and parallel to the opposite side. It then turns out as a consequence of a previously established theorem on parallel lines that the angles 1 and 2 are equal, as are the angles 3 and 4. However the angles 1, 3, and the angle A of the triangle itself do add up to 180°, and so the same is true for the angles of the triangle. This method of proof is not routine. The idea of drawing the line through A must be supplied by the mind. Some methods of proof seem so devious and artificial that they have provoked critical comments. The philosopher Arthur Schopenhauer called Euclid's proof of the Pythagorean theorem "a mouse-trap proof" and "a proof walking on stilts, nay, a mean, underhand proof."

The above example has been offered to emphasize the fact that ingenious mathematical work must be done in finding methods of proofs even after the question of what to prove is disposed of. In the search for a method of proof, as in finding what to prove, the mathematician must use audacious imagination, insight, and creative ability. His mind must see possible lines of attack where others would not. In the domains of algebra, calculus, and advanced analysis especially, the first-rate mathematician depends upon the kind of inspiration that we usually associate with the creation of music, literature, or art. The composer feels that he has a theme which when properly developed will produce true music. Experience and a knowledge of music aid him in arriving at this conviction. Similarly, the mathematician surmises that he has a conclusion which will follow from the axioms of mathematics. Experience and knowledge may guide his thoughts into the proper channels. Modifications of one sort or another may be required before a correct proof and a satisfactory statement of the new theorem are achieved. But essentially both mathematician and composer are moved by an afflatus which enables them to see the final edifice before a single stone is laid.

We do not know just what mental processes may lead to correct insight any more than we know how it was possible for Keats to write fine poetry or why Rembrandt was able to turn out fine paintings. One might say of mathematical creation what P. W. Bridgman, the noted physicist, has said of scientific method, that it consists of "doing one's damnedest with one's mind, no holds barred." There is no logic or infallible guide which tells the mind how to think. The very fact that many great mathematicians have tackled a problem and failed and that another comes along and solves it shows that the mind has something to contribute.

The preceding discussion of the creation of mathematics should correct several mistaken popular impressions. When creating a mathematical proof, the mind does not see the cold, ordered arguments which one reads in texts, but rather it perceives an idea or a scheme which when properly formulated constitutes the deductive proof. The formal proof, so to speak, merely sanctions the conquest already made by the intuition. Secondly, the deductive proof is not the preferable form by which to grasp the idea or method employed. In fact the deductive argument often conceals the idea because the logical form is not perspicuous to the intuition. At the very least the details of the arguments obscure the main threads. The value of the deductive organization of the proof is that it enables the creator and the reader to test the arguments by the standards of exact reasoning. Thirdly, there is the prevalent but mistaken notion that scientists and mathematicians must keep their minds open and unbiased in pursuing an investigation. They are not supposed to prejudge the conclusion. Actually the mathematician must first decide *what* to prove, and this conclusion not only does but must precede the search for the proof, or else he would not know where to head. This is not to say that the mathematician may not sometimes make a false conjecture. If he does, his search for a proof will fail or in the course of the search he will realize that he cannot prove what he is after, and he will correct his conjecture. But in any case he knows what he is trying to prove.

EXERCISES

1. Consider the parallelogram $ABCD$ (Fig. 3–8). By definition, the opposite sides are parallel. Now introduce the diagonal BD. Does observation suggest a possible theorem relating the triangles ABD and BDC?

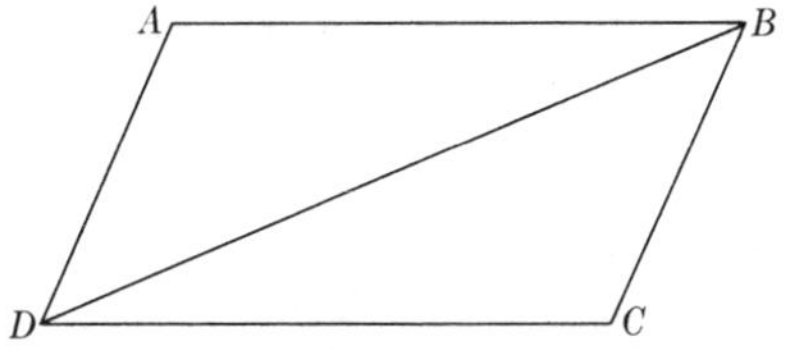

Fig. 3–8

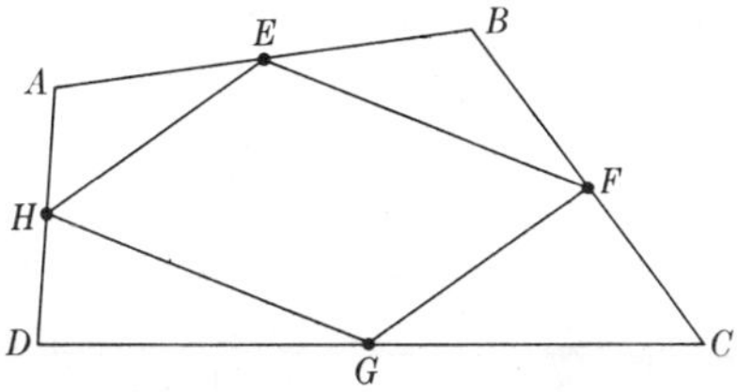

Fig. 3–9

2. Consider any quadrilateral $ABCD$ (Fig. 3–9) and the figure formed by joining the mid-points E, F, G, H of the sides of the quadrilateral. Does observation or intuition suggest any significant fact about the quadrilateral $EFGH$?
3. The formula $n^2 - n + 41$ is supposed to yield primes for various values of n. Thus when $n = 1$,

$$1^2 - 1 + 41 = 41,$$

and this is a prime. When $n = 2$,

$$2^2 - 2 + 41 = 43,$$

and this is a prime. Test the formula for $n = 3$ and $n = 4$. Are the resulting values of the formula primes? Have you proved, then, that the formula always yields primes?
4. Can you specify conditions under which two quadrilaterals will be congruent, that is, have the same size and shape?
5. The following lines show some calculations with the sum of the cubes of whole numbers:

$$1^3 = 1,$$
$$1^3 + 2^3 = 1 + 8 = 9 = 3^2 = (1 + 2)^2,$$
$$1^3 + 2^3 + 3^3 = 1 + 8 + 27 = 36 = 6^2 = (1 + 2 + 3)^2$$
$$1^3 + 2^3 + 3^3 + 4^3 = 1 + 8 + 27 + 64 = 100 = 10^2 = (1 + 2 + 3 + 4)^2.$$

What generalization do these few calculations suggest?

REVIEW EXERCISES

1. What basis did the Egyptians and Babylonians have for believing in the correctness of their mathematical conclusions?
2. Compare Greek and pre-Greek understanding of the concepts of mathematics.
3. What was the Greek plan for establishing mathematical conclusions?

4. In what sense is mathematics a creation of the Greeks rather than of the Egyptians and the Babylonians?
5. Suppose we accept the premises that all professors are intelligent people and all professors are learned people. Which of the following conclusions is validly deduced?
   a) Some intelligent people are learned.
   b) Some learned people are intelligent.
   c) All intelligent people are learned.
   d) All learned people are intelligent.
6. Suppose we accept the premises that all college students are wise, and no professors are college students. Which of the following conclusions is validly deduced?
   a) No professors are wise.
   b) Some professors are wise.
   c) All professors are wise.
7. Is the following argument valid?
   All parallelograms are quadrilaterals, and figure *ABCD* is a quadrilateral. Hence figure *ABCD* is a parallelogram.
8. What conclusion can you deduce from the premises,

   Every successful student must work hard,

   and

   John does not work hard?
9. Smith says,

   If it rains I go to the movies.

   If Smith went to the movies, what can you conclude deductively?
10. Smith says,

    I go to the movies only if it rains.

    If Smith went to the movies, what can you conclude deductively?

Topics for Further Investigation

To pursue any of these topics use the books listed below under Recommended Reading.

1. The life and work of the Pythagoreans
2. The life and work of Euclid

Recommended Reading

Bell, Eric T.: *The Development of Mathematics*, 2nd ed., Chaps. 2 and 3, McGraw-Hill Book Co., N.Y., 1945.

Bell, Eric T.: *Men of Mathematics*, Simon and Schuster, New York, 1937.

Clagett, Marshall: *Greek Science in Antiquity*, Chap. 2, Abelard-Schuman, Inc., New York, 1955.

Cohen, Morris R. and E. Nagel: *An Introduction to Logic and Scientific Method*, Chaps. 1 through 5, Harcourt Brace and Co., New York, 1934.

Coolidge, J. L.: *The Mathematics of Great Amateurs*, Chap. 1, Dover Publications, Inc., New York, 1963.

Hamilton, Edith: *The Greek Way to Western Civilization*, Chaps. 1 through 3, The New American Library, New York, 1948.

Jeans, Sir James: *The Growth of Physical Science*, 2nd ed., Chap. 2, Cambridge University Press, Cambridge, 1951.

Neugebauer, Otto: *The Exact Sciences in Antiquity*, Princeton University Press, Princeton, 1952.

Smith, David Eugene: *History of Mathematics*, Vol. I., Dover Publications, Inc., New York, 1958.

Struik, Dirk J.: *A Concise History of Mathematics*, Dover Publications, Inc., New York, 1948.

Taylor, Henry Osborn: *Ancient Ideals*, 2nd ed., Vol. I, Chaps. 7 through 13, The Macmillan Co., New York, 1913.

Wedberg, Anders: *Plato's Philosophy of Mathematics*, Almqvist and Wiksell, Stockholm, 1955 (for students of philosophy).

CHAPTER 4

NUMBER: THE FUNDAMENTAL CONCEPT

A marvelous neutrality have these things Mathematical, *and also a strange participation between things supernatural, immortal, intellectual, simple, and indivisible, and things natural, mortal, sensible, compounded and divisible.*

JOHN DEE (1527–1608)

4-1 INTRODUCTION

Just as we are inclined to accept the sun, moon, and stars as our birthright and do not appreciate the grandeur, the mystery, and the knowledge which can be gleaned from the contemplation of the heavens, so are we inclined to accept our number system. There is, however, this difference. Many of us would not claim the latter and would gladly sell it for a mess of pottage. Because we are forced to learn about numbers and operations with numbers while we are still too young to appreciate them—a preparation for life which hardly excites our interest in the future—we grow up believing that numbers are drab and uninteresting. But the number system warrants attention not only as the basis of mathematics, but because it contains weighty and beautiful ideas which lend themselves to powerful applications.

Among past civilizations, the Greeks best appreciated the wonder and power of the concept of number. They were, of course, a people with great intellectual perception, but perhaps because they viewed numbers abstractly, they saw more clearly their true nature. The very fact that one can abstract from many diverse collections of objects a property such as "fiveness" struck the Greeks as a marvelous discovery. If one may use the ridiculous to accentuate the sublime, one may say that the Greek delight in numbers was the rational counterpart of the hysteria which many young and old Americans experience when they encounter numbers in the form of baseball scores and batting averages.

4-2 WHOLE NUMBERS AND FRACTIONS

The first Greeks who, to our knowledge, expressed their satisfaction with numbers and propounded a philosophy based on numbers which is extremely alive and vital today were the Pythagoreans. This group was founded in the middle of the sixth century B.C. by Pythagoras. We know rather little that is certain about this man. However, it seems very likely that he was born in 569 B.C. in a Greek settlement on the island of Samos in the Aegean Sea. Like many other Greeks he traveled to Egypt and to the Near East to learn what these older civilizations had to offer, and then settled in Croton, another Greek city in southern Italy. Pythagoras and his followers were among the early founders of the great Greek civilization, and so it is not surprising to find that the rational attitude which characterizes the Greeks was still surrounded in his times with mystical and religious doctrines prevalent in Egypt and its eastern neighbors. In fact the Pythagoreans were a religious sect as well as students of philosophy and nature.

Membership in the group was restricted, and the members were pledged to secrecy. Among their religious doctrines was the belief that the soul was tainted by the body. To purify the soul they maintained celibacy; their religious practices were also supposed to be efficacious in purifying the soul. At death the soul was reincarnated in another human or an animal. Like most mystics they observed certain taboos. They would not touch a white cock, walk on the highways, use iron to stir a fire, or leave the marks of ashes on a pot.

The secrecy of the group, its aloofness, and an attempted interference in the political affairs of Croton finally aroused the people of this city to drive out the Pythagoreans. We do not know for certain what happened to Pythagoras. One story has it that he fled to Metapontum, another Greek city in southern Italy, and was murdered there. However, the Pythagoreans continued to be influential in Greek intellectual life. One of their notable members was the philosopher Plato.

The Pythagoreans were impressed with numbers and, because they were mystics, attached to the whole numbers meanings and significances which we now regard as childish. Thus, they considered the number "one" as the essence or very nature of reason, for reason could produce only one consistent body of doctrines. The number "two" was identified with opinion, clearly because the very meaning of opinion implies the possibility of an opposing opinion, and thus of at least two. "Four" was identified with justice because it is the first number which is the product of equals. Of course, one can also be thought of as 1 times 1, but to the Pythagoreans one was not a number in the full sense because it did not represent quantity. The Pythagoreans represented numbers as dots in sand or by pebbles, and for each number the dots or pebbles had a special arrangement. Thus the number "four" was pictured as four dots suggesting a square, and so the square and justice were also linked. Foursquare and square shooter still mean a person who acts justly. "Five" signified marriage because it was the union of the first masculine number, three, and the first feminine number, two. (Odd numbers were masculine and even numbers feminine.) The number "seven" represented health and "eight," friendship or love.

We shall not pursue all the ideas which the Pythagoreans developed about numbers. What is significant about their work is that they were the first to study properties of whole numbers. As we shall see in a later chapter, they also possessed the vision of deep mystics and saw that numbers could be used to represent and even embody the essence of natural phenomena.

The speculations and results obtained by the Pythagoreans about whole numbers and ratios of whole numbers, or fractions as we prefer to call them, were the beginning of a long and involved development of arithmetic as a science as opposed to arithmetic as a tool for daily applications. During the 2500 years since the Pythagoreans first called attention to the importance of numbers, man has not only learned to better appreciate the idea but has invented excellent methods of writing quantity and of performing the four operations of arithmetic, i.e., ambition, distraction, uglification, and derision, as Lewis Carroll called them. While these methods of writing and operating with numbers are largely familiar, there are a few facts which are worthy of comment.

One of the most important members of our present number system is the mathematical representation of no quantity, that is, zero. We are accustomed to this number and yet usually fail to appreciate two facts about it. The first is that this member of our number system came rather late. The idea of using zero was conceived by the Hindus and, like other of their ideas, reached Europe through the Arabs. It had not occurred to earlier civilizations, even to the Greeks, that it would be useful to have a number which represents the absence of any objects. Connected with this late appearance of the number is the second significant fact, namely, that zero must be distinguished from nothing. Undoubtedly it was the inability of earlier peoples to perceive this distinction which accounts for their failure to introduce the zero. That zero must be distinguished from nothing is easily seen from several examples. A student's grade in a course he never took is no grade or nothing. He may, however, have the grade of zero in a course he has taken. If a person has no account in a bank, his balance is nothing. If he has a bank account, he may very well have a balance of zero.

Because zero is a number, we may operate with it; for example, we may add zero to another number. Thus $5 + 0 = 5$. By contrast $5 +$ nothing is meaningless or nothing. The only restriction on zero as a number is that one cannot divide by zero. Division by zero does, so to speak, produce nothing. Because so many false steps in mathematics result from division by zero, it is well to understand clearly why we cannot do this. The answer to a problem of division, say $\frac{6}{2}$, is some number which when multiplied by the divisor yields the dividend. In our example, 3 is the answer because $3 \cdot 2 = 6$. Hence the answer to $\frac{5}{0}$ should be a number which when multiplied by 0 gives 5. However, any number multiplied by 0 gives 0 and not 5. Thus, there is no answer to the problem of $\frac{5}{0}$. In the case of $\frac{0}{0}$ the answer should be some number which when multiplied by 0 yields the dividend 0. However, any number may then serve as a quotient because any number multiplied by 0 gives 0. But mathematics cannot tolerate such an ambiguous situation. If $\frac{0}{0}$ arises and any number may serve as an answer, we do not know what number to take and hence are not aided. It is as if we asked a person for directions to some place and he replied, Take any direction.

With the availability of zero, mathematicians were finally able to develop our present method of writing whole numbers. First of all we count in units and represent large quantities in tens, tens of tens, tens of tens of tens, etc. Thus we represent two hundred and fifty-two by 252. The left-hand 2 means, of course, two tens of tens; the 5 means 5 times 10; and the right-hand 2 means 2 units. The concept of zero makes such a system of writing quantities practical since it enables us to distinguish 22 and 202. Because ten plays such a fundamental role, our number system is called the *decimal system,* and ten is called the *base.* The use of ten resulted most likely from the fact that man counted on his fingers and, when he had used the fingers on his two hands, considered the number arrived at as a larger unit.

Because the position of an integer determines the quantity it represents, the principle involved is called *positional notation.* The decimal system of positional notation is due to the Hindus; however, the same scheme was used two millenniums earlier by the Babylonians, but with base 60 and in more limited form since they did not have zero.

The operations of arithmetic, addition, subtraction, multiplication, and division, are of course familiar to us, but it is perhaps not recognized that these operations are quite sophisticated and remarkably efficient. They date back to Greek times and gradually evolved, as improvements in the methods of writing numbers and the concept of zero were introduced. The Europeans picked up the methods from the Arabs. Previously the Europeans had used the Roman system of writing numbers, and the operations were based on that system. Partly because these latter methods were relatively cumbersome and partly because education was limited to a few people, those who acquired the art of calculation were regarded as skilled mathematicians. In fact the processes defied the average man so much that it seemed to him that those possessed of the ability must have magical powers. Good calculators were called practitioners of the "Black Art."

To appreciate the efficiency of our present methods we would have to learn the older ones and even acquire some facility in them, to make the comparison a fair one. But we cannot spare the time and effort. Perhaps the one point we should emphasize is how much our methods of arithmetic depend upon positional notation. This can be seen even in a simple problem of addition. To add 387 and 359 say, the written work is

$$\begin{array}{r} 387 \\ \underline{359} \\ 746 \end{array}$$

However, in performing this work, we think as follows. We add the units 7 and 9, the "tens" quantities 8 and 5, and the "hundreds" 3 and 3, separately. When we add the 7 and 9, we obtain 16. We recognize that 16 is $1 \cdot 10 + 6$, and so we add the $1 \cdot 10$ to the $13 \cdot 10$ already obtained from the 8 and 5. We say that we "carry" the $1 \cdot 10$ over, and instead of $13 \cdot 10$ we obtain $14 \cdot 10$. However, $14 \cdot 10$ is $(10 + 4) \cdot 10$ or $1 \cdot 10^2 + 4 \cdot 10$. Thus we write 4 in the tens' column, add the $1 \cdot 10^2$ to the $6 \cdot 10^2$ already obtained from the 3 and 3, and arrive at $7 \cdot 10^2$. All these steps are usually executed rather mechanically by writing the appropriate numbers in the units', tens', and hundreds' places and by using the process called carrying. Were we to analyze the processes of subtraction, multiplication, and division, we would again see how the steps which we learn mechanically in elementary school are just the skeletal processes of thinking suited to positional notation in base ten.

A word about fractions may also be in order. The natural method of writing fractions, for example, $\frac{2}{3}$ or $\frac{7}{5}$, to express parts of a whole presents no difficulties of comprehension. However the operations with fractions do seem to be somewhat arbitrary and mysterious. To add $\frac{2}{3}$ and $\frac{7}{5}$, say, we go through the following process:

$$\frac{2}{3} + \frac{7}{5} = \frac{10}{15} + \frac{21}{15} = \frac{31}{15}.$$

What we have done is to express each fraction in an equivalent form such that the denominators are now alike, and then add the numerators. We are not required by law to add fractions in this manner. It would, of course, be much simpler if we agreed to add fractions by adding the numerators and adding the denominators so that

$$\frac{2}{3} + \frac{7}{5} = \frac{9}{8}.$$

As a matter of fact, when we multiply two fractions, we do multiply the numerators and multiply the denominators so that it does seem as though the mathematicians prefer to be unnecessarily complicated about the addition of fractions.

The explanation of this seeming mathematical idiosyncrasy is simple: the operations with fractions are formulated to fit experience. When one has $\frac{2}{3}$ of a pie and $\frac{7}{5}$ of a pie, he has in all not $\frac{9}{8}$ but $\frac{31}{15}$ of a pie. In other words, if mathematical concepts and operations are to fit experience, the nature of the operations is forced upon us. In the case of multiplication of fractions it is correct that multiplying the numerators and multiplying the denominators will yield the fraction which represents the physical result. Thus suppose we had to find $\frac{2}{3}$ of $\frac{7}{5}$, that is $\frac{2}{3} \cdot \frac{7}{5}$. We think of $\frac{2}{3}$ as $2 \cdot \frac{1}{3}$. Now

$$\tfrac{1}{3} \cdot \tfrac{7}{5} = \tfrac{1}{3} \cdot \tfrac{21}{15} = \tfrac{7}{15}.$$

Then

$$2 \cdot \tfrac{1}{3} \cdot \tfrac{7}{5} = 2 \cdot \tfrac{7}{15} = \tfrac{14}{15}.$$

The same result is obtained by multiplying the original numerators and multiplying the original denominators.

The operation of dividing one fraction by another presents a little more difficulty. To see how we arrive at the correct process let us start with some simple examples. Suppose we had to answer the question of how many one-thirds of a pie are in 2 pies. Mathematically this question is formulated as how much is

$$\frac{2}{\frac{1}{3}}?$$

We should note that one bar is larger than the other and the longer bar separates the numerator, the 2, from the denominator $\frac{1}{3}$. Now, we know on physical grounds that we can obtain 6 one-thirds from 2 pies. We can obtain this answer arithmetically by inverting the denominator $\frac{1}{3}$ and multiplying the inverse into the numerator 2. That is,

$$\frac{2}{\frac{1}{3}} = \tfrac{2}{1} \cdot \tfrac{3}{1} = \tfrac{6}{1} = 6.$$

Now let us complicate the problem slightly. How many two-thirds of a pie are contained in 2 pies? Again this question is formulated mathematically as

$$\frac{2}{\frac{2}{3}}.$$

We know on physical grounds that there are 3 two-thirds of a pie in 2 pies. We can obtain this answer arithmetically by inverting the denominator and multiplying this inverse fraction into the numerator. Thus,

$$\frac{2}{\frac{2}{3}} = \tfrac{2}{1} \cdot \tfrac{3}{2} = \tfrac{6}{2} = 3.$$

Now let us complicate the problem still more. We would certainly agree that 2 pies are the same as $\frac{10}{5}$ pies. If therefore we had to answer the question of how many two-thirds of a pie are contained in $\frac{10}{5}$ pies, we would know from the preceding example that the answer is 3. How could we obtain this answer directly? The question is, how much is

$$\frac{\frac{10}{5}}{\frac{2}{3}}?$$

Let us invert the denominator and multiply the inverse into the numerator. Thus

$$\frac{\frac{10}{5}}{\frac{2}{3}} = \frac{10}{5} \cdot \frac{3}{2} = \frac{30}{10} = 3.$$

Again we see that the process of inverting the denominator and multiplying it into the numerator gives the result which we know on physical grounds is correct.

The significant point, then, is that the rule "to divide one fraction by another, invert the denominator and multiply this inverse into the numerator" is designed to make the mathematical operation give a result which fits experience. This is, of course, the same principle which applies to the other operations. Logically, we may say that we define the operations to be what we have just illustrated for addition, multiplication and division, and in our purely mathematical definitions we do not say anything about agreement with physical facts. But, of course, the definitions would be pointless if they did not give physically correct results.

Fractions, like the whole numbers, can be written in positional notation. Thus

$$\frac{1}{4} = \frac{25}{100} = \frac{20}{100} + \frac{5}{100} = \frac{2}{10} + \frac{5}{100}.$$

If we now agree to suppress the powers of 10, that is 10, 100, and higher powers where they occur, then we can write $\frac{1}{4} = 0.25$. The decimal point reminds us that the first number is really $\frac{2}{10}$, the second $\frac{5}{100}$, and so forth. The Babylonians had employed positional notation for fractions, but they used base 60 rather than base 10, just as they had for whole numbers. The decimal base for fractions was introduced by sixteenth-century European algebraists. Of course, the operations with fractions can also be carried out in decimal form.

The disappointing feature of the decimal representation of fractions is that some simple fractions cannot be represented as decimals with a finite number of digits. Thus when we seek to express $\frac{1}{3}$ as a decimal, we find that neither 0.3, nor 0.33, nor 0.333, and so on, suffices. All one can say in this and similar cases is that by carrying more and more decimal digits one comes closer and closer to the fraction, but no finite number of digits will ever be the exact answer. This fact is expressed by the notation

$$\frac{1}{3} = 0.333\ldots,$$

where the dots indicate that we must keep on adding threes to approach the fraction $\frac{1}{3}$ more and more closely.

From the standpoint of applications the fact that some fractions cannot be expressed as decimals with a finite number of digits does not matter because we can always carry enough digits to obtain an answer as accurate as the application requires.

EXERCISES

1. What is the principle of positional notation?
2. Why is the number zero almost indispensable in the system of positional notation?
3. What is the meaning of the statement that zero is a number?
4. What two methods are there of representing fractions?
5. What principle determines the definitions of the operations with fractions?

4-3 IRRATIONAL NUMBERS

The Pythagoreans, as we noted earlier, were the first to appreciate the very concept of number, and sought to employ numbers to describe the basic phenomena of the physical and social worlds. Numbers to the Pythagoreans were also interesting in and for themselves. Thus they liked square numbers, that is, numbers such as 4, 9, 16, 25, 36, and so on, and observed that the sums of certain pairs of square numbers, or perfect squares, are also square numbers. For example, $9 + 16 = 25$, $25 + 144 = 169$, and $36 + 64 = 100$. These relationships can also be written as

$$3^2 + 4^2 = 5^2, \qquad 5^2 + 12^2 = 13^2, \qquad \text{and} \qquad 6^2 + 8^2 = 10^2.$$

The three numbers whose squares furnish such equalities are today called Pythagorean triples. Thus 3, 4, 5 constitute a Pythagorean triple because $3^2 + 4^2 = 5^2$.

The Pythagoreans liked these triples so much because, among other features, they have an interesting geometrical interpretation. If the two smaller numbers are the lengths of the sides or arms of a right triangle, then the third one is the length of the hypotenuse (Fig. 4–1). Just how the Pythagoreans knew this geometrical fact is not clear, but assert it they did. They also claimed that in *any* right triangle, the square of the length of one arm added to the square of the length of the other gives the square of the length of the hypotenuse. This more general assertion is still called the Pythagorean theorem and

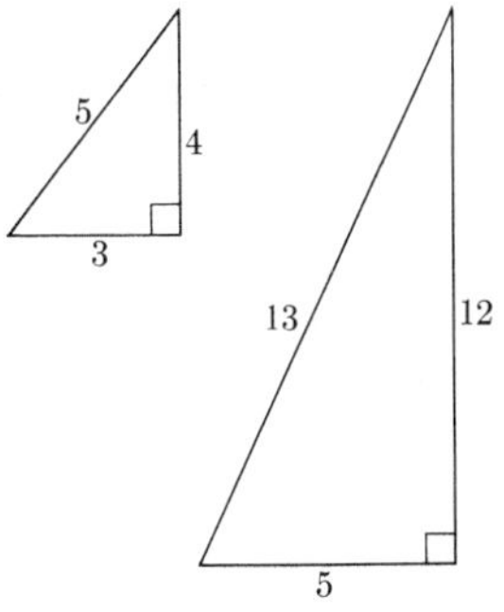

Fig. 4–1

a proof of it, such as we learn in high-school geometry, was given about 200 years later by Euclid. Pythagoras is said to have been so overjoyed with this theorem that he sacrificed an ox to celebrate its discovery.

This theorem proved to be the undoing of a central doctrine in the Pythagorean philosophy and caused woe and misery to many mathematicians. But before we pursue this story, we should look into a few simple properties of the whole numbers which are embodied in the following exercises.

EXERCISES

1. Prove that the square of any even number is an even number. [*Suggestion:* By definition every even number contains 2 as a factor.]
2. Prove that the square of any odd number is an odd number. [*Suggestion:* Every odd number ends in 1, 3, 5, 7, or 9.]
3. Let a stand for a whole number. Prove that if a^2 is even, then a is even. [*Suggestion:* Use the result in Exercise 2.]
4. Establish the truth or falsity of the assertion that the sum of any two square numbers is a square number.

There are tragedies in mathematics also, and one of these struck the very group of mathematicians who deserved a better fate. The Pythagoreans had constructed, at least to their own satisfaction, a philosophy which asserted that all natural phenomena and all social and ethical concepts were in essence just whole numbers or relationships among whole numbers. But one day it occurred to a member of the group to examine the seemingly simplest case of the Pythagorean theorem. Suppose each arm of a right triangle (Fig. 4–2) is 1 unit in length; how long, he asked, is the hypotenuse? The Pythagorean theorem says that the square of (the length of) the hypotenuse equals the sum of the squares of the arms. Hence if we call c the unknown length of the hypotenuse, then the theorem says that

$$c^2 = 1^2 + 1^2$$

or

$$c^2 = 2.$$

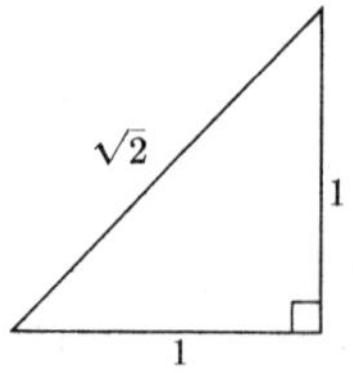

Fig. 4–2

Now 2 is not a square number, that is, a perfect square, and so c is not a whole number. But it certainly seemed reasonable to this Pythagorean that c should be a fraction; that is, there should be a fraction whose square is 2. Even the simple fraction $\frac{7}{5}$ comes close to being the correct value because $(\frac{7}{5})^2 = \frac{49}{25}$, and this is almost 2. However, simple trial does not easily yield a fraction whose square is 2. Hence this Pythagorean became worried, and he decided to investigate the question of whether there is a fraction whose square is 2. We shall examine his reasoning which, as far as we know, is the same as that given in Euclid's famous work on geometry, the *Elements*.

The number c which we seek to determine is one whose square is 2. Let us denote it by $\sqrt{2}$. All we mean by this symbol is that it represents a number whose square is 2. And now let us suppose that $\sqrt{2}$ is a fraction a/b, where a and b are whole numbers. Moreover, to make matters simpler, let us suppose that any factors common to a and b are cancelled. Thus if a/b were $\frac{8}{6}$, for example, we would cancel the common factor 2 and write it as $\frac{4}{3}$. Hence we have assumed so far that

$$\sqrt{2} = \frac{a}{b} \tag{1}$$

and that a and b have no common factors.

If equation (1) is correct, then by squaring both sides, a step which utilizes the axiom that equals multiplied by equals give equals (because we multiply the left side by $\sqrt{2}$ and the right side by a/b), we obtain

$$2 = \frac{a^2}{b^2}.$$

Again by employing the axiom that equals multiplied by equals yield equals, we may multiply both sides of this last equation by b^2 and write

$$2b^2 = a^2. \tag{2}$$

The left side of this equation is an even number because it contains 2 as a factor. Hence the right side must also be an even number. But if a^2 is even, then, according to Exercise 3 above, a must be even. If a is even, it must contain 2 as a factor. That is, $a = 2d$, where d is some whole number. If we substitute this value of a in (2) we obtain

$$2b^2 = (2d)^2 = 2d \cdot 2d = 4d^2. \tag{3}$$

Since then

$$2b^2 = 4d^2,$$

we may divide both sides of this equation by 2 and obtain

$$b^2 = 2d^2. \tag{4}$$

We now see that b^2 is an even number and so, by again appealing to the result in Exercise 3, we find that b is an even number.

What we have shown in the above argument is that if $\sqrt{2} = a/b$, then a and b must be even numbers. But at the very outset we had cancelled any common factors in a and b; yet we find that a and b still contain 2 as a common factor. This result contradicts the fact that a and b have no common factors.

Why do we arrive at a contradiction? Since our reasoning is correct, the only possibility is that the assumption that $\sqrt{2}$ equals a fraction is not correct. In other words, $\sqrt{2}$ cannot be a ratio of two whole numbers.

This proof is so neat that one can almost believe the legend that Pythagoras sacrificed an ox in honor of its creation. But there are at least two reasons for discrediting this tale. The first is that if all the legends telling of Pythagoras sacrificing an ox were true, he could not have had time for mathematics. The second reason is that the above proof was not a triumph for the Pythagoreans but a disaster. The symbol $\sqrt{2}$ is a number because it represents the length of a line, namely the hypotenuse of the triangle in Fig. 4–2. But this number is not a whole number or a fraction. The Pythagoreans had, however, developed an embracing philosophy which asserted that everything in the universe reduced to whole numbers. Clearly, then, this philosophy was inadequate. Indeed the existence of numbers such as $\sqrt{2}$ was such a serious threat to the Pythagorean philosophy that another legend, more credible, states that the Pythagoreans, who were at sea when the above discovery was made, threw overboard the member who made it, and pledged to keep the discovery secret.

But secrets will out, and later Greeks not only learned that $\sqrt{2}$ is neither a whole number nor a fraction, but they discovered that there is an indefinitely large collection of other numbers which are not whole numbers or fractions. Thus $\sqrt{3}$, $\sqrt{5}$, $\sqrt{7}$, and, more generally, the square root of any number which is not a perfect square, the cube root of any number which is not a perfect cube, and so on, are numbers which are not whole numbers or fractions. The number π, which is the ratio of the circumference of any circle to its diameter, is also neither a whole number nor a fraction. All these new numbers are called irrational numbers, the word "irrational" now meaning that these numbers cannot be expressed as ratios of whole numbers, although in Pythagorean times it meant unmentionable or unknowable.

If these irrational numbers are really so common and represent lengths of sides of triangles and circumferences of circles in terms of the diameters, why weren't they encountered before? Didn't the Babylonians and Egyptians run across them? They did. But since they were concerned only with having numbers serve their practical purposes, they used convenient approximations. Thus, when they encountered a length such as $\sqrt{2}$, they were content to use a value such as 1.4 or 1.41. For π, as we noted in an earlier chapter, they used values even as crude as 3. Not only did these peoples use such approximations, but they never realized that the most complicated fraction or decimal could never represent an irrational number exactly. The Egyptians and Babylonians treated irrational numbers and their mathematics in general rather lightheartedly. We may hail their blithe spirits, but mathematicians they never were.

The Greeks, as we know, were of a different intellectual breed and could not be content with approximations, but they also exhibited a weakness. Although they recognized that quantities exist which are neither whole numbers nor fractions, they were so convinced that the concept of number could not comprise anything else than whole numbers or fractions that they did not accept irrationals as numbers. Instead they thought of such quantities only as geometrical lengths or areas. Thus the Greeks never did develop an *arithmetic* of irrational numbers. In their astronomical work, for example, they used only whole numbers and fractions. The difficulty which the Greeks experienced also baffled all mathematicians up to modern times. The greatest mathematicians refused to accept irrationals as numbers and followed the Greek procedure of thinking about such quantities as lengths or areas. All these people wished that the Pythagoreans had thrown all irrational numbers overboard rather than the man who discovered them.

But the needs of society often oblige even mathematicians to face unpleasantnesses. In the seventeenth century, science began to develop at an amazing rate, and science needs quantitative results. It may be nice to know that $\sqrt{2}$ is a certain length and that $\sqrt{2} \cdot \sqrt{3}$ is an area, but this knowledge does not suffice when one needs numerical results. And so finally mathematicians had to accept the fact that if they were to treat numerically all the quantities that arise in scientific work, they must handle irrational numbers as numbers. The mathematicians' refusal over centuries to grant irrationals the status of numbers illustrates one of the surprising features of the history of mathematics. New ideas are often as unacceptable in this field as they are in politics, religion, and economics.

The situation, then, which must be faced squarely is that there are other numbers besides whole numbers and fractions. It is, of course, quite understandable that whole numbers and fractions should have been created and used first, for these numbers arise in the simplest physical situations man encounters. The irrational numbers on the other hand are not commonly encountered. Only the application of a theorem such as the Pythagorean theorem brings them to our attention, and even then one must go through a proof such as that examined above, to see that they are not whole numbers or fractions. But the fact that irrational numbers are late-comers does not mean that they are less acceptable or less genuine numbers. Just as we gradually add to our knowledge of the varieties of human beings and animals which exist in our physical world, so must we broaden our knowledge of the varieties of numbers and with true liberality accept these strangers on the same basis as the already familiar numbers.

However, if we are to use irrational numbers, we must know how to operate with them, that is, how to add, subtract, multiply, and divide them. We have already noted with whole numbers and fractions that if we wish the operations to fit experience, we must formulate the operations accordingly. So it is with the irrational numbers. We could define addition, multiplication, and the other operations as we please. But if we wish these operations to represent physical situations, we must define them properly. However, there is no real difficulty here. Since irrational numbers are quantities, as are whole numbers and fractions, we may use the latter as a guide to the proper operations with irrational numbers.

Let us consider a few examples which will be sufficient to indicate the general principles. Should we say that

$$\sqrt{2} + \sqrt{3} = \sqrt{5}?$$

To answer this question let us consider the analogous question: May we say that

$$\sqrt{4} + \sqrt{9} = \sqrt{13}?$$

It is clear in the latter case that $2 + 3$ does not equal $\sqrt{13}$, for $\sqrt{13}$ is certainly less than 4. Hence we should *not* add the radicands, that is, the 2 and the 3, in the preceding equation. One might then ask, How much is $\sqrt{2} + \sqrt{3}$? Since both summands are numbers, the sum is also a number, but it cannot be written more compactly than $\sqrt{2} + \sqrt{3}$. This inability to combine the summands is not something new or troublesome. When we add 2 and $\frac{1}{2}$, for example, the answer continues to be $2 + \frac{1}{2}$. We usually omit the plus sign and write $2\frac{1}{2}$, but the summands are really not combined.

Let us consider next whether

$$\sqrt{2} \cdot \sqrt{3} = \sqrt{6}.$$

Here too we shall see what the analogous operation with whole numbers suggests. Is it true that

$$\sqrt{4} \cdot \sqrt{9} = \sqrt{36}?$$

The answer is clearly yes, and so we shall agree that to multiply square roots we shall multiply the radicands. That is,

$$\sqrt{2} \cdot \sqrt{3} = \sqrt{6}.$$

The definitions of the operations of subtraction and division are also readily determined. Thus $\sqrt{3} - \sqrt{2}$ yields a definite number, but the difference cannot be written any more compactly than $\sqrt{3} - \sqrt{2}$.

For division, say $\sqrt{3}/\sqrt{2}$, the procedure, as in the case of multiplication, is suggested by observing that

$$\frac{\sqrt{9}}{\sqrt{4}} = \sqrt{\frac{9}{4}},$$

for this equation simply says that $\frac{3}{2} = \frac{3}{2}$. Hence we shall agree that

$$\frac{\sqrt{3}}{\sqrt{2}} = \sqrt{\frac{3}{2}}.$$

The general principle which these examples illustrate is that operations with irrational numbers are defined so as to agree with the same operations on whole numbers when the latter are expressed as roots. We could state our definitions in general form, but there is no need to do so.

In applications we often approximate irrational numbers by fractions or decimals because actual physical objects cannot be constructed exactly anyway. Thus if we had to construct a length which strictly should be $\sqrt{2}$, we would approximate $\sqrt{2}$. Since $(1.4)^2 = 1.96$ and 1.96 is nearly 2, we could approximate $\sqrt{2}$ by 1.4. If we desired a more accurate approximation, we might determine to the nearest hundredth the number whose square approximates 2. Thus, since

$$(1.41)^2 = 1.988 \qquad \text{and} \qquad (1.42)^2 = 2.016,$$

we see that 1.41 is a good two-decimal approximation of $\sqrt{2}$. We could, of course, improve still more on the accuracy of the approximation. We should, however, realize that no matter how many decimal places we employed, we would never obtain a number which is exactly $\sqrt{2}$ because any decimal with a finite number of digits or a whole number plus such a decimal is just another way of writing a fraction, whereas $\sqrt{2}$, as the above proof showed, can never equal a quotient of two whole numbers.

The fact that we often approximate an irrational number when we wish to construct something raises a question which merits an answer. The question is, Why don't we approximate irrational numbers wherever they arise and forget about operations with irrationals as such? For example, to calculate $\sqrt{2} \cdot \sqrt{3}$, we could approximate $\sqrt{2}$ by, say 1.41, approximate $\sqrt{3}$ by 1.73, and then multiply 1.41 by 1.73. The answer is 2.44, and since $(2.44)^2$ is 5.95, we see that we have a good approximation to $\sqrt{6}$. If we wanted a more accurate answer, we could approximate $\sqrt{2}$ and $\sqrt{3}$ more closely and then multiply. One reason we do not approximate in mathematics proper is that mathematics is an exact science. It insists on reasoning as rigorous as human beings can perform. We pay a price for this rigor by expending more thought and effort, but we shall see that mathematics has made its contributions just because it insists on exactness.

There is also a practical advantage in working with irrational numbers as such. Let us suppose that some problem required us to calculate $(\sqrt{3})^4$, that is, $\sqrt{3} \cdot \sqrt{3} \cdot \sqrt{3} \cdot \sqrt{3}$. The person who insists on approximating would now approximate $\sqrt{3}$ to some number of decimal places, for example, 1.732, and then calculate $(1.732)^4$ While the practical person takes an hour to calculate and check his arithmetic, the mathematician would see at once that

$$\sqrt{3}\cdot\sqrt{3}\cdot\sqrt{3}\cdot\sqrt{3}=(\sqrt{3}\cdot\sqrt{3})(\sqrt{3}\cdot\sqrt{3})=3\cdot 3=9,$$

and could spend the rest of the hour in refreshing sleep. Moreover, the mathematician's answer is exact, whereas the practical man's answer is not accurate even to the four demical places with which he started, because the product of two approximate numbers is less accurate than either factor. To achieve an answer accurate to four decimal places, the practical man would have to use an approximation of $\sqrt{3}$ containing seven decimal places and then multiply.

The irrational number is the first of many sophisticated ideas which the mathematician has introduced to think about and cope with the real world. The mathematician creates these concepts, devises ways of working with them which fit real situations, and then uses his abstractions to think about the phenomena to which the ideas apply.

EXERCISES

1. Express the answers to the following problems as compactly as you can:

a) $\sqrt{3}+\sqrt{5}$ b) $\sqrt[3]{2}+\sqrt[3]{7}$ c) $\sqrt[3]{2}+\sqrt{7}$ d) $\sqrt{7}+\sqrt{7}$

e) $\sqrt{3}\cdot\sqrt{7}$ f) $\sqrt[3]{2}\cdot\sqrt[3]{5}$ g) $\sqrt[3]{2}\cdot\sqrt[3]{4}$ h) $\sqrt{12}\cdot\sqrt{3}$

i) $\dfrac{\sqrt{5}}{\sqrt{2}}$ j) $\dfrac{\sqrt{8}}{\sqrt{2}}$ k) $\dfrac{\sqrt[3]{10}}{\sqrt[3]{2}}$

2. Simplify the following:

a) $\sqrt{50}$ b) $\sqrt{200}$ c) $\sqrt{75}$

[*Suggestion:* $\sqrt{50}=\sqrt{25\cdot 2}=\sqrt{25}\cdot\sqrt{2}$.]

3. Criticize the following argument: No irrational number can be expressed as a decimal with a finite number of decimal places. The number $\frac{1}{3}$ cannot be expressed as a decimal with a finite number of decimal places. Hence $\frac{1}{3}$ is an irrational number.

4-4 NEGATIVE NUMBERS

One more addition to the number system which has considerably extended the power of mathematics comes from far-off India. Numbers are commonly used to represent an amount of money, in particular the amount of money which a person owns. Perhaps because the Hindus were in debt more often than not, it occurred to them that it would also be useful to have numbers which represent the amount of money one owes. They therefore invented what are now called negative numbers, while the previously known numbers are called positive numbers. Thus numbers which we denote by -3, $-\frac{5}{2}$, and $-\sqrt{2}$ came into existence. Where necessary to distinguish clearly positive from negative numbers or to emphasize what is positive as opposed to what is negative, one writes $+3$ or $+\frac{5}{2}$ instead of 3 or $\frac{5}{2}$.

It is not necessary, incidentally, to use such symbols as -3 to represent the negative counterpart to 3. Modern banks and large commercial corporations, which deal with negative numbers continually, often write these in red ink, whereas positive numbers are written in black ink. However, we shall find that placing the minus sign in front of a number to indicate a negative number is a convenience.

The use of positive and negative numbers is not limited to the representation of assets and debts. One represents temperatures below 0° as negative temperatures, while temperatures above 0° are positive. Likewise heights above and below sea level can be represented by positive and negative numbers, respectively. It is sometimes convenient to represent time after and before a specified event by positive and negative numbers. For example, using the birth of Christ as the event, the year 50 B.C. can just as well be described as the year -50.

To derive more use from the concept of negative numbers it must be possible to operate with them just as we operate with positive numbers. The operations with negative numbers and with negative and positive numbers together are easy to understand if one keeps in mind the physical significance of these operations. For example, suppose a man has assets of 3 dollars and debts of 8 dollars. What is his net wealth? Clearly the man is 5 dollars in debt. The same calculation is represented in terms of positive and negative numbers by stating that the amount 8 dollars must be taken from 3 dollars, that is, $3 - 8$, or that a debt of 8 dollars must be added to assets of 3 dollars, that is, $+3 + (-8)$. The answer is obtained by subtracting the smaller numerical value (that is, the smaller number without regard to sign) from the larger numerical value and giving the answer the sign attached to the larger numerical value. That is, we subtract 3 from 8 and call the answer negative because the larger numerical value, namely 8, has the minus sign attached to it.

Since negative numbers represent debts and subtraction usually has the physical meaning of "taking away" or "removing," then the subtraction of a negative number means the removal of a debt. Thus, if a person has assets of, say 3 dollars, but this figure already takes into account a debt of 8 dollars, the removal or cancellation of the debt leaves the person with assets of 11 dollars. Mathematically we say $+3 - (-8) = +11$. In words, to subtract a negative number we add the corresponding positive number.

Suppose a man goes into debt at the rate of 5 dollars per day. Then in 3 days after a given date he will be 15 dollars in debt. If we denote a debt of 5 dollars as -5, then going into debt at the rate of 5 dollars per day for 3 days can be stated mathematically as $3 \cdot (-5) = -15$. That is, the multiplication of a positive and a negative number yields a negative number whose numerical value is the product of the two given numerical values.

In the very same situation in which a man goes into debt at the rate of 5 dollars per day, his assets three days *before* a given date are 15 dollars more than they are at the given date. If we represent time before the given or zero date by -3 and the loss per day as -5, then his relative financial position 3 days *ago* can be expressed as $-3 \cdot (-5) = +15$; that is, to consider his assets three days ago, we would multiply the debt per day by -3, whereas to calculate the financial status three days in the future, we multiply by $+3$. Hence the result is $+15$ in the former case compared to -15 in the latter.

There is one more definition concerning negative numbers which is readily seen to be sensible. For the positive numbers and zero we say for obvious reasons that 3 is greater than 2, that 2 is greater than $\frac{1}{2}$, and that any positive number is greater than zero. The negative numbers are said to be less than the positive numbers and zero. Moreover, we say that -5 is less than -3, or that -3 is greater than -5. If one thinks of these various numbers as representing people's wealth, then the agreement concerning their order fits our usual understanding of relative wealth. A person whose financial status is -3 is wealthier than one whose status is -5; one is better off to be 3 dollars than 5 dollars in debt. Incidentally, the symbol $>$ is used to denote "greater than" as in $5 > 3$, and the symbol $<$ denotes "less than" as in $-5 < -3$.

The relative position of the various positive and negative numbers and zero is readily remembered if one visualizes these numbers as points on a line as shown in Fig. 4–3. The figure is really not different from that obtained by moving a thermometer scale into a horizontal position.

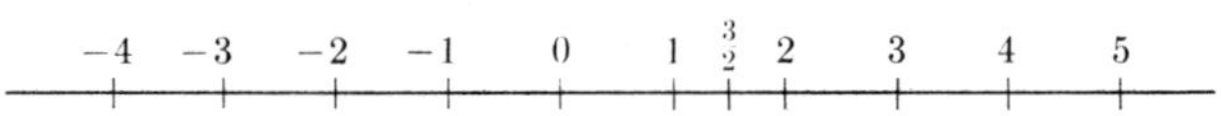

Fig. 4–3

The above situations, which illustrate how the definitions of the operations with positive and negative numbers were suggested, are of course by no means the only ones in which positive and negative numbers are employed. Indeed the usefulness of negative numbers would hardly be great were this the case. However, these simple financial transactions show not only how mathematicians arrived at the definitions, but that there is no more mystery about negative numbers than about positive ones. The definitions represent in abstract form what takes place physically, and, as with all numbers, we can think in terms of the abstractions to arrive at a knowledge of physical happenings.

It may be of some comfort to the reader to know that the concept of negative numbers, like the concept of irrational numbers, was resisted by mathematicians for several hundred years. The history of mathematics illustrates the rather significant observation that it is more difficult to get a truth accepted than to discover it. The mathematicians to whom "number" meant whole numbers and fractions found it hard to accept negative numbers as true numbers. They, too, failed to realize for centuries that mathematical concepts are man-made abstractions which can be introduced at will if they can serve useful purposes.

EXERCISES

1. Suppose a man has \$3 and incurs a debt of \$5. What is his net worth?
2. Suppose a man owes \$5 and then incurs a new debt of \$8. Use negative numbers to calculate his financial condition.
3. Suppose a man owes \$5 and earns \$8. Use positive and negative numbers to calculate his net worth.
4. Suppose a man owes \$13, and a debt of \$8 is cancelled. Use negative numbers to calculate his net worth.
5. A man loses money in business at the rate of \$100 per week. Let us denote this change in his assets by -100 and let us denote time in the future by positive numbers and time in the past by negative numbers. How much will the man lose in 5 weeks? How much more was the man worth 5 weeks ago?

4-5 THE AXIOMS CONCERNING NUMBERS

In the preceding chapter we said that mathematics proceeds by deductive reasoning from explicitly stated axioms. Yet thus far in this chapter we have said nothing about axioms. The reason is simply that the axioms concerning numbers are such obvious properties that we use them automatically without realizing that we are doing so.

This situation may perhaps be better understood by means of an analogy. Whenever a child at play throws a ball up into the air, he expects that the ball will come down. He is really assuming that all balls thrown up will come down. Of course, this assumption is well founded in experience; nevertheless, the child's expectation that the ball will come down is a deduction from the assumption just stated and the additional premise that he is throwing a ball up into the air. Recognition of the fact that he has made an assumption makes clear the reasoning, conscious or unconscious, behind the act.

To understand the deductive process in the mathematics of numbers, as well as in geometry, we must recognize the existence and use of the axioms. We do not hesitate to say that $275 + 384 = 384 + 275$. Surely we did not add 384 objects to 275, count the total, then add 275 objects to 384, count that total, and check that the two totals agree. Rather, whenever in our experience we combined two groups of objects, we found that we obtained the same total collection regardless of whether we put the first group with the second or the second with the first. Of course, our evidence to the effect that the order of addition is immaterial is limited to a small number of cases, whereas in practice we use this fact with all numbers. Hence, we are really making an assumption, namely, that for *any* two numbers a and b, integral, fractional, irrational, and negative, the order of addition will not affect the result. Thus our assumption also includes the affirmation that $\sqrt{3} + \sqrt{5} = \sqrt{5} + \sqrt{3}$. It is important for another reason to recognize that this assumption is being made. Numbers are not apples or cows. They are abstractions from physical situations. Mathematics works with these abstractions in order to deduce information about physical situations. However, if the axioms are not well chosen, the deductions will not apply. Hence it is well to note what assumptions are being employed and to ascertain that they are well founded in experience.

Let us, therefore, note the axioms which we have been using and will continue to use. The first axiom is the one discussed in the preceding paragraph:

AXIOM 1. For any two numbers a and b,

$$a + b = b + a.$$

The axiom is called the *commutative axiom of addition* because it says that we can commute or interchange the order of the two numbers to be added. We note that subtraction is not commutative, that is, $3 - 5$ does not equal $5 - 3$.

If we had to calculate $3 + 4 + 5$, we could first add 4 to 3 and then add 5 to this result, or we could add 5 to 4, and then add this result to 3. Of course, the result is the same in the two cases, and this is exactly what our second axiom says.

AXIOM 2. For any numbers a, b, and c,

$$(a + b) + c = a + (b + c).$$

This axiom is called the *associative axiom of addition* because we can associate the three numbers in two different ways in performing the addition.

The two axioms we have just discussed have their analogues for the operation of multiplication.

AXIOM 3. For any two numbers a and b,

$$a \cdot b = b \cdot a.$$

This axiom is called the *commutative axiom of multiplication*. Incidentally, the dot which is used to denote multiplication is omitted if there is no danger of misunderstanding. Thus, we could as well write $ab = ba$. The axiom is clearly a property of numbers; yet we sometimes fail to recognize that it is applicable. Many a student hesitates to write $5 \cdot a$ instead of $a \cdot 5$. But the commutative axiom says that the two expressions are equal. We might note in this context that the operation of division is not commutative, for $4 \div 2$ does not equal $2 \div 4$.

AXIOM 4. For any three numbers a, b, and c,

$$(ab)c = a(bc).$$

This axiom is called the *associative axiom of multiplication.* Thus $(3 \cdot 4)5 = 3(4 \cdot 5)$.

We also find in our work with numbers that it is convenient to use the number 0. To recognize formally that there is such a number and that it has the properties which its physical meaning requires, we state another axiom.

AXIOM 5. There is a unique number 0 such that

a) $0 + a = a$ for every number a,
b) $0 \cdot a = 0$ for every number a,
c) if $ab = 0$ then either $a = 0$ or $b = 0$ or both are 0.

The number 1 is another whose properties are somewhat special. Again, we know from the physical meaning of 1 just what its properties are, but if one is to justify the operations with 1 by appealing to axioms rather than to physical meaning, there must be a statement which tells us just what these properties are. In the case of 1, it is sufficient to specify a sixth axiom.

AXIOM 6. There is a unique number 1 such that

$$1 \cdot a = a$$

for every number a.

In addition to adding and multiplying any two numbers, we also have physical uses for the operations of subtraction and division. We know that given any two numbers a and b, there is a number c which results when b is subtracted from a. From a practical standpoint, it is helpful to recognize that subtraction is the inverse operation to addition. What this means is simply that if we have to find the answer to $5 - 3$ we can and, in fact, do ask ourselves what number added to 3 gives 5. If we know addition, we can then answer the subtraction problem. Even if we obtain the answer by a special subtraction process, and we do in the case of large numbers, we check it by adding the result to what we subtracted to see if it gives the original number or minuend. Hence a subtraction problem such as $5 - 3 = x$ really asks for what number x added to 3 gives 5, that is, $x + 3 = 5$.

In our logical development of the number system we wish to affirm that we can subtract any number from any other number and we phrase this statement so that the meaning of subtraction is precisely what it is, the inverse of addition.

AXIOM 7. If a and b are any two numbers, there is a unique number x such that

$$a = b + x.$$

Of course, the quantity x is what we usually denote by $a - b$.

The relation of division to multiplication is also that of an inverse operation. When we seek the answer to $\frac{8}{2}$ we may happen to know directly from experience that the answer is 4. But if we don't, we can reduce the division problem to a multiplication problem and ask what number, x, multiplied by 2 gives 8, and if we know multiplication we can find the answer. Here too, as in the case of subtraction, even if we use a special division process such as long division to find the answer, we check the answer by multiplying the divisor by the quotient to see if the product is the dividend. The reason for doing this is simply that the basic meaning of a/b is to find some number x such that $bx = a$.

In our logical development of the number system we affirm that we can divide any number by any other number (except 0) and we phrase the assertion so that the meaning of division is precisely what it actually is, the inverse of multiplication.

AXIOM 8. If a and b are any two numbers, except that $b \neq 0$, then there is a unique number x such that

$$bx = a.$$

Of course x is the number usually denoted by a/b.

The next axiom is not quite so obvious. It says, for example, that $3 \cdot 6 + 3 \cdot 5 = 3(6 + 5)$. In this example we can perform the calculation to see that the left and right sides are equal, but this is really not necessary. Suppose we had 157 cows in one herd and 379 in another, and each herd increased sevenfold. The total number of cattle is then $7 \cdot 157 + 7 \cdot 379$. But if the original two herds were one herd with $157 + 379$ cows, and this single herd increased sevenfold, we would have $7(157 + 379)$ cows. It is physically clear that we have the same number of cows now as before, that is, that $7 \cdot 157 + 7 \cdot 379 = 7(157 + 379)$. Stated in general terms, the axiom is:

AXIOM 9. For any three numbers a, b, and c,

$$ab + ac = a(b + c).$$

This axiom, called the *distributive axiom*, is very useful. For example, to calculate $571 \cdot 36 + 571 \cdot 64$ we can apply the axiom to state that this quantity is $571(36 + 64)$ or $571 \cdot 100$ or 57,100. We say often that we have *factored* the quantity 571 out of the sum $571 \cdot 36 + 571 \cdot 64$.

We should note that from

$$ab + ac = a(b + c)$$

we can also state that

$$ba + ca = (b + c)a,$$

because in each term of the first equation we can apply the commutative axiom of multiplication to change the order of the factors.

We often use the second form of the distributive axiom. Thus, suppose a is some number and we wish to calculate $5a + 7a$. We can replace this sum by $(5 + 7)a$, and obtain $12a$.

The distributive axiom is also applicable in the following situation. Suppose we have to calculate

$$\frac{296 + 148}{296}.$$

One might be tempted to cancel the two numbers 296. But this is incorrect. The given fraction means

$$\tfrac{1}{296}(296 + 148),$$

and the distributive axiom tells us that we may write instead

$$\tfrac{1}{296} \cdot 296 + \tfrac{1}{296} \cdot 148, \quad \text{or} \quad 1 + \tfrac{1}{2}.$$

In addition to the above axioms, we have the following evident properties of numbers:

AXIOM 10. Quantities equal to the same quantity are equal to each other.

AXIOM 11. If equal quantities are added to, subtracted from, multiplied with, or divided into equal quantities, the results are equal. However, division by zero is not permitted.

The set of axioms we have just given is not complete; that is, it does not form the logical basis for *all* of the properties of the positive and negative whole numbers, fractions, and irrational numbers. However, the set does provide the logical basis for what is usually done with numbers in ordinary algebra. Moreover, it does give some idea of what the axiomatic basis for mathematical work with numbers amounts to.

Now that we have the axioms, what do we do with them? We can prove theorems about numbers. Let us consider a few examples. Negative numbers were introduced to represent physical happenings such as debts or time before a given event. When we examined the physical situation in which we wished to use these numbers, we found that if the numbers are to be useful then we should agree that, for example,

$$-2 \cdot 3 = -6 \quad \text{and} \quad -2 \cdot (-3) = 6.$$

There we agreed to operate with positive and negative numbers so as to make the results fit the physical situation. In the deductive approach to numbers we prove on the basis of our axioms that certain theorems are correct. Let us prove that a positive number times a negative number is negative.

Let a and b be positive numbers. Then $-b$ is a negative number; for example, $-b = -5$. We shall prove that $a(-b) = -ab$. We know by Axiom 7, wherein we let a be 0, that if b and 0 are given, then there is a number x such that $b + x = 0$. This number x is denoted by $0 - b$ or $-b$. Then

$$b + (-b) = 0.$$

Now we can multiply both sides of this equation by a, and since equals multiplied by equals give equals, we have

$$a[b + (-b)] = a \cdot 0.$$

Now $a \cdot 0 = 0 \cdot a$ by Axiom 3, and $0 \cdot a = 0$ by Axiom 5. By applying the distributive axiom, Axiom 9, to the left-hand side of our equation, we obtain

$$a \cdot b + a(-b) = 0, \tag{1}$$

and now we see that $a(-b)$ is the number which added to ab gives 0. But Axiom 7 says that given ab and 0 (these are the a and b of Axiom 7), there is a *unique* number x such that

$$ab + x = 0. \tag{2}$$

This number x is denoted by $0 - ab$ or $-ab$. But equation (1) says that $a(-b)$ is the number which added to ab gives 0. Since there is just one such number which when added to ab gives 0, and that number we know is $-ab$, it must be be that $a(-b) = -ab$.

The proof is now complete and yet may not be convincing. The reason is simply that we are so accustomed to operating with numbers on the basis of physical arguments and experience with them that we have not accustomed ourselves to reasoning with numbers on an axiomatic basis.

Let us consider another proof. In Section 4–2 we gave a physical argument to show that if we divide a/b by c/d then we can get the answer by inverting c/d to obtain d/c and then multiplying; that is,

$$\frac{\frac{a}{b}}{\frac{c}{d}} = \frac{a}{b} \cdot \frac{d}{c} = \frac{ad}{bc}.$$

We can prove, on the basis of our axioms, that this rule of inverting the denominator and then multiplying is correct.

To divide a/b by c/d is to find a number x such that

$$\frac{a}{b} = x \cdot \frac{c}{d}. \tag{3}$$

Now Axiom 8 tells us that there is a unique number x which satisfies such an equation. We do know that

$$\frac{a}{b} = \frac{ad}{bc} \cdot \frac{c}{d}$$

because if we cancel common factors in the numerator and denominator of the right side we obtain a/b. Hence one number which can serve as the x which satisfies equation (3) is ad/bc. But since the value of x is unique, $x = ad/bc$. Thus the result of dividing a/b by c/d is ad/bc. We note that the answer ad/bc is obtained by multiplying a/b by the inverse of c/d. Hence to divide one fraction by another, we invert the denominator and multiply.

This proof, like the preceding one, may not be convincing, and the reason is the same. We are not accustomed to reasoning about numbers on an axiomatic basis. Rather, we have relied upon the physical meaning of numbers and operations. Historically, the mathematicians did the same thing. They learned to operate with numbers by noting the uses to which numbers were put, and they constructed the axiomatic basis long afterward, just to satisfy themselves that deductive proofs of the properties of numbers could be made.

Since we, too, are accustomed to the properties and operations with numbers since childhood, and we are sure of these properties, we shall not often cite axioms to justify our steps. Thus if we write $3a$ in place of $a \cdot 3$, we shall not cite the commutative axiom of multiplication as the justification for this step. In fact, it would be pedantic to do so. The axioms are useful, rather, in helping us to determine what is correct when our experience fails us or leaves us in doubt. However, we should not lose sight of the fact that the mathematics built upon the number system is a deductive system. This point needs emphasis because we begin to learn arithmetic at an early age by rote and thereafter we tend to operate with numbers mechanically without perceiving that we are constantly using axioms of numbers.

EXERCISES

1. Do you believe that

$$256(437 + 729) = 256 \cdot 437 + 256 \cdot 729?$$

Why?

2. Is it correct to assert that

$$a(b - c) = ab - ac?$$

[*Suggestion:* $b - c = b + (-c)$.]

3. Perform the operations called for in the following examples:

a) $3a + 9a$ b) $a \cdot 3 + a \cdot 9$ c) $a(5 + \sqrt{2})$ d) $7a - 9a$

e) $3(2a + 4b)$ f) $(4a + 5b)7$ g) $a(a + b)$ h) $a(a - b)$

i) $2(8a)$ j) $a(ab)$

4. Carry out the multiplication:

$$(a + 3)(a + 2).$$

[*Suggestion:* Regard $(a + 3)$ as a single quantity and apply the distributive axiom.]

5. Calculate $(n + 1)(n + 1)$.
6. If $3x = 6$, is $x = 2$? Why?
7. If $3x + 2 = 7$, is $3x = 5$? Why?
8. Is the equality $x^2 + xy = x(x + y)$ correct?
9. Is it correct to assert that

$$a + (bc) = (a + b)(a + c)?$$

* 4-6 APPLICATIONS OF THE NUMBER SYSTEM

Something of the power, methodology, and subtlety of mathematical reasoning can already be seen in the applications which have been made of the several types of number. Indeed, we shall see that these resulted in significant scientific discoveries.

Let us begin with some rather simple matters. Suppose a man drives a car for one mile at 60 miles per hour and for another mile at 120 miles per hour. What is his average speed? We tend to answer this question by applying the common procedure for finding an average. Thus, if a man buys one pair of shoes for \$5 and another for \$10, the average price is \$5 + \$10 divided by 2, or \$7.50. Hence it would seem as though the average speed in the above problem should be 60 + 120 divided by 2, or 90 miles per hour. However, this answer is not correct. The number 90 is an average in the arithmetic sense, but it is not the average we seek. The average speed should be that speed which would enable the man to drive the two miles *in the same time* as it took him to drive that distance at the two different speeds. Now, it took the man 1 minute to drive the first mile and it took him $\frac{1}{2}$ minute to drive the second mile. Hence it took him $1\frac{1}{2}$ minutes to drive 2 miles. We now ask, what average

---

* Starred sections and chapters can be omitted without disrupting the mathematical continuity.

speed maintained for $1\frac{1}{2}$ minutes would cover 2 miles? Since the average speed multiplied by the total time should give the total distance, the average speed is the total distance divided by the total time, that is,

$$\text{average speed} = \frac{2}{\frac{3}{2}} = \frac{4}{3}.$$

The average speed is then $\frac{4}{3}$ miles per minute or 80 miles per hour.

The point of this example, not a momentous one to be sure, is merely that the unthinking, blind application of arithmetic does not produce the correct result. The notion of average speed serves a physical purpose, and unless we are clear about what average speed is supposed to mean, we shall not profit by the use of arithmetic.

EXERCISES

1. A man can row a boat in still water at 6 mi/hr. He plans to row upstream for 12 mi and then back in a river whose current flows at 2 mi/hr. Thus his speed upstream is 4 mi/hr and his speed downstream is 8mi/hr. He reasons that his average speed is 6 mi/hr and that the entire trip of 24 mi should therefore take 4 hr. Is this reasoning correct?
2. Suppose that a merchant sells apples at a price of 2 for 5¢ and oranges at 3 for 5¢. To make his arithmetic simpler he decides to sell any 5 pieces of fruit for 10¢ or at the average price of 2¢ per piece. Thus, if he sells 2 apples and 3 oranges, he sells 5 pieces of fruit at 2¢ each and receives the same 10¢ as if he had sold them at the original separate prices. Is the merchant's average price correct? [*Suggestion:* Consider what results if he sells 12 apples and 12 oranges.]
3. Suppose the merchant wishes to sell a apples and b oranges to some customer at the prices given in Exercise 2. What should the average price be?
4. Given the data of Exercise 2, is there an average price which would be correct no matter how many apples and how many oranges are sold?
5. One man can dig a certain ditch in 2 days and another can dig the same ditch in 3 days. What is their average rate of ditch-digging per day?

Let us consider next an application of simple arithmetic to genetics. Suppose we have before us 2 red aces and 2 red kings from the usual deck of 52 cards. How many different pairs consisting of one ace and one king can be put together? Since each ace can be paired with either of 2 kings, there are 2 different pairs for any one ace. Since we have 2 aces, there are $2 \cdot 2$ or 4 different pairs.

Now let us suppose that we have 2 red aces, 2 red kings, and 2 red queens. How many different sets consisting of one ace, one king, and one queen can we form with the given cards? We saw above that there are 4 different pairs of aces and kings. With each of these 4 pairs we can place 2 different queens. Hence there are $4 \cdot 2$ or 8 different sets of 3 cards. We note that $4 \cdot 2 = 2 \cdot 2 \cdot 2 = 2^3$.

If we have 2 red aces, 2 red kings, 2 red queens, and 2 red jacks, the number of different sets, each consisting of one ace, one king, one queen, and one jack, can also be readily calculated. Each of the 8 choices of ace, king, and queen can be paired with each of the 2 jacks. Hence there are $8 \cdot 2$ or 16 choices in all. Now,

$$8 \cdot 2 = 4 \cdot 2 \cdot 2 = 2 \cdot 2 \cdot 2 \cdot 2 = 2^4.$$

Clearly, if we had 10 different pairs of cards and had to make all possible choices of 10 cards, one from each pair, the number of all possible sets of 10 cards would be

$$2 \cdot 2 \cdot 2 \cdot 2 \cdot 2 \cdot 2 \cdot 2 \cdot 2 \cdot 2 \cdot 2 = 2^{10} = 1024.$$

This simple reasoning about cards has an important application to genetics. The reproductive cells (as well as ordinary cells) of the human male contain 24 pairs of chromosomes. When a sperm cell is formed from the reproductive cell, it contains 24 chromosomes, each coming from one of the 24 pairs. Hence a sperm cell can be formed in 2^{24} possible combinations. The reproductive cells of the human female also contain 24 pairs of chromosomes. An ovum formed from the female reproductive cell contains 24 chromosomes, each coming from one of the 24 pairs of the reproductive cell. Hence there are 2^{24} possible ways in which an ovum can be formed. In conception, any one sperm joins, or fertilizes, any one ovum. Since there are 2^{24} possible sperms and 2^{24} possible ova, the number of possible chromosome combinations for the fertilized ovum is then

$$2^{24} \cdot 2^{24} = 16{,}777{,}216 \cdot 16{,}777{,}216 = 281{,}474{,}976{,}710{,}656.$$

This is the number of possible variations in the genetic make-up of any one child a man and wife may have. Actually, the number of variations is somewhat larger. Each chromosome contains genes, and these determine the hereditary qualities. Biologists have found that any two paired chromosomes in a reproductive cell may exchange some genes, and this exchange gives rise to new varieties of sperm cells and ova.

EXERCISES

1. The usual deck of 52 cards contains 4 different aces and 4 different kings. How many different pairs of cards, each pair consisting of one ace and one king, can be formed from the aces and kings?
2. A manufacturer offers his automobile in 3 different colors, with or without a heater, and with or without a radio. How many different choices can a purchaser make?

3. A girl has 3 hats, 2 dresses, and 2 pairs of shoes. How many different costumes does she have?
4. There are six numbers on a die (singular of dice). How many different pairs of numbers can show up on a throw of a pair of dice? The two dice are to be marked so that a throw of a 2 on one die (say A) and of a 5 on the other (say B) can be distinguished from the reverse arrangement (5 on A and 2 on B).

We have already discussed the fact that our method of writing quantities uses the idea of positional notation in base ten (see Section 4–2). However, some civilizations used other numbers as a base. For example, the Babylonians, for reasons that are obscure, selected 60. This system was taken over by the Greek astronomers and was used in Europe for many mathematical and all astronomical calculations as late as the seventeenth century. It still survives in our practice of dividing hours and angles into 60 minutes and 60 seconds. In adopting ten as a base, Europe followed the practice of the Hindus. Let us challenge history and see whether we can derive some advantage from a change to a new base.

We shall choose base six. The quantities from zero to five would be designated by the symbols 0, 1, 2, 3, 4, 5, as in base ten. The first essential difference comes up when we wish to denote six objects. Since six is to be the base, we would no longer use the special symbol 6, but place the 1 in a new position to denote 1 times the base, just as in base ten the 1 in 10 denotes one times the base, or the quantity ten. Hence, to write six in base six, we would write 10, but now the symbols 10 means 1 times six plus 0. Thus the symbols 10 can denote two different quantities, depending upon the base employed. Seven in base six would be written 11, because in base six these symbols mean 1 times six + 1, just as 11 in base ten means 1 times ten + 1. Again the symbols 11 represent different quantities, depending upon the base implied. As another example, to denote twenty-two in base six we write 34, because these symbols now mean 3 times six + 4.

In base ten, to write numbers larger than ninety-nine, we use a third position, the hundreds' place, to indicate tens of tens. Similarly in base six, when we reach numbers larger than thirty-five, we use a third position to denote sixes of sixes. Thus thirty-eight would be written in base six as 102, wherein the one means 1 times six times six, the 0 means 0 times six, and the 2 denotes just 2 units. To express very large numbers we would use four-place numbers, five-place numbers, and so forth.

We can perform the usual arithmetic operations in base six. However, we would have to learn new addition and multiplication tables. For example, in base ten we write $5 + 3 = 8$, whereas in base six, eight must be written 12. Hence our addition table would have to state that $5 + 3 = 12$. Likewise, our new multiplication table would have to list $3 \cdot 5 = 23$, because fifteen is $2 \cdot 6 + 3$ or, in base six, 23. When we learned to use base ten, we had to memorize the result of adding each number from 0 to 9 to every number from 0 to 9, and the result of multiplying each number from 0 to 9 with every number from 0 to 9. For base six we would have to learn to add (and multiply) only numbers from 0 to 5 to (or with) the numbers of this set. Thus our addition and multiplication tables would be shorter, and we would learn arithmetic sooner as youngsters. We might even pass the hurdle of arithmetic so easily that we might get to like mathematics. The only disadvantage of base six would be that to represent large quantities we would have to use more digits. For example, the quantity fifty-four, written as 54 in base ten, must be written as 130 in base six, because fifty-four equals $1 \cdot 6^2 + 3 \cdot 6 + 0$.

There are people who campaign for the adoption of base twelve, because it offers special advantages. For one thing, more fractions can be written as finite decimals in base twelve. Thus $\frac{1}{3}$ must be written as the unending decimal 0.333 . . . in base ten, but can be written as 0.4 in base twelve, since in this base 0.4 means $\frac{4}{12}$. Also, since the English system of denoting length calls for 12 inches in one foot, we could, for example, express 3 feet and 6 inches as 36 inches in base twelve, whereas to express this number of inches in base ten, we must first calculate $3 \cdot 12 + 6$ and then write 42 in base ten. To a limited extent we could use base twelve in our method of recording time. In the United States the day has two sets of twelve hours, and in base twelve the hours of the day would run from 0 to 20. Whereas determining what 7 hours after 6 o'clock will be requires at present some computation, under the addition table for base twelve we would state at once that $7 + 6 = 11$. However, base ten is now so widely used that a change to another base for ordinary daily use or commerce is hardly likely.

In the subject of bases we have an idea that was pursued for centuries, largely as an interesting and amusing speculation, but which suddenly became highly important in science and even in the commercial world. For several centuries mathematicians worked on the design of machines which would perform arithmetical computations quickly and thus remove a good deal of the drudgery of arithmetic. Although some types of computing machines were invented and used, mathematicians saw their golden opportunity in the electronic devices developed by modern radio engineers. The key is the radio vacuum tube, which can be made to pass current by applying a voltage to it or can be kept inactive. Two maneuvers are thus possible. In base two all numbers require only two symbols, the 0 and the 1. A typical number would be 1011, which means

$$1 \cdot 2^3 + 0 \cdot 2^2 + 1 \cdot 2 + 1.$$

This number can be recorded by the machine by employing four tubes, one for the units' place, another for multiples of 2, a third for multiples of 2^2, and a fourth for multiples of 2^3. To record 1011, the first, second, and fourth tubes can be activated, and the third, which records the third place in the number, kept inactive. The currents passed by the tubes which "fire" are recorded by the machine in special circuits. Another number can then be fed into the machine. Let us suppose that it is to be added to the first number. The result of having two 1's in the same place means, in base 2, that the sum is to be 0 and a 1 carried over to the next place. This operation is readily performed by the circuits. While this description of an electronic computer certainly doesn't begin to present the ingenious ideas which engineers and mathematicians have incorporated, we may perhaps see that the workings of a vacuum tube are ideally suited to operations in base two.

To take advantage of the fact that computers can perform calculations in base two, the numbers to be worked on are converted beforehand from base ten to base two and then fed to the machine along with other instructions. The machine then operates in this latter base. The result is, of course, reconverted to base ten.

Because computers work with microsecond speed, they are exceedingly valuable in any commercial or industrial organization which must process a lot of numerical data. Calculations in banks, insurance companies, and in industry, which used to require an immense amount of human labor, are now performed by machines. Computers are the first in a new series of machines which keep track of great quantities of data, select information from millions of cards on which data are recorded, plan factory operations, direct machinery, and may soon provide translations of foreign-language publications.

Electronic computing machines are an enormous boon to science and mathematics also. The arithmetic required to extract concrete information from mathematical formulas is often so lengthy that it would take years to perform these calculations. Computing machines do such work in hours. Moreover, mathematicians no longer hesitate to work on problems which will lead to extensive computations, because they now know that their work will not be in vain.

Computing machines may help us to learn more about the action of the human brain. According to biologists, the nerve cells in the nerve chains and the cells in our brains respond to electrical impulses much as a vacuum tube does. Just as a tube will "fire" when it receives electrical current beyond a certain minimum value and remain inactive otherwise, so do the nerve cells in the nerve chains transmit an electrical impulse to whatever organ they may lead when this impulse exceeds a threshhold value; otherwise they are inactive. Computing machines also have a memory; that is, partial results of calculations are stored automatically in a special device, called the memory. When these results are needed, the memory device releases them. Thus the result of an addition process might be stored in the memory device until the result of some multiplication process is obtained and then, if so instructed, the machine will add these two results. The machine's memory device, then, functions somewhat like the human memory. Hence, in two respects at least, electronic computing machines simulate the actions of human nerves and memory. Though machines are, in speed, accuracy, and endurance, superior to the human brain, one should not infer, as many popular writers are now suggesting, that machines will ultimately replace brains. Machines do not think. They perform the calculations which they are directed to perform by people who have the brains to know what calculations are wanted. Nevertheless, we undoubtedly have in the machine a useful model for the study of some functions of the human brain and nerves.

EXERCISES

1. Construct an addition table for base six.
2. Construct a multiplication table for base six.
3. Construct an addition and multiplication table for base two.
4. The following numbers are in base ten:

   9, 10, 12, 36, 48, 100.

   Write the respective quantities in base six.
5. The following numbers are in base six:

   5, 10, 12, 20, 100.

   Write the respective quantities in base ten.
6. Write the fraction $\frac{1}{2}$ as a decimal in base six.
7. The quantity 0.2 is in base six. Write the corresponding quantity in base ten.
8. The number 101 is written in some unknown base and equals ten. What is the base?
9. Find the least number of weights needed to weigh, to the nearest pound, objects weighing from 0 to 63 lb. The scale to be used contains two pans and the weights are to be put in one pan. [*Suggestion:* Consider the problem of representing all numbers from 0 to 63 in base 2.]

Some of the most remarkable uses of numbers, which have led to profound discoveries, are found in the study of the structure of matter. During the early and middle part of the nineteenth century, certain basic experimental facts about the varieties of matter found in nature led, after some inevitable fumbling on the part of John Dalton, Amadeo Avogadro, and Stanislav Cannizzaro, to the theory that all matter is made up of atoms. Thus hydrogen, oxygen, chlorine, copper, aluminum, gold, silver, and all other varieties of matter are composed of atoms. Experimental techniques were developed to measure the relative weights of the atoms of different elements. The convenient unit of weight chosen was $\frac{1}{16}$ the weight of the oxygen atom, so that the weight of the atom of oxygen is 16. The weight of the hydrogen atom then proved to be 1.0080, that of copper 63.54, that of gold 197.0, and so on.

By this time, too, a number of chemical properties of these various elements, such as their melting temperatures, boiling temperatures, and their ability to combine with other elements to form compounds, had been determined.

The question which had begun to stir the chemists was whether there existed any law or principle which utilized the atomic weights of these elements. The crowning discovery was made in 1869 by Dimitri Ivanovich Mendeléev (1834–1907). He found, as he began to arrange the elements in order of increasing atomic weight, that every eighth element, starting from a given one, had chemical properties similar to the first one. Thus, the gases fluorine and chlorine, the latter the eighth element starting from fluorine, both combine readily with metals. However, as he continued to place each of the 63 different elements known in his time in the eighth position after the one with similar chemical properties, he saw that he had to leave blank spaces. Mendeléev was so much impressed with the periodicity of chemical properties that he did not hesitate to leave the blank spaces and to affirm that there must be elements to fill these spaces. Since each of these missing elements should have chemical properties similar to those of the element found in the eighth position preceding or succeeding the missing element, he could even predict some of the properties of the unknown elements. Mendeléev described the properties of three of the missing elements, and his immediate successors discovered them. These are now called scandium, gallium, and germanium. Still later, others were found. The interesting fact about Mendeléev's work from the mathematical standpoint is that he had no physical explanation of why elements eight positions removed from one another should have similar chemical properties. He knew only that the number eight was the key to the arrangement, and he followed this mathematical guide faithfully. Long after Mendeléev's time, other elements, for example helium, were found, which do not fit into this arrangement, but his periodic table is still the basic one which all students of chemistry learn today.

Simple arithmetic continued to play a leading role in subsequent developments of atomic theory. The continuing study of the atomic weights of various elements and their chemical properties showed that elements formerly regarded as pure were really not so. Thus, there are two kinds of hydrogen. These have similar chemical properties, but different atomic weights; in fact, one is twice as heavy as the other. Since both were previously called hydrogen, and since they do, in any case, have similar chemical properties, these two forms of hydrogen are called isotopes of hydrogen. Likewise, there is not one substance, oxygen, but there are three, of atomic weights 16, 17, and 18. Uranium, a very important element today, has two isotopes of atomic weights 238 and 235.

The startling fact which emerged from the discovery of isotopes is that when all isotopes are distinguished and the relative weights of the distinct elements determined, the weight of any one element is within 1% of a whole number. Such a fact can hardly be accidental. The explanation would seem to be that all these elements are really multiples of a single element, namely the lighter isotope of hydrogen, which has the least weight of all elements. In other words, the various elements which previously appeared to be entirely different substances, now were seen to be just smaller or larger collections of the same element, but arranged in special ways peculiar to the substance. (Strictly speaking, the fundamental building block is not the lighter isotope of hydrogen, but what is now called the proton. The lighter hydrogen isotope also has an electron whose weight is insignificant by comparison.)

If all of the different elements are really just aggregates of the lighter hydrogen atom, it should be possible to remove some atoms and convert one substance into another. Thus, we should be able to convert mercury, which is the next heavier element after gold, into gold. And we can. What the medieval alchemists hoped to do on mystical and superficial grounds, we can now do on the basis of far better scientific knowledge. Unfortunately, the cost of converting mercury into gold is too great to make it worth while. But we do have uses for the transmutation of elements which are, in our age, more valued, and which we shall describe in a moment.

A scientist who has a theory cannot afford to overlook even one detail, trivial as it may seem, which does not square with his theory. If all elements are merely combinations of the lighter hydrogen atom, then their weights should be exact multiples of the weight of this atom instead of having values within 1% of such weights. (The electrons in the atoms do not account for the difference.) This discrepancy must be explained. The lightest isotope of oxygen had, somewhat arbitrarily, been given weight 16. With this rather arbitrary standard, the lighter hydrogen atom has weight 1.008 rather than exactly 1. But helium, which consists of 4 hydrogen atoms, proves to have weight 4.0028. However, if it consists of 4 hydrogen atoms, its atomic weight should be 4 times 1.008, or 4.032. The difference, $4.032 - 4.0028$, or about 0.03, is the discrepancy which must be accounted for. Now it so happens that Einstein, working in an entirely different field, the theory of relativity, had already shown that mass can be converted into energy. Energy can take different forms. It can be the heat created by burning coal or wood, or it can be radiation such as comes to us from the sun. At the moment the precise form of it does not matter. What does matter is the thought which occurred to scientists that perhaps, when 4 hydrogen atoms are fused to form helium, the missing 0.03 of matter is converted into energy in the process. Hence the fusion of elements should release energy. And experiments showed that this is indeed what happens. The energy which is released is called the binding energy, and it is this energy which is released when a hydrogen bomb is exploded.

In this brief account of the role of arithmetic in chemistry and atomic theory, we have said almost nothing about the great thinking and brilliant experiments which physicists and chemists contributed. Our interest has been to show how the use of simple numbers supplies scientists with a powerful tool. Of course, the mathematics of numbers remains to be developed, and we shall learn how much more can be accomplished with slightly more advanced tools. But we can already see something of what the Pythagoreans envisioned when they spoke of numbers as the essence of reality.

REVIEW EXERCISES

1. Calculate:

   a) $\frac{3}{5} + \frac{4}{7}$ b) $\frac{3}{5} - \frac{4}{7}$ c) $\frac{4}{7} - \frac{3}{5}$ d) $\frac{2}{9} + \frac{5}{12}$

   e) $\frac{2}{9} - \frac{5}{12}$ f) $\frac{2}{9} - \frac{-5}{12}$ g) $-\frac{2}{9} + \frac{-5}{12}$ h) $\frac{a}{b} + \frac{c}{d}$

   i) $\frac{a}{b} - \frac{c}{d}$ 0 j) $\frac{a}{b} - \frac{-c}{d}$ k) $\frac{1}{x} + \frac{1}{2}$

2. Calculate:

   a) $\frac{3}{5} \cdot \frac{4}{9}$ b) $\frac{3}{5} \cdot \frac{-4}{9}$ c) $\frac{-3}{5} \cdot \frac{-4}{9}$ d) $(-\frac{3}{5}) \cdot (-\frac{4}{9})$

   e) $\frac{a}{b} \cdot \frac{c}{d}$ f) $\frac{a}{b} \cdot \frac{c}{a}$ g) $\frac{a}{b} \cdot \frac{b}{a}$ h) $\frac{a}{b} \cdot \frac{-c}{d}$

   i) $\frac{2}{5} \div \frac{1}{5}$ j) $\frac{2}{3} \div \frac{3}{7}$ k) $\frac{3}{5} \div \frac{6}{10}$ l) $\frac{21}{6} \div \frac{7}{4}$

   m) $\frac{a}{b} \div \frac{c}{d}$ n) $\frac{21}{8} \div 5\frac{1}{2}$ o) $-8 \div -2$

3. Calculate:

   a) $(2 \cdot 5)(2 \cdot 7)$ b) $2a \cdot 2b$ c) $2a \cdot 3b$

   d) $2x \cdot 3y$ e) $2x \cdot 3y \cdot 4z$

4. Calculate:

   a) $(\frac{3}{4} \cdot \frac{5}{7}) \div \frac{3}{2}$ b) $\frac{3 + 6a}{3}$ c) $\frac{3a + 6b}{3}$

   d) $\frac{4x + 8y}{2}$ e) $\frac{ab + ac}{a}$

5. Calculate:

   a) $\sqrt{49}$ b) $\sqrt{121}$ c) $\sqrt{\frac{9}{4}}$ d) $\sqrt{\frac{81}{16}}$

e) $\sqrt{3}\sqrt{3}$ f) $\sqrt{2}\sqrt{8}$ g) $\sqrt{2}\sqrt{4}$ h) $\sqrt{2}\sqrt{\frac{5}{3}}$

6. Simplify:

a) $\sqrt{32}$ b) $\sqrt{48}$ c) $\sqrt{72}$ d) $\sqrt{8}$

e) $\sqrt{\frac{9}{4}}$ f) $\sqrt{\frac{18}{4}}$ g) $\sqrt{\frac{27}{4}}$ h) $\sqrt{\frac{27}{8}}$

7. Write as a fraction:

a) 0.294 b) 0.3742 c) 0.08 d) 0.003

8. Approximate by a number which is correct to 1 decimal place:

a) $\sqrt{3}$ b) $\sqrt{5}$ c) $\sqrt{7}$

9. Are the following equations true for all values of a and b? [*Suggestion:* If you wish to disprove a general statement, it is sufficient to show one instance where it does not hold.]

a) $2(a + b) = 2a + 2b$ b) $2ab = 2a \cdot 2b$

c) $\frac{a + b}{2} = \frac{a}{2} + \frac{b}{2}$ d) $\frac{ab}{2} = \frac{a}{2} \cdot \frac{b}{2}$

e) $\sqrt{a^2 + b^2} = \sqrt{a^2} + \sqrt{b^2}$ f) $\sqrt{a^2 + b^2} = a + b$

g) $\sqrt{ab} = \sqrt{a}\sqrt{b}$ h) $\sqrt{a^2 - b^2} = a - b$

i) $\sqrt{a + b} = \sqrt{a} + \sqrt{b}$

10. Write the following numbers in positional notation but in base 2. The only digits one can use in base 2 are 0 and 1.

a) 1 b) 3 c) 5 d) 7 e) 8 f) 16 g) 19

11. The following numbers are in base 2. Write the corresponding quantities in base 10.

a) 1 b) 101 c) 110 d) 1101 e) 1001

Topics for Further Investigation

1. The Egyptian method of writing whole numbers and fractions.
2. The Babylonian method of writing whole numbers and fractions.
3. The Roman method of writing whole numbers and fractions.
4. The fundamental arithmetical laws of atomic theory. (Use the references to Holton and Roller and to Bonner and Phillips).
5. Pythagorean number theory.

Recommended Reading

BALL, W. W. ROUSE: *A Short Account of the History of Mathematics,* Chaps. 1 and 2, Dover Publications, Inc., New York, 1960.

BONNER, F. T. and M. PHILLIPS: *Principles of Physical Science,* Chap. 7, Addison-Wesley Publishing Co., Inc., Reading, Mass., 1957.

COLERUS, EGMONT: *From Simple Numbers to the Calculus,* Chaps. 1 through 8, Wm. Heineman Ltd., London, 1954.

DANTZIG, TOBIAS: *Number, the Language of Science,* 4th ed., Chaps. 1 through 6, The Macmillan Co., New York, 1954 (also in a paperback edition).

DAVIS, PHILIP J.: *The Lore of Large Numbers,* Random House, New York, 1961.

EVES, HOWARD: *An Introduction to the History of Mathematics,* Rev. ed., pp. 29–64, Holt, Rinehart and Winston, Inc., New York, 1964.

GAMOW, GEORGE: *One Two Three . . . Infinity,* Chap. 9, The New American Library, New York, 1953.

HOLTON, G. and D. H. D. ROLLER: *Foundations of Modern Physical Science,* Chaps. 22 and 23, Addison-Wesley Publishing Co., Inc., Reading, Mass., 1958.

JONES, BURTON W.: *Elementary Concepts of Mathematics,* Chaps. 2 and 3, The Macmillan Co., New York, 1947.

SMITH, DAVID EUGENE: *History of Mathematics,* Vol. I, pp. 1–75, Vol. II, Chaps. 1 through 4, Dover Publications, Inc., New York, 1953.

ALGEBRA, THE HIGHER ARITHMETIC

Algebra is the intellectual instrument which has been created for rendering clear the quantitative aspect of the world.

ALFRED NORTH WHITEHEAD

5-1 INTRODUCTION

Mathematics is concerned with reasoning about certain special concepts, the concepts of number and the concepts of geometry. Reasoning about numbers —if one is to go beyond the simplest procedures of arithmetic—requires the mastery of two facilities, vocabulary and technique, or one might say, vocabulary and grammar. In addition, the entire language of mathematics is characterized by the extensive use of symbolism. In fact, it is the use of symbols and of reasoning in terms of symbols which is generally regarded as marking the transition from arithmetic to algebra, though there is no sharp dividing line.

The task of learning the vocabulary and techniques of algebra may be compared with that which faces the prospective musician. He must learn to read music and he must develop the technique for playing an instrument. Since our goal in mathematics is far more the acquisition of an understanding than the attainment of professional competence, the problem of learning the vocabulary and techniques will hardly be a severe one.

5-2 THE LANGUAGE OF ALGEBRA

The nature and use of the language of algebra are readily illustrated, although the illustration is at the moment a trivial one. Most readers have encountered parlor number games, of which the following is an especially simple example. The leader of the game says to any member of the group: Take a number; add 10; multiply by 3; subtract 30; and give me your answer. And now, says the leader, I shall tell you the number you chose originally. To the amazement of the audience he does so immediately. The secret of his method is absurdly simple. Suppose the subject chooses the number a. Then adding 10 yields $a + 10$. Multiplication by 3 means $3(a + 10)$. By the distributive axiom, this quantity is $3a + 30$. Subtraction of 30 yields $3a$. The leader has only to divide by 3 the number given to him to tell the subject what his original choice was.

If the leader wishes to be especially impressive, he can ask the subject to perform many more computations that will yield a simple and known multiple of the original number, and he can give the original number just as readily. By representing in the language of algebra the operations which he asks the subject to perform and by noting what the operations amount to, the leader can easily see how the final result is related to the original number chosen.

The language of algebra involves more than the use of a letter to represent a number or a class of numbers. The expression $3(a + 10)$ contains, in addition to the usual plus sign of arithmetic, the parentheses which denote that the 3 multiplies the entire quantity $a + 10$. The notation b^2 is a shorthand expression for $b \cdot b$ and is read b-square. The word *square* enters here because b^2 is the area of a square whose side is b. Likewise, the notation b^3 means $b \cdot b \cdot b$ and is read b-cube. The word *cube* is suggested by the fact that b^3 is the volume of a cube whose side is b. The expression $(a + b)^2$ means that the entire quantity $a + b$ is to be multiplied by itself. An expression such as $3ab^2$ means 3 times some quantity a and that product multiplied by the quantity b^2. In addition, the notation uses the convention that numbers and letters following one another with no symbol in between any two are to be multiplied together. Another important convention stipulates that if a letter is repeated in an expression, it stands for the same number throughout. For example, in $a^2 + ab$ the value of a must be the same in both terms. Thus algebra uses many symbols and conventions to represent quantities and operations with quantities.

Why do mathematicians bother with such special symbols and conventions? Why must they place hurdles in the way of would-be students of their subject? The answer is not that mathematicians are trying to introduce hurdles; nor are they seeking to impress people by making their subject look awesome. Rather the symbolism of algebra and the symbolism of mathematics in general are an unfortunate necessity. The most weighty reason is comprehensibility. Symbolism enables the mathematician to write lengthy expressions in a compact form so that the eye can see quickly and the mind can retain what is being said. To describe in words even the simple expression $3ab^2 + abc$ would require the phrase, "The product of 3 times a number multiplied by a second number which is multiplied into itself added to the product of the first number, the second number, and still a third number." It is unfortunate that our eyes and minds are limited. The long and complicated sentences that would be required if ordinary language were used could not be remembered and, in fact, can be so involved as to be incomprehensible.

In addition to comprehensibility, there is the advantage of brevity. The expression in ordinary language of what is covered in typical texts on mathematics would require tomes of two to ten or fifteen times the customary size of such books.

Still another advantage is clarity. Ordinarily English or, for that matter, any other language is ambiguous. The statement, "I read the newspaper," can mean that one reads newspapers regularly, once in a while, or often, or that one has read the newspaper, presumably the paper of the day. One must judge by the context just what this sentence means. Such ambiguity is intolerable in exact reasoning. By using symbols for specific ideas mathematics avoids ambiguity or, to put the matter positively, each symbol has its own precise meaning, and so the resulting expressions are clear.

Symbolism is one of the sources of the remarkable power of algebra. Suppose that one wished to discuss equations of the form $2x + 3 = 0$, $3x + 7 = 0$, $4x - 9 = 0$, and the like. The particular numbers which appear in these equations do not happen to be important in the discussion; in fact, one wishes to include all equations in which the product of some number and x is added to some other number. The way to represent all possible equations of this form is

$$ax + b = 0. \qquad (1)$$

Here a stands for any number, and so does b. These numbers are known, but their precise value is not stated. The letter x stands for some unknown number. By reasoning about the general form (1) the mathematician covers the millions of separate cases which arise when a and b have specific values. Thus, by means of symbolism, algebra can handle a whole class of problems in one bit of reasoning.

Of course, it is unfortunate that one must learn the elements of a new language to master some mathematics. But one could with much justice complain that the French people insist on their language, the Germans on theirs, and so on. Obviously English is the best language, and the French and Germans are exhibiting provincialism by insisting on holding on to their respective languages. The language of mathematics has the additional merit of being universal.

There are justifiable criticisms of the symbolism of algebra, although they are hardly major ones. Mathematicians are greatly concerned about the accuracy of their reasoning, but pay little attention to the aesthetics or appropriateness of their symbolism. Very few symbols suggest their meaning. The signs $+$, $-$, $=$, $\sqrt{\ }$ are easy to write, but they are historical accidents. No mathematician has bothered to replace these by at least prettier ones, perhaps $\diamond$ for plus. The seventeenth-century mathematician Gottfried Wilhelm Leibniz, who did spend days on the choice of symbols in an effort to make them suggestive, was an exception. There are even inconsistencies in symbolism which, once recognized, fortunately do not impair the clarity. For example, when two letters, such as ab, are written together with no symbol between them, then it is understood that multiplication is meant. However two numbers, such as $3\frac{1}{2}$, with no symbol between them, mean $3 + \frac{1}{2}$.

Symbolism entered algebra rather late. The Egyptians, Babylonians, Greeks, Hindus, and Arabs knew and applied a great deal of the algebra which we learn in high school. But they wrote out their work in words. Their algebraic style is in fact called rhetorical algebra because, except for a few symbols, they used ordinary rhetoric. It is significant that symbolism entered mathematics in the sixteenth and seventeenth centuries when pressure to improve the efficiency of mathematics was applied by science. The idea of using symbols was no longer new, but mathematicians were undoubtedly stimulated to extend the application of symbolism and to adopt it readily.

EXERCISES

1. Why does mathematics use symbols?
2. Criticize the statement that all men are created equal.
3. In the following symbolic expressions the letters stand for numbers. Write out in words what the expressions state.

   a) $a + b$ b) $a(a + b)$ c) $a(a^2 + ab)$ d) $3x^2y$

   e) $(x + y)(x - y)$ f) $\dfrac{x + 3}{7}$ g) $\frac{1}{7}(x + 3)$

4. Does

$$\frac{x + 3}{7} = \frac{1}{7}(x + 3)?$$

   [*Suggestion:* What do these symbolic expressions say in words?]
5. Write in symbols: (a) three times a number plus four; (b) three times the square of a number plus four.

5-3 EXPONENTS

One of the simplest examples of the convenience of algebraic symbolism or algebraic language is found in the use of exponents. We have already used frequently such expressions as 5^2. In this expression the number 2 is an *exponent*, and 5 is called the *base*. The exponent is placed above and to the right of the base to indicate that the quantity to which it applies, 5 in this example, is to be multiplied by itself, so that $5^2 = 5 \cdot 5$. Of course, there would be no great value in the use of exponents if their use were limited to such instances. Suppose, however, that we wished to indicate

$$5 \cdot 5 \cdot 5 \cdot 5 \cdot 5 \cdot 5.$$

Here 5 occurs as a factor six times. We can indicate this quantity by means of exponents thus: 5^6. That is, when the exponent is a positive whole number it indicates how many times the quantity to which it is applied occurs as a factor in a product of this quantity and itself. In such instances as 5^6 the use of exponents saves a lot of writing and counting of factors.

Exponents are even more useful than we have thus far indicated. Suppose we wished to write

$$5 \cdot 5 \cdot 5 \cdot 5 \cdot 5 \cdot 5 \text{ times } 5 \cdot 5 \cdot 5 \cdot 5.$$

With exponents we can write

$$5^6 \cdot 5^4.$$

Moreover, the original product calls for 5 multiplied by itself to a total of 10 factors. This product can be written as 5^{10}. We see, however, that if we add the exponents in $5^6 \cdot 5^4$, we also get 5^{10}. That is, it is correct to write

$$5^6 \cdot 5^4 = 5^{6+4} = 5^{10}.$$

More generally, when m and n are positive whole numbers,

$$a^m \cdot a^n = a^{m+n}.$$

This statement is really a theorem on exponents. Its proof is trivial. All that the theorem says is that if one quantity contains a as a factor m times, and another quantity contains a as a factor n times, the product of these two quantities contains a as a factor $m + n$ times.

And now suppose we wished to write

$$\frac{5 \cdot 5 \cdot 5 \cdot 5 \cdot 5 \cdot 5}{5 \cdot 5 \cdot 5 \cdot 5}.$$

With the use of exponents we can write

$$\frac{5^6}{5^4}.$$

Moreover, if we were to calculate the value of the original quotient, we know that we could cancel 5's in the numerator and denominator. We would be left with

$$\frac{5 \cdot 5}{1} \quad \text{or} \quad 5^2.$$

We can obtain the same result if in $5^6/5^4$ we subtract the 4 from the 6 and so arrive at 5^2. Here, too, as in the case of multiplication, the exponents keep track of the number of 5's which occur in the numerator and denominator, and the subtraction of the 4 from the 6 tells us the net number of 5's remaining as factors.

In more general language, we can say that if m and n are positive whole numbers, and if m is greater than n, then

$$\frac{a^m}{a^n} = a^{m-n}.$$

This result, too, is a theorem on exponents, and the proof is again trivial because all the theorem states is that if we cancel the a's common to numerator and denominator, we shall have $m - n$ factors left over.

We might find that we have to deal with

$$\frac{5 \cdot 5 \cdot 5 \cdot 5}{5 \cdot 5 \cdot 5 \cdot 5 \cdot 5 \cdot 5}.$$

In exponent form this quotient is

$$\frac{5^4}{5^6}.$$

This time, if we cancel the 5's common to the numerator and denominator, we are left with 5 occurring twice in the denominator; that is, we are left with

$$\frac{1}{5 \cdot 5} = \frac{1}{5^2}.$$

We can obtain this result at once by subtracting the exponent 4 in the numerator from the exponent 6 in the denominator. In general form we have the theorem: If m and n are positive whole numbers and if n is greater than m, then

$$\frac{a^m}{a^n} = \frac{1}{a^{n-m}}.$$

There is the possibility of encountering

$$\frac{5 \cdot 5 \cdot 5 \cdot 5}{5 \cdot 5 \cdot 5 \cdot 5}.$$

In exponent form this quotient is written

$$\frac{5^4}{5^4}.$$

It would be nice to be able to simplify this expression, too, by using exponents. However, here, unlike the two previous cases, the two exponents are equal. If we were to, say, subtract the exponent 4 of the denominator from the exponent 4 of the numerator, we would have

$$\frac{5^4}{5^4} = 5^{4-4} = 5^0.$$

Now 5^0 has no meaning. However we know that $5^4/5^4$ has the value 1. If we agree to give a meaning to a zero exponent and, in fact, agree that a number to the 0 exponent is to be 1, then we can use the symbol 5^0. Further, with this meaning we can properly write

$$\frac{5^4}{5^4} = 5^{4-4} = 5^0 = 1.$$

In general, if m is a positive whole number,

$$\frac{a^m}{a^m} = a^{m-m} = a^0 = 1.*$$

EXERCISES

1. Simplify the following expressions by using the theorems on exponents:

a) $5^4 \cdot 5^6$ b) $6^3 \cdot 6^7$ c) $10^5 \cdot 10^4$ d) $x^2 \cdot x^3$ e) $\frac{5^7}{5^4}$ f) $\frac{10^7}{10^4}$

g) $\frac{x^4}{x^2}$ h) $\frac{10^4}{10^7}$ i) $10 \cdot 10^4$ j) $\frac{5^4}{5^7}$ k) $\frac{7^4}{7^4}$ l) $\frac{5^7 \cdot 5^2}{5^8}$

2. Could we apply the above theorems on exponents to negative numbers as bases? Then, is it true that $(-3)^5(-3)^4 = (-3)^9$?
3. Which of the following equations are correct?

a) $3^2 + 3^4 = 3^6$ b) $3^2 \cdot 3^4 = 3^6$ c) $3^2 + 3^4 = 6^6$

d) $\frac{6^5}{6^7} = \frac{1}{6^2}$ e) $3^4 + 3^4 = 3^8$

We can use exponents even more effectively than has thus far been indicated. Suppose that in the course of algebraic work, there occurred the expression

$$5^3 \cdot 5^3 \cdot 5^3 \cdot 5^3.$$

* It is necessary to add that a must not be 0, because then the original quotient has no meaning.

Could we write this more briefly? By the very meaning of an exponent we certainly can write

$$5^3 \cdot 5^3 \cdot 5^3 \cdot 5^3 = (5^3)^4.$$

We can go further. The left-hand side of this equation contains 5 as a factor 12 times. We can recognize the same fact if we multiply the two exponents on the right-hand side. That is,

$$(5^3)^4 = 5^{12}.$$

This example is the essence of another theorem on exponents, namely, if m and n are positive integers, then

$$(a^m)^n = a^{mn}.$$

There is one more commonly useful theorem on exponents. Suppose we wished to denote

$$2 \cdot 2 \cdot 2 \cdot 2 \cdot 3 \cdot 3 \cdot 3 \cdot 3$$

briefly, by taking advantage of exponents. We certainly could write this quantity as

$$2^4 \cdot 3^4.$$

However, we know that the order in which we multiply numbers does not matter. Hence it is correct that

$$2 \cdot 2 \cdot 2 \cdot 2 \cdot 3 \cdot 3 \cdot 3 \cdot 3 = 2 \cdot 3 \cdot 2 \cdot 3 \cdot 2 \cdot 3 \cdot 2 \cdot 3,$$

and now if we use exponents, we can say that

$$2^4 \cdot 3^4 = (2 \cdot 3)^4.$$

What this fact amounts to, in general terms, is that if m is a positive whole number, then

$$a^m \cdot b^m = (a \cdot b)^m.$$

EXERCISES

1. Use the theorems on exponents to simplify the following expressions:

a) $3^4 \cdot 3^4 \cdot 3^4$ b) $(3^4)^3$ c) $(5^4)^2$ d) $10^2 \cdot 10^2 \cdot 10^2$

e) $(10^4)^3$ f) $5^4 \cdot 2^4$ g) $3^7 \cdot 3^3$ h) $10^4 \cdot 3^4$

2. Calculate the values of the following quantities:

a) $2^5 \cdot 5^5$ b) $\dfrac{2^4 \cdot 3^4}{6^3}$ c) $\dfrac{4^5 \cdot 2^5}{8^6}$ d) $\dfrac{(ab)^3}{a^2}$ e) $\dfrac{a^3b^3}{(ab)^2}$

3. Which of the following equations are correct?

a) $(3 \cdot 10)^4 = 3^4 \cdot 10^4$ b) $(3 \cdot 10^2)^3 = 3^3 \cdot 10^6$ c) $(3 + 10)^4 = 3^4 + 10^4$

d) $(3^2 \cdot 5^3)^4 = 3^8 \cdot 5^{12}$ e) $(3^4)^3 = 3^7$ f) $(3^2)^3 = 3^9$

All the above theorems deal with positive integral (whole-numbered) exponents or zero as an exponent. Though we shall not deal with other types of numbers as exponents, it is significant to know that the exponent notation can be more valuable than we have thus far indicated. Let us consider $\sqrt{3}$. We know that

$$\sqrt{3} \cdot \sqrt{3} = 3.$$

Let us suppose that we would like to investigate whether the exponent notation could be used to simplify work with irrational numbers. (Of course, this is the kind of problem that one takes up when he has nothing better to do.) Now the right-hand side of the above equation can be written as 3^1. No matter what exponent notation we do adopt for $\sqrt{3}$, say 3^a, the equation would have to read

$$3^a \cdot 3^a = 3^1.$$

Moreover, we would like, if possible, to maintain the validity of our previous theorems on exponents. In the present case, we would like to be able to say that

$$3^a \cdot 3^a = 3^{a+a} = 3^{2a},$$

and since $3^{2a} = 3^1$, we would have to have $2a = 1$ or $a = \frac{1}{2}$. What these exploratory thoughts suggest is that if we denote

$$\sqrt{3} \quad \text{by} \quad 3^{1/2},$$

then we would be able to use at least the first theorem on exponents to state that

$$3^{1/2} \cdot 3^{1/2} = 3^{1/2+1/2} = 3^1 = 3.$$

As a matter of fact, this example typifies what is done. Thus we use the notation

$$\sqrt{3} = 3^{1/2}, \qquad \sqrt[3]{3} = 3^{1/3}, \qquad \sqrt[5]{4} = 4^{1/5},$$

and so on.

5-4 ALGEBRAIC TRANSFORMATIONS

Symbolism is a means to an end. The function of algebra is not to display symbols but to convert or transform expressions from one form to another which may be more useful for the problem in hand.

Let us consider an example. Suppose that in the course of some mathematical work we encounter the expression

$$(x + 4)(x + 3). \tag{2}$$

The letter x in this expression may stand for some number whose value we do or do not happen to know, or it may stand for any one of some class of numbers. What matters is that x stands for a number. If x is a number, then $x + 4$ is a number. We may now apply the distributive axiom, which states that for any numbers a, b, and c,

$$a(b + c) = ab + ac. \tag{3}$$

If we compare (2) and (3) we see that (2) has the form of (3) if we think of $x + 4$ in (2) as the a of (3). Then by applying the distributive axiom to (2) we may assert that

$$(x + 4)(x + 3) = (x + 4)x + (x + 4)3. \tag{4}$$

We also know that there is another form of the distributive axiom, namely

$$(b + c)a = ba + ca.$$

If we apply this axiom to each of the terms on the right side of (4), we see that

$$(x + 4)x = x^2 + 4x \qquad \text{and} \qquad (x + 4)3 = 3x + 12.$$

If we substitute these last two results on the right side of (4), we have

$$(x + 4)(x + 3) = x^2 + 4x + 3x + 12 = x^2 + (4 + 3)x + 12 \tag{5}$$

or

$$(x + 4)(x + 3) = x^2 + 7x + 12. \tag{6}$$

Before discussing what this example illustrates, let us note that we do not usually carry out the multiplication of $(x + 4)$ by $(x + 3)$ in this long and rather cumbersome fashion. Instead we write

$$\begin{array}{r} x + 4 \\ x + 3 \\ \hline 3x + 12 \\ x^2 + 4x \\ \hline x^2 + 7x + 12. \end{array}$$

The partial product $3x + 12$ results from multiplying $x + 4$ by 3, and the partial product $x^2 + 4x$ results from multiplying $x + 4$ by x. The two partial products are then added. This manner of carrying out the multiplication is faster, but fails to indicate explicitly that we have used the distributive axiom several times.

The main point of the above example is that we have transformed the expression $(x + 4)(x + 3)$ into the expression $x^2 + 7x + 12$. We do not maintain that the latter expression is more attractive than the former, but it may be more useful in a particular mathematical application. On the other hand, we might, in some situation, find ourselves with the expression $x^2 + 7x + 12$ and, by recognizing that it is equal to $(x + 4)(x + 3)$, be able to make progress toward some significant conclusion. In this latter transformation we say that we have *factored* $x^2 + 7x + 12$ into $(x + 4)(x + 3)$. Which of the two forms is more useful depends upon the application in hand. At the moment we should merely see that algebra is concerned with the technique of such transformations, and that a skilled mathematician should be able to perform them rapidly. Since we shall not become too involved in complicated technical processes, we shall not spend much time in developing skills.

The problem of factoring to which we referred in the preceding paragraph does arise reasonably often. For example, one usually starts with an expression such as $x^2 + 6x + 8$ and seeks to transform it into a product of factors of the form $(x + a)(x + b)$. The original expression is said to be of second degree because it contains x^2 but no higher *power* of x. The factors are first-degree expressions because each contains x but no higher power of x. The problem is to find the correct values of a and b so that the product $(x + a)(x + b)$ will equal the original expression. We know from our work on multiplication [see equation (5)] that

$$x^2 + (a + b)x + ab = (x + a)(x + b).$$

Hence to factor the second-degree expression, we should look for two numbers a and b whose sum is the coefficient of x and whose product is the constant. Thus to factor $x^2 + 6x + 8$, we look for two numbers whose sum is 6 and whose product is 8. By mere trial of the possible factors of 8 we see that $a = 4$ and $b = 2$ will meet the requirement; that is:

$$x^2 + 6x + 8 = (x + 4)(x + 2).$$

EXERCISES

1. Transform to an equal expression:

a) $3x \cdot 5x$ b) $(x + 4)(x + 5)$ c) $(3x + 4)(x + 5)$

d) $(x - 3)(x + 3)$ e) $(x + \frac{5}{2})(x + \frac{5}{2})$ f) $(x + \frac{5}{2})(x - \frac{5}{2})$

2. Factor the following expressions. Experiment with numbers to find the correct factors.

a) $x^2 + 9x + 20$ b) $x^2 + 5x + 6$ c) $x^2 - 5x + 6$

d) $x^2 - 9$ e) $x^2 - 16$ f) $x^2 + 7x - 18$

3. Prove that $x(x^2 + 7x) = x^3 + 7x^2$.
4. Can you think of a way of testing or verifying (not proving) that

$$x^2 + 5xy + 6y^2 = (x + 3y)(x + 2y)$$

for all values of x and y?
5. Write out in words the equivalent of

$$(x - 3)(x + 3) = x^2 - 9.$$

6. A high school girl had to simplify $(a^2 - b^2)/(a - b)$. She reasoned that a^2 divided by a gives a. Minus divided by minus gives plus. And b^2 divided by b gives b. Hence the answer is $a + b$. Is the answer correct? Is the argument correct?
7. There is a well known "proof" that $2 = 1$. The proof runs as follows. Suppose a and b are two numbers such that

$$a = b.$$

We may multiply both sides of this equation by a and obtain

$$a^2 = ab.$$

Now we may subtract b^2 from both sides and obtain

$$a^2 - b^2 = ab - b^2.$$

By factoring we may replace the left and right sides of this equation by

$$(a - b)(a + b) = b(a - b).$$

Division of both sides of this equation by $a - b$ yields

$$a + b = b.$$

Since $a = b$, we may as well write

$$2b = b.$$

But now we can divide both sides of this last equation by b, and there results

$$2 = 1.$$

Find the flaw in this proof.

5-5 EQUATIONS INVOLVING UNKNOWNS

The study of algebraic transformations as such is not very interesting. It is much like the grammar of a language. The significant uses of these transformations occur in larger investigations which we shall undertake later. However, a direct use of the processes of algebra does arise in the problem of finding unknown quantities, a problem not without some interest in itself and one which also arises in the course of broader investigations.

A somewhat practical, though by no means vital, example is the following. The radiator of a car contains 10 gallons of liquid 20 per cent of which is alcohol. The owner wishes to draw off a quantity of liquid and replace it by pure alcohol so that the resulting mixture contains 50 per cent alcohol. How many gallons of liquid should he draw off?

Now the very practical person who refuses to use mathematics can handle this situation very readily. He can draw off 5 gallons of the mixture and replace it by 5 gallons of alcohol. Then the mixture will certainly contain at least 50 per cent alcohol because even the remaining 5 gallons contain some alcohol. However, if the final mixture need contain only 50 per cent alcohol, then the practical person has wasted alcohol and therefore money. If he draws off 4 gallons, 6 gallons will be left, and since 20 per cent of this is alcohol, the alcoholic content is $1\frac{1}{5}$ gallons. If he now adds 4 gallons of alcohol he will have $5\frac{1}{5}$ gallons of alcohol, or more than 50 per cent, in the 10 gallons. On the other hand, if he draws off only 3 gallons, 20 per cent of the remaining 7 gallons is $\frac{7}{5}$ gallons of alcohol, and the addition of 3 more will yield $4\frac{2}{5}$ gallons of alcohol out of 10, or less than 50 per cent. The correct answer lies somewhere between 3 and 4, but where? Instead of continuing to guess let's use a little algebra.

Let x be the number of gallons of the mixture to be drawn off and to be replaced by an equal amount of pure alcohol. Then the number of gallons remaining of the original mixture is $10 - x$. Of this 20 per cent, or $\frac{1}{5}$, is alcohol, so that of the $10 - x$ gallons, $(\frac{1}{5})(10 - x)$ is alcohol. After the x gallons are replaced with pure alcohol, the amount of alcohol in the tank will be $(\frac{1}{5})(10 - x) + x$. We should like to fix x so that the amount of alcohol should be 50 per cent of 10 gallons, or 5 gallons. Hence we seek the value of x which satisfies the equation

$$\frac{1}{5}(10 - x) + x = 5. \tag{7}$$

Now we can apply the distributive axiom to start off our transformations and write

$$\frac{1}{5} \cdot 10 - \frac{1}{5}x + x = 5. \tag{8}$$

The terms $-\frac{1}{5}x + x$ or $x - \frac{1}{5}x$ amount to $\frac{4}{5}x$. Hence (8) is equivalent to

$$2 + \frac{1}{5}x = 5. \tag{9}$$

If we now subtract 2 from both sides of this equation, the result will still be an equality because equals subtracted from equals give equals. Then

$$\tfrac{4}{5}x = 3.$$

We now multiply both sides of this equation by $\frac{5}{4}$, and since equals multiplied by equals give equals, we have

$$x = 3 \cdot \tfrac{5}{4}. \tag{10}$$

Hence the answer is that the owner of the car should draw off $3\frac{3}{4}$ gallons of the original liquid. We knew before we applied algebra that the answer lies between 3 and 4, and we now know exactly where.

The more significant point made by this example, however, is that we started with equation (7) which expresses the condition to be satisfied by the unknown quantity x and that, by executing a series of almost mechanical steps justified by axioms about numbers, we arrived at a new equation, (10), which tells us what we wish to know. In other words, we performed a series of transformations which carried us from one equation to another and we profited thereby. The answer is not sensational, but we see how the manipulation of symbols gives us new information.

There is another point which the above example illustrates, at least in a minor way. Once we formulate equation (7) we forget all about the physical situation and concentrate solely on the equation. Nothing that is not relevant to the problem, i.e., to the problem of determining the number x, interferes with our thinking. Ernst Mach, a famous scientist of the late nineteenth century, said that mathematics is characterized by "a total disburdening of the mind," and we can now see what he meant. The make of the car, the shape of the radiator, the fact that the owner may be concerned with protecting the liquid in the radiator from freezing, and any other facts which have nothing to do with determining x can be forgotten. We disburden our minds of everything but the quantitative facts expressed in equation (7), and proceed to handle quantitative relationships only.

Equation (7) is rather simple. It is called a *linear* or *first-degree equation* because the unknown x occurs to the first power only. Let us consider a second example which will again illustrate the transformation value of algebra, but which also has other interesting features. Suppose that one ship is at A (Fig. 5–1) and another is at B, exactly 10 miles north of A. The ship at B is steaming east at the rate of 2 miles per hour. The ship at A is capable of traveling at a speed of 5 miles per hour and wishes to intercept the other ship. To set his course properly the captain of the ship at A must know where the two will meet.

Let us suppose that C is the point where they will meet. If the captain can determine the distance BC, he will head along the hypotenuse of a right triangle whose arms are AB and BC. Let us therefore denote the distance BC by x. Now that we seem to have labeled all relevant quantities, we encounter the first puzzling aspect of this problem, namely, that we do not have any equation to find x. Without this, of course, we can only sit and do some wishful thinking. Yet we do have enough information to set up such an equation.

What we have overlooked is a physical fact which is implied by the given information: The *time that the ship at B will take to travel to C must be the same as the time it will take the ship at A to reach C.* Since the ship at B travels at 2 miles per hour, it will take $x/2$ hours to reach C. To calculate the time required by the ship at A to reach C, we need the distance AC. We do not know AC, but we can at least express its value by means of the Pythagorean theorem of geometry. This theorem says in the present instance that

$$AC^2 = 100 + x^2.$$

Then

$$AC = \sqrt{100 + x^2}.$$

The time required for the ship at A to travel the distance AC at 5 miles per hour is

$$\frac{\sqrt{100 + x^2}}{5}.$$

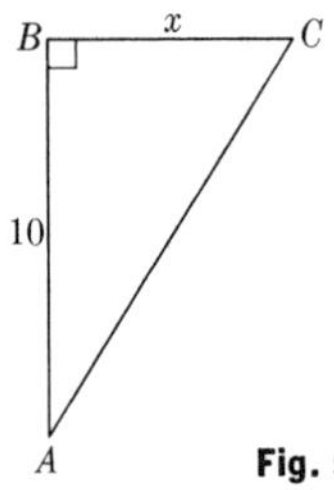

Fig. 5–1

We next equate the time required for the ship at B to travel the distance BC and the time required for the ship at A to travel the distance AC. This equation is

$$\frac{x}{2} = \frac{\sqrt{100 + x^2}}{5}. \qquad (11)$$

We now have an equation to work with. Let us see whether we can transform it so that it will yield a value for x. Since the square root is annoying, let us square both sides, i.e., multiply the left side by itself and the right side by itself. Since the left side equals the right side we are in effect multiplying equals by equals, and so the step is justified. Squaring both sides, we obtain

$$\frac{x^2}{4} = \frac{100 + x^2}{25}. \qquad (12)$$

Since fractions are also annoying, let us multiply both sides by 100. We choose 100 because both 25 and 4 divide evenly into 100. Thus

$$100 \cdot \frac{x^2}{4} = 100 \cdot \frac{100 + x^2}{25}.$$

We may apply our operations with fractions to write

$$25x^2 = 4(100 + x^2).$$

Application of the distributive axiom yields

$$25x^2 = 400 + 4x^2. \tag{13}$$

Now we subtract $4x^2$ from both sides, and because equals subtracted from equals yield equals, we obtain

$$21x^2 = 400. \tag{14}$$

Division of both sides by 21, which is a division of equals by equals, yields

$$x^2 = \frac{400}{21}\cdot \tag{15}$$

Now we ask ourselves what number squared yields 400/21. Certainly $\sqrt{400/21}$ is one possibility. But a negative number squared or multiplied by itself is also positive. Hence there are two possible answers:

$$x = \sqrt{400/21} \qquad \text{and} \qquad x = -\sqrt{400/21}. \tag{16}$$

Let us accept both of these for the moment and dispose first of a purely arithmetical question. How much is $\sqrt{400/21}$? Well, we can divide 21 into 400 and obtain 19.05 to two decimal places. We must now find $\sqrt{19.05}$. There is an arithmetic process for finding the square root of a number, but for our purposes it will be sufficient to estimate the answer. Clearly 4 is too small and 5 is too large. By sheer trial we find that $(4.3)^2 = 18.49$ and $(4.4)^2 = 19.36$. Hence the correct value lies between 4.3 and 4.4. If we wished to have a more accurate answer, we could now try 4.31, 4.32, and so on, until we found a result which came as close to 19.05 as possible, and so obtain an answer to the nearest hundredths' place. We shall accept 4.4 as good enough for our purposes and thus we may say that

$$x = 4.4 \qquad \text{and} \qquad x = -4.4. \tag{17}$$

And now we have more than we want; we have two answers, whereas we sought only one. Of course, we wish to use the positive answer because the x we seek stands for a length which is positive. This is the value which has the proper physical meaning in our problem. But the question, How did the negative value of x get into the picture, remains open. The answer involves a rather important point about the nature of mathematics and its relation to the physical world. The mathematician starts with concepts and axioms which express some idealized facts about the world, and proceeds to apply these concepts and axioms to solve physical problems. In the present case the methods used lead to two solutions. Hence the methods may involve new elements which are not present in the physical world, even though the intent was to stay close to it. Thus, squaring both sides of equation (11), a justifiable mathematical step, introduced a new solution, for, if our original equation had been

$$\frac{x}{2} = -\frac{\sqrt{100 + x^2}}{5}, \tag{18}$$

we would have obtained the same equation, (12) and everything we did thereafter would have applied to (18) as well as (11). Hence, in this case, we can see specifically where mathematics departs from the physical situation.

The main point to be noted is then that, although mathematical concepts and operations are formulated to represent aspects of the physical world, mathematics is not to be identified with the physical world. However, it tells us a good deal about that world if we are careful to apply it and interpret it properly. We shall find that this point, which eluded the best thinkers until the late nineteenth century, will acquire increasing importance as we proceed.

There is another valuable lesson to be learned from the solution of the problem we have just examined. When we arrived at step (13), we combined terms in x^2 and then proceeded to find x. The subsequent work led to a fair amount of arithmetic. An engineer working with the same problem and perhaps satisfied with an approximate answer might argue that the term $4x^2$ is small compared with the term $25x^2$ and so disregard it. Instead of our next equation, (14), his new equation would then read

$$25x^2 = 400,$$

and by dividing both sides of this equation by 25, he would obtain

$$x^2 = 16.$$

It now follows that

$$x = 4 \quad \text{and} \quad x = -4.$$

Thus 4 is an approximate answer. Engineers often are satisfied with such approximations because, in constructing actual objects of wood and steel, they cannot meet a specified value exactly. Not only can't one measure exactly, but tools and machines also introduce errors. By neglecting $4x^2$ in (13) the engineer gained the advantage of finding the approximate answer much more readily than we were able to determine the correct answer even to one decimal place only.

In the present problem the saving is trivial, but approximation may make a lot of difference in more difficult problems. Whereas the mathematician, who seeks exact answers, will work months and years on a problem, the engineer will often settle for an approximate answer and obtain it far more easily. The point we are making is not that the engineer is smarter. To get on with his job the engineer must arrive at an answer quickly, whereas the mathematician's job is to obtain a correct answer, no matter how long it takes. Both are true to the objectives and spirit of their own work. Moreover, in making approximations, the engineer raises a question which he may not be able to answer. How good is his approximation? After all, while physical constructions and measurements are not exact, beams must fit. Hence the engineer should really ascertain that the approximation is good enough for his purposes. If he can tolerate an error of only 0.1 of an inch, he must make sure that his approximations do not introduce a larger error.

In really difficult problems the engineer will make approximations and, usually with the aid of a mathematician, determine the error introduced. If he cannot do so, he will often overdesign; that is, if the approximate result shows that a beam supporting a building need be only one inch thick, he may make it two inches thick and thereby hope that he has more than allowed for the error. Is he certain even with this precaution that his beam will hold up? No. Big bridges have collapsed because such calculations and additional precautionary measures were not enough. A recent example was the Tacoma bridge in the State of Washington. The bridge did not withstand the force of the wind and collapsed.

EXERCISES

1. The speed of sound in an iron rod is 16,850 ft/sec, and the speed in air is 1100 ft/sec. If a sound originating at one end of the rod is heard one second sooner through the rod than through the air, how long is the rod?
2. A bridge AB is 1 mi (5280 ft) long in winter and expands 2 ft in the summer. For simplicity suppose that the shape in summer is the triangle ACB shown in Fig. 5–2. How far does the center of the bridge drop in summer, that is, how long is CD? Before calculating the answer, estimate it. To calculate, use the Pythagorean theorem and estimate the square root to the nearest foot.

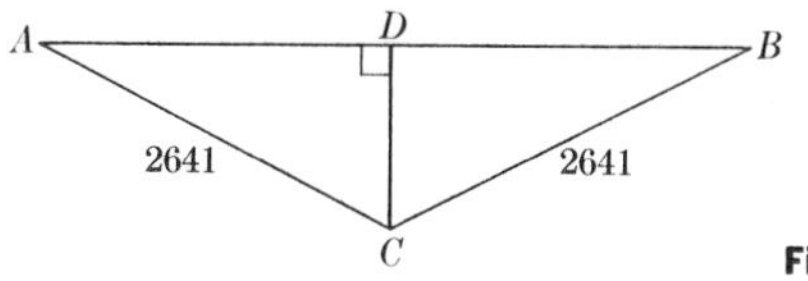

Fig. 5–2

3. An airplane which can fly at a speed of 200 mi/hr in still air flies a distance of 800 mi with the wind in the same time as it flies 640 mi against the wind. What is the

speed of the wind? [*Suggestion:* If x is the speed of the wind, then the speed of the plane when flying with the wind is $200 + x$; the speed of the plane when flying against the wind is $200 - x$.]

4. The population of town A is 10,000 and is increasing by 600 each year. The population of town B is 20,000 and is increasing by 400 each year. After how many years will the two towns have the same population?
5. A rope hanging from the top of a flagstaff is 2 ft longer than the staff. When pulled out taut, it reaches a point on the ground 18 ft from the foot of the staff. How high is the staff?
6. A publisher finds that the cost of preparing a book for printing and of making the plates is \$5000. Each set of 1000 printed copies costs \$1000. He can sell the books at \$5 per copy. How many copies must he sell to at least recover his costs?
7. We may certainly say that

$$\tfrac{1}{4}\text{ dollar} = 25\text{ cents.}$$

We take the square root of both sides and obtain

$$\tfrac{1}{2}\text{ dollar} = 5\text{ cents.}$$

What is wrong?

8. A glass which is half full certainly contains as much liquid as a glass which is half empty. Then

$$\tfrac{1}{2}\text{ full} = \tfrac{1}{2}\text{ empty.}$$

If we multiply both sides by 2 we obtain

$$1\text{ full} = 1\text{ empty,}$$

or a full glass contains as much as an empty glass. What is wrong?

5-6 THE GENERAL SECOND-DEGREE EQUATION

Our discussion of the solution of equations in the preceding section dealt with two types of equations, first-degree equations illustrated by equation (7) and second-degree equations illustrated by equation (14). No difficulties can arise in the process of solving first-degree equations, i.e., equations which, by proper algebraic operations, can be expressed in the form

$$ax + b = 0, \tag{19}$$

where a and b are definite numbers and x is the unknown. Equation (19) can readily be solved for x.

The case of second-degree equations is not so simple. We were fortunate that equation (14) led to (15), and that by taking the square root of both sides we obtained the two solutions, or roots as they are called. However we might have to solve an equation such as

$$x^2 - 6x + 8 = 0. \tag{20}$$

This equation is more complicated than (14) because (20) also contains the first-degree term in x.

In solving equation (20), we still do not encounter much trouble. We know from our work on transforming algebraic expressions that the left-hand side of (20) can be factored; that is, the equation can be written as

$$(x - 2)(x - 4) = 0. \tag{21}$$

We now see that when $x = 2$, the left side is zero because

$$(2 - 2)(2 - 4) = 0.$$

When $x = 4$, the left side is again zero because

$$(4 - 2)(4 - 4) = 0.$$

Hence the solutions or roots are

$$x = 2 \quad \text{and} \quad x = 4.$$

Now suppose we had to solve the second-degree equation

$$x^2 + 10x + 8 = 0. \tag{22}$$

This time it is not possible to find simple factors of the left side. Equations such as (22) do arise in real problems. Hence the mathematician considers the question, Is there a method which will solve such second-degree equations? Naturally he studies those he can solve to see whether they furnish any clue to such a method.

Examination of equation (20) reveals an interesting fact. The roots are 2 and 4. The sum of these two numbers is 6, and the coefficient, or multiplier, of x is -6. The product of 2 and 4 is 8, and 8 is the constant term, that is, the term free of x. These facts might be a coincidence, and so the mathematician would investigate whether they hold for other simple equations. Consider the very simple equation:

$$x^2 - 4 = 0. \tag{23}$$

Here the roots are $+2$ and -2. Their sum is 0, and we note that the term in x is missing, which means it is $0 \cdot x$. The product of the roots is -4, precisely the constant term in (23). Presumably we have some facts about the roots, but how can we use them?

Equations of the form (23) are easy to solve, since one only has to take a square root. Perhaps the method we should seek is one which reduces all equations of the type (20) to the type (23). But how do we do this? The sum of the roots in (23) is zero. The sum of the roots in (20) is 6, and this is the negative of the coefficient of x. If we added to each root of (20) one-half the coefficient of x, that is, -3, the sum of the roots would be zero. What this suggests, then, is to *form a new equation whose roots are the roots of the old one, each increased* by one-half the coefficient of x.* Since the coefficient of x is -6, we let

$$y = x + (-3) = x - 3$$

or

$$x = y + 3. \tag{24}$$

If we substitute this value of x in (20), we obtain

$$(y + 3)^2 - 6(y + 3) + 8 = 0.$$

We now calculate the square in the first term, carry out the multiplication in the second term, and find that

$$y^2 + 6y + 9 - 6y - 18 + 8 = 0$$

or

$$y^2 - 1 = 0.$$

Then

$$y^2 = 1$$

and

$$y = 1 \quad \text{and} \quad y = -1.$$

But from (24) we see that

$$x = 1 + 3 \quad \text{and} \quad x = -1 + 3$$

or

$$x = 4 \quad \text{and} \quad x = 2.$$

Thus we obtain *without factoring* the very same roots of equation (20) that we found previously by factorization.

Now let us reconsider equation (22), namely

$$x^2 + 10x + 8 = 0. \tag{22}$$

Since the roots cannot be obtained by any apparent method of factoring, let us see whether the idea just tried works here also; that is, let us form a new equation whose roots are the roots of (22) increased by one-half the coeffi-

---

* We use the term "increased" here, even though in the example we add a negative quantity to each root and really decrease the value of the roots.

cient of x. The roots of (22) are represented by x. Then we shall form a new equation whose roots y are:

$$y = x + \tfrac{10}{2} = x + 5. \tag{25}$$

From (25) we have

$$x = y - 5. \tag{26}$$

We substitute this value of x in (22) and obtain

$$(y - 5)^2 + 10(y - 5) + 8 = 0.$$

We perform the indicated multiplications and obtain

$$y^2 - 10y + 25 + 10y - 50 + 8 = 0.$$

By combining terms we find that

$$y^2 - 17 = 0$$

or

$$y^2 = 17.$$

Then

$$y = \sqrt{17} \quad \text{and} \quad y = -\sqrt{17}.$$

We now use (26) to state that

$$x = \sqrt{17} - 5 \quad \text{and} \quad x = -\sqrt{17} - 5. \tag{27}$$

We have found the two roots of (22) without factoring.

EXERCISES

1. Find the roots of the following equations by factoring the left-hand side:

   a) $x^2 - 8x + 12 = 0$ b) $x^2 + 7x - 18 = 0$

2. Find the roots of each of the equations in Exercise 1 by forming a new equation whose roots are "larger" than those of the original equation by one-half the coefficient of x.

3. Solve the following equations by the method of forming a new equation whose roots are "larger" than those of the original equation by one-half the coefficient of x.

   a) $x^2 + 12x + 9 = 0$ b) $x^2 - 12x + 9 = 0$

The method of solving second-degree equations by forming a new equation seems to work, but we have no proof that it will always work. To secure a general proof we shall use one of the basic devices of algebra; that is, instead of working with particular equations, we shall consider the general second-degree equation

$$x^2 + px + q = 0. \tag{28}$$

Here p and q are letters, each of which can stand for *any* given real number. The use of the letters p and q must be distinguished from the use of x to stand for the specific unknown roots of the equation. Now we follow the method employed to solve equations (20) and (22); that is, we form a new equation whose roots are the roots of (28), each increased by one-half the coefficient of x. This means that we introduce the expression

$$y = x + \frac{p}{2}.$$

Then

$$x = y - \frac{p}{2}. \tag{29}$$

We substitute this value of x in (28) and obtain

$$\left(y - \frac{p}{2}\right)^2 + p\left(y - \frac{p}{2}\right) + q = 0.$$

By squaring the first term and multiplying through by p in the second one, we obtain

$$y^2 - py + \frac{p^2}{4} + py - \frac{p^2}{2} + q = 0.$$

The terms involving py cancel. Moreover, $p^2/4 - p^2/2 = -p^2/4$. Hence

$$y^2 - \frac{p^2}{4} + q = 0.$$

By adding $p^2/4$ to both sides and subtracting q from both sides, we obtain

$$y^2 = \frac{p^2}{4} - q.$$

Hence

$$y = \sqrt{\frac{p^2}{4} - q} \quad \text{and} \quad y = -\sqrt{\frac{p^2}{4} - q}.$$

In this general case, we cannot determine the numerical value of the square root, but we can leave the result in this form. We now see from equation (29) that

$$x = \sqrt{\frac{p^2}{4} - q} - \frac{p}{2} \quad \text{and} \quad x = -\sqrt{\frac{p^2}{4} - q} - \frac{p}{2}. \tag{30}$$

This result is remarkable.* We have shown that the roots of any equation of the form (28) (that is, no matter what p and q are) are given by the expressions (30).

We really have accomplished more than we sought to accomplish. We sought a method of solving an equation such as (22). We not only have found such a method, but, since the result (30) holds for *any* such equation, we do not have to go through the entire process each time; we proceed by simply substituting the proper value of p and q in (30). Thus if we compare equations (22) and (28), we see that the p in (22) is 10 and q is 8. Hence let us substitute 10 for p and 8 for q in (30). We find

$$x = \sqrt{\tfrac{100}{4} - 8} - \tfrac{10}{2} \quad \text{and} \quad x = -\sqrt{\tfrac{100}{4} - 8} - \tfrac{10}{2},$$

or

$$x = \sqrt{17} - 5 \quad \text{and} \quad x = -\sqrt{17} - 5. \tag{31}$$

This is exactly the result obtained in (27).

By working with the general form $x^2 + px + q = 0$ instead of equations with specific numbers as coefficients, we have shown how to solve *any* second-degree equation. This general result could never be derived from equations with numerical coefficients because there are infinitely many such equations, and one could not investigate them all. Thus the use of letters to represent any one of a class of numbers gives mathematics a power and generality which achieves what could not be accomplished in many lifetimes of effort with particular equations. Of course, to people who do not care to solve one quadratic the ability to solve all is no boon. But even these people have benefited indirectly. The preceding theory illustrates how the mathematician, when called upon to solve the same type of problem repeatedly, seeks a general method which will handle all of them.

---

* In many books a method is given for solving the general second-degree equation $ax^2 + bx + c = 0$. If we divide this equation by a, we obtain $x^2 + (b/a)x + (c/a) = 0$. This equation is now of the same form as (28), where $p = b/a$ and $q = c/a$. If we enter these values of p and q in (30), we get the roots

$$x = -\frac{b}{2a} + \frac{\sqrt{b^2 - 4ac}}{2a} \quad \text{and} \quad x = -\frac{b}{2a} - \frac{\sqrt{b^2 - 4ac}}{2a}.$$

The use of letters such as p and q, which has made an enormous difference in the effectiveness of mathematics, seems like a small idea once understood, and yet it is a rather recent development. From the time of the Babylonians and Egyptians to about 1550, all the equations solved had numerical coefficients. Although many algebraists realized that the method they used for one set of numerical coefficients would work for any other, they had no general proof. The idea of employing general coefficients in algebraic equations, an idea which, as we shall see, was taken over into other domains of mathematics, is due to François Vieta (1540–1603), a great French mathematician. The remarkable fact about Vieta is that he was a lawyer who worked for the kings of France. Mathematics was just a hobby to him, but one at which he "worked" extensively. Vieta was fully conscious of what he had done by introducing literal coefficients. He said that he was introducing a new kind of algebra which he called *logistica speciosa*, that is calculation with whole species, as opposed to the numerical work of his predecessors which he called *logistica numerosa*.

We could consider other examples of how the processes of algebra permit us to solve equations involving unknowns, but we shall not devote more time to the subject. What is important is the recognition that by means of algebra we can extract information from some given facts. It is also important to see how readily and mechanically the processes of solving equations yield the desired information. In fact, one of the curious things about mathematics that clearly emerges even from our brief work in algebra is that mathematics which is concerned with reasoning nevertheless creates processes which can be applied almost mechanically, that is, without reasoning. The thinking is, so to speak, mechanized and this mechanization enables us to solve complicated problems in no time. We think up processes so that we don't have to think.

It may be necessary to caution the reader again that while the techniques of transformations are necessary to perform useful and interesting mathematical work, they are not the substance of mathematics. If all that one learns in mathematics is the ability to execute these techniques, however quickly and accurately, he will not see the real purpose, nature, and accomplishments of mathematics. To a large extent, techniques are a necessary evil, like practicing scales on a piano, in order to be able to play grand and beautiful compositions. Naturally those who wish to be professional mathematicians must learn as many of these techniques as possible.

EXERCISES

1. Solve by means of (30) the following equations:

a) $x^2 - 8x + 10 = 0$ b) $x^2 + 8x + 10 = 0$ c) $x^2 - 6x - 9 = 0$

d) $2x^2 + 8x + 6 = 0$ e) $x^2 - 8x + 16 = 0$

* 5-7 THE HISTORY OF EQUATIONS OF HIGHER DEGREE

The search for generality in mathematics began in the sixteenth century. One type of generality became possible when Vieta showed how to treat a whole class of equations by means of literal coefficients. Another direction which the search for generality took was the investigation of equations of degree higher than the second.

The first of the notable mathematicians to pursue the mathematics of equations of higher degree and certainly the greatest combination of mathematician and rascal is Jerome Cardan. He was born in Pavia, Italy, in 1501 to somewhat disreputable parents, although his father was a lawyer, doctor, and minor mathematician. Cardan had no upbringing worth speaking about and was sickly during the first half of his life. Despite these handicaps, he studied medicine and became so celebrated a physician that he was invited to treat prominent people in many countries of Europe. At various times he was professor of medicine, and he also lectured on mathematics at several Italian universities.

He was aggressive, high-tempered, disagreeable, and even vindictive, as if anxious to make the world suffer for his early deprivations. Because illnesses continued to harass him and prevented him from enjoying life, he gambled daily for many years. This experience undoubtedly helped him to write a now famous book, *On Games of Chance*, which treats the probabilities in gambling. He even gives advice on how to cheat, which was also gleaned from experience.

A product of his age in many respects, Cardan collected and published prolifically legends, false philosophical and astrological doctrines, folk cures, methods of communion with spirits, and superstitions. Apparently he himself believed in spirits and in astrology. He cast horoscopes, many of which proved to be false. Toward the end of his life he was imprisoned for casting the horoscope of Christ, but was soon pardoned, pensioned by the Pope, and lived peacefully until his death in 1576. In his *Book of My Life*, an autobiography, he says that despite his years of trouble he has to be grateful, for he had acquired a grandson, wealth, fame, learning, friends, belief in God, and he still had fifteen teeth.

Part of Cardan's rascality concerns our present subject. The mathematicians of the sixteenth century had undertaken to solve higher-degree equations, for example, equations of the third degree such as

$$x^3 - 6x = 8.$$

Among them was another famous man, Nicolò of Brescia, better known as Tartaglia (1499–1557), whom we shall meet occasionally in other contexts. Tartaglia had discovered a method for solving third-degree equations, and Cardan wished to publish this method in a book he was writing on algebra, which later appeared under the title *Ars Magna*, the first major book on algebra in modern times. After refusing to divulge the method, Tartaglia finally acquiesced, but asked Cardan to keep it secret. However, Cardan wished his book to be as important as possible and so published the method, though acknowledging that it was Tartaglia's. From this book, which appeared in 1545, the mathematical world learned how to solve third-degree equations. In this same book Cardan also published a method of solving fourth-degree equations discovered by one of his own pupils, Lodovico Ferrari (1522–1565). Although general coefficients were not in use as yet, it was clear that all third- and fourth-degree equations could be solved. In other words, the solutions could be expressed in terms of the coefficients by means of the ordinary operations of algebra, i.e., addition, subtraction, multiplication, division, and roots (though not necessarily square roots), in just about the manner in which (30) expresses the solutions of a second-degree equation in terms of the coefficients p and q.

And now the mathematicians' interest in generality took over. Since the *general* equations of the first, second, third, and fourth degree could be solved, what about fifth-, sixth- and higher-degree equations? It seemed certain that these equations could also be solved. For three hundred years many mathematicians worked on this basic problem and made almost no progress. And then a young Norwegian mathematician, Niels Henrik Abel (1802–1829), showed at the age of 22 that fifth-degree equations could not be solved by the processes of algebra. Another youth, Évariste Galois (1811–1832), who failed twice to pass the entrance examinations for the École Polytechnique and spent just one year at the École Normale, demonstrated that all general equations of degree higher than the fourth cannot be solved by means of the operations of algebra. In a letter he wrote the night before he was killed in a duel, Galois explained his ideas and showed how a new and general theory of the solution of equations could be developed. Galois' ideas gave algebra a totally new turn. Instead of being a tool, a series of techniques for the transformation of expressions into more useful ones, it became a beautiful body of knowledge which can be of interest in itself. Unfortunately we cannot undertake to study Galois' ideas, or the Galois theory as it is called, because there are more basic things to be learned first.

This brief account of the search for generality in the solution of equations has been given here because it illustrates many important features of mathematics. One is the persistence, stubbornness if you will, of mathematicians over hundreds of years. Another is the experience that the search for generality leads to new and important developments, even though at the outset the generality is sought for its own sake. Today, the solution of higher-degree equations is a most practical matter, and we owe to Galois the most revealing insight into this subject. We also find in this history of the theory of equations a major example of how mathematicians find problems on which to work, problems of significance drawn from other problems which have humble and practical origins such as simple equations involving unknowns.

REVIEW EXERCISES

1. Carry out the indicated multiplication:

   a) $3(2x + 6)$ b) $(x + 3)(x + 2)$ c) $(x + 7)(x - 2)$
   d) $(x + 3)(x - 3)$ e) $(x + \frac{7}{2})(x + \frac{3}{2})$ f) $(2x + 1)(x + 2)$
   g) $(x + y)(x - y)$

2. Factor the following expressions. You may have to experiment to find the correct factors.

   a) $x^2 - 9$ b) $x^2 - 16$ c) $x^2 - a^2$
   d) $a^2 - b^2$ e) $x^2 + 6x + 9$ f) $x^2 + 7x + 6$
   g) $x^2 + 5x + 4$ h) $x^2 - 6x + 9$ i) $x^2 - 7x + 6$
   j) $x^2 - 5x + 4$ k) $x^2 - 7x + 12$ l) $x^2 + 6x - 16$
   m) $x^2 + 6x - 27$

3. If $2x + 7 = 5$, what does $2x$ equal, and what does x equal?
4. Solve the following equations. State what you do in each step.

   a) $2x + 9 = 12$ b) $2x + 12 = 9$ c) $\frac{x}{10} + 3 = 4$ d) $\frac{x}{10} + \frac{3}{5} = \frac{4}{5}$
   e) $\frac{x}{10} + \frac{3}{5} = \frac{6}{7}$ f) $3x + \frac{3}{10} = \frac{4}{5}$ g) $\frac{5x}{2} + \frac{3}{5} = 6$ h) $ax + 2 = b$
   i) $ax - b = c$

5. A solution of acid and water contains 75% water. How many grams of acid would you add to 50 grams of the solution to make the percentage of water 60%?
6. A student has grades of 60 and 70 on two examinations. What grade must he earn on a third examination to attain an average of 75%?
7. Solve the following equations by factoring:

   a) $x^2 - 6x + 5 = 0$ b) $x^2 - 6x - 7 = 0$ c) $x^2 - 7x + 6 = 0$
   d) $x^2 + 6x - 27 = 0$ e) $x^2 - 7x + 12 = 0$ f) $x^2 - 5x - 14 = 0$

8. Solve the following equations by the method of forming a new equation whose roots are "larger" than those of the original equation by one-half the coefficient of x.

   a) $x^2 + 10x + 9 = 0$ b) $x^2 - 10x + 9 = 0$ c) $x^2 + 10x + 6 = 0$

d) $x^2 - 10x + 6 = 0$ e) $x^2 - 12x + 15 = 0$ f) $x^2 + 12x + 15 = 0$

9. Solve the following equations by applying formula (30) of the text:

a) $x^2 + 12x + 6 = 0$ b) $x^2 - 12x + 6 = 0$ c) $x^2 + 12x - 6 = 0$

d) $x^2 - 12x - 6 = 0$ e) $2x^2 + 12x + 6 = 0$ f) $3x^2 + 27x + 15 = 0$

g) $t^2 + 10t = 8$

10. In Section 5–5 of the text, we solved a problem wherein one ship sets its course properly so as to overtake another ship. To set up the equation which solved the problem, equation (11), we started by letting x be the distance which the ship traveling east covers. Solve the same problem by letting t be the time that both ships travel until they meet. Then $x = 2t$. The algebra of this alternative solution is easier to handle. However it is not so obvious that we should let our unknown be the time of travel.

Topics for Further Investigation

1. The rise of symbolism in algebra.
2. The history of the solution of equations.

Recommended Reading

BALL, W. W. ROUSE: *A Short Account of the History of Mathematics*, pp. 201–243, Dover Publications Inc., New York, 1960.

COLERUS, EGMONT: *From Simple Numbers to the Calculus*, Chaps. 9 through 13, Wm. Heinemann Ltd., London, 1954.

ORE, OYSTEIN: *Cardano, The Gambling Scholar*, Chaps. 1 through 5, Princeton University Press, Princeton, 1953.

SAWYER, W. W.: *A Mathematician's Delight*, Chap. 7, Penguin Books Ltd., Harmondsworth, England, 1943.

SMITH, DAVID E.: *History of Mathematics*, Vol. II, pp. 378–470, Dover Publications Inc., New York, 1958.

WHITEHEAD, ALFRED N.: *An Introduction to Mathematics*, Chaps. V and VI, Holt, Rinehart and Winston, Inc., New York, 1939 (also in paperback).

CHAPTER 6

THE NATURE AND USES OF EUCLIDEAN GEOMETRY

Circles to square and cubes to double
Would give a man excessive trouble.

MATTHEW PRIOR

6–1 THE BEGINNINGS OF GEOMETRY

Just as the study of numbers and its extensions to algebra arose out of the very practical problems of keeping track of property, trading, taxation, and the like, so did the study of geometry develop from the desire to measure the area of pieces of land (or geodesy in general), to determine the volumes of granaries, and to calculate the dimensions and amount of material needed for various structures.

The physical origin of the basic figures of geometry is evident. Not only the common figures of geometry but the simple relationships, such as perpendicularity, parallelism, congruence, and similarity, derive from ordinary experiences. A tree grows perpendicular to the ground, and the walls of a house are deliberately set upright so that there will be no tendency to fall. The banks of a river are parallel. A builder constructing a row of houses according to the same plan wishes them to have the same size and shape, that is, to be congruent. A workman or machine producing many pieces of a particular item makes them congruent. Models of real objects are often similar to the object represented, especially if the model is to be used as a guide to the construction of the object.

The science of geometry, indeed, the science of mathematics, was founded by the Greeks of the classical period. We have already described the major steps: the recognition that there are abstract concepts or ideas such as point, line, triangle, and the like, which are distinct from physical objects, the adoption of axioms which contained the surest knowledge about these abstractions man can obtain, and the decision to prove deductively any other facts about these concepts. The Greeks converted the disconnected, empirical, limited geometrical facts of the Egyptians and Babylonians into a vast, systematic, and thoroughly deductive structure.

Although the Greeks also studied the properties of numbers, they favored geometry. The reasons are pertinent. First of all, the Greeks liked exact thinking, and found that this faculty was more readily applied to geometry. Possible theorems are rather easily gleaned from the visualization of geometrical configurations. The neat correspondence between deductively established conclusions and intuitive understanding further increases this appeal of geometry. That one can draw pictures to represent what one is thinking about in geometry has its drawbacks. One is prone to confuse the abstract concept with the picture and to accept unconsciously properties of the picture. Of course, the idea of a triangle must be distinguished from the triangle drawn in chalk or pencil, and no properties of the picture may be used unless they are contained in the axioms or in some previously proved theorem. The Greeks were careful to make this distinction.

Secondly, the Greek philosophers who founded mathematics were intrigued with the design and structure of the universe, and they studied the heavens, certainly the most impressive spectacle in nature, to fathom the design. The shapes and paths of the heavenly bodies and the over-all plan of the solar system were of interest. On the other hand, they hardly saw any value in the ability to describe the exact locations of the moon, sun, and planets and to predict their precise locations at a given time, information of importance in calendar reckoning and in navigation.

Thirdly, since commerce and daily business were handled in large part by slaves, and were in any case in low regard, the study of numbers, which served such purposes, was subordinated. Why worry about the uses of numbers for measurement and trade if one does not measure or trade? One does not need the dimensions of even one rectangle to speculate about the properties of all rectangles.

The Greek philosophers emphasized an aspect of reality which is today, at least in scientific circles, neglected. To the Greeks of the classical period the reality of the universe consisted of the forms which matter possessed. Matter as such was formless and therefore meaningless. But an object in the shape of a triangle was significant by the very fact that it was triangular.

Finally, there were purely mathematical grounds for the Greek emphasis on geometry. The Greeks were the first to recognize that quantities such as $\sqrt{2}$, $\sqrt{3}$, $\sqrt[3]{2}$, etc., are neither whole numbers nor fractions, but they failed to recognize that these were new types of numbers, and that one could reason with them. To handle all types of quantities, they conceived the idea of treating them as line segments. As line segments, the hypotenuse of a right triangle (Fig. 4–2) and the arms have the same character, despite the fact that if the arms are each 1 unit long, the hypotenuse has the irrational length $\sqrt{2}$. To execute their plan of treating all quantities geometrically, the Greeks converted the algebraic processes developed in Egypt and Babylonia into geometrical ones. We could illustrate how the Greeks solved equations geometrically, but their methods are no longer favored. For science and engineering, the knowledge that a certain line segment solves an equation is not nearly so useful as a numerical answer which can be calculated to as many decimal places as needed. But the classical Greeks, who regarded exact reasoning as paramount in importance and who deprecated practical applications, found the solution of their difficulty in geometry and were content with this solution. Geometry remained the basis for all exact mathematical reasoning until the seventeenth century, when the needs of science forced the shift to number and algebra and the ultimate recognition that these could be built up as logically as geometry. In the intervening centuries arithmetic and algebra were regarded as practical disciplines.

Of course, the Greek conversion of exact mathematics to geometry was, from our present viewpoint, a backward step. Not only are the geometrical methods of performing algebraic processes insufficient for science, engineering, commerce, and industry, but they are by comparison clumsy and lengthy. Moreover, because Greek geometry was so complete and so admirable, mathematicians following in the Greeks' footsteps continued to think that exact mathematics must be geometrical. As a consequence, the development of algebra was unnecessarily delayed.

6–2 THE CONTENT OF EUCLIDEAN GEOMETRY

The major book on geometry of the classical Greek era is Euclid's *Elements*, a work on plane and solid geometry. Written about 300 B.C., it contains the best results produced by dozens of fine mathematicians during the period from 600 to 300 B.C. The work of Thales, the Pythagoreans, Hippias, Hippocrates, Eudoxus, members of Plato's Academy, and many others furnished the material which Euclid organized. His text was not the first to be written, but unfortunately we do not have copies of the earlier ones. It is quite certain that the particular axioms one finds in the *Elements*, the arrangement of the theorems, and many of the proofs are all due to Euclid. The geometry texts used in high schools today in essence reproduce Euclid's work, although these contemporary versions usually contain only a small part of the 467 theorems and many corollaries found in the *Elements*. Euclid's version is so marvelously knit together that most readers are amazed to see so many profound theorems deduced from the few self-evident axioms.

Though the reader may already be familiar with the basic theorems of Euclidean geometry, we shall take a few moments to review some features of the subject and the nature of the accomplishment. We might note first the structure of Euclid's *Elements*. He begins with some definitions of the basic concepts: point, line, circle, triangle, quadrilateral, and the like. Although modern mathematicians would make some critical comments about these definitions, we shall not discuss them at present. (See Chapter 20.)

Euclid then states ten axioms on which all subsequent reasoning is based. We shall note these merely to see that they do indeed describe apparently unquestionable properties of geometric figures. The first five axioms are:

AXIOM 1. Two points determine a unique straight line.

AXIOM 2. A straight line extends indefinitely far in either direction.

AXIOM 3. A circle may be drawn with any given center and any given radius.

AXIOM 4. All right angles are equal.

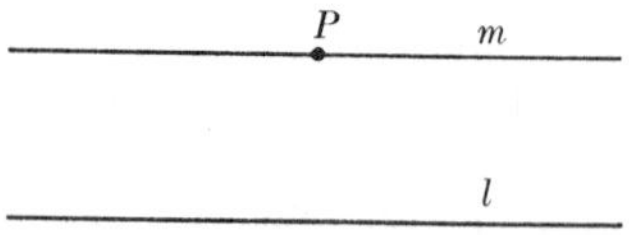

Fig. 6–1.
The parallel axiom.

AXIOM 5. Given a line l (Fig. 6–1) and a point P not on that line, there exists in the plane of P and l and through P one and only one line m, which does not meet the given line l.

In a separate *definition* Euclid defines parallel lines to be any two lines in the same plane which do not meet, that is, do not have any points in common. Thus, Axiom 5 asserts the existence of parallel lines.

The remaining five axioms are:

AXIOM 6. Things equal to the same thing are equal to each other.

AXIOM 7. If equals be added to equals, the sums are equal.

AXIOM 8. If equals be subtracted from equals, the remainders are equal.

AXIOM 9. Figures which can be made to coincide are equal (congruent).

AXIOM 10. The whole is greater than any part.

The formulations of these axioms are not quite the same as those prescribed by Euclid. Axiom 5 is, in fact, different from Euclid's, but is stated here in the form which is most likely to be familiar to the reader. The differences between Euclid's versions and those introduced by later mathematicians are not important for our present purposes, and so we shall not take time now to note them. (See Chapter 20.)

After stating his axioms, Euclid proceeded to prove theorems. Many of these theorems are indeed simple to prove and obviously true of the geometrical figures involved. But Euclid's purpose in proving them was to play safe. As we shall see in later chapters, many a conclusion seems obvious but is false. Of course, the major proofs are those which establish conclusions that are not at all obvious and, in some cases, even come as a surprise.

Partly to refresh our memories about some theorems of Euclidean geometry and partly to note once again the deductive procedure of mathematics, let us review one or two proofs. A basic theorem of Euclidean geometry asserts the following:

THEOREM 1. An exterior angle of a triangle is greater than either remote interior angle of the triangle.

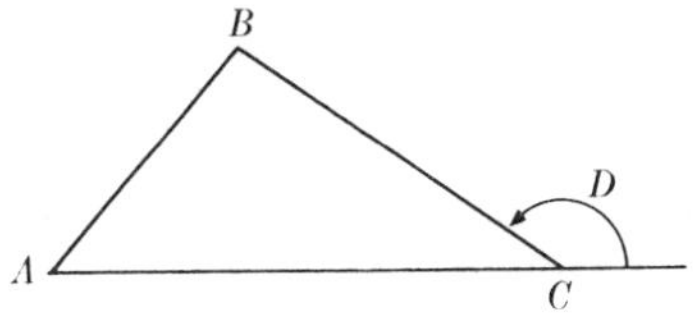

Fig. 6–2.
An exterior angle of a triangle is greater than either remote interior angle.

Before proving this theorem, let us be clear about what it says. Angle *D*, in Fig. 6–2, is called an exterior angle of triangle *ABC* because it is outside the triangle and is formed by one side, *BC*, and an extension of another side, *AC*. With respect to angle *D*, angles *A* and *B* are remote interior angles of triangle *ABC*, whereas angle *C* is an adjacent interior angle. Hence we have to prove that angle *D* is larger than angle *A* and larger than angle *B*. Let us prove that angle *D* is larger than angle *B*.

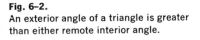

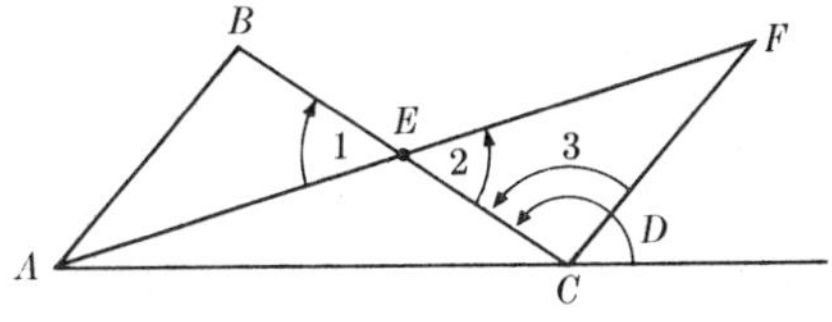

Fig. 6–3

The problem before us is a tantalizing one because, while it does seem visually obvious that angle *D* is greater than angle *B*, there is no apparent method of proof. An idea is needed, and this is supplied by Euclid. He tells us to bisect side *BC* (Fig. 6–3), to join the mid-point *E* of *BC* to *A*, and to extend *AE* to the point *F*, so that $AE = EF$. He then proves that triangle *AEB* is congruent to triangle *CEF*, that is, that the sides and angles of one triangle are equal, respectively, to the sides and angles of the other. This congruence is easy to prove. Euclid had previously proved that vertical angles are equal, and we see from Fig. 6–3 that angles 1 and 2 are vertical angles. Further, the fact that *E* is the mid-point of *BC* means that $BE = EC$. Moreover, we constructed *EF* to equal *AE*. Hence, in the two triangles in question, two sides and the included angle of one triangle are equal to two sides and the included angle of the other. But Euclid had previously proved that two triangles are congruent if merely two sides and the included angle of one are equal to two sides and the included angle of the other. Since these facts are true of our triangles, the two triangles must be congruent.

Because triangles *AEB* and *CEF* are congruent, angle *B* of the first triangle equals angle 3 of the second one. We know that angle 3 is the angle to choose in the second triangle as the angle which corresponds to *B*, because angle *B* is opposite *AE*, and angle 3 is opposite the corresponding equal side *EF*. The proof is practically finished. Angle *D* is larger than angle 3 because the whole, angle *D* in our case, is greater than the part, angle 3. Hence angle *D* is also greater than angle *B* because angle *B* has the same size as angle 3.

We have now proved a major theorem, and we should see that a series of simple deductive arguments leads to an indubitable result.

And now let us prove another, equally important theorem which will exhibit one or two other features of Euclid's work:

> THEOREM 2. If two lines are cut by a transversal so as to make alternate interior angles equal, then the lines are parallel.

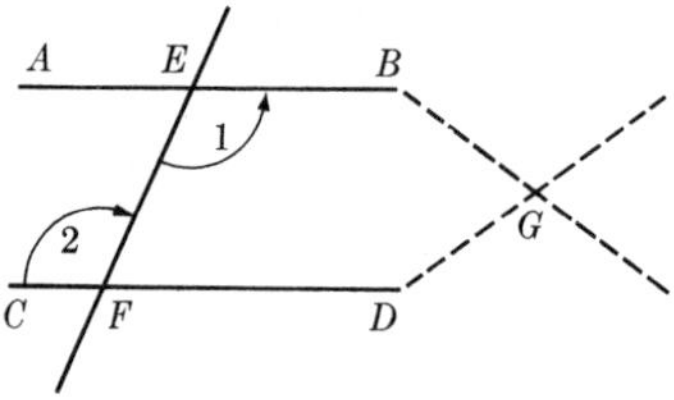

Fig. 6–4

Again let us see what the theorem means before we consider its proof. In Fig. 6–4, *AB* and *CD* are two lines cut by the transversal *EF*. The angles 1 and 2 are called alternate interior angles, and we are told that they are equal. The theorem asserts that, under this condition, *AB* must be parallel to *CD*. As in the case of the preceding theorem, the assertion is seemingly correct, and yet the method of proof is by no means apparent.

Here Euclid uses what is usually called the indirect method of proof; that is, he supposes that *AB* is not parallel to *CD*. Two lines that are not parallel must, by definition, meet somewhere. Thus *AB* and *CD* meet, let us say, in the point *G*. But now *EG*, *GF*, and *FE* form a triangle. Angle 2 is an exterior angle of this triangle and angle 1 is a remote interior angle. Since we have the theorem that in *any* triangle an exterior angle is greater than either remote interior angle, it follows that angle 2 must be greater than angle 1. But, in the above figure, we were given as fact that angle 2 equals angle 1. We have arrived at a contradiction which, if we did not make any mistakes in reasoning, has only one explanation: somewhere we introduced a false premise. We find that the only questionable fact is the assumption that *AB* is not parallel to *CD*. But there are only two possibilities, namely, that *AB* is parallel to *CD* or that it is not parallel to *CD*. Since the latter supposition led to a contradiction, it must be that AB is parallel to CD. Thus the theorem is proved.

We should be sure to note that the indirect method of proof is a deductive argument. The essence of the argument is that if AB is not parallel to CD, then angle 2 must be greater than angle 1. But angle 2 is not greater than angle 1. Hence it is not true that AB is not parallel to CD. But AB is or is not parallel to CD. If nonparallelism is not true then parallelism must hold.

Though we shall use a few other theorems of Euclidean geometry in subsequent work, we shall not present their proofs. We are now reasonably familiar with the nature of proof in geometry, and so we shall merely state the theorems when we wish to use them.

Perhaps one other point about the contents of the *Elements* warrants attention. A superficial survey of the many different theorems may leave one with the impression that the Greek geometers proved what they could and produced merely a mélange. But there are broad themes in Euclidean geometry, and these are pursued systematically. The first major theme is the study of conditions under which geometric figures must be congruent. This is a highly practical subject. Suppose, for example, that a surveyor has two triangular pieces of land and wishes to show that they are equal or congruent. Must he measure all the sides and all the angles of the first piece and show that they are of the same size, respectively, as the sides and angles of the second piece? Not at all! There are several Euclidean theorems which can aid the surveyor. If he can show, for example, that two angles and the included side of the first triangle equal, respectively, the two angles and the included side of the second one, then Euclid's theorem tells him that the triangular pieces of land must be equal.

A second major theme in Euclid's work is the similarity of figures, that is, figures with the same shape. We have already mentioned that models of houses, ships, and other large structures are often built to assist in planning. One may wish to know what conditions will guarantee the similarity of the model and the actual structure. Let us suppose that the model or some part of it is triangular in shape. One of Euclid's theorems tells us that if the corresponding sides of two triangles have the same ratio, then the two triangles will be similar. Thus, if the model is constructed so that each side of the model is $\frac{1}{100}$ of the corresponding side of the actual structure, we know that the model will be similar to the structure. This similarity is useful because, by definition, two triangles are similar if the angles of one equal the corresponding angles of the other. Hence, an engineer can measure the angles of the model and know precisely what the angles of the actual structure will be.

Suppose that two figures are neither congruent nor similar. Could they have some other significant property in common? One answer, clearly, is area. And so Euclid considers conditions under which two figures may have the same area, or, in Euclid's language, be equivalent.

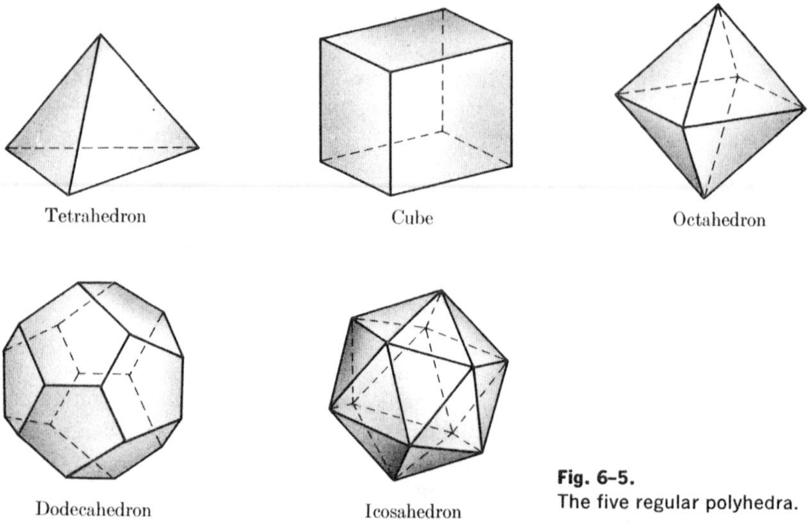

Fig. 6-5.
The five regular polyhedra.

There are many other themes in Euclid, such as interesting properties of circles, quadrilaterals, and regular polygons. He also considers the common solid figures such as pyramids, prisms, spheres, cylinders, and cones. Finally, Euclid devotes considerable space to a class of figures which all Greeks favored, the regular polyhedra (Fig. 6–5).

EXERCISES

1. What essential fact distinguishes axioms from theorems?
2. Why were the Greeks willing to accept the statements 1 through 10 above as axioms?
3. Use the indirect method of proof to show that if two angles of a triangle are equal, then the opposite sides are equal. [*Suggestion:* Suppose that angle A (Fig. 6–6) equals angle C, but that BC is greater than BA. Lay off $BC' = BA$ and draw AC'. Use the theorem that the base angles of an isoceles triangle are equal and Theorem 1 above.]

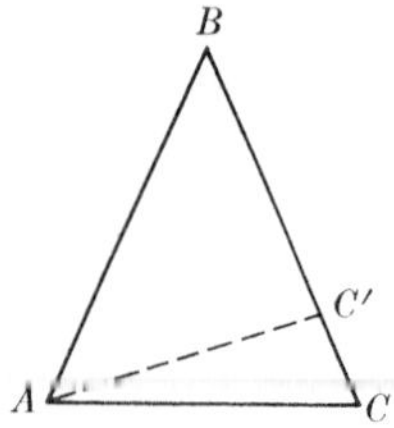

Fig. 6–6

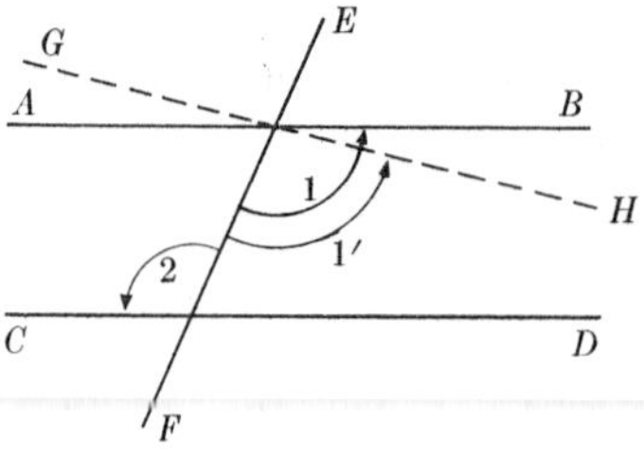

Fig. 6–7

4. Use the indirect method of proof to show that if two lines are parallel, alternate interior angles must be equal. [*Suggestion:* Suppose angle 1 in Fig. 6–7 is greater than angle 2. Then draw GH so that angle 1′ equals angle 2. Now use Theorem 2 and Axiom 5.]
5. In Section 3–7, we have briefly outlined the proof that the sum of the angles of a triangle is 180°. Write out the full proof.
6. Under what conditions would two parallelograms be congruent?
7. What conditions would ensure the similarity of two rectangles?
8. A right triangle has an arm 1 mi long and a hypotenuse 1 mi plus 1 ft long. How long is the other arm? Before you apply mathematics, use your imagination to estimate the answer. To work out the problem, use the Pythagorean theorem which says that the square of the hypotenuse equals the sum of the squares of the arms.
9. A farmer is offered two triangular pieces of land. The dimensions are 25, 30, and 40 ft and 75, 90, and 120 ft, respectively. Since the dimensions of the second one are 3 times the dimensions of the first, the two triangles are similar. The price of the larger piece is 5 times the price of the smaller one. Use intuition, measurement, or mathematical proof to decide which is the better buy in the sense of price per square foot.

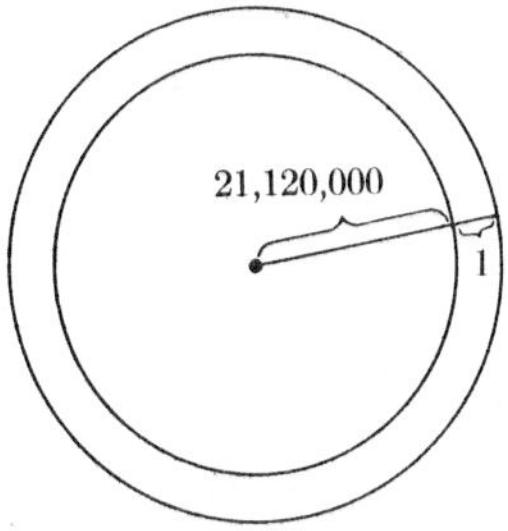

Fig. 6–8

10. Suppose a roadway is to be built around the earth and each point on the surface of the roadway is to be 1 ft above the surface of the earth (Fig. 6–8). Given that the radius of the earth is 4000 mi or 21,120,000 ft, estimate by how much the length of the roadway would exceed the circumference of the earth. Then use the fact that the circumference of a circle is 2π times the radius and calculate how much longer the roadway would be.
11. Criticize the statement: Euclid assumes that two parallel lines do not meet.

6–3 SOME MUNDANE USES OF EUCLIDEAN GEOMETRY

The creation of Euclidean geometry was motivated by the desire to learn the properties of figures in the world about us. Let us see now whether the knowledge can be applied to the world to good advantage.

Suppose a farmer has 100 feet of fencing at his disposal and he wishes to enclose a rectangular piece of land. Since the perimeter will be 100 feet, the farmer can enclose a piece of land 10 feet by 40 feet, 15 feet by 35 feet, 20 feet by 30 feet, or of still other dimensions, all of which yield a perimeter of 100 feet. The farmer plans to garden in the enclosed plot and therefore wishes the enclosed area to be as large as possible. He notes that the dimensions 10 by 40 would yield an area of 400 square feet; the dimensions 15 by 35 enclose 525 square feet; and the dimensions 20 by 30 enclose 600 square feet. Evidently the area can vary considerably despite the fact that the perimeter in each case is 100 feet. The question then arises, What dimensions would yield the maximum area?

Our first task in seeking to answer this question is to make some reasonable conjecture about these dimensions. We might then be able to prove that the conjecture is correct. Since in the present instance it is easy to play with the numbers involved, let us make a little table of dimensions (always yielding a perimeter of 100 feet) and the corresponding area.

Dimensions, in feet	*Area, in square feet*
1 by 49	49
5 by 45	225
10 by 40	400
15 by 35	525
20 by 30	600

Study of the table suggests that the more nearly equal the dimensions are, the larger is the area. Hence one might readily conjecture that if the dimensions were equal, that is, if the rectangle were a square, the area would be a maximum.

We can see at once that the dimensions 25 by 25 give an area of 625 square feet, and this area is larger than any of the areas in the table. So far our conjecture is confirmed. However, we could not be sure that some other dimensions, perhaps $24\frac{1}{2}$ by $25\frac{1}{2}$, would not do even better. Moreover, even if we could be certain that the square furnishes the largest area among all rectangles with a perimeter of 100 feet, the question would arise whether the square would continue to be the answer for some other perimeter. Hence, let us see whether we can prove the general theorem that *of all rectangles with the same perimeter, the square has maximum area.*

Figure 6–9 shows the rectangle *ABCD*. Since this rectangle is not a square, let us erect on the longer side a square which has the same perimeter. Thus, the square *EFGD* has the same perimeter as *ABCD*. We now denote equal segments by the same letters. The perimeter of the rectangle is then $2x + 2u + 2y$, and the perimeter of the square is $2x + 2v + 2y$. Since the two figures have the same perimeter, we have

$$2x + 2v + 2y = 2x + 2u + 2y.$$

If we subtract $2x$ and $2y$ from both sides of this equation and then divide both sides by 2, we obtain

$$v = u. \tag{1}$$

Moreover, because the square has equal sides,

$$y = x + v. \tag{2}$$

If we now multiply the left side of equation (2) by the left side of equation (1), and do the same for the right sides, the results must be equal. Hence,

$$yv = u(x + v),$$

or, by the distributive axiom,

$$yv = ux + uv.$$

Since $yv = ux$ plus an additional area, it must be that yv is greater than ux. Now yv is area B in the figure, and ux is area A. Thus B is greater than A, and so $B + C$ is greater than $A + C$. But $B + C$ is the area of the square, and $A + C$ is the area of the rectangle. Hence the square has more area than the rectangle.

We have proved that a square has more area than a rectangle of the same perimeter, *no matter what this perimeter may be.* A little thinking proves in a few minutes what may have taken man hundreds of years to learn through trial and error.

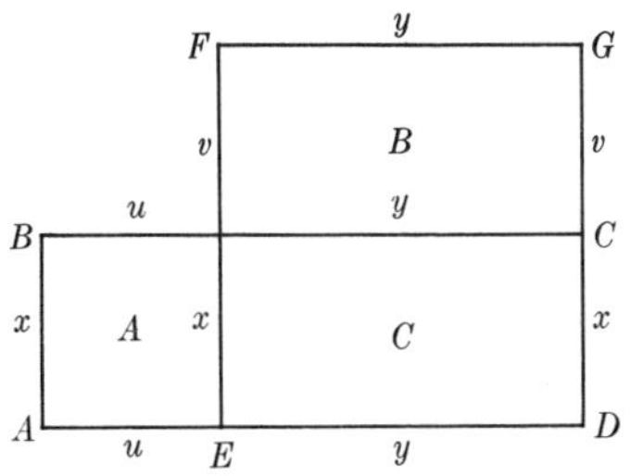

Fig. 6–9.
Of all rectangles of the same perimeter the square has the greatest area.

The result is far more useful than may appear at first sight. Suppose a house is to be built. The major consideration is to have as much floor area or living space as possible. Now the perimeter of the floor determines the number of feet of wall that will be needed and hence the cost of the walls. To obtain the maximum floor area for a given cost of walls, the shape of the floor should be square.

A farmer who seeks the rectangle of maximum area with given perimeter might, after finding the answer to his question, turn to gardening, but a mathematician who obtains such a neat result would not stop there. He might ask next, Suppose we were free to utilize any quadrilateral rather than just rectangles, which one of all quadrilaterals with the same perimeter has maximum area? The answer happens to be a square, though we shall not prove it. The mathematician might then consider the question, Which pentagon of all pentagons with the same perimeter has maximum area? One can show that the answer is the regular pentagon, that is, the pentagon whose sides are all equal and whose angles are all equal. Now the square also has equal sides and equal angles. Hence it would seem that if one compares all polygons of the same perimeter and same number of sides, then the one with equal sides and equal angles, i.e., the regular polygon, should have maximum area. This general result can also be proved.

But now an obvious question comes to the fore. The square has maximum area among all quadrilaterals of the same perimeter. The regular pentagon has maximum area among all pentagons of the same perimeter. Suppose that we compared the regular pentagon with the square of the same perimeter. Which would have more area? The answer, perhaps surprising, is the regular pentagon. And now the conjecture seems reasonable that of two regular polygons with the same perimeter, the one with more sides will have more area. This is so. Where does this result lead? One can form regular polygons of more and more sides, which all have the same perimeter. As the number of sides increases, the area increases. But as the number of sides increases, the regular polygon approaches the circle in shape. Hence the circle should have more area than any regular polygon of the same perimeter. And since the regular polygon has more area than an arbitrary polygon, *the circle has more area than any polygon with the same perimeter*. This result is a famous theorem.

Now the sphere, among surfaces, is the analogue of the circle among curves. Hence, a reasonable conjecture would be that the spherical surface bounds more volume than any other surface with the same area. This conjecture can be proved. Nature obeys this mathematical theorem. For example, if one blows up a rubber balloon, the balloon assumes a spherical shape. The reason is that the rubber must enclose the volume of air blown into the balloon and the rubber must be stretched. But rubber contracts as much as possible. The spherical figure requires less surface area to contain a given volume of gas than does any other shape. Hence, with the spherical shape, the rubber is stretched as little as possible.

The problem of bounding the greatest possible area with a perimeter of given length has a variation whose solution shows how ingenious mathematical reasoning can be. Suppose that a person has a fixed amount of fencing at his disposal and wishes to enclose as much area as possible along a river front in such a way that no fencing is required along the shore itself. The question now is, What should the shape of the boundary curve be? According to a legend, which may or may not have a factual basis, this problem was solved thousands of years ago by Dido, the founder of the city of Carthage on the Mediterranean coast of Africa. Dido, the daughter of the king of the Phoenician city of Tyre, ran away from home. She took a fancy to this land on the Mediterranean, and made an agreement to pay a definite sum of money for as much land as "could be encompassed by a bull's hide." Dido thereupon took a bull's hide, cut it up into thin long strips, tied the strips together, and used this length to "encompass land." She chose an area along the shore, because she was smart enough to realize that no hide would be needed along the shore. But there still remained the question of what shape to use for the boundary formed by the hide, that is, for *ABC* of Fig. 6–10. Dido decided that the most favorable shape was a semicircle, enclosed that shape, and built a city there.

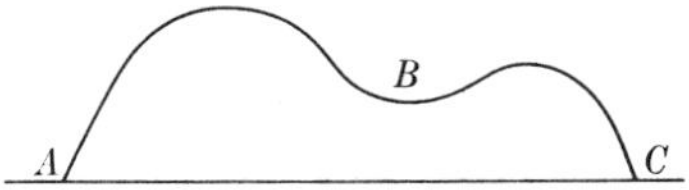

Fig. 6–10

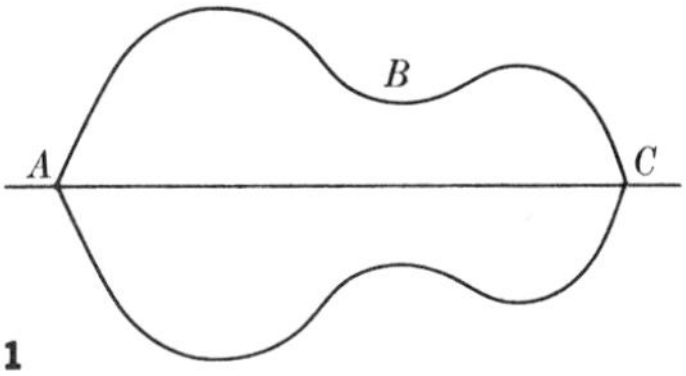

Fig. 6–11

A sequel to this story, which has nothing to do with the mathematics of Dido's problem, is not without relevance to the history of mathematics. Shortly after she founded Carthage, Aeneas, a refugee from Troy, intent on getting to Italy to found his own city, was blown ashore along with his compatriots. Dido took a fancy to Aeneas also, and did her best to persuade him to remain at Carthage, but despite the best of hospitality, Aeneas could not be diverted from his plan, and soon sailed away. Rejected and scorned, Dido was so despondent that she threw herself on a blazing pyre just as Aeneas sailed out of the harbor. And so an ungrateful and unreceptive man with a rigid mind caused the loss of a potential mathematician. This was the first blow to mathematics which the Romans dealt.

Dido's fate was a tragic end to a brilliant beginning, for her solution to the geometrical problem described above was correct. The answer is a semicircle. We do not know how Dido found the answer, but it can be obtained very neatly. The way to prove it is by complicating the problem. Suppose that, instead of bounding an area on one side of the seashore, which we idealize as the line *AC* (Fig. 6–11), we try to solve the problem of enclosing an area on both sides of *AC* with *double* the length of hide Dido had for one side, i.e., now we seek to solve the problem by determining the maximum area which can be completely enclosed by a perimeter of given length. The answer to this problem is a circle. If, therefore, we choose a semicircle for arc *ABC*, it will contain maximum area on one side of the shore. For if there were a more favorable shape than the semicircle, the mirror image in *AC* of that shape would, together with the original, do better than the circle and yet have the same perimeter as the circle. But this is impossible.

Our last few pages have dealt with problems which grew out of determining the rectangle of maximum area with given perimeter. We can see from the lines of thought pursued how the mathematician can raise one question after another on this same theme of figures with maximum area and given perimeter and will find the answers to these questions. Moreover, many of these answers prove to be applicable to physical problems.

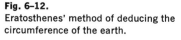

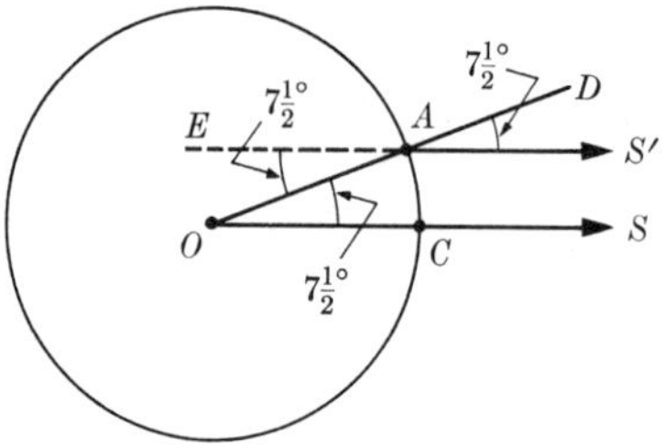

Fig. 6–12.
Eratosthenes' method of deducing the circumference of the earth.

The first reasonably accurate calculation of the size of the earth was made by a simple application of Euclidean geometry. One of the most learned men of the Alexandrian Greek world, Eratosthenes (275–194 B.C.), a geographer, mathematician, poet, historian, and astronomer, used the following plan. At the summer solstice, the sun shone directly down into a well at Syene (C in Fig. 6–12). As Eratosthenes well appreciated, this meant that the sun was directly overhead. At the same time, at the city of Alexandria, 500 miles north of Syene, the direction of the sun was AS', whereas the overhead direction was OAD. Now the sun is so far away that the lines AS' and CS could be taken to be parallel. Eratosthenes measured the angle DAS' and found it to be $7\frac{1}{2}°$. But this angle equals the vertical angle OAE, and the latter and angle AOC are alternate interior angles of parallel lines. Hence angle AOC is also $7\frac{1}{2}°$, or $7\frac{1}{2}/360$, or $1/48$ of the entire angle at O. Then arc AC is $1/48$ of the entire circumference. Since AC is 500 miles, the entire circumference is $48 \cdot 500$ or 24,000 miles.

Strabo, a Greek geographer who lived in the first century B.C., tells us that after Eratosthenes obtained this result, he realized that one might sail from Greece past Spain across the Atlantic Ocean to India. This is, of course, what Columbus attempted. Fortunately or unfortunately, the geographers who lived after Eratosthenes, notably Poseidonius (first century B.C.) and Ptolemy (second century A.D.), gave other results which were interpreted by Columbus (because of some uncertainty about the units of distance used by these early scientists) to mean that the circumference of the earth is 17,000 miles. Had he known the correct value, he might never have undertaken to sail to India because the greater distance might have daunted him.

EXERCISES

1. Suppose that DF (Fig. 6–13) is the course of a railroad, and A and B are two towns. It is desired to build a station somewhere on DF so that the station will be equally distant from A and B. Where should the station be built? One draws the line AB and, at its mid-point, erects the perpendicular CE. The point E on DF is equidistant from A and B. Prove this statement.

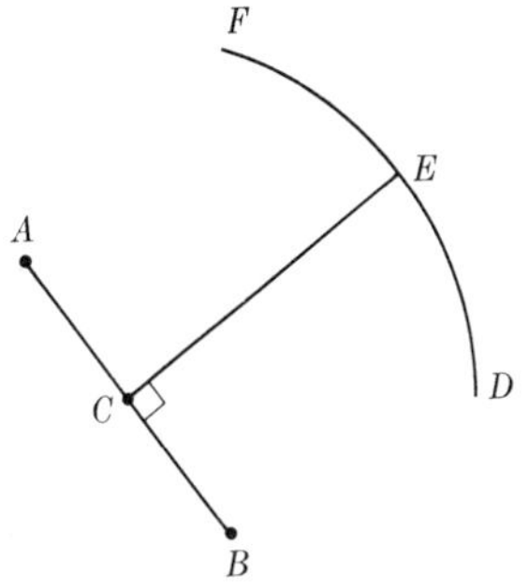

Fig. 6–13

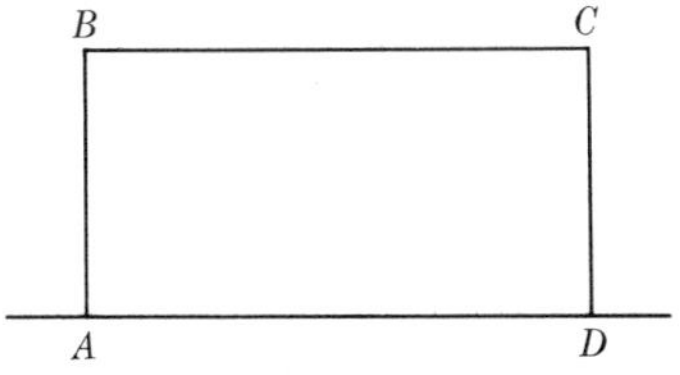

Fig. 6–15

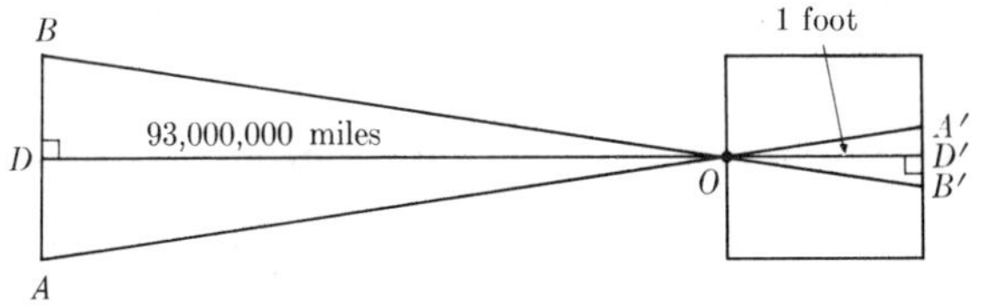

Fig. 6–14

2. A pinhole camera is a practical device if a long exposure time is possible. In fact, one of the best pictures of the scene following the explosion of the first atomic bomb was made with a pinhole camera. The principle involves similar triangles. The object AB being photographed (Fig. 6–14) appears on the film inside the box as $A'B'$. If one draws OD perpendicular to AB, the extension of OD to D' will be perpendicular to $A'B'$. Then triangles OAD and $OA'D'$ are similar. Now suppose the sun, whose radius is AD, is photographed. We know that OD is 93,000,000 mi. Suppose that OD', the width of the box, is 1 ft. The length $A'B'$ is readily measured and is found to be 0.009 ft. What is the radius of the sun?
3. A farmer has 400 yd of fencing and wishes to enclose a rectangle of maximum area. What dimensions should he choose?
4. A farmer has p yd of fencing and wishes to enclose a rectangle of maximum area. What dimensions should he choose?
5. A farmer plans to enclose a rectangular piece of land alongside a lake; no fencing is required along the shoreline AD (Fig. 6–15). He has 100 ft of fence and wishes the area of the rectangle to be as large as possible. What dimensions should he choose?

6. Of any two numbers whose sum is 12, the product is greatest for 6 and 6; that is, $6 \cdot 6$ is greater than $5 \cdot 7$, $4 \cdot 8$, $3\frac{1}{2} \cdot 8\frac{1}{2}$, and so forth. Can you explain why this is so? [*Suggestion:* Think in geometrical terms.]
7. Suppose h is the known height of a mountain, and R is the radius of the earth (Fig. 6–16). How far is it from the top of the mountain to the horizon; that is, how long is x? [*Suggestion:* Use the fact that the line of sight from the top of the mountain to the horizon is tangent to the circle shown, and that a radius of a circle drawn to the point of tangency is perpendicular to the tangent.]

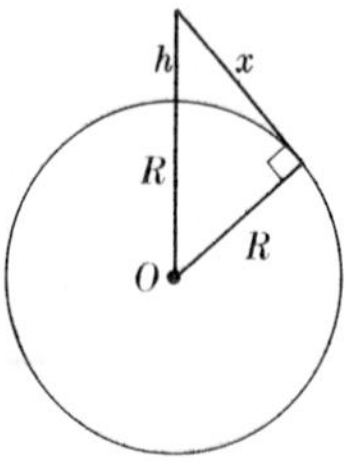

Fig. 6–16

8. Having obtained the exact answer to Problem 7, can you suggest a good approximate answer which would suffice for many applications and yet make calculation easier?
9. A boy stands on a cliff $\frac{1}{2}$ mi above the sea. How far away is the horizon?
10. Knowing that of all rectangles with the same perimeter, the square has maximum area, prove that of all rectangles with the same area, the square has the least perimeter. [*Suggestion:* Use the indirect method of proof. Suppose, then, that the square has more perimeter than the rectangle of the same area and consider the square which has the same perimeter as the rectangle.]

* 6–4 EUCLIDEAN GEOMETRY AND THE STUDY OF LIGHT

Light is certainly a pervasive phenomenon. Man and the physical world are subject daily to the light of the sun, and the process of vision of course is dependent upon light. Hence it is to be expected that the Greeks, the first great students of nature, would investigate this phenomenon. Plato and Aristotle had much to say on the nature of light, and the Greek mathematicians also tackled the subject. It has continued to be a primary concern of mathematicians and physicists right down to the present day. Despite man's continuous experience with light, the nature of this occurrence is still largely a mystery. Through mathematics and through Euclidean geometry in particular, man obtained his first grip on the subject. Two books by Euclid were the beginning of the mathematical attack.

In ordinary air, light is observed to travel along straight lines. This preference of light for the simplest and shortest path is in itself of significance. But Euclid proceeded beyond this point to study the behavior of light under reflection in a mirror, and discovered a now famous mathematical law of light.

Suppose light issuing from A (Fig. 6–17) takes the path AP to the point P on the mirror m. As we all know, the light is reflected and takes a new direction, PA'. The significant fact about this reflection, which was pointed out by Euclid, is that the reflected ray, i.e., the line PA' along which the reflected light travels, always takes a direction such that angle 1 equals angle 2. Angle

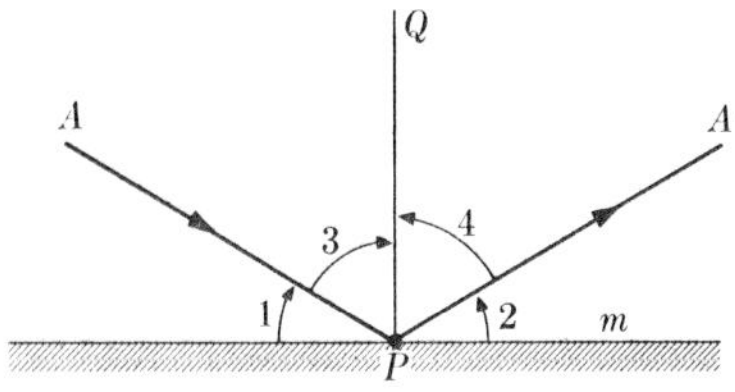

Fig. 6-17.
The law of reflection of light.

1 is called the angle of incidence and angle 2 the angle of reflection.* It is, of course, very obliging of light to follow such a simple mathematical law. As a consequence, we are able to prove other facts rather readily.

Assume there is a source of light at A (Fig. 6-18), and rays of light spread out in all directions from A. Many of these will strike the mirror. But through a definite point A' only one of these rays will pass, namely the ray PA' for which angle 1 equals angle 2. To prove that only one ray from A will pass through A', let us suppose that another ray, AQ, is also reflected to A'. Now angle 2 is an exterior angle of triangle $A'QP$. Hence

$$\angle 2 > \angle 4.$$

Angle 3 is an exterior angle of triangle AQP, and so

$$\angle 3 > \angle 1.$$

Since angle 1 equals angle 2, we see from the two preceding inequalities that

$$\angle 3 > \angle 4.$$

Then QA' cannot be the reflected ray corresponding to the incident ray AQ because the reflected ray must make an angle with the mirror which equals angle 3.

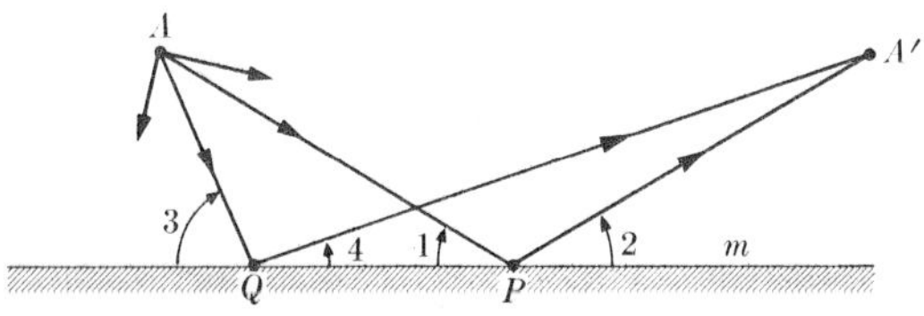

Fig. 6-18

The more interesting point, which was first observed and proved by the Greek mathematician and engineer Heron (first century A.D.), is that the

* It is more common to introduce the perpendicular PQ to the mirror and to call angle 3 the angle of incidence and angle 4 the angle of reflection. However, if angle 1 equals angle 2, then angle 3 equals angle 4.

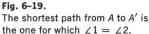

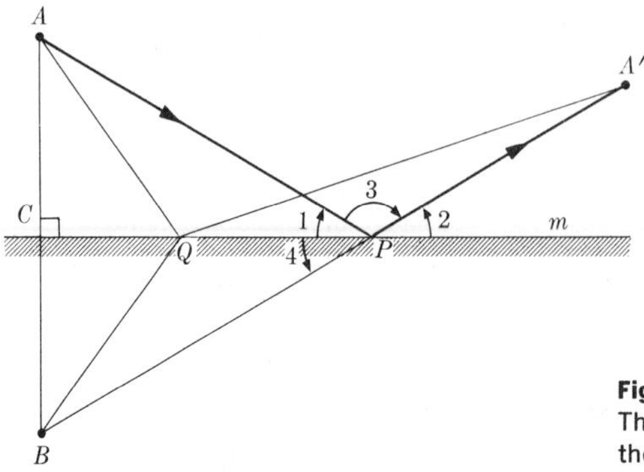

Fig. 6–19.
The shortest path from A to A' is the one for which $\angle 1 = \angle 2$.

unique ray from A (Fig. 6–19) which does reach A' after reflection in the mirror travels the shortest possible path in going from A to the mirror and then to A'. In other words, $AP + PA'$ is less than $AQ + QA'$, where Q is any point on the mirror other than P, the point at which the angle of incidence equals the angle of reflection.

How can we prove this theorem? Nature not only sets problems for us, but often solves them too, if we are but keen enough in our observations. If a person at A' sees in the mirror the reflection of an object at A, he must be looking in the direction $A'P$ and actually sees the image of A at B. Hence, perhaps we should bring B into our thinking. Closer observation shows that the mirror image of an object is on the perpendicular from A to the mirror and, moreover, seems to be as far behind the mirror as the object is in front. That is, AB seems to be perpendicular to the mirror and AC seems to equal CB.

Let us use this suggestion. We construct the perpendicular from A to the mirror, thus obtaining AC, and extend AC by its own length to B. Now it is not hard to see that triangles ACQ and BCQ are congruent because QC is common to both triangles, the angles at C are right angles, and $AC = CB$. Hence, $AQ = BQ$, because they are corresponding parts of congruent triangles. Likewise triangles ACP and BCP are congruent and $AP = BP$. We wish to prove that

$$(AP + PA') < (AQ + QA').$$

But now, since $AP = BP$ and $AQ = BQ$, it will be enough to prove that

$$(BP + PA') < (BQ + QA'.) \tag{3}$$

Well, we have exchanged one difficulty for another, but perhaps this second one is easier to overcome. Physically one looks directly along $A'P$ and sees B. If we could prove that BPA' is a straight line, then, of course, the inequality (3) would be proved because BQ and QA' are the other two sides of triangle $A'BQ$, and the sum of these two sides must be greater than the third side. Our goal, then, is to prove that $A'PB$ is a straight line.

We know that

$$\angle 1 + \angle 3 + \angle 2 = 180° \tag{4}$$

because m is a straight line. But angle 1 equals angle 4 because triangle PCA and PCB are congruent. Also, according to the law of reflection, angle 2 equals angle 1. If, therefore, in (4) we replace angle 1 by angle 4 and angle 2 by angle 1, we have

$$\angle 4 + \angle 3 + \angle 1 = 180°. \tag{5}$$

Hence, $A'PB$ is a straight line and the inequality (3) is proved. Then the light ray, in going from A to m to A', really travels the shortest path.

This behavior of light rays is striking. It seems to show that nature is interested in accomplishing its ends by the most efficient means. We shall find this theme to be a recurring one, and it will be seen to have broad applicability.

We have proved a theorem about light rays, but we have also proved somewhat more. As far as the mathematics is concerned, the lines AP and PA' are any lines which make equal angles with m, and the fact that they are light rays plays no role. What we have proved, then, is a theorem of geometry, namely:

Of all the broken line paths from a point A to a point on a line and then to a point A' on the same side as A, the shortest path is the one fixed by the point P on m for which AP and $A'P$ make equal angles with m.

This theorem has applications in quite different domains (see the exercises). It is worth noting how the study of light gives rise to purely mathematical theorems. The converse of this theorem is, incidentally, equally true and is presented in the exercises.

EXERCISES

1. Where is the mirror image of a point A which is in front of a plane mirror?
2. Suppose that m (Fig. 6–20) is the shore of a river and a pier is to be built somewhere along m so that merchandise can be trucked from the pier to two inland towns, A and A'. Where should the pier be built so that the total trucking distance from the pier to A and from the pier to A' is a minimum?

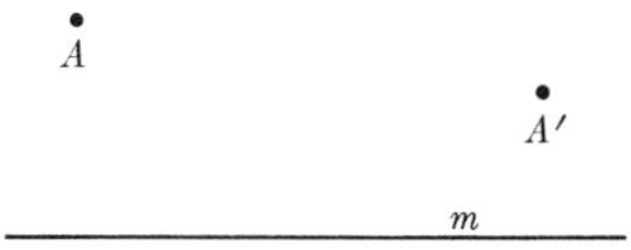

Fig. 6–20

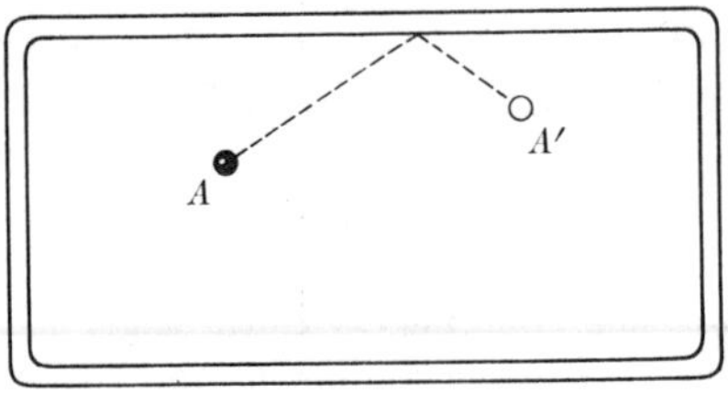

Fig. 6–21

3. A billiard player wishes to hit the ball at A (Fig. 6–21) in such a way that it will strike side m of the table and then hit the ball at A'. Now billiard balls behave like light rays, that is, the angle of reflection equals the angle of incidence. At what point on m should the billiard player aim?
4. A billiard ball starting from a point A on the table (Fig. 6–21) strikes two successive sides and then travels along the table. What can you say about the final in relation to the original direction of travel?

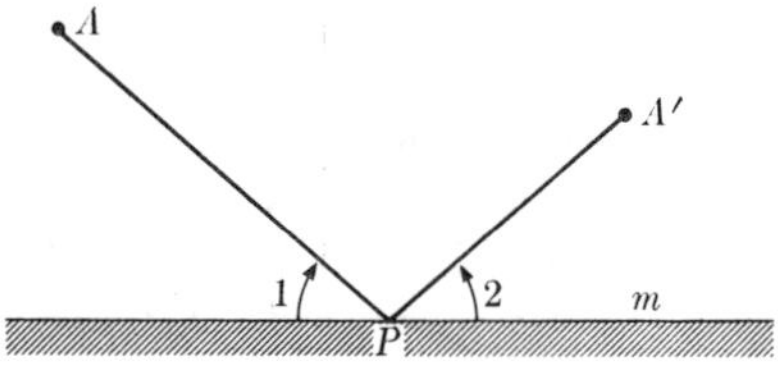

Fig. 6–22

5. In the text we proved that if angle 1 equals angle 2 (Fig. 6–22), then $AP + PA'$ is the shortest path from A to any point on the mirror to A'. Prove the converse, namely, that if $AP + PA'$ is the shortest path, then $\angle 1$ must equal $\angle 2$. [*Suggestion:* Use the indirect method of proof. If $\angle 1$ does not equal $\angle 2$, then one can find another point, P', on m for which the angles made by AP' and $A'P'$ with m are equal.]

6–5 CONIC SECTIONS

The *Elements* of Euclid dealt with plane figures which can be built up with line segments and circles, with the corresponding solid figures which can be built up with pieces of a plane, such as prisms and the regular polyhedra, and with the sphere. But the classical Greeks also studied another class of curves which they called conic sections because they were originally obtained by slicing a cone with a plane. The resulting curves, the parabola, ellipse, and hyperbola, were treated by Euclid in a separate book. Unfortunately, no copies of this book have survived. But a little after Euclid's time another famous Greek geometer, Apollonius, wrote a book entitled *Conic Sections*, which is known to us and which is about as exhaustive in its treatment of these curves as the *Elements* are about figures formed by lines and circles.

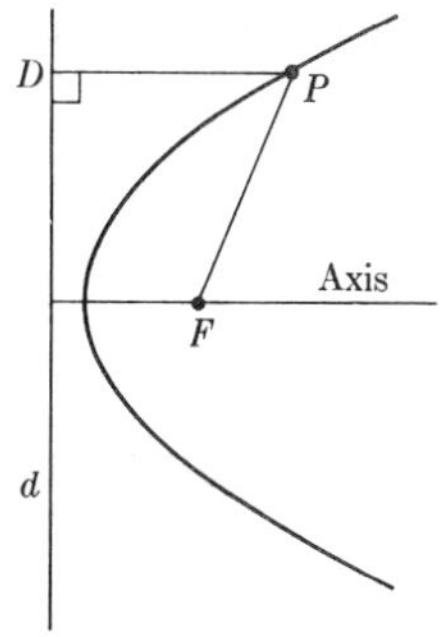

Fig. 6–23.
The parabola.

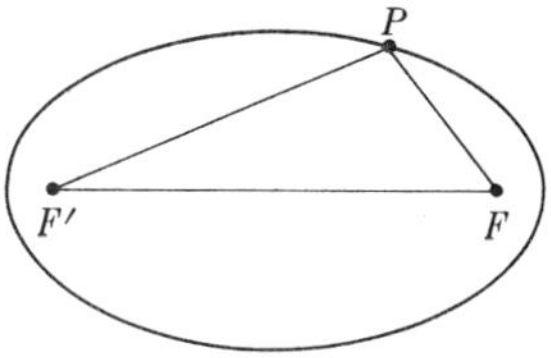

Fig. 6–24.
The ellipse.

Conic sections were introduced, as already noted, by cutting a conical surface with a plane. However, the curves themselves can be considered apart from the surface on which they lie. For example, the circle is also one of the conic sections. Yet we know that the circle can be defined as the set of all points which are at a fixed distance from a given point, and this definition does not involve the cone at all. Indeed, insofar as properties and applications of these curves are concerned, it is far more convenient to disregard the conical surface and concentrate on the curves themselves.

Let us consider, therefore, the direct definitions of conic sections. To define the parabola, we start with a fixed point F and a fixed line d (Fig. 6–23). We then consider the set of all points, each of which is equally distant from F and d. Thus the point P in Fig. 6–23 is such that $PF = PD$. The collection of all points, each of which is equidistant from F and d, fills out a curve called the parabola. The point F is called the *focus* of the parabola, and the line d is called the *directrix*.

Each choice of a point F and line d determines a parabola. Hence there are infinitely many different parabolas. The general shape of all such curves is, however, about the same. Each is symmetric about the line which passes through F and is perpendicular to d. This line is called the *axis* of the parabola. Each parabola passes between its focus and directrix and opens out as it extends farther and farther from the directrix.

The direct definition of the ellipse is also simple. We start with two fixed points F and F' (Fig. 6–24) and consider any constant quantity greater than the distance F to F'. If, for example, the distance from F to F' is 6, we may choose 10 as the constant quantity. One then determines all points for each of which the distance from F and the distance from F' add up to 10. This collection of points is called an ellipse. Thus, if P is a point for which $PF + PF'$ equals 10, then P lies on the ellipse determined by F, F', and the quantity 10. The points F and F' are called the *foci* of the ellipse.

By changing the distance FF' or the quantity 10, one obtains another ellipse. Some ellipses are long and narrow; others are almost circular. All are symmetric about the line FF' and about the line perpendicular to and midway between F and F'.

The direct definition of the hyperbola also calls for choosing two fixed points F and F', called foci, and a constant quantity which, however, must be less than the distance from F to F'. If FF' is 6, then the constant quantity can, for example, be 4. We now consider any point P for which the difference $PF' - PF$ equals 4. All such points lie on the right-hand portion of Fig. 6–25, whereas the points for which $PF - PF' = 4$ lie on the left-hand portion of the figure. The two portions together are the hyperbola; each portion is a branch of the hyperbola.

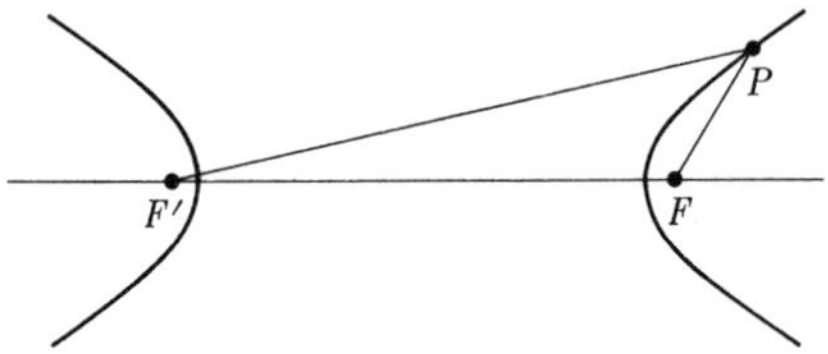

Fig. 6–25. The hyperbola.

As for the ellipse, each choice of the distance FF' and the constant quantity determines a hyperbola. Here, too, the curve is symmetric about the line FF' and about a line perpendicular to and midway between F and F'. One branch opens to the right and the other to the left.

We shall not prove that the curves we have defined by means of focus and directrix or by means of foci and constant quantities are the same as those obtainable by slicing a conical surface. In our future work we shall use the direct definitions.

EXERCISES

1. Since the circle is also a conic section, it should be included among one of the three types—parabola, ellipse, and hyperbola. From the shapes of these curves it would appear that the circle falls among the ellipses. Can you see how the circle may arise as a special kind of ellipse?
2. Suppose that we have an ellipse for which $F'F$ is 6 and the constant quantity is 10. If the point P of the ellipse lies on the line $F'F$ to the right of F, how much is PF?
3. For the ellipse, why must the constant quantity be chosen greater than the distance $F'F$?
4. Given a parabola for which the distance from focus to directrix is 10, how far from the focus is that point on the parabola which lies on the axis?

* 6–6 CONIC SECTIONS AND LIGHT

Next to straight line and circle, conic sections are the most valuable curves mathematics has to offer for the study of the physical world. We shall examine here the uses of the parabola in the control of light.

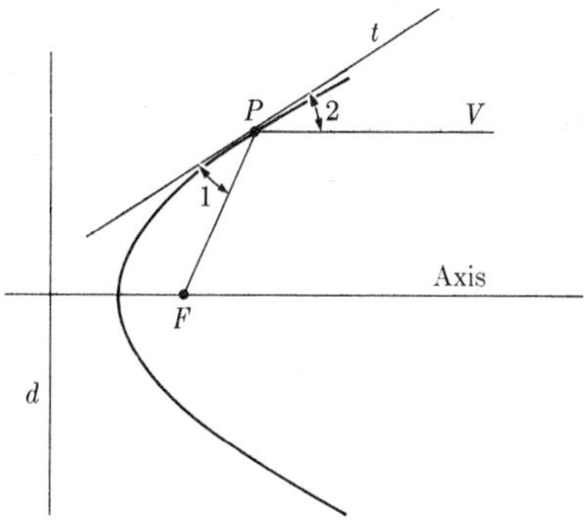

Fig. 6–26.
The reflecting property of the parabola.

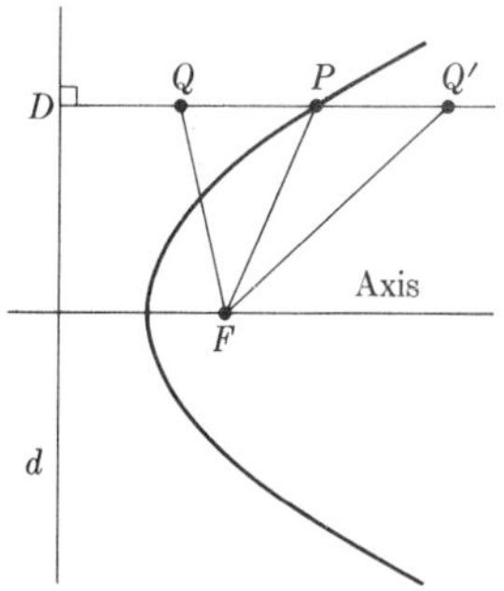

Fig. 6–27

Let P be any point on the parabola (Fig. 6–26). By the tangent to the parabola at P we mean the line through P which meets the parabola in just that one point and lies entirely outside the curve. From the standpoint of the control of light, the curve possesses a most pertinent property. If P is any point on the curve and F is the focus, then the line FP and PV, the line through P parallel to the axis, where V is any point on this parallel, make equal angles with the tangent t at P. That is, angle 1 equals angle 2.

Before proving the geometrical property just stated, let us see why it is significant. If a light ray issues from some source of light at F and strikes a parabolic mirror at P, it will be reflected in accordance with the law that the angle of incidence equals the angle of reflection. The curve acts at P as though it had the direction of the tangent. Then angle 1 is the angle of incidence. Because angle 1 equals angle 2, the reflected ray will be PV. Hence the reflected ray will travel out parallel to the axis of the parabola. Now, P is any point on the parabola. Hence any ray leaving F and striking the parabola will, after reflection, travel out parallel to the axis of the parabola, and the reflected light will form a powerful beam in one direction. We thus obtain a concentration of light.

Let us now prove that PF and PV make equal angles with the tangent at P. We shall prove first that every point outside of the parabola is farther from the focus than from the directrix, and every point inside is closer to the focus than to the directrix. Consider the point Q (Fig. 6–27) outside the parabola. We wish to show $QF > QD$, where QD is the distance from Q to the directrix. We continue the line QD until it strikes the parabola at P. Now

$$QF > PF - PQ$$

because any side of a triangle is greater than the difference of the other two sides. Since P is on the parabola, by the very definition of the curve, $PF = PD$. then

$$QF > PD - PQ = QD.$$

We may use the same figure to show that Q', any point inside the parabola, is closer to F than to the directrix, that is, that $Q'F < Q'D$. First,

$$Q'F < PF + PQ'$$

because any side of a triangle is less than the sum of the other two sides. Since $PF = PD$,

$$Q'F < PD + PQ' = Q'D.$$

And now let us prove that PF and PV of Fig. 6–26 make equal angles with the tangent t at P. We shall invert our approach just to make the proof easier. Let us draw a line t through the point P (Fig. 6–28), which makes equal angles with PF and PV, and we shall prove that this line is the tangent to the parabola at P. We shall make the proof by showing that any point Q on this line lies outside the parabola. Since this line does have one point P in common with the parabola, it must, by the definition of tangent, be the tangent.

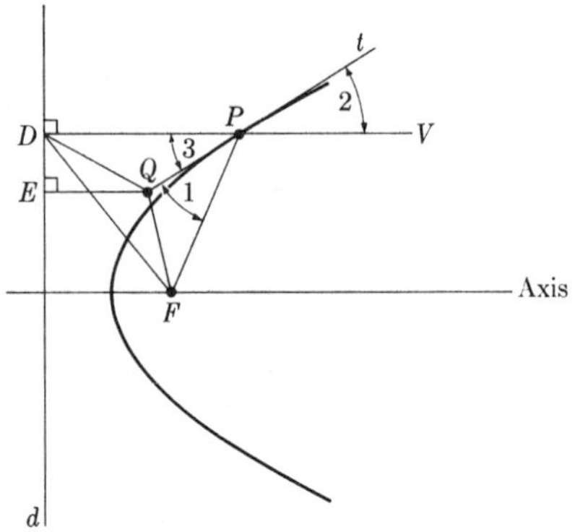

Fig. 6–28

Consider the triangles PDQ and PFQ. We know that $PD = PF$ because P is a point on the parabola. Further, since $\angle 1 = \angle 2$ by the very choice of the line t, and since $\angle 2 = \angle 3$ because they are vertical angles, then $\angle 1 = \angle 3$. Finally, PQ is common to the two triangles. Then the triangles are congruent and $QD = QF$ because they are corresponding sides of the congruent triangles. Now QE is the distance from Q to the directrix, and $QE < QD$ because the hypotenuse of a right triangle is longer than either arm. Then $QF > QE$. According to the preceding proof, Q must lie outside the parabola. Since Q is any point on t (except P, of course), the line t must be the tangent at P. Thus the line through P which makes equal angles with PF and PV is the tangent at P.

We now know, then, that any light ray issuing from F and striking the parabola at P will be reflected along PV, that is, parallel to the axis. The parabolic mirror's power to concentrate light in one direction is very useful. The commonest application is found in automobile headlights. In each headlight there is a small bulb. Surrounding this bulb is a surface (Fig. 6–29), called a paraboloid, which is formed by rotating a parabola about its axis. (The surface is, of course, silvered so that it will reflect.) Light issuing in millions of directions from the bulb, which is placed at the focus of the paraboloidal mirror, strikes the mirror, is reflected along the axis of the paraboloid, and illuminates strongly whatever lies in that direction. The effectiveness of this arrangement may be judged from the fact that the light thrown forward by bulb and mirror is about 6000 times as intense as that thrown in the same direction by the bulb alone. The reflecting property of the paraboloidal mirror is also utilized in searchlights and flashlights.

The reflecting property of a paraboloidal mirror can be used in reverse. If a beam of parallel light rays enters such a mirror while traveling parallel to the axis, each ray will be reflected by some point on the surface in accordance with the law of reflection. But since FP (Fig. 6–29) and VP make equal angles with the tangent, the reflected ray will travel along PF and all reflected rays will arrive at the focus F. Hence there will be a great concentration of light at F.

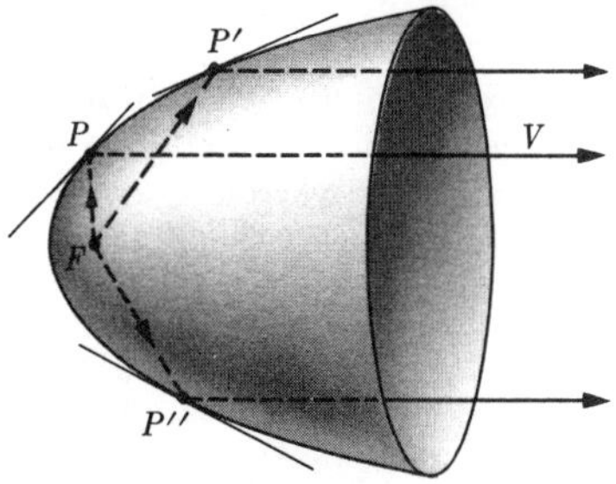

Fig. 6–29.
The paraboloidal mirror.

This concentration of light is used effectively in telescopes. The light emitted by stars is so faint that it is necessary to collect as much as possible in order to obtain a clear image. The axis of the telescope is therefore directed toward the star, and, because this source is so far away, the rays enter the telescope practically parallel to the axis, travel down the telescope to a paraboloidal mirror at the back, and are reflected to the focus of the mirror.

Radio waves behave very much like light rays. Hence paraboloidal reflectors made of metal are used to concentrate radio waves issuing from a small source into a powerful beam. Conversely, a paraboloidal antenna can pick up faint radio signals and produce a relatively strong signal at the focus. Since radio is used today for hundreds of purposes, the paraboloidal radio antenna is a very common instrument.

We see from this brief account that the conic sections are immensely valuable. Some of the most momentous applications have yet to be described and will be taken up in later chapters.

How did the Greeks come to study these curves? As far as we know, the conic sections were discovered in attempts to solve the famous construction problems of Euclidean geometry, i.e., to trisect any angle, to construct a square equal in area to a given circle, and to construct the side of a cube whose volume is twice that of a cube of given side. The constructions were to be performed subject to the restriction that only a straight edge (not a ruler) and a compass be used. Having obtained the curves, the Greeks continued to work on them, partly because they were interested in geometrical forms and partly because they discovered the uses of these curves in the control of light. Apollonius himself wrote a book entitled *On Burning Glasses*, whose subject was the parabola as a means of concentrating light and heat, and there is a story that Archimedes constructed a huge paraboloid which focused the sun's rays on the Roman ships besieging his city of Syracuse, and thus set them on fire.

We see in the history of conic sections one more example of how mathematicians, pursuing a subject far beyond the immediate problems which give rise to it, come to make important contributions to science.

EXERCISES

1. Let Q be any point outside of an ellipse (Fig. 6–30). Prove that $F_2Q + F_1Q$ is greater than a, where a is the sum of the distances of any point on the ellipse from the foci. [*Suggestion:* Introduce the point P where F_2Q cuts the ellipse.]

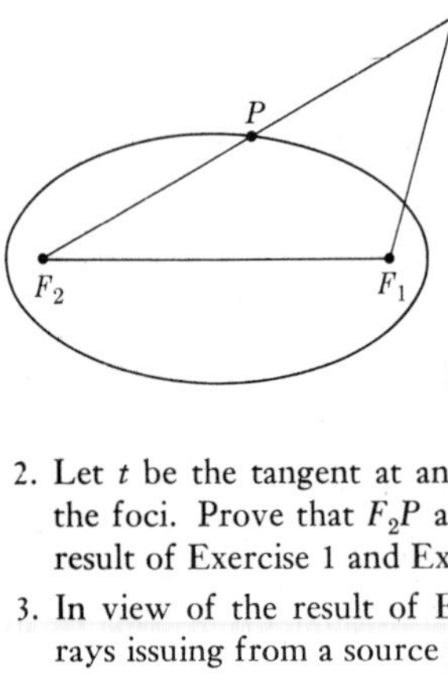

Fig. 6–30

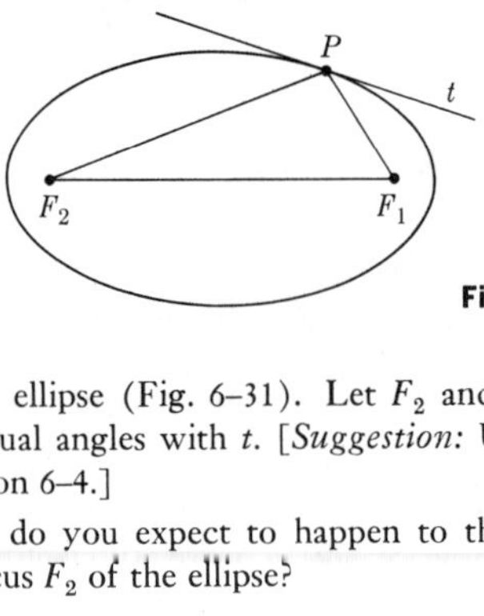

Fig. 6–31

2. Let t be the tangent at any point P of an ellipse (Fig. 6–31). Let F_2 and F_1 be the foci. Prove that F_2P and F_1P make equal angles with t. [*Suggestion:* Use the result of Exercise 1 and Exercise 5 of Section 6–4.]

3. In view of the result of Exercise 2, what do you expect to happen to the light rays issuing from a source placed at the focus F_2 of the ellipse?

4. When the distance between the two foci F_2 and F_1 of an ellipse approaches 0, the ellipse approaches a circle in shape. What do the lengths F_2P and F_1P become when F_2 and F_1 coincide? What theorem about circles follows as a special case of the result in Exercise 2?

* 6-7 THE CULTURAL INFLUENCE OF EUCLIDEAN GEOMETRY

If the development of mathematics had ceased with the creation of Euclidean geometry, the contribution of the subject to the molding of Western civilization would still have been enormous, for Euclidean geometry was and still is an overwhelming demonstration of the power and effectiveness of our reasoning faculty. The Greeks loved to reason and applied it to philosophy, political theory, and literary criticism. But philosophy breaks down into philosophies whose relative merits become the object of much dispute between the adherents of one school and those of another. Plato's *Republic* may indeed be the perfect answer to the quest for a satisfactory political system, but we must still be convinced of this fact. And literary criticism certainly does not lead to universally accepted standards and the creation of universally acclaimed literature. In Euclidean geometry, however, the Greeks showed how reasoning which is based on just ten facts, the axioms, could produce thousands of new conclusions, mostly unforeseen, and each as indubitably true of the physical world as the original axioms. New, unquestionable, thoroughly reliable, and usable knowledge was obtained, knowledge which obviated the need for experience or which could not be obtained in any other way.

The Greeks, therefore, demonstrated the power of a faculty which had not been put to use in other civilizations, much as if they had suddenly shown the world the existence of a sixth sense which no one had previously recognized. Clearly, then, the way to build sound systems of thought in any field was to start with truths, apply deductive reasoning carefully and exclusively to these basic truths, and thus obtain an unquestionable body of conclusions and new knowledge.

The Greeks themselves recognized this broader significance of Euclidean geometry, and Aristotle stressed that the Euclidean procedure must be the aim and goal of all sciences. Each science must start with fundamental principles relevant to its field and proceed by deductive demonstrations of new truths. This ideal was taken over by theologians, philosophers, political theorists, and the physical scientists. We shall see later on how widely and how deeply it influenced subsequent thought.

By teaching mankind the principles of correct reasoning, Euclidean geometry has influenced thought even in fields where extensive deductive systems could not be or have not thus far been erected. Stated otherwise, Euclidean geometry is the father of the science of logic. We pointed out in Chapter 3 that certain ways of combining statements lead to unquestionable conclusions, provided the original premises are unquestionable. These ways are called principles or methods of deductive reasoning. Where did we get these principles? The answer is that the Greeks learned to recognize them in their work on Euclidean geometry and then appreciated that these principles apply to all concepts and relationships. If one argues from the premises that all bankers are wealthy and some bankers are intelligent to the conclusion that some intelligent men are wealthy, he is using a principle of valid reasoning discovered in the work on Euclidean geometry. The indirect method of proof which we applied earlier in this chapter owes its recognition to the same source. Toward the end of the classical Greek period, Aristotle formulated the valid principles of reasoning and created the science of logic. In particular, he called attention to some basic laws of logic, such as the principle of contradiction, which says that no proposition can be both true and false, and the principle of the excluded middle, which states that any proposition must be either true or false.

It is because Euclidean geometry applies these principles of reasoning so clearly and so repeatedly that this subject is often taught as an approach to reasoning. The Greeks themselves stressed the value of mathematics as a preparation for the study of philosophy. Whether this is the best way of learning to reason may perhaps be disputable, but there is no doubt that historically this is the way in which Western man learned. And it is pertinent that even current texts on logic use mathematical examples quite freely because these illustrate the principles clearly, unobscured by irrelevant implications or by vagueness in the concepts and relations employed.

The most portentous fact about Euclidean geometry is that it inspired a large-scale mathematical investigation of nature. From the outset the geometrical studies were an investigation of nature. But as the Greeks proved more and deeper theorems and these theorems continued to agree perfectly with observations and measurements, the Greeks became convinced that through mathematics they were learning some of the secrets of the design of this world. It became clear that mathematics was the instrument for this investigation, and the results fostered the expectation that the further application of mathematics would reveal more and more of that design. Just how far the Greeks were emboldened to carry this venture will be apparent in the next two chapters. From the Greeks the Western world learned that mathematics was the extraordinarily powerful instrument with which to explore nature.

REVIEW EXERCISES

1. One of the basic theorems of Euclidean geometry is that the base angles of an isoceles triangle are equal. Euclid's proof proceeded thus. Given triangle ABC (Fig. 6–32) with $AB = AC$, prolong AB to D and AC to E so that $BD = CE$.

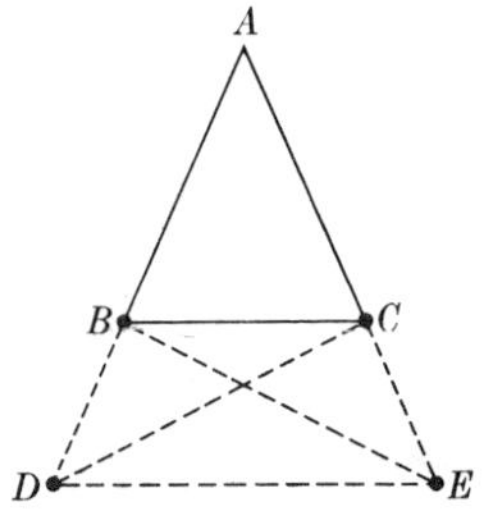

Fig. 6-32

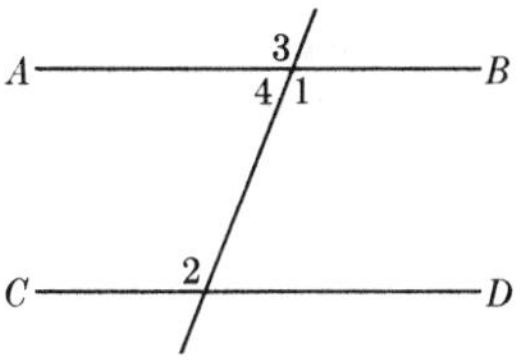

Fig. 6-33

Now draw DE, BE, and DC. Complete the proof by first proving that $BE = DC$ and then that $\angle\, CBD = \angle\, BCE$.

2. In the text we proved that if the alternate interior angles 1 and 2 of Fig. 6–33 are equal, the lines AB and CD are parallel. Prove

   a) if the corresponding angles 2 and 3 are equal, the lines are parallel;

   b) if the angles 2 and 4 are supplementary, that is, if their sum is 180°, then the lines are parallel.

3. Suppose that m in Fig. 6–34 represents a road, and a telephone central is to be built somewhere on this road to serve towns at A and A'. Where along the road should the central be built to minimize the total distance from the central to A and the central to A'?

Fig. 6-34

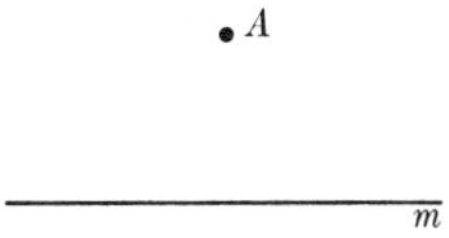

Fig. 6-35

4. A ship must pass between the guns of a fort at A (Fig. 6–35) and equally powerful guns along the shore m. What path should it take to be as safe as possible from all the guns?

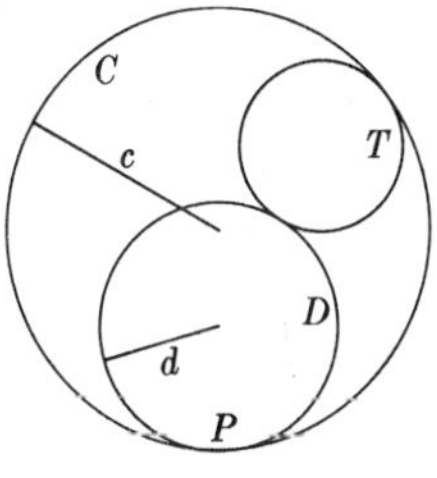

Fig. 6-36

5. Let C and D be two fixed circles (Fig. 6–36) with radii c and d, respectively, and $c > d$. Moreover D and C are tangent internally at P. Now let T be a third circle which is tangent externally to D and internally to C. Show that the positions of the centers of all possible circles T is an ellipse whose foci are the centers of C and D.

Topics for Further Investigation

1. The use of Euclidean geometry in the design of spherical mirrors. Use the first reference to Kline or the reference to Taylor below or any college physics text.
2. The use of Euclidean geometry in the design of optical lenses. Use the references to Taylor below or any college physics text.
3. The contents of Euclid's *Elements*. Use the reference to Heath.
4. Euclidean geometry as a manifestation of Greek culture. Use the second reference to Kline.

Recommended Reading

BALL, W. W. R.: *A Short Account of the History of Mathematics,* pp. 13–63, Dover Publications, Inc., New York, 1960.

BOYS, C. VERNON: *Soap Bubbles,* Dover Publications, Inc., New York, 1959.

COURANT, R. and H. ROBBINS: *What is Mathematics?,* pp. 329–338, pp. 346–361, Oxford University Press, New York, 1941.

EVES, HOWARD: *An Introduction to the History of Mathematics,* Rev. ed., pp. 52–130, Holt, Rinehart and Winston, Inc., New York, 1964.

HEATH, SIR THOMAS L.: *A Manual of Greek Mathematics,* Chaps. 8, 9 and 10, Dover Publications, Inc., New York, 1963.

KLINE, MORRIS: *Mathematics: A Cultural Approach,* Sections 6–8, Addison-Wesley Publishing Co., Reading, Mass., 1962.

KLINE, MORRIS: *Mathematics and the Physical World,* Chaps. 6 and 17, T. Y. Crowell Co., New York, 1959. Also in paperback, Doubleday and Co., N.Y., 1963.

SAWYER, W. W.: *Mathematician's Delight,* Chaps. 2 and 3, Penguin Books, Harmondsworth, England, 1943.

SCOTT, J. F.: *A History of Mathematics,* Chap 2, Taylor and Francis, Ltd., London, 1958.

SMITH, DAVID EUGENE: *History of Mathematics,* Vol. I., Chap. 3, Vol. II, Chap. 5, Dover Publications, Inc., New York, 1958.

TAYLOR, LLOYD WM.: *Physics, The Pioneer Science,* Chaps. 29–32, Dover Publications, Inc., New York, 1959.

CHAPTER 7

CHARTING THE EARTH AND THE HEAVENS

Thrice happy souls! to whom 'twas given to rise
To truths like these, and scale the spangled skies!
Far distant stars to clearest view they brought,
And girdled ether with their chains of thought.
So heaven is reached:—not as old they tried
By mountains piled on mountains in their pride.

OVID

7-1 THE ALEXANDRIAN WORLD

The course of mathematics is very much dependent upon the caprices of man. What more the classical Greeks might have produced had they been able to continue their way of life uninterruptedly, we shall never know. In 352 B.C. Philip II of Macedonia, a province to the north of Athens and outside the pale of Greek culture, started out to conquer the world. He defeated Athens in 338 B.C. In 336 B.C. Alexander the Great, Philip's son, took over the Macedonian armies, completed the conquest of Greece, conquered Egypt, and penetrated Asia as far east as India, and Africa as far south as the cataracts of the Nile. For a new capital he chose a site in Egypt which was central in his empire. Too big a man to be hampered by modesty, he called the capital Alexandria. Alexander drew up plans for the city and for populating it, and the work was begun. Alexandria did become the center of the Hellenistic world, and even 700 years later was still called the noblest of all cities.

Alexander was the most cosmopolitan of men and sought to break down barriers of race and creed. Hence he encouraged and invited Greeks, Egyptians, Jews, Romans, Ethiopians, Arabs, Indians, Persians, and Negroes to settle in the city. At that time the Persian culture was flourishing, and so Alexander made special efforts to fuse Greek and Persian ways of life. He himself married Statira, daughter of Darius, in 325 B.C. and compelled 100 of his generals and 10,000 of his soldiers to marry Persians. After his death written orders were found to transport large groups of Asians to Europe and vice versa.

Alexander died while still engaged in reconstructing the world, and his empire split into three parts. Of these Egypt proved to be the most significant from the standpoint of mathematical progress. Alexander had indeed chosen a good site for his capital. Located at the junction of Asia, Africa, and Europe, it became the center of trade, which brought wealth to the city. The successors of Alexander who ruled Egypt and who adopted the title of Ptolemy were wise men. They appreciated the cultural greatness of classical Greece and decided to make Alexandria a great cultural center. Under their direction part of the wealth was used to beautify the city with splendid buildings, baths, parks, theaters, temples, libraries, and a national archive. They also erected a famous building devoted to the Muses of literature, art, and science, called the Museum, and adjacent to it, an enormous library to house manuscripts. At its height this library was said to contain 750,000 works, an enormous number in view of the fact that in those days "books" were written and reproduced by hand. The Ptolemies invited scholars from all over the world to work there and supported them. Euclid and Eratosthenes, to speak for the moment of men we have already met, lived and worked at this center; Apollonius was educated there; and we shall meet other luminaries shortly. These men, coming from all over the world, brought knowledge of their lands, people, animals, and vegetation to Alexandria, and this in itself helped to make Alexandria cosmopolitan.

The scholars set to work in the fields of mathematics, science, philosophy, philology, astronomy, history, geography, medicine, jurisprudence, natural history, poetry, and literary criticism. Fortunately, Egyptian papyrus, cheaper than parchment, was available for books, and so many more works could not only be written but copied. Alexandria became in fact the center of the book-copying trade of the ancient world. The scholars undertook not only to create and write, but they sent expeditions all over the world to gather knowledge. At Alexandria they built a huge zoological garden and a botanical garden to house the species of animals and plants brought back by these expeditions.

Alexander had planned to fuse cultures in his new empire, and at Alexandria his goal was realized. The culture which developed there was indeed different from that of classical Greece, for reasons that are of interest because they account for the kind of mathematics the Alexandrians produced. First of all, the rather sharp segregation between free men and slaves which existed in Athens was destroyed. The scholars came from all parts of the world and from all economic levels and took a natural interest in the scientific, commercial, and technical problems of commerce, industry, engineering, and navigation. Although Athens also was primarily a sea power and lived on trade, Alexandrian commerce and navigation were far more widespread. Hence there developed an intense interest in astronomy and in geography, i.e., in the subjects enabling man to tell time, navigate over land and sea, build roads, and determine boundaries of the empire. Free men engaged in commerce are naturally more concerned with materials, methods of production, and new ventures. Finally, though the nucleus of the scholars gathered at Alexandria was Greek, it was exposed to the influence of the practical Egyptians to whom mathematics, to the extent that it was used in ancient Egypt, was a tool for engineering, commerce, and state administration.

The results of the new outlook and interests are readily detected. First of all, there was a sharp increase in mechanical devices, which of course aid men in their work. Even training schools to educate young people in mechanics were established. Pulleys, wedges, tackles, geared devices, and a mileage-measuring instrument such as is found in the modern automobile were invented. Archimedes, the greatest intellect of the Alexandrian world, constructed a planetarium which reproduced the motions of the heavenly bodies and designed a pump for raising water from a river to land. He used pulleys to launch a heavy galley for King Hiero of Syracuse. Instruments to improve astronomical measurements were also invented.

Another science whose beginnings may be found in Alexandria is the study of gases. The Alexandrians, notably Heron (about first century A.D.), a famous mathematician and engineer, learned that the steam created by heating water seeks to expand and that compressed air can also exert force. Heron is responsible for many inventions which used these forces. Temple doors opened automatically when a coin was deposited. Inside the temple another coin inserted in a machine blessed the donor by automatically sprinkling holy water upon him. Fires lit under the altar created steam, and the mystified and awe-struck audience observed gods who raised their hands to bless the worshippers, gods shedding tears, and statues pouring out libations. Doves rose and descended under the unobservable action of steam. Guns similar to the toy bee-bee gun were operated by compressed air. Steam power was used to drive automobiles in the annual religious parade along the streets of Alexandria.

The Alexandrians also studied water power and applied it. They invented improved water clocks (used in the courts to limit the time allowed to lawyers), fountains in which figures moved under water pressure, pumps to bring water from wells and cisterns, musical organs worked by water pressure, and a water-spraying device operating on exactly the same principle as that applied in contemporary lawn-sprinklers.

The study of sound and light was intensified. We have already mentioned Euclid's and Heron's studies on the reflection of light by mirrors. Books on optics were written not only by Euclid and Apollonius but also by Heron, the astronomer Ptolemy (whom we shall discuss shortly), and others. Indeed the Alexandrians were the first to concern themselves with a second basic phenomenon of light, refraction, which we shall encounter in this chapter.

Chemical and medical skills, if not a science of chemistry, show a marked advance in Alexandria. The Egyptians had previously acquired some knowledge in these areas, as we know from their ability to embalm. However, metallurgical studies, including the first text on the subject, and the investigation of chemicals, including poisons and their uses, were essentially new developments. Dissection of bodies, forbidden in classical Greece, was permitted, and the Alexandrian world produced the beginnings of anatomy and the most famous doctor of the ancient world, Galen.

Where was mathematics in this scheme of things? The Greeks brought to Alexandria a fully formed, mature, and philosophically oriented mathematics which had little bearing on practical problems. Although the great Alexandrian mathematicians continued to display the Greek genius for theory and abstraction, they combined with that an interest in the world about them and in practical problems. To the classical Greek concern with qualitative properties such as congruence, the Alexandrians added a new theme, *quantitative* results which are useful in a variety of ways.

To illustrate the combination of old and new we might note that while Euclid chronologically belongs to the Alexandrian period, his mathematical work is in essence a recapitulation of the work done in the classical period. Thus Euclid tells us, for example, that the ratio of the area of any circle to the square of its radius is the same for all circles. In symbols, if A is the area of any circle of radius r, then

$$\frac{A}{r^2} = k,$$

where k is the same number for *all* circles. But now suppose that we wish to find the area of a particular circle. Does Euclid's theorem help us? Not directly. We know from the preceding equation that for any circle

$$A = kr^2,$$

where k is a constant. But how much is k? This quantity which we usually denote by π is an irrational number. It is not readily computed and, because it is irrational, can be expressed as a decimal only approximately. One of Archimedes' great achievements, which also illustrates the interest in quantitative knowledge, is his determination that π lies between $3\frac{1}{7}$ and $3\frac{10}{71}$. The achievement is all the more remarkable because neither the classical nor the Alexandrian Greeks had an efficient system for writing and operating with numbers.

As a matter of fact, Archimedes (287–212 B.C.) is the man whose work best illustrates the character of Alexandrian Greek mathematics. He derived many formulas for the areas and volumes of geometric figures, and his results, as opposed to those of Euclid and Apollonius, made actual computations possible. At the same time, Archimedes also pursued the classical Greek interest in proof and in beautiful mathematical results. In this area, he was proudest of his proof that the ratio of the volume of a sphere inscribed in a cylinder (Fig. 7–1) is to the volume of the cylinder as 2 is to 3. He also proved that the same ratio holds for the areas of the sphere and the cylinder. Archimedes was so pleased with this result that he asked that it be inscribed on his tombstone. After Archimedes was killed by a Roman soldier during the Roman conquest of Syracuse, the Romans built an elaborate tomb on which they inscribed this theorem. It was this inscription which enabled Cicero to recognize the tomb on a visit to Syracuse two hundred years later.

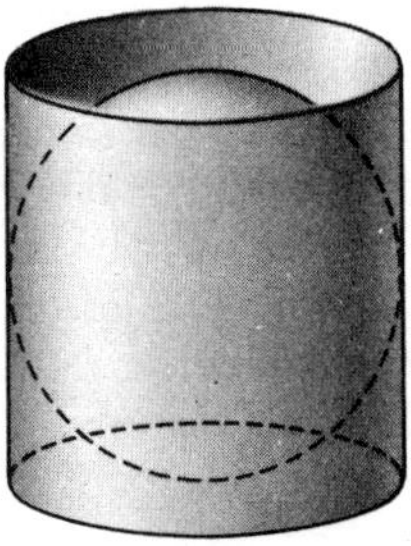

Fig. 7–1.
The volume of a sphere inscribed in a cylinder is two-thirds the volume of the cylinder.

Even in his physical studies Archimedes displayed this combination of theoretical and practical interests. He took up the subject of the lever, a device which had been used in Egypt and Babylonia for thousands of years. Like a true Greek, he produced a scientific work, *On the Lever*, along the lines of Euclidean geometry; that is, he started from axioms and proved theorems about the lever. He did the same with the subjects of floating bodies and centers of gravity of various surfaces and volumes. To these achievements must be added his inventions, of which we have already spoken.

The work of some other giants of the Alexandrian civilization also illustrates the combination of theoretical and practical interests. Eratosthenes (273 B.C.–192 B.C.), director of the library at Alexandria, was distinguished in mathematics, poetry, philology, philosophy, and history. He was the first outstanding mathematical geographer and geodesist. The calculation of the circumference of the earth, which we studied in the preceding chapter, is one of his great achievements. He collected and integrated all available geographical knowledge, introduced methods of surveying, made maps, and compiled all of this information in his *Geographica*.

Eratosthenes was also an astronomer. He constructed some new instruments, made many astronomical measurements, and, among other applications, used his astronomical knowledge to improve the calendar. As a result of his work, an old Greek calendar based on a year of 12 months each containing 30 days was replaced by the Egyptian year of 365 days, to which Eratosthenes added an extra day every fourth year. This calendar was adopted by the Romans when Julius Caesar called in Sosigenes, an Alexandrian, to reform the calendar. Julius contributed his name. The Julian calendar was taken over by the Western world with the slight modification that we omit the leap year in three out of every four century years.

The work in geography and astronomy, continued by such famous men as Strabo (*ca.* 63 B.C.–*ca.* 15 B.C.), Poseidonius (first century B.C.), and many others, was crowned by the achievements of two of the greatest men of the Alexandrian world, Hipparchus and Ptolemy. Hipparchus (second century B.C.), about whom we know rather little, lived at Rhodes, but was in close touch with the developments in Alexandria. After criticizing Eratosthenes' *Geographica*, he refined the method of locating places on the earth by systematically employing latitude and longitude. He improved astronomical instruments, measured irregularities in the moon's motion, catalogued about 1000 stars, and estimated the length of the solar year as 365 days, 5 hours, and 55 minutes, i.e., he overestimated by about 6½ minutes. One of his notable astronomical discoveries was the precession of the equinoxes, a slow change in the time of occurrence of the spring and fall equinoxes. Hipparchus is the creator of the most famous and most useful astronomical theory of antiquity, about which we shall learn more later.

The work of Hipparchus is known to us largely through the writings of the mathematician, astronomer, geographer, and cartographer Claudius Ptolemy. Ptolemy, who is believed to be Egyptian—he was no relation to the Greek rulers of Egypt—lived from about 100 to 178 A.D. One of his influential achievements was his *Guide to Geography*, or *Geographica*, the most comprehensive work of antiquity on this subject. This book, which contains the latitude and longitude of 8000 places, almost every place on the earth then known, estimates of the size and extent of the habitable world, and methods of map-making, summarized the geographical knowledge of the ancient world and became the standard atlas for over a thousand years. Better known is Ptolemy's great work on astronomy, the *Mathematical Syntaxis* or *The Mathematical Collection*, which the Arabs called *Al Megiste* (an Arabic and Greek combination meaning "the greatest")—later Anglicized as *The Almagest.* This book contains the full development of Hipparchus' and Ptolemy's astronomical theory, generally known as Ptolemaic theory, which dominated astronomy until about 1600 A.D. when it was superseded by the work of Copernicus and Kepler.

7-2 BASIC CONCEPTS OF TRIGONOMETRY

The theoretical sciences of geography and astronomy require their own mathematical tool, trigonometry. Hipparchus and Ptolemy created this branch of mathematics whose first presentation is found in Ptolemy's *Almagest.* With this simple branch of mathematics it is possible to calculate the sizes and distances of the heavenly bodies as easily as one calculates the area of a rectangle. In presenting the trigonometry of Hipparchus and Ptolemy, we shall not use their notation and proofs; however, the modern approach is not essentially different.

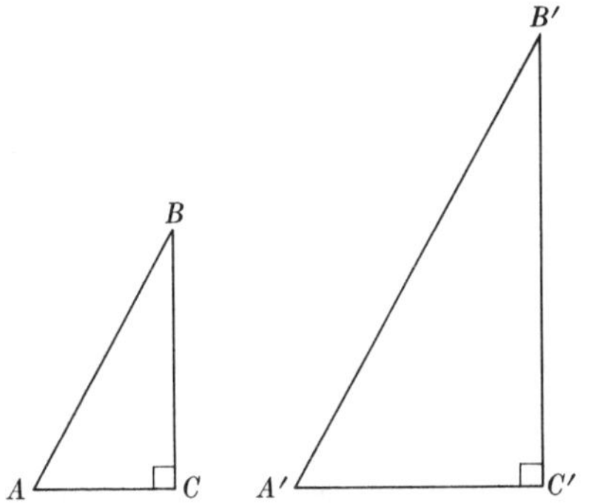

Fig. 7-2.
Similar right triangles.

Let us consider the two right triangles shown in Fig. 7–2 and let us suppose that angle A equals angle A'. Since all right angles are equal, angle C equals angle C'. One of the key theorems in Euclidean geometry states that the sum of the angles in any triangle is 180°. Since all three angles in each triangle add up to the same amount and two angles of one are equal to two of the other, the third angles must be equal; i.e., angle B equals angle B'.

Now another theorem of Euclidean geometry states that if two triangles are similar, the ratio of any two sides in one equals the ratio of the corresponding sides in the other. Thus, for example,

$$\frac{BC}{AB} = \frac{B'C'}{A'B'} .$$

Let us note here that triangle $A'B'C'$ is *any* other right triangle which has an acute angle, A', equal to angle A. Hence for any such triangle the ratio $B'C'/A'B'$ must equal BC/AB. Therefore, if we could compute this ratio—it is a number of course—for any one right triangle containing a given angle A, we would know it for all right triangles having an acute angle equal to A.

Before we pursue this idea, let us observe that what we said about the ratio BC/AB applies to any other ratio of two sides of triangle ABC. Of the many ratios we can form three are especially useful and are given names. These ratios are:

$$\text{sine } A = \frac{\text{side opposite angle } A}{\text{hypotenuse}} = \frac{BC}{AB},$$

$$\text{cosine } A = \frac{\text{side adjacent to angle } A}{\text{hypotenuse}} = \frac{AC}{AB},$$

$$\text{tangent } A = \frac{\text{side opposite angle } A}{\text{side adjacent to angle } A} = \frac{BC}{AC} .$$

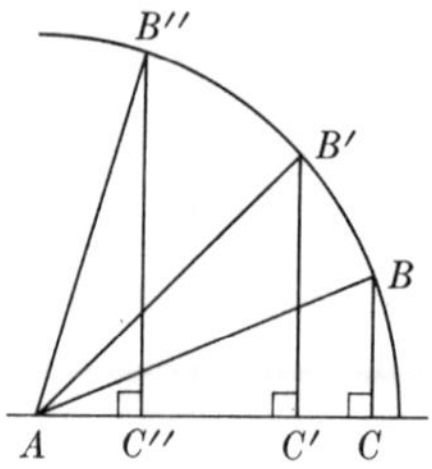

Fig. 7–3.
The variation of sin *A* with angle *A*.

The angle *A* is written alongside the name of each ratio. This practice is necessary not only because the very use of such words as opposite and adjacent depends upon which angle of the triangle we are talking about, but also because the values of the ratios depend upon the size of the angle. It is very common to abbreviate these names as sin, cos, and tan, respectively.

Since we intend to employ these ratios, our first task should be to see whether we can compute them for angles of various sizes. First of all let us get some general notion of how these ratios vary with the angle. Let us consider sin *A* as an example. We have already pointed out that the values of these ratios for a given angle *A* are the same in any right triangle containing *A*. To study the variation of sin *A* as *A* changes, we can then take right triangles whose hypotenuse is 1. We know from the very definition of sin *A* that it is the ratio of the side opposite angle *A* to the hypotenuse. Since sin *A* equals BC/AB and $AB = 1$, then $\sin A = BC$. When *A* is small (Fig. 7–3), *BC* or sin *A* is small. We should expect, then, that for an angle close to 0°, the sine of that angle should be close to 0. On the other hand, as Fig. 7–3 shows, when angle *A* increases and the hypotenuse is kept one unit in length, the opposite side must increase; hence the sine ratio must increase. When angle *A* is very close to 90°, as in triangle $AC''B''$, the side $B''C''$ is almost as large as AB''; hence sin *A* must be close to 1. When angle *A* is 90°, it can no longer be an acute angle of a right triangle, but because sin *A* approaches 1 as *A* approaches 90°, it is agreed that in this special case we shall take sin *A* to be 1. Likewise, we take sin 0° to be 0. The general point of this discussion is that sin *A* varies from 0 to 1 as *A* varies from 0° to 90°.

Next let us take a particular angle and let us see whether we can calculate the three ratios. We shall choose 30°. Consider the equilateral triangle *ABD* (Fig. 7–4). We know that in such a triangle each angle is 60°. If we now draw the angle bisector *AC*, then angle *BAC* is 30°. Moreover, triangle *ACB* is a right triangle because triangles *ACB* and *ACD* are congruent, and hence the two angles at *C* must be equal. Since the sum of these two angles is 180°, each must be 90°. Triangle *ACB* is, then, a right triangle containing an acute angle of 30°.

Now it does not matter how long we take *AB* to be, for we saw earlier that we may compute the ratios in any right triangle containing the given acute angle. Let us therefore choose a convenient number, say 2, for the length of AB. Since ABD is equilateral, side $BD=2$. But because triangles ACB and ACD are conguent, $CB=CD$. Hence $CB=1$. We now find the length of AC. The Pythagorean theorem says that

$$(AC)^2 + (CB)^2 = (AB)^2;$$

therefore

$$(AC)^2 = (AB)^2 - (CB)^2.$$

Since $AB=2$ and $CB=1$,

$$(AC)^2 = 4 - 1,$$

or

$$AC = \sqrt{3}.$$

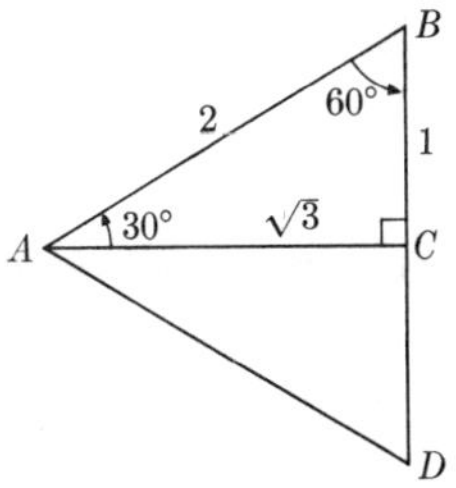

Fig. 7–4

We can now use the definitions of sine, cosine, and tangent to state at once that

$$\sin 30° = \frac{\text{side opposite}}{\text{hypotenuse}} = \frac{1}{2};$$

$$\cos 30° = \frac{\text{side adjacent}}{\text{hypotenuse}} = \frac{\sqrt{3}}{2};$$

$$\tan 30° = \frac{\text{side opposite}}{\text{side adjacent}} = \frac{1}{\sqrt{3}}.$$

As a dividend for our patience we get more information from the above reasoning than we sought. Let us note that angle B, which is 60°, is also an acute angle in a right triangle, and we know the lengths of the sides. Hence, since $\sin B$ is the side opposite angle B divided by the hypotenuse, we have

$$\sin 60° = \frac{\sqrt{3}}{2}.$$

Similarly, by applying the definitions of cosine and tangent we obtain

$$\cos 60° = \tfrac{1}{2}, \qquad \tan 60° = \sqrt{3}.$$

We must admit that in undertaking to find the ratios belonging to 30° we selected a simple case. For most angles the ratios are not so easily found, and a good deal of geometry must be applied. The process of determining the ratios for angles from 0° to 90° is not particularly fascinating. Fortunately these values were obtained by Hipparchus and Ptolemy and compiled in a table to be found in Ptolemy's *Almagest*. (These tables were checked and extended by many later mathematicians.) Hence let us take over their results which appear in the "Table of Trigonometric Ratios" (in the Appendix).

The table gives the sine, cosine, and tangent values for each angle from 0° to 90°. For angles from 0° to 45° we use the left-hand column and the headings across the top of the page. For example, alongside of 30° and under tangent we find 0.5774. This number is the approximate decimal value of $1/\sqrt{3}$. To find the sine, cosine, or tangent of an angle from 45° to 90° we use the right-hand column and the column designations at the *bottom* of the page. For example, to find sin 60° we look for 60° in the right-hand column and above the word sine we find 0.8660. This number is the approximate decimal value of $\sqrt{3}/2$.

Our table does not give the ratios for angles which contain minutes and seconds as well as degrees. There are tables which do so, but we shall not bother with them because the idea is the same. Where we need the value of a ratio for an angle not in the table, it will be supplied in the text proper.

Let us note that we can use these tables in reverse. If, for example, we are given $\tan A = 1.7321$, we can look down one tangent column and up the other until we come to 1.7321. We can then look to the left (or to the right, depending upon where we locate this number), and find the angle which has the given tangent value. In the present case we must choose the angle at the right, namely 60°. If the table does not contain the exact value given, it will suffice, for our purposes, to choose the one nearest to it.

EXERCISES

1. Use the isosceles right triangle shown in Fig. 7–5 to compute sin 45°, cos 45°, and tan 45°.
2. Use the Table of Trigonometric Ratios to find

a) sin 20°	b) sin 70°	c) cos 35°
d) cos 55°	e) tan 15°	f) tan 80°

3. Use Fig. 7–3 in the text to determine the range of cosine values as angle A varies from 0° to 90°.
4. Use Fig. 7–3 in the text to determine the range of tangent values as angle A varies from 0° to 90°.
5. Show that when A and B are the two acute angles of a right triangle, then $\sin A = \cos B$ and $\cos A = \sin B$.
6. Prove that $\cos(90° - A) = \sin A$ and that $\sin(90° - A) = \cos A$.
7. Show that $\sin^2 A + \cos^2 A = 1$ for any acute angle A. Here the notation $\sin^2 A$ means $(\sin A)(\sin A)$ or the square of $\sin A$. Can the result be used to compute trigonometric ratios? If so, how?
8. State the definitions of sine, cosine, and tangent of angle D in terms of the sides DE, EF, and FD of the triangle shown in Fig. 7–6.

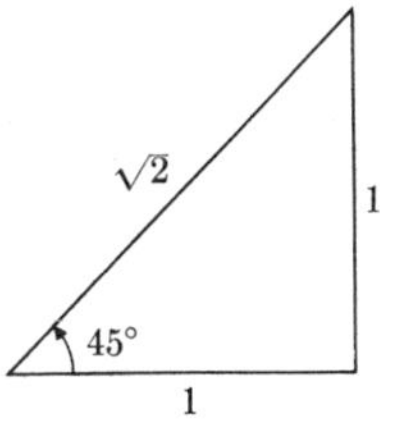

Fig. 7-5

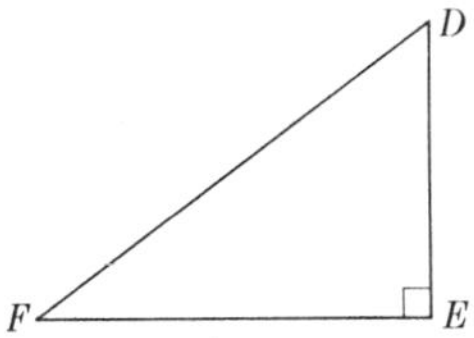

Fig. 7-6

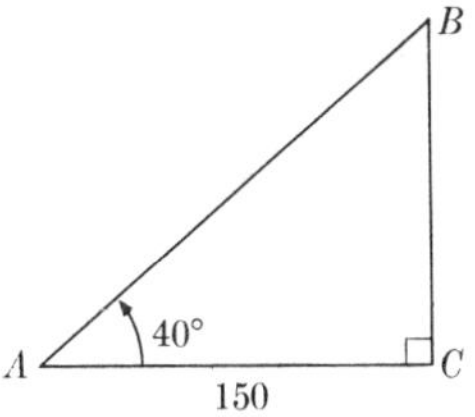

Fig. 7-7

7-3 SOME MUNDANE USES OF TRIGONOMETRIC RATIOS

Before we venture onto vast stretches of the earth's surface or into the heavens, let us see what we can do with trigonometric ratios in rather simple, homely situations. Suppose we had to find the height of the cliff BC in Fig. 7-7. Of course, we could climb the cliff, let a rope down from point B until it just reaches C, pull up the rope, and measure the length which stretched from B to C. There is, however, an easier method which is especially recommended to people who do not like heights.

Instead of climbing the cliff, one can walk along the ground from C to any convenient point A. The distance from C to A is then measured; let us suppose it proves to be 150 feet. At A, a person measures the angle between the horizontal AC and the line of sight from A to B. A surveyor would use a transit for this purpose, but there are simpler devices, called protractors, which one can carry in his pocket. Suppose that angle A turns out to be 40°. We are interested in side BC and we know side AC. The fact that these two sides are the side opposite angle A and the side adjacent to angle A suggests that we use the tangent ratio and write

$$\tan 40° = \frac{BC}{150}.$$

This equation involves numbers, and we can therefore apply the axiom that equals multiplied by equals give equals, to justify multiplying both sides by 150. We obtain

$$150\,(\tan 40°) = BC.$$

Now tan 40° can be found in the table which Hipparchus and Ptolemy so considerately prepared, and proves to be 0.8391. Hence

$$BC = 150(0.8391) = 126.$$

The answer, then, is 126 feet. We ignore the decimals because the given information is presumably accurate only to the nearest foot.

EXERCISES

1. To measure the width BC of a canyon (Fig. 7–8), a surveyor at C walks along the edge (preferably alongside the edge) to some convenient point A. He then measures AC and the angle A. Suppose AC is 300 ft and angle A is 56°. How large is BC?
2. At some point on the ground, located at a distance from the Empire State Building in New York City, an observer finds that the angle between the horizontal and the line of sight to the top is 5° (Fig. 7–9). The building is 1248 ft high. How far away is the observer?

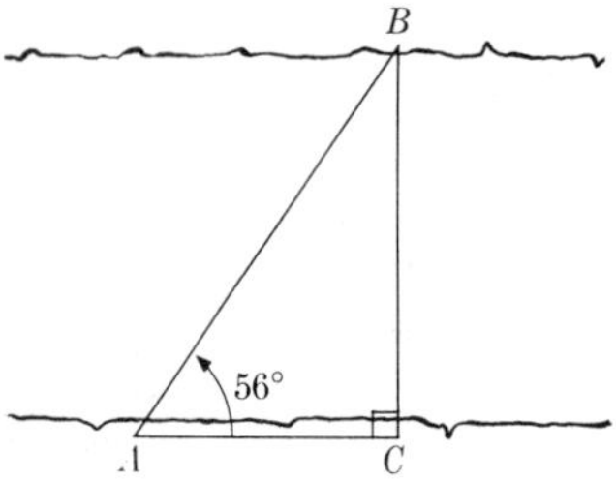

Fig. 7–8

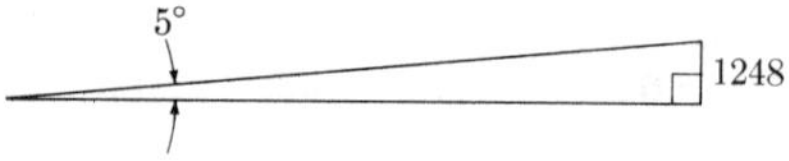

Fig. 7–9

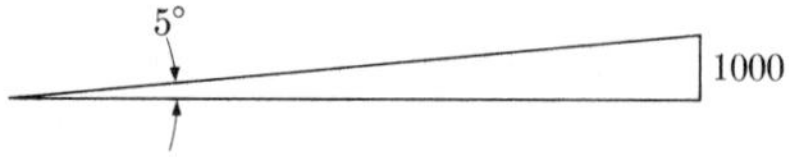

Fig. 7–10

3. A railroad line is being planned which must rise to 1000 ft (Fig. 7–10) at a "grade" of 5°. How long must the line be?
4. A lighthouse beacon is 400 ft above sea level (Fig. 7–11), and the sea around it is obstructed by rocks extending as far as 300 ft from the base of the lighthouse. A sailor on a ship's deck 20 ft above sea level measures the angle between his horizontal and the line of sight to the top of the beacon and finds it to be 50°. Is his ship clear of the rocks?
5. The Alexandrian Greek mathematician and engineer Heron showed how one could dig a tunnel under a mountain by working from both ends simultaneously and have the borings meet. He chose a convenient point A on one side, a con-

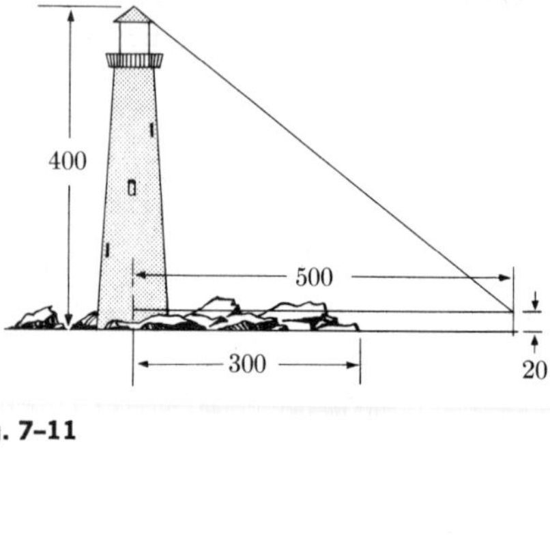

Fig. 7–11

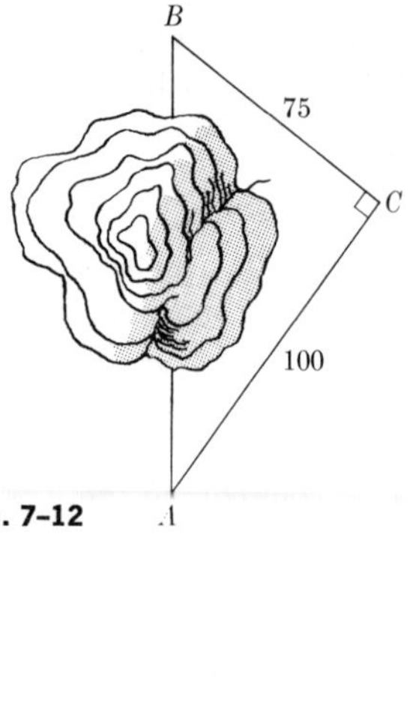

Fig. 7–12

venient point B on the other, and finally point C for which angle ACB is 90° (Fig. 7–12). He next measured AC and BC and found their lengths to be 100 ft and 75 ft, respectively. Now, said Heron, it is possible to calculate angles A and B. He then instructed the workers at A to follow a line which made the calculated angle with AC, and gave analogous directives to the workers at B. How did he calculate angle A and angle B?

6. Trigonometric ratios can be used to compute the radius of the earth, whence, of course, the circumference can be determined by plane geometry. The method is an alternative to Eratosthenes' procedure. From a point A which is 3 mi above the surface of the earth (A can be the top of a mountain or an airplane), an observer looks to the horizon. His line of sight, AC in Fig. 7–13, is just tangent to the earth's surface. According to a theorem of Euclidean geometry, the radius OC of the earth is perpendicular to the tangent at C. Hence triangle ACO is a right triangle. Suppose that the size of angle A is 87° 46′. Let us denote the length of OC by r. Then OD is also r. We can now say that

$$\sin 87°46' = \frac{r}{r+3}.$$

Given that sin 87° 46′ is 0.99924, calculate r.

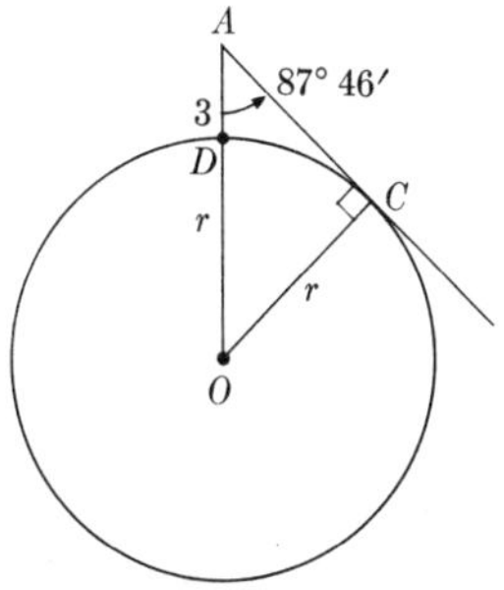

Fig. 7–13

✻ 7–4 CHARTING THE EARTH

We have already related that geography was one of the major interests of the Alexandrians. Here Hipparchus and Ptolemy, helped by the trigonometry they had created, made great strides. Let us see how they determined the locations of important places and how they calculated the distances between such places.

Hipparchus proceeded by employing systematically an idea already advanced prior to his time, namely the scheme of latitude and longitude. The earth is, of course a sphere. Let us consider circles with center O, which is the center of the earth, each going through the North and South Poles, N and S in Fig. 7–14. Thus NWS is one half of such a circle; the other half runs in back of our figure and is therefore invisible. Likewise NVS is one half of another such circle. Obviously we can think of such a circle through N, S, and any other point on the earth's surface. Each half circle from N to S is called a longitude line or a meridian of longitude.

To distinguish among these many lines, we introduce another circle, $XWVU$, which is perpendicular to the longitude lines and halfway between the two poles. This circle is called the equator. Now one of the longitude lines, say NWS, is chosen as the starting line, so to speak. (Today this line goes through the city of Greenwich, England.) We consider next any other line,

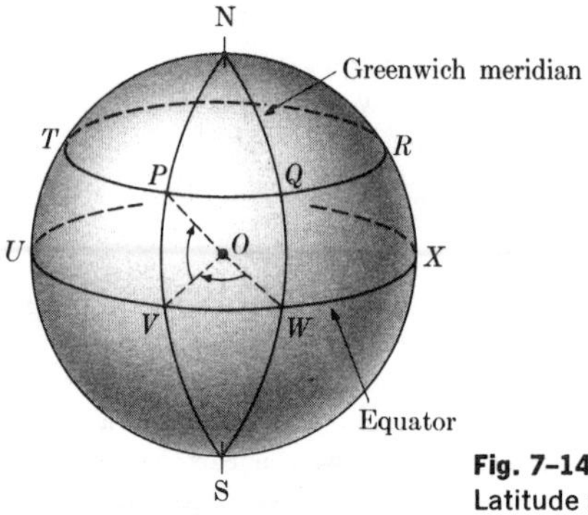

Fig. 7–14.
Latitude and longitude.

such as NVS in our figure. The angle VOW formed at the earth's center, O, by the lines VO and OW is called the longitude of any point on NVS. Thus longitude is an angle. To distinguish the meridians of longitude on the left of NWS from those on the right, we use the term "west longitude" to designate the angles determined by the former, and apply the term "east longitude" to those formed by the latter.

Thus any point on the earth's surface has a definite longitude. However all points on the half circle NVS have the same longitude. How shall we distinguish any one of these points from the others? The answer is: by introducing horizontal circles going around the earth. The equator is one such circle, and the circle $TPQR$ of our figure is another. Clearly we can introduce many such circles lying in planes parallel to the equator. These circles are called circles of latitude. Again we have the problem of distinguishing among these circles. This is solved by introducing angles formed at the center of the earth, such as POV of Fig. 7–14, where P is any point on a circle of latitude, O is the center of the earth, and V is on the equator and on the meridian of longitude through P. The angle POV is called the latitude of P. If P is north of the equator, it is said to have north latitude; if it is south of the equator, it is said to have south latitude. Thus points on the same meridian of longitude are distinguished by their differing latitudes.

The point P is a typical point on the earth's surface, and its position is now described by its latitude and longitude. For example, it might have 30° north latitude and 50° west longitude. In this case, angle POV is 30°, and angle VOW is 50°. Any point north or south of P, that is, on the same meridian, will have the same longitude as P but a different latitude. Any point east or west of P, that is, on the same circle of latitude, will have the same latitude as P but a different longitude.

We have described what is meant by the latitude and longitude of any point on the earth's surface, but how do we determine the latitude and longitude for any given point P? (After all, we cannot penetrate to the center of

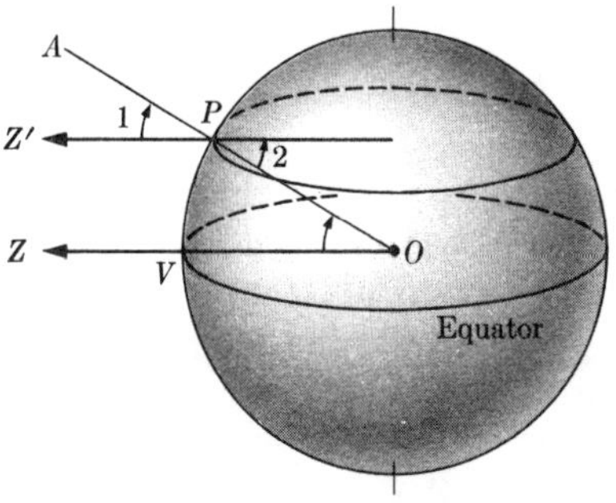

Fig. 7-15.
The determination of latitude at a point on the earth's surface.

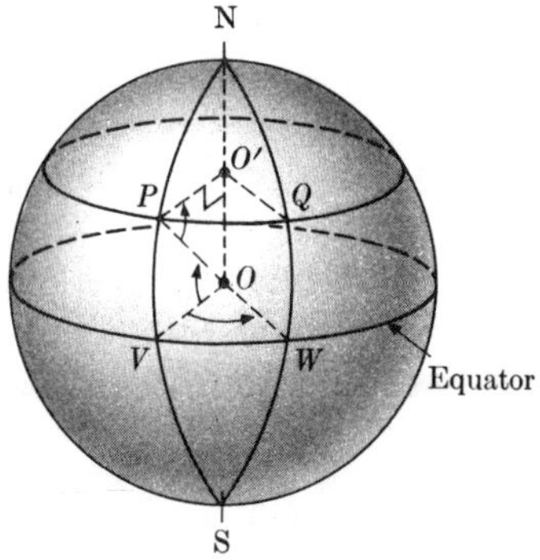

Fig. 7-16

the earth to measure the angles POV and VOW.) There are numerous methods available. We shall describe a simple one just to see that the latitude and longitude of places on the earth can be determined. Suppose we seek the latitude of some point P (Fig. 7-15). On the day of the spring equinox, that is about March 21, the sun is in the plane of the equator and, at noon on that day, it is also in the plane of the meridian of longitude. For a person at P, the overhead direction is PA, and the direction to the sun is PZ'. Now the sun is so far away that PZ' and VZ can be taken to be parallel lines. Then angle 2 equals the angle of latitude POV because they are alternate interior angles of parallel lines. But angle 1 equals angle 2 because they are verticle angles. Hence angle 1 equals the latitude of P. But angle 1 can be measured. It is the angle between the direction of the sun and the overhead direction at P. Thus the latitude of P can be determined.

There are other methods of measuring latitude as well as methods for finding the longitude of places on the earth. It is of interest that the methods of measuring latitude are more readily applied. The problem of determining longitude accurately aboard a ship at sea was not resolved until the middle of the eighteenth century. We shall have more to say about this later.

We may suppose, then, that the latitude and longitude of places on the earth can be determined. Can we now determine how far apart two places are? We can and we shall illustrate the process. Suppose P (Fig. 7-16) is New York City, which has a north latitude of 41° and a west longitude of 74°. Hence angle POV is 41°, and angle VOW is 74°. Let us answer first the question, How far north of the equator is New York City? This question is easy to answer. The distanee we seek is the arc PV. But POV is 41° and arc PV is the arc of a circle whose radius is the radius of the earth. Hence arc PV is that part of the circumference of the earth which 41° is of 360°; that is, if we take the circumference of the earth to be 25,000 miles, then

$$PV = \tfrac{41}{360} \cdot 25{,}000 = 2847.$$

Thus New York City is 2847 miles north of the equator.

Now let us calculate how far west New York City is of the point Q which has the same latitude and has longitude 0°. This point Q is actually the location of Morella, Spain, a small town about 200 miles east of Madrid. Since the longitude of New York City is 74°, angle VOW is 74°. But the distance we seek is not arc VW but arc PQ. Now arc PQ is on the circle of latitude through P. This circle has its center at O' on the straight line through NS, and its radius, $O'P$, is not the radius of the earth. If we could calculate $O'P$, then we could calculate the circumference of the circle of latitude, and since angle $PO'Q$ is also 74°, we could calculate arc PQ.

Our problem then reduces to finding $O'P$. We can find it. The radius $O'P$ is a side of the triangle $OO'P$. Moreover, OO' is perpendicular to $O'P$. Hence we have a right triangle. Since $O'P$ and OV are parallel, angle $O'PO$ equals the latitude of P because $O'PO$ and POV are alternate interior angles of parallel lines. Hence in triangle $O'PO$,

$$\cos 41° = \frac{O'P}{OP}$$

or

$$O'P = OP \cos 41°.$$

Now OP is the radius of the earth, or 4000 miles. From our table we find that cos 41° is 0.7547. Hence

$$O'P = 4000 \cdot 0.7547 = 3019.$$

We may now calculate arc PQ. This arc is $\tfrac{74}{360}$ of the circumference of the circle whose radius is 3019 miles. Hence

$$PQ = \tfrac{74}{360} \cdot 2\pi \cdot 3019.$$

Using the approximate value of 3.14 for π, we find that

$$PQ = 3897.$$

Thus New York City is 3897 miles west of Morella, Spain.

We have computed the distance between two points on the same meridian of longitude and the distance between two points on the same circle of latitude. We could investigate how to calculate the distance between two points on the earth's surface which have neither the same longitude nor the same latitude. However, we have seen enough of the method to comprehend how the trigonometric ratios can be used. Only one point may be worthy of note here. Suppose that P and Q (Fig. 7–17) are two points on the surface of the earth, and we now consider the question, What is the distance between them? We cannot mean the straight-line distance between P and Q because this does not lie on the earth's surface. The distance along the surface of the earth from P to Q must then be an arc of a curve. Which shall we choose? If we choose a circle *whose center O is the center of the earth* and which passes through P and Q, then we shall have what is called a *great circle*. The shorter of the two arcs from P to Q along this great circle is the shortest distance from P to Q along the surface of the sphere. This theorem of spherical geometry, which we shall not prove, is noteworthy because it tells us what route ships and planes should take if they are to save time and expense.

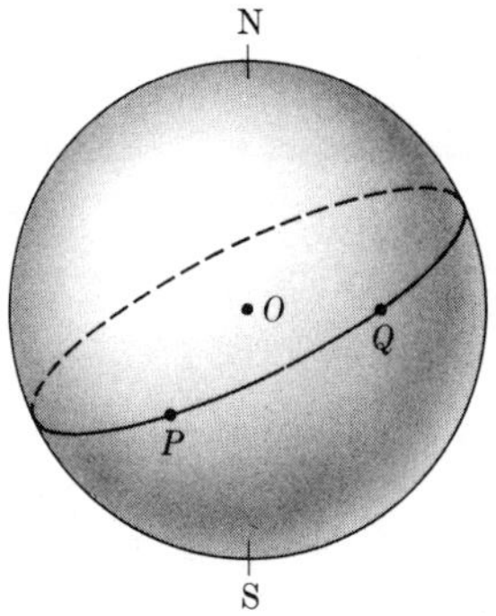

Fig. 7–17.
A great circle on the earth's surface.

Let us consider this theorem in connection with travel by the shortest route between two points such as New York City and Morella, Spain, which are on the same circle of latitude. Although in this case, one wishes to reach a point due east or west (depending on the direction of the trip) of a given point, the circle of latitude is *not* the shortest route because it is not a great circle. We saw in fact that the circle of latitude has O' as its center (Fig. 7–16), whereas the center of the earth is O.

Determining the latitude and longitude of places on the earth and their distances apart is valuable not only for navigation but for map-making. Both Hipparchus and Ptolemy made maps of the ancient world. Although we shall not describe their mathematical methods, we would like to call attention to the problem of making a map. A map is supposed to be a reproduction on flat paper of the relative locations of places on the earth. Now the earth is a sphere, and the one deficiency of this most prized figure of the Greeks is that one cannot take a sphere, cut it open, and lay it flat without creasing, folding, stretching, or tearing the material. One can see this readily if he peels an orange and then tries to flatten out the skin.

Since it is not possible to flatten a sphere without distorting it, any attempt to reproduce on flat paper the relationships that exist on the sphere must involve a distortion of areas, or the relative directions of one place from another, or distances. Hipparchus and Ptolemy therefore invented several methods of map making each of which has features useful for one or more purposes. Thus some methods preserve area, others direction, and still others project great circles into straight lines so that the shortest distance on the sphere between two points is represented by the shortest distance on the map. No map can be a true representation in all respects.

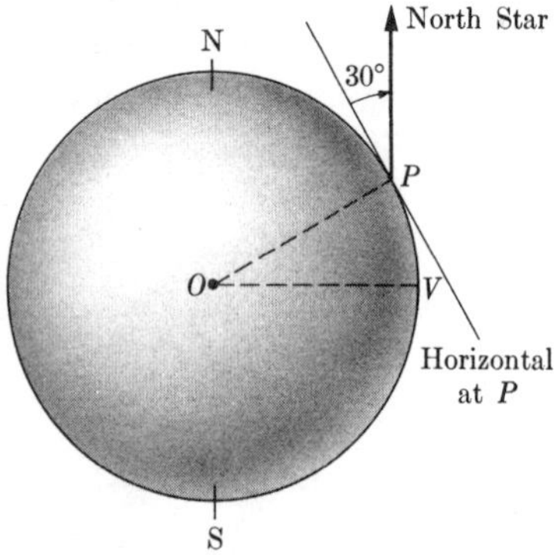

Fig. 7–18

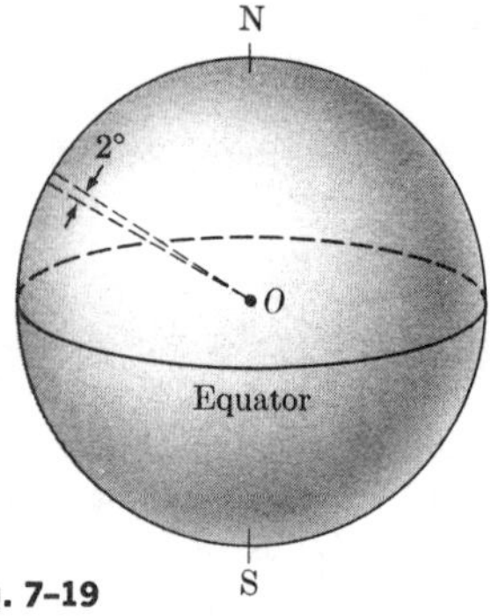

Fig. 7–19

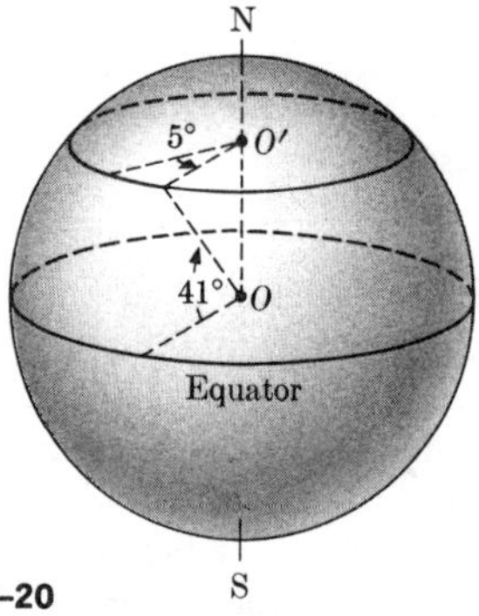

Fig. 7–20

EXERCISES

1. To determine the latitude of a point P (Fig. 7–18) on the surface of the earth, an observer at P measures the angle between the horizontal at P and the direction of the North Star. He finds this angle to be 30°. What is the latitude of P?
2. As one travels north along a meridian from the South Pole to the North Pole, how does his latitude change?
3. Suppose that one travels west from some point on the 0° meridian. How does his longitude change?

4. Of the two circles of latitude, 30° north and 40° north, which has the larger radius?
5. If a man changes his latitude by 2° in traveling along a meridian (Fig. 7–19), how far does he travel?
6. Suppose a man travels due west along the 41° circle of latitude and changes his longitude by 5° (Fig. 7–20). How far does he travel?
7. In one day (24 hr) the earth rotates through 360°. Hence a person has in effect moved around in a complete circle. How far has a person traveled who is at 41° latitude?

* 7-5 CHARTING THE HEAVENS

From the determination of the latitude and longitude of places on the earth's surface and of distances between places, Hipparchus and Ptolemy proceeded to the far more ambitious problem of calculating the sizes and distances of the heavenly bodies. The classical Greeks had indeed speculated about these sizes and distances, but since they relied far more upon aesthetically pleasing principles than upon keen observation, measurement of angles, and numerical calculation, their conclusions were often absurd.

The Alexandrian Greeks made the decisive step in quantitative astronomy. They were, as we have noted, more disposed to measure. Moreover many of them, including Hipparchus himself, had improved the astronomical instruments and the sundials and water clocks which helped to fix more accurately the time at which observations were made. Hipparchus and Ptolemy also had at their disposal in Alexandria a wealth of astronomical data which the Egyptians, Babylonians, and Alexandrians had compiled over many centuries. Let us see how these men "triangulated" the heavens. We shall not reproduce their exact procedures but merely show the essential principles.

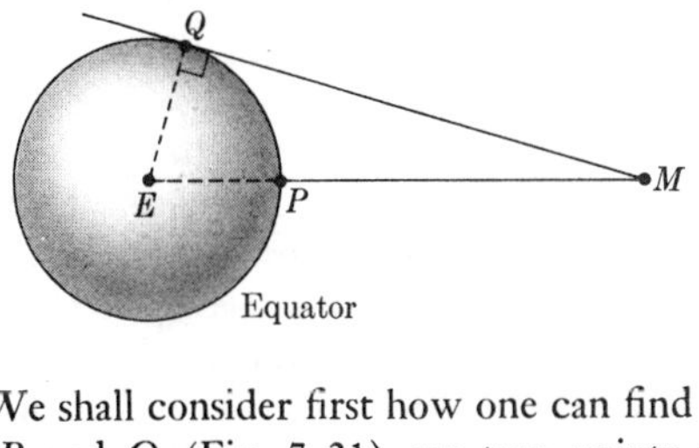

Fig. 7–21.
Finding the distance to the moon.

We shall consider first how one can find the distance to the moon. Suppose that P and Q (Fig. 7–21) are two points on the earth's equator which are chosen to satisfy the following conditions: The moon is to be directly overhead at P; that is, the moon, M, regarded as a point, is to be on the line from the center of the earth, E, through P. The moon is in this position at certain times each month. The point Q is chosen such that the moon is just visible from it. This means that the moon is clearly visible from points closer to P but not visible from points farther away from P and along the equator. Another way of saying the same thing is that the line MQ is tangent to the equator at Q. Let us draw the line EQ. The angle EQM is a right angle because the radius of a circle drawn to the point of contact of a tangent is perpendicular to the tangent.

We now have a right triangle. Moreover, EQ is the radius of the earth and this is known. The angle at E is the difference in longitude between the points P and Q, and since the longitudes of places on the earth are known, so is angle E. A modern value for it is 89° 4′, a value far more accurate than Hipparchus or Ptolemy could have obtained with their instruments. The calculation of EM is now child's play, for

$$\cos E = \frac{EQ}{EM}.$$

The value of cos E or cos 89° 4′, taken from a larger trigonometric table than ours, is 0.0163. Moreover EQ is 4000 miles. Then

$$0.0163 = \frac{4000}{EM}.$$

If we multiply both sides of this equation by EM and then divide by 0.0163, we obtain

$$EM = \frac{4000}{0.0163} = 245{,}000.$$

Our data yield $EM = 245{,}000$ miles, and if we now subtract EP, the radius of the earth, we find that PM, the distance from the surface of the earth to the moon, is 241,000 miles. Hipparchus arrived at the figure of about 280,000 miles because his angular measure of E was not so accurate.*

Precisely the same method can be used to find the distance to the sun. The point M (Fig. 7–21) would now represent the sun. However, because the distances PM and QM are much larger in the case of the sun, angle E is larger and very close to 90°. Moreover, the angle must be measured very accurately because a small error in the angle will cause a large error in the value of PM. For this reason the result of Hipparchus and Ptolemy, of the order of millions of miles, was, as they realized, not very accurate (see Exercise 1).

Let us now find the radius of the moon. Whereas in the preceding calculation we regarded the moon as a point, this idealization will obviously not

* The distance of the earth from the moon varies over the year. The above value is about an average value.

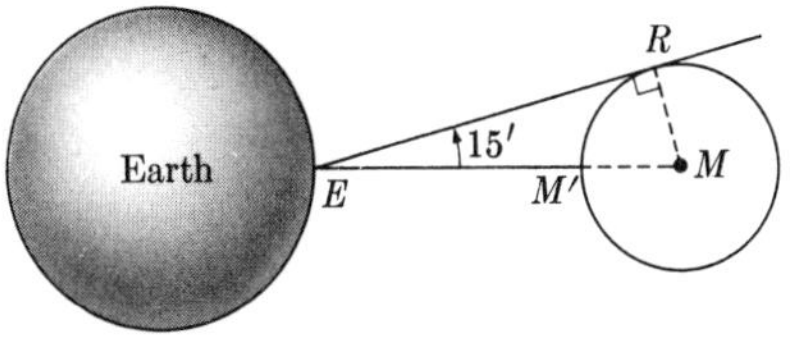

Fig. 7–22.
Determining the radius of the moon.

do in finding the radius. Instead let us regard the moon as a small sphere with center M and radius MR (Fig. 7–22). At a point E on the earth's surface, one measures the angle between the line EM which has the direction from E to the center of the moon, and the line ER which is tangent to the moon's surface. This angle proves to be 15′. We know the distance from earth to moon, at least when the moon is idealized as a point. Let us use this distance, even though it is not exactly EM in our figure. We shall see that the error introduced is minor. Hence EM for us is 241,000 miles. We shall use again the Euclidean theorem that a radius of a circle drawn to the point of tangency of a tangent is perpendicular to the tangent. For our figure this theorem says that MR is perpendicular to ER. Then in the right triangle EMR we have

$$\sin E = \frac{MR}{EM}.$$

Now angle $E = 15'$ and $\sin 15'$, taken from a table giving sine values for angles in minutes, is 0.0044. Moreover EM is 241,000. Hence

$$0.0044 = \frac{MR}{241{,}000}.$$

Then

$$MR = 241{,}000 \cdot 0.0044 = 1060.$$

Thus the radius of the moon is 1060 miles. We can see now that the error introduced by using 241,000 miles as the distance to the center of the moon cannot be great because the radius of the moon is only 1060 miles. The distance of 241,000 miles is really the distance EM' since, in determining the distance to the moon, we could observe only the surface. (For a more accurate calculation of the moon's radius see Exercise 3.) It is of interest that Hipparchus obtained the result of 1,333 miles; his measurement of angle E was not as accurate as the modern one.

The method just used to find the radius of the moon can also be applied to find the radius of the sun. The point M in Fig. 7–22 becomes the center of the sun, and the distance EM becomes the distance to the sun (see Exercise 2). Angle E is about the same for this case as for the moon, as one might

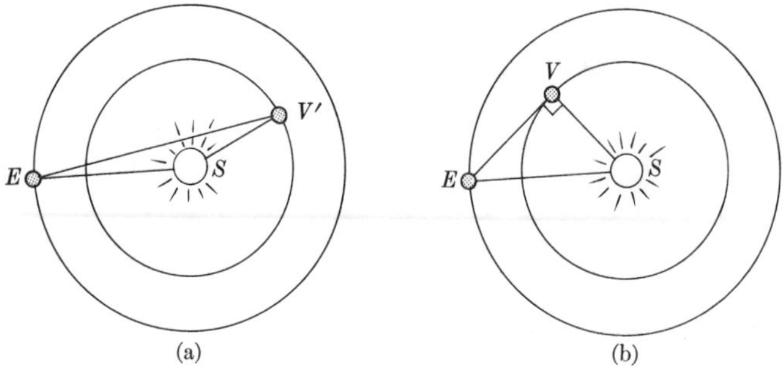

Fig. 7–23.
Determining the distance of Venus from the sun.

expect from the fact that when the moon is between the earth and the sun, the moon just about eclipses the sun.

We can find the distances to the moon and sun and the radii of the moon and sun by making measurements on the surface of the earth. But now suppose that we wish to calculate the distance from Venus to the sun. If we were to use the preceding methods we should have to make measurements on the surface of Venus. Of course, we all expect to be able to make the trip to Venus shortly, and can then make the measurements. In the meantime, to satisfy our curiosity, we shall employ a somewhat less direct method.

Let us regard all three bodies, the earth, the sun, and Venus, as points and let us suppose that the paths of earth and Venus are circular. At any time the three bodies are the vertices of triangle ESV' in Fig. 7–23(a). From the earth we can observe the size of angle E, which, of course, changes as the earth and Venus move around the sun.

A neat fact, which emerges from a study of Fig. 7–23(b), is that when angle E is a maximum, then the line from earth to Venus is tangent to the path of Venus around the sun. For when the angle at E is a maximum, the line from the earth to Venus is farthest from ES and still meets the circle on which Venus travels. But such a line must be tangent to the circle. A tangent to a circle is perpendicular to the radius drawn to the point of contact. Hence the radius SV (Fig. 7–23b) is perpendicular to EV. What we should do, then, is measure the angle E at various times of the year and find out when it is largest. At this time EV is perpendicular to SV.

Measurements show that the largest value of angle E is 47°. If in Fig. 7–23(b), we use 47° for E and the fact that angle V must then be a right angle, we have

$$\sin 47° = \frac{SV}{ES}.$$

From our tables we find that sin 47° = 0.7314. The distance ES is 93,000,000 miles. Hence

$$0.7314 = \frac{SV}{93{,}000{,}000}.$$

Then

$$SV = 93{,}000{,}000 \cdot 0.7314 = 68{,}000{,}000.$$

Thus the distance from Venus to the sun is 68,000,000 miles.

We can begin to see from these examples how Hipparchus and Ptolemy gave mankind its first reasonable values for the dimensions of our solar system. The figures they produced were staggering to the Greeks because these people believed that our solar system and universe were far smaller.

The crowning achievement of Hipparchus and Ptolemy was the creation of a new astronomical theory which described the paths of the heavenly bodies and enabled man to predict their positions. We shall consider their theory in the next chapter.

EXERCISES

1. Let us use the method given in the text to find the distance to the sun. We know that in Fig. 7–24, QE is the radius of the earth, or 4000 mi. The angle at E is the difference in longitude between P and Q and in our case is 89° 59′ 51″. Given that $\cos E = 0.000043$, find ES.

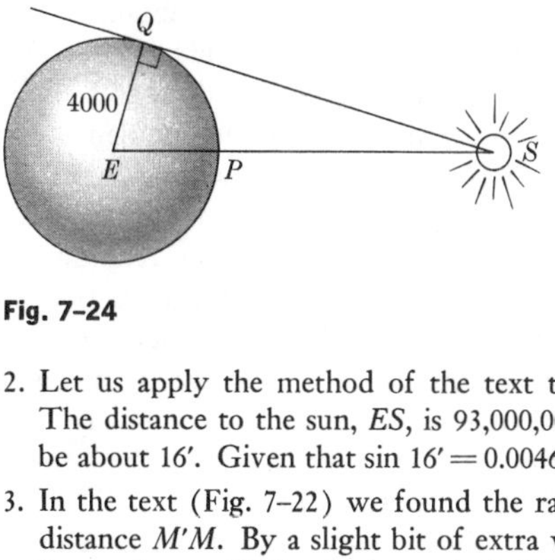

Fig. 7–24

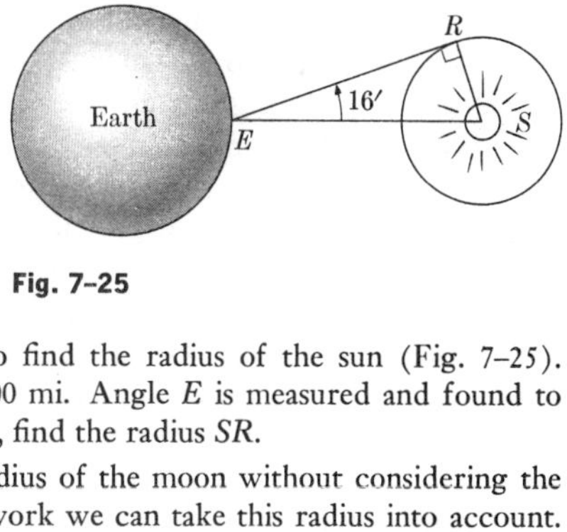

Fig. 7–25

2. Let us apply the method of the text to find the radius of the sun (Fig. 7–25). The distance to the sun, ES, is 93,000,000 mi. Angle E is measured and found to be about 16′. Given that sin 16′ = 0.0046, find the radius SR.
3. In the text (Fig. 7–22) we found the radius of the moon without considering the distance $M'M$. By a slight bit of extra work we can take this radius into account. Let us denote $M'M$, which equals RM, by r. Then, since EM' is 241,000 miles and angle E equals 15′, we have

$$\sin 15' = \frac{r}{241{,}000 + r}.$$

Use the value of sin 15′ given in the text to find r.

4. Use the method in the text to find the distance from Mercury to the sun. The relevant angle E in this case is 23°.

✱ 7-6 FURTHER PROGRESS IN THE STUDY OF LIGHT

We saw in the preceding chapter that Euclid had already formulated one basic law for the behavior of light, namely the law of reflection. The Alexandrians undertook to study a second basic phenomenon of light, namely the change in the direction of light as it passes from one medium to another.

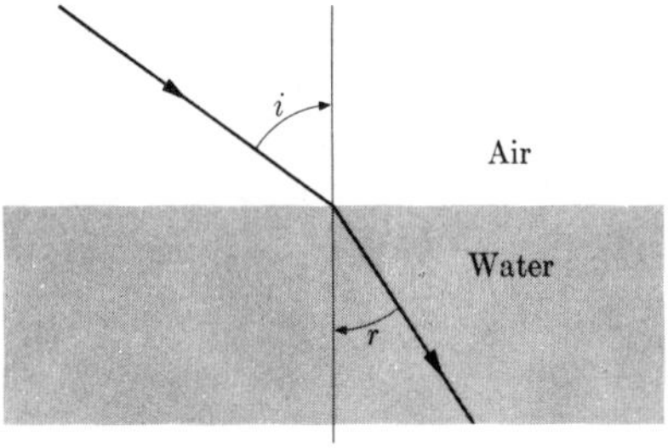

Fig. 7-26.
Refraction of light.

We often recognize that something strange does happen when light goes from air to water, say, because a straight rod when partially immersed in water seems to bend sharply at the water level. Also if one should shine a flashlight beam into water, he would observe the sudden change in the direction of the beam as it enters the water. This bending of light is called refraction. The Alexandrians sought to determine the extent of this change in direction. Specifically, if i is the angle (Fig. 7–26) which the direction of the incident light ray makes with the perpendicular to the surface which separates the two media, say air and water, and if r is the angle which the refracted light makes with this same perpendicular, then what the Alexandrians sought is the relationship between angle i and angle r. But the Alexandrians and Ptolemy, in particular, who worked very hard on this problem, were baffled. They did observe that as i increased, r increased, but the increase did not occur in any simple manner. Moreover, the r which corresponds to a given i is not the same for any two different media. Thus, if the first medium should be air, then for the same i, the value of r for glass would be different from that for water.

Ptolemy did not succeed in arriving at the correct law, but he developed the mathematical tool which finally enabled the Dutchman Willebrord Snell and the Frenchman René Descartes to discover and express it. It was found in the seventeenth century that light travels with a finite velocity and that this velocity is different in different media. Let us suppose that we have two media bordering each other as shown in Fig. 7–26, and let v_1 be the velocity of light in the upper medium and v_2 the velocity in the lower medium. Then Snell and

Descartes demonstrated by arguments we shall not reproduce that

$$\frac{\sin i}{\sin r} = \frac{v_1}{v_2}. \tag{1}$$

Thus it is the sine ratio which proves to be the key to this phenomenon of light. We see here, as we shall see many times later, how—as more mathematical ideas are placed at our disposal—we can take hold of more natural phenomena.

To familiarize ourselves with the law of refraction, we shall consider a concrete example. Let us suppose that the two media in question are air and water. The ratio of v_1 to v_2 in this case is 4 to 3. If we are then given a value of i, say 30°, we can find r. Since $\sin 30° = \frac{1}{2}$, it follows from (1) that

$$\frac{\frac{1}{2}}{\sin r} = \frac{4}{3}.$$

Therefore

$$\tfrac{1}{2} = \tfrac{4}{3} \sin r.$$

If we multiply both sides of this equation by $\frac{3}{4}$, we obtain

$$\sin r = \tfrac{3}{8} = 0.3750.$$

We have now but to find the angle whose sine value is 0.3750. From our table we see that $r = 22°$ to the nearest degree. Hence, as the light enters the water, the angle between its direction and the perpendicular will change from 30° to 22°.

We now know how light refracts. Can we put this knowledge to use? Let us assume that the sun is close to the horizon. Of course, light from the sun streams out in all directions, but some rays will travel horizontally. Suppose that the surface over which the light travels is the surface of a large body of water (Fig. 7–27). Then some rays will enter the water at a very large angle of incidence i, in fact, so close to 90° that we shall consider angle i to be 90°. The question we shall discuss is, How large is angle r in this case? To obtain an answer, we follow the procedure described in the preceding paragraph. This time, angle i is 90°, and sin 90° is 1. If we substitute this value in Formula (1), we have

$$\frac{1}{\sin r} = \tfrac{4}{3},$$

or

$$\sin r = \tfrac{3}{4}.$$

Reference to the table shows that $r = 49°$. Here the change in direction is

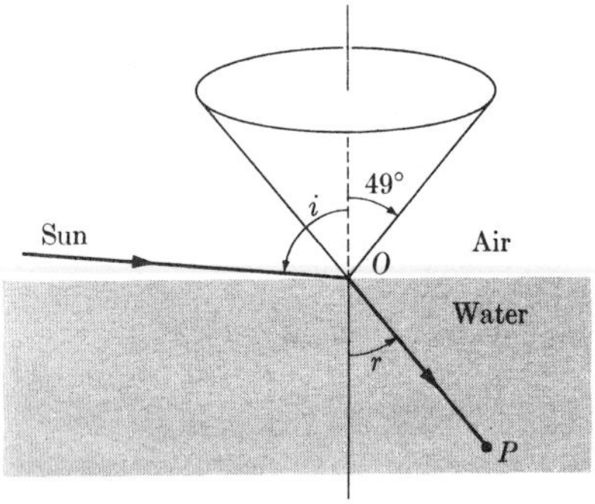

Fig. 7–27.
The fish-eye view of the world.

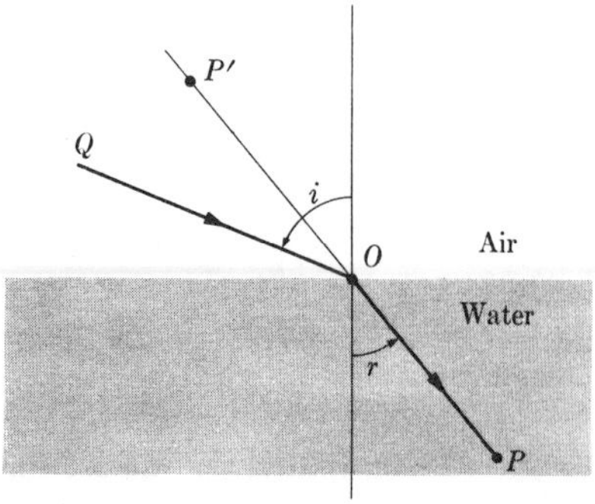

Fig. 7–28

considerable. Before we draw any further conclusions, let us note that 90° is the largest angle of incidence possible. If angle i should be less than 90°, angle r will be less than 49°. Hence any light entering the water will take a direction which makes an angle between 0° and 49° with the perpendicular.

Now suppose that a person located at the point P (Fig. 7–27) in the water sees the light ray OP coming toward him. Since the light travels along the direction OP, he will conclude that the sun is located above the water in the direction PO. Moreover, if light from any source enters the water at an angle of incidence less than 90°, the angle of refraction will be even less than 49°. In such cases a person in the water observing this light will conclude that the light comes from a source whose angle of incidence is less than 49°. The point of this discussion is that a person at P will believe that all objects in the air are situated within a 49°-angle from the perpendicular because the light from them will seem to come from a direction within this range. The region in the air extending in all directions within 49° of the perpendicular at O is the interior of a cone. Hence to a person in the water all objects seem to lie within this cone. We have of course presumed that the person in the water does not know the refractive effect of light or that he at least does not know how much light is bent under refraction. It is fairly certain that fish do not know mathematics, and so the inference that all objects above water must be within the 49°-cone around the perpendicular is called the fish-eye view of the world.

To be a little clearer about the possible error into which one may be led, suppose that light comes to a person in a submarine at P (Fig. 7–28) and that the direction of the light is OP. The object which emits the light will then appear to lie along the line PO. If, to hit the object, he shoots a bullet in the direction PO, the bullet will enter the air and follow the direction POP'. But the object at which he believes to be shooting is located along the direction OQ.

Let us now reverse the roles of air and water and let us suppose that the light originates in the water. Assume, in fact, that a beam of light is shot in the direction of *QO* of Fig. 7–29. The angle of incidence is now the angle marked *i* in the figure. The angle of refraction is the angle *r*. Since the ratio of the two velocities of light, that is v_1/v_2, is now $\frac{3}{4}$, the law of refraction (1) becomes

$$\frac{\sin i}{\sin r} = \frac{3}{4}$$

or

$$\sin r = \tfrac{4}{3} \sin i. \qquad (2)$$

We see that sin *r* is greater than sin *i* and that therefore *r* must be greater than *i*. This is as it should be because light in going from water to air will bend *away* from the perpendicular.

Let us now suppose that angle *i* is greater than 49°, say 60°, and let us seek to determine the corresponding angle of refraction, *r*. Then, knowing that sin 60° = 0.8660, we find by (2) that

$$\sin r = \tfrac{4}{3}(0.8660) = 1.155.$$

We see then that sin *r* is greater than 1. Unfortunately, there is no angle whose sine value is greater than 1, and so there is no angle of refraction. Thus mathematics predicts that the light cannot leave the water. Is this the case physically? Well, no decently behaving light ray would wish to disobey the mathematical law of refraction. And none does. The light remains in the water. But what does it do? The answer is that the light is *reflected* from the boundary between air and water. Since the light must return to the water, it may as well do what it has already learned how to do under the process of reflection, namely, be reflected at an angle equal to the angle of incidence which in the present case is 60° (Fig. 7–30).

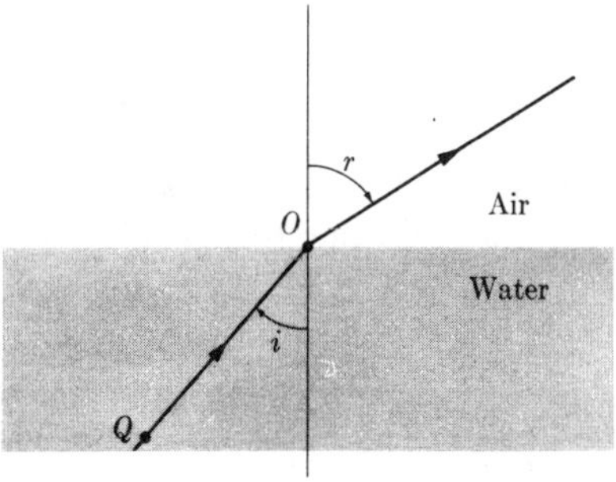

Fig. 7–29.
Refraction from water into air.

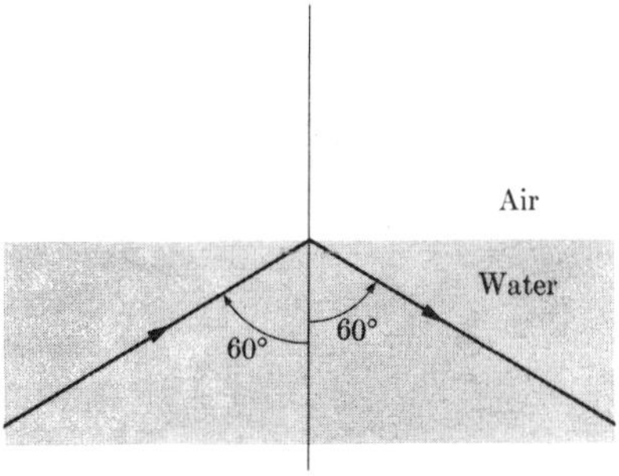

Fig. 7–30.
Total reflection.

Thus we have discovered that if light seeks to pass from one medium to a second one in which the velocity is greater, then for all angles larger than a certain angle (49° for water and air), the light is not refracted but reflected. This particular angle, the largest for which refraction is still possible, is called the *critical angle*, and the phenomenon that for all greater angles of incidence the light is reflected is called *total reflection*.

This phenomenon is indeed a surprising one. It means that a surface, such as the surface of the water in the above example, serves as a mirror for some angles of incidence. Now mirrors are very useful devices. Generally they are made by silvering the back of a glass plate. Would there be any use for the phenomenon of total reflection in view of the fact that here too we encounter a reflecting surface? As a matter of fact, the phenomenon is put to use in a number of familiar instruments.

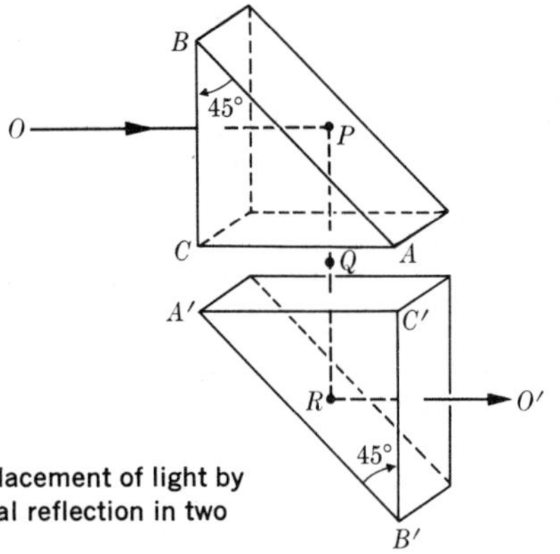

Fig. 7–31.
Parallel displacement of light by means of total reflection in two prisms.

Let us consider the following situation (Fig. 7–31), where *ACB* and *A′C′B′* are two prism-shaped pieces of glass with the faces *AC* and *A′C′* parallel to each other. Both prisms are shaped as isosceles right triangles. Suppose that *OP* is a light ray which first strikes the face *BC* perpendicularly. Here the angle of incidence is 0°. Hence the angle of refraction is also 0°, and the light therefore goes through unchanged in direction. The light ray strikes the face *BA* at an angle of 45°. Now if the prisms are made of flint glass, the critical angle is 37°. Thus the light ray strikes the face *BA* at an angle of incidence greater than the critical angle. In accordance with the phenomenon of total reflection, the light is reflected at an angle of 45° to *AB* and follows the direction *PQ*. The light strikes the faces *AC* and *A′C′* at an angle of incidence of 0° and so goes right through unchanged. It then strikes the face *A′B′* at an angle of 45°. Since this angle of incidence also is greater than the critical angle, the light is again totally reflected at an angle of 45° with *B′A′*

and takes the direction RO'. Thus the final ray, RO', has the same direction as the original ray, OP, but is displaced by the distance PR.

We might well ask, Does this combination of prisms have any practical value? One application is the periscope. The two prisms are at opposite ends of a long vertical tube. Now OP is the light received above water and RO' is the light received below. One could very well use two silvered mirrors at BA and $A'B'$ and obtain the same result. But silvered mirrors tarnish with age and lose their effectiveness. Moreover, well-made glass prisms reflect almost all the light that falls on a face such as BA, whereas a silvered mirror reflects only about 70% of the incident light; the rest is absorbed or scattered in all directions. Hence the prism not only outlasts the silvered mirror but is much more efficient.

Another application of the above combination of two prisms is made in binoculars. The two tubes which first receive the light are deliberately placed rather far apart so that the field of vision is large. But the eye pieces of the binoculars cannot be farther apart than the distance between a person's eyes. In each half of a binocular, the incident light is displaced as OP is displaced to RO'. Then the two incoming rays, one in each of the main tubes, can be far apart, whereas the two emerging rays are no farther apart than the eyes of a person.

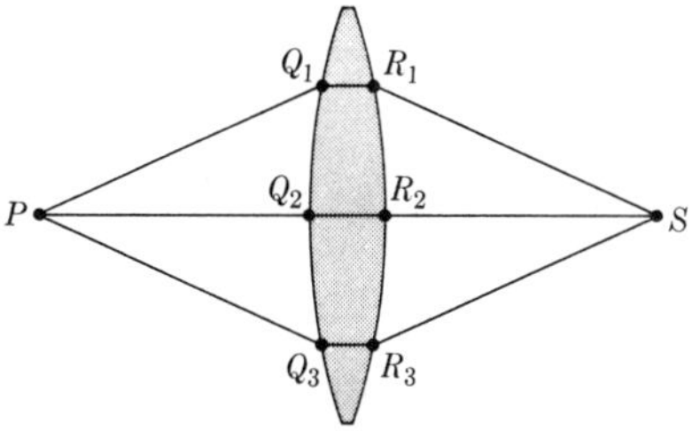

Fig. 7–32. Refraction by a lens.

Total reflection is but one phenomenon of the refractive effect of light. The most common use of the refractive effect of light is in lenses. If light streams out from an object at P (Fig. 7–32) in all directions, some of the rays will strike the lens at points such as Q_1, Q_2, and Q_3. There their directions will change because they are entering glass. Thus the rays PQ_1, PQ_2, and PQ_3 may take the directions Q_1R_1, Q_2R_2, and Q_3R_3, respectively. At the right-hand surface of the glass, the rays re-enter the air and, since the medium in which the light is traveling changes, the light rays bend again. By properly shaping the lens surfaces, that is $Q_1Q_2Q_3$ on the left and $R_1R_2R_3$ on the right, the light from P may be made to concentrate at S. All optical instruments, such as telescopes, microscopes, binoculars, and cameras, contain lenses of this kind.

The eye itself is a complicated refracting device. When light enters the eye (Fig. 7–33), it passes through a liquid (denoted by A in the figure), called the aqueous humor, then through the lens, *L*, which is made of a fibrous jelly, and finally it enters another liquid, *V*, called the vitreous humor. Although all three media have some refractive effect upon the light, most of the refraction occurs when it encounters the aqueous humor. To be perceived, light rays that enter the eye must strike the retina, *R*, in the rear. The eye has a ciliary muscle which changes the shape of the lens and therefore the direction of the light rays passing through the eye so that the rays are directed toward the retina. Eyes which for one reason or another cannot direct the rays to the retina must be aided by additional lenses in eyeglasses. Clearly the science of medicine profits immensely from the mathematical and physical knowledge acquired about the action of the eye.

In the camera, the lens or lenses are fixed in shape. The film acts as does the retina in the eye. Since the shapes of the lenses are fixed, the distances of the lenses from the film can be varied to enable the refracted light to reach the proper places on the film.

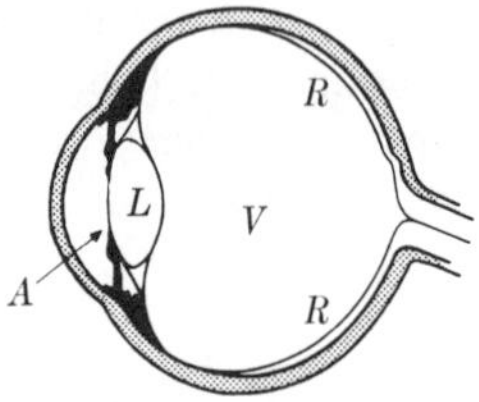

Fig. 7–33.
A sketch of the eye.

We have been discussing the law of refraction and some of the remarkable effects which take place at a sharp boundary between the two media. But the refractive effect of light is equally striking and important when there is a gradual change in the nature of the medium through which the light passes. Let us consider the passage of light through air, which is not a uniform medium. Generally it is more dense near the ground and thinner at higher altitudes. Hence, when light comes to a person at *P* (Fig. 1–1) from the sun at *O*, the light ray follows a curved path as it travels through the earth's atmosphere because it is continually refracted. For the observer at *P* the direction of the incoming light is *O′P*, and hence he thinks that the source lies along the direction *PO′*. This is the reason that we are often deceived about the true position of the sun (see Chapter 1).

The refractive effect of light is, as we can see, a peculiar phenomenon. Why does light behave this way? We do not understand what light is and so cannot analyze the substance itself to learn why it refracts, but we have another kind of explanation which sheds light on nature's operations. The clue lies in the law of refraction. We note that refraction depends upon the velocity of light in the medium. The seventeenth-century mathematician Pierre de Fermat, whom we shall meet again, pondered on this fact and, after analyzing the law of refraction, found an important principle. Suppose that

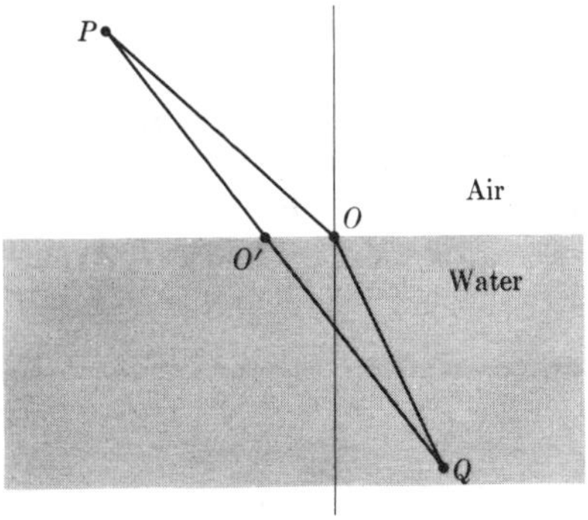

Fig. 7–34.
Light takes the path requiring least time.

light travels from the point P in air (Fig. 7–34) to the point Q in water and bends at O in accordance with the law of refraction. Were the light to follow the straight-line path from P to Q instead of the broken-line path POQ, it would travel a shorter distance. Let us note, however, that the distance $O'Q$ in water would be longer than OQ. Because the velocity in water is smaller than in air, the light might lose more time in traveling the path $O'Q$ instead of OQ than it might save by traveling the shorter distance PO' instead of PO in air. By a mathematical argument Fermat showed that light takes the path which requires *least time*.

But is this fact true for other phenomena of light? When light travels from one point to another in a uniform medium, it takes the straight-line path. It would seem as though in this case light chooses the criterion of shortest path and not that of least time. But in a uniform medium the velocity of light is constant, and so the shortest path requires the least time. Let us consider next what happens when light goes from a point P to a mirror and then to a point Q. We proved in Chapter 6 that light takes the shortest path. But here too the light travels in one medium and, because the medium is uniform, the velocity is constant; hence the shortest path again means least time. It would appear from Fermat's analysis that nature is wise. It knows mathematics and employs it in the interest of economy.

We have gotten a little ahead of our story by presenting the mathematical law of refraction and Fermat's analysis of the deeper implications of this law. The Alexandrian Greeks had grappled with the phenomenon of refraction and, as we noted earlier, supplied the key in the concept of the trigonometric ratios, but did not attain the law itself or see its meaning in terms of least time. But by providing these ratios and by charting the earth and heavens, the Alexandrians extended enormously man's mathematical understanding of the physical world. The power of mathematics to describe and analyze nature's ways was advanced well beyond the stage at which Euclid and Apollonius had left it. The crowning achievement of the Alexandrians is yet to be related.

EXERCISES

1. Given that the ratio of the velocity of light in air to that in water is 4 to 3 and that the angle of incidence of a light ray originating in the air and striking the surface of the water is 45°, what is the angle of refraction?
2. Suppose a light ray traveling in glass strikes the boundary of the glass and seeks to enter the air beyond the boundary. The velocity of light in the glass is two-thirds of its velocity in air. What angles of incidence can the light ray have and still penetrate into the air?

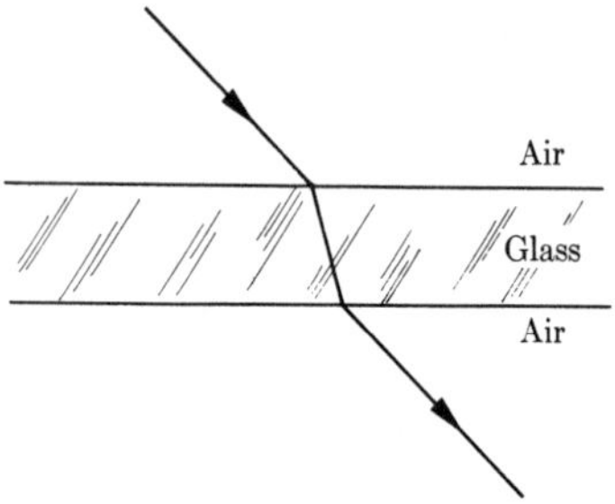

Fig. 7–35

3. Prove that a light ray passing through a plate of glass (Fig. 7–35) emerges parallel to its original direction but is somewhat displaced.
4. Suppose that one measures the angle of incidence, i, and the angle of refraction, r, for a light ray passing from air into a plate of glass, and assume that angle i proves to be 50° and angle r, 45°. The velocity of light in air is 186,000 mi/sec. What is the velocity of light in the glass?
5. What is the mathematical theme of this chapter?
6. Is it correct to say that the trigonometry of the Alexandrian Greeks is an extension of Euclidean geometry?
7. Contrast the classical and the Alexandrian Greek activities in mathematics.

REVIEW EXERCISES

1. Use the Table of Trigonometric Ratios to find the angle
   a) whose sine is 0.3256,
   b) whose tangent is 0.5317,
   c) whose cosine is 0.3256,
   d) whose tangent is 1.8807.
2. It is possible to find the sine, cosine, and tangent of 45° in somewhat the same manner as we found the corresponding values of 30° and 60°. Take a right triangle whose arms are each 1. Calculate the length of the hypotenuse by means of the Pythagorean theorem. Now write the values of sin 45°, cos 45°, and tan 45°.

3. Find the sine, cosine, and tangent of the acute angle A of a right triangle
   a) when the opposite side is 5 and the hypotenuse is 13,
   b) when the opposite side is 12 and the adjacent side is 5,
   c) when the opposite side is $\sqrt{3}$ and the adjacent side is 2,
   d) when the opposite side is $\sqrt{3}$ and the adjacent side is $\sqrt{6}$,
   e) when the opposite side is 1 and the hypotenuse is $\sqrt{10}$.
4. If $\sin A = \frac{3}{5}$, find $\cos A$ and $\tan A$.
5. If $\cos A = \frac{1}{2}$, find $\sin A$ and $\tan A$.
6. If $\tan A = \frac{3}{5}$, find $\sin A$ and $\cos A$.
7. To find the width AB of a river, a line segment AC perpendicular to AB is measured along one bank and found to be 100 ft. By sighting along CA and CB, the angle ACB is found to be 40°. How wide is the river?
8. The shadow on the horizontal ground of a vertical pole is 15 ft. At the end of the shadow the angle between the horizontal and the line of sight to the top of the pole is 20°. Find the height of the pole.
9. A wire 60 ft long reaches from the top of a 40-ft pole to the ground. What angle does the wire make with the pole?
10. From the top of a lighthouse 60 ft high, the angle between the vertical and the line of sight to a ship at sea is 35°. How far is the ship from the foot of the lighthouse?
11. An observer in an airplane 2000 ft directly above a gun observes that the angle between his vertical and the line of sight to an enemy target is 50°. How far is the target from the gun?
12. Find the radius and the circumference of the circle of latitude 23° north.
13. Suppose a man changes his longitude by 5° while traveling along the circle of latitude 23° north. How far does he travel?
14. Find the radius of the circle of latitude 67° north.
15. Suppose a light ray traveling in air strikes the water at an angle of incidence of 45°. What is the angle of refraction?
16. A ray of light starts from a point P in water and strikes the surface at an angle of incidence of 30° and emerges into air. What is the angle of refraction of the light ray?

Topics for Further Investigation

1. The mathematics of lenses. Use the references to Taylor, or to Sears and Zemansky, or look up any elementary physics book.
2. The mathematics of map-making. Use the references to Brown, Raisz, Deetz, or Chamberlin.
3. The history of mathematics during the Alexandrian period. Use the references to Smith, Ball, Eves, or Scott.
4. The creation of trigonometry. Use the references to Aaboe.
5. The life and work of Archimedes. Use any history.

Recommended Reading

AABOE, ASGER: *Episodes from the Early History of Mathematics*, Chap. 4, Random House, New York, 1964.

BALL, W. W. ROUSE: *A Short Account of the History of Mathematics*, 4th ed., Chaps. 4 and 5, Dover Publications, Inc., New York, 1960.

BROWN, LLOYD A.: *The Story of Maps*, Little, Brown and Co., Boston, 1944.

CHAMBERLIN, WELLMAN: *The Round Earth on Flat Paper*, National Geographic Society, Washington, D.C., 1947.

DEETZ, CHARLES H. and OSCAR S. ADAMS: *Elements of Map Projection*, pp. 1–52. U.S. Department of Commerce, Special Publication No. 68, 1938.

GREENHOOD, DAVID: *Mapping*, The University of Chicago Press, Chicago, 1964.

HEATH, SIR THOMAS L.: *A Manual of Greek Mathematics*, Chap. 14, Dover Publications Inc., New York. 1963.

PARSONS, EDWARD A.: *The Alexandrian Library*, The Elsevier Press, Amsterdam, 1952.

RAISZ, E.: *General Cartography*, McGraw-Hill Book Co., New York, 1948.

SAWYER, W. W.: *Mathematician's Delight*, Chap. 13, Penguin Books, Harmondsworth, England, 1943.

SCOTT, J. F.: *A History of Mathematics*, Chap. 3, Taylor and Francis, Ltd., London, 1958.

SEARS, FRANCIS W. and MARK ZEMANSKY: *University Physics*, 3rd ed., Chaps. 39–43, Addison-Wesley Publishing Co., Inc., Reading, Mass., 1964.

SMITH, DAVID E.: *History of Mathematics*, Vol. I, Chap. 4, Dover Publications, Inc., New York, 1958.

TAYLOR, LLOYD W.: *Physics, The Pioneer Science*, pp. 442–470, Dover Publications, Inc., New York, 1959.

CHAPTER 8

THE MATHEMATICAL ORDER OF NATURE

Great men! elevated above the common standard of human nature, by discovering the laws which celestial occurrences obey, and by freeing the wretched mind of man from the fears which the eclipses inspired.

PLINY

8-1 THE GREEK CONCEPT OF NATURE

The Greeks, as we now know, molded the nature of mathematics, constructed Euclidean geometry and trigonometry, and applied their theoretical results to objects in space, to the behavior of light, to mapping the earth, and to determining the sizes and distances of heavenly bodies. But these extensive and magnificent achievements within mathematics proper and in its applications do not exhibit the full greatness of the Greek genius, and are indeed dwarfed by the Greeks' grand conception of the universe itself.

Possessed with insatiable curiosity and courage, they asked and answered the questions which occur to many, are tackled by few, and are resolved only by individuals of the highest intellectual caliber. Is there any plan underlying the workings of the entire universe? Are planets, men, animals, plants, light, and sound merely physical accidents or are they part of a grand plan? Because they were dreamers enough to arrive at new points of view, the Greeks fashioned a conception of the universe which has dominated all subsequent Western thought. They affirmed that nature is rationally and indeed mathematically designed. All phenomena apparent to the senses, from the motions of planets in the heavens to the stirrings of leaves on a tree, can be fitted into a precise, coherent, intelligible pattern. The Greeks were the first people with the audacity to conceive of such law and order in the welter of phenomena and the first with the genius to uncover a pattern to which nature conforms. They dared to ask for and they found a design underlying the greatest spectacle man beholds, the motion of the brilliant sun, the changing shapes of the many-hued moon, the piercing shafts of the planets, the broad panorama of lights from the canopy of stars, and the seemingly miraculous eclipses of the sun and moon.

8-2 PRE-GREEK AND GREEK VIEWS OF NATURE

To appreciate the originality and boldness of the steps which the Greeks took in this direction, one must compare their attitude with what preceded. To all pre-Greek civilizations and later ones which lay beyond the Greek pale, nature appeared arbitrary, capricious, mysterious, and even terrifying. The ancient Egyptians and Babylonians did note the periodic motions of the sun and moon. But the motions of the planets made no sense at all. These bodies moved with varying speeds at different times of the year; at times they stood still; and often they reversed their courses. They appeared and disappeared. The few regularities which were observed in these motions were beclouded by the many irregularities.

If these two ancient peoples had any expectation at all that the universe would continue to function in the future as it had in the past, it was because they believed that sun, moon, and planets were gods who would most likely behave in a gentlemanly and beneficent manner. In the complex actions of nature, they saw no glimpse of plan, order, or law. They scarcely dreamed of design and certainly conceived no embracing theories.

Even the Greeks of about 1000 B.C. accepted fanciful accounts of the universe, accounts which are found in Homer and Hesiod. There were many gods, each of whom played some role in the creation and maintenance of the universe. Indeed the names Jupiter, Saturn, Venus, Mercury, and Mars are merely the Roman names for the Greek gods, and the Greek names, such as Aphrodite for Venus and Hermes for Mercury, were replacements for Babylonian names. These gods not only determined but even intervened in the affairs of man.

Rather suddenly, or so at least our knowledge of history indicates, rational accounts of the structure of the universe and of the motions of heavenly bodies appeared in the Greek city of Miletus located in Ionia, a region of Asia Minor. There is the theory that the Miletans, far from home and therefore free of the tyranny of beliefs which a society imposes on its members and yet repelled by the strange doctrines they encountered among the peoples of the Near East, were propelled into thinking for themselves. Certainly from 600 B.C. onward rational views dominate the picture. These Greeks and their successors were the first to reveal the passionate desire for knowledge, the love of reason, and the conviction that nature not only is rational but that an examination of nature's ways would reveal the order inherent in the physical world. The new thesis is proclaimed by the Ionian Anaxagoras: "Reason rules the world." The early rational theories are crude from a modern standpoint, but the new outlook is evident.

The decisive step leading to the construction of precise and verifiable scientific theories in place of vague and largely speculative accounts was the involvement of mathematics. This step was made by the Pythagoreans. We have already noted the prepossession of these people with the concept of number, though admixed with mystical and religious doctrines. In their philosophy of nature the Pythagoreans began with the principle that number is the essence of all substance. Unlimited space furnishes the material for particular forms of matter. But to the Pythagoreans any form was a pattern of discrete points arranged, as small pebbles might be, to build up the form. Hence the forms reduced to numbers. Since number is the essence of any object, the explanation of natural phenomena could be achieved only through number.

The natural philosophy of the Pythagoreans is hardly very substantial. Aesthetic principles commingled with an obsession to find number relationships certainly led to assertions transcending observational evidence. Nor did the Pythagoreans develop any one branch of physical science very far. One can justifiably call their theories superficial. But whether by a lucky stroke or by intuitive genius the Pythagoreans did hit upon two doctrines which later proved to be all important. The first is that nature is built in accordance with mathematical principles, and the second that number relationships reveal the order in nature. They underlie and unify the seeming diversity exhibited by nature. The Pythagoreans said in fact that numbers and number relationships are the essence of nature. This statement will assume deeper meaning when we get to modern times.

Perhaps because mathematics developed considerably in the intervening century, the principle that nature is mathematically designed emerged more sharply and was applied more substantially in Plato's time. Plato was indeed a Pythagorean but a master in his own right who influenced Greek thought in a most important century, the fourth century B.C. He was the founder of an academy in Athens, a university which attracted the leading thinkers of his day and which, in fact, endured for nine hundred years.

Plato's own doctrines were extreme. Reality to him was not to be found in the physical world but in a system of ideas and in an ideal plan of the universe which God himself had created and contemplates. The visible and sensible world is just a vague, dim, and imperfect realization of these ideas. Moreover, the ideas were perfect and eternal, whereas the physical world is imperfect and decays. One might say that, unlike the Pythagoreans, Plato did not wish to comprehend the physical world through mathematics but aimed at understanding the mathematical plan itself which observation of the physical world suggested very imperfectly.

For example, Plato describes the real science of astronomy. The visible figures in the heavens are far inferior to the true objects, namely those objects that are to be apprehended by reason and mental conceptions. The varied configurations which the sky presents to the eye are to be used only as diagrams to assist in the study of higher truths. We must treat astronomy, like geometry, as a series of problems suggested by visible things. True astronomy deals with the laws of motion of true stars in a mathematical heaven of which the visible heaven is but an imperfect expression. True astronomy must leave the actual heavens alone. It is clear, incidentally, that Plato, like the classical Greeks in general, was indifferent to the practical problems of navigation, calendar reckoning, and the measurement of time.

Although the planets, at least as seen from the earth, do not appear to follow any regular course (the word "planet" means in fact "wanderer," and the planets were referred to as the vagabonds of the sky), Plato was sure—because "God eternally geometrizes"—that there was a mathematical pattern underlying and governing the motions of all heavenly bodies. Plato's own attempts to find such a plan were crude, largely because he would not devote himself to a careful study of the actual motions. But he did pose to his colleagues and students the problem of devising a mathematical scheme that would call for regular motions and yet account for the irregular motions we see, the problem he described as "saving the appearances."

8-3 GREEK ASTRONOMICAL THEORIES

One of Plato's pupils, Eudoxus (408–355 B.C.), who later became one of the most famous of Greek mathematicians, did take on this problem and, by creating the first major astronomical theory known to history, made one of the great and ingenious contributions to the demonstration of the mathematical design of nature. We shall not present the details of his theory. It was constructed before Hipparchus and Ptolemy calculated the sizes and distances of the heavenly bodies, and so Eudoxus did not have the data on which to build an accurate system. The defects in the theory were soon recognized.

The problem of finding the design of planetary motions continued to engage the minds of the Greeks, possibly because they were not distracted by the "heavenly" stars of stage, screen, and radio with whom many modern minds seem to be preoccupied. One of the solutions advanced but rejected is worthy of mention. Aristarchus, who lived about 270 B.C. and who had made many estimates of the sizes and distances of heavenly bodies, though with methods cruder than those developed later by Hipparchus and Ptolemy, proposed the theory that the planets move in circles about the sun. Aristarchus, to our knowledge, did not attempt to show that such a theory would fit the data known to his time. But the theory was not acceptable to his contemporaries and successors because it was totally at variance with Greek conceptions of the universe and Greek physics. For one thing, the Greeks already knew that simple circular motion would not do because the distance of the earth from the sun was known not to be constant. One piece of evidence was that the apparent diameter of the sun varied with the seasons. Another objection to Aristarchus' plan arose from the knowledge that the earth consisted of heavy matter; it was inconceivable that such a heavy body could be in motion. The planets, on the other hand, were supposed to be made of some light substance and so their motion was feasible. This distinction between the physical constitution of the earth and that of the planets was almost universally accepted up to the seventeenth century. Moreover, if the earth were in motion, why did objects on the earth not fall behind? Greek physics had no answer to this argument.

The supreme achievement of all Greek efforts aimed at exhibiting the mathematical design of the universe is the astronomical theory of Hipparchus and Ptolemy. These two men, as we noted in the preceding chapter, had created the mathematical method that enabled them to determine the sizes and distances of the sun, moon, and several planets, the method which, as Ptolemy put it, gave them the tool needed to base astronomy "on the incontrovertible ways of arithmetic and geometry." They also had older Egyptian and Babylonian observations at their disposal as well as innumerable others made by Hipparchus himself at Rhodes and by the observatory in Alexandria. They tackled the plan of organizing all this knowledge into one comprehensive scheme.

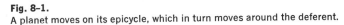

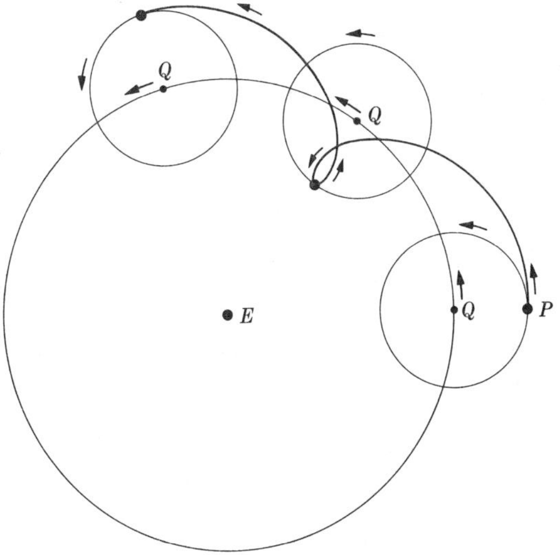

Fig. 8–1.
A planet moves on its epicycle, which in turn moves around the deferent.

In the astronomy of Hipparchus and Ptolemy, which we now refer to as the Ptolemaic theory, the earth is the center of the universe and stationary. To account for the motion of a planet P (Fig. 8–1), these men assumed that P moves at a constant speed along a circle whose center is Q. At the same time that P moves around Q, Q is supposed to be moving in a circle and at a constant speed around the earth, E. The circle on which P moves is called an epicycle, and the circle on which Q moves is called the deferent. Hipparchus and Ptolemy could, of course, choose the radii of the two circles and the speeds at which P and Q move on their respective circles so that the motion of P agreed with the observed positions of the particular planet. For each planet the choice of radii and speeds was different.

Actually the above scheme did not give these men enough latitude. Hence their astronomical system contained also some minor devices which enabled them to fit a system of such circles to the motion of any one heavenly body, but the essential principle is the use of deferent and epicycle. It should be noted that the motion of a planet as viewed from the earth is actually quite complicated and yet, by the above scheme, is readily understood in terms of a combination of circular motions. This theory accounted for planetary motions within the accuracy of observations attained in Alexandrian times. From the time of Hipparchus an eclipse of the moon could be predicted to within an hour or two. Predictions of the sun's motion were not so precise, but we must recall here a point made in the preceding chapter, namely, that calculations of the sun's distances at various times were not exact because the requisite angles were too small to be measured accurately.

The scheme we have just described is contained in Ptolemy's *Almagest*, the book mentioned earlier (Chapter 7). This theory was quantitatively so precise that it was accepted as the true design of the heavens until the work of Copernicus and Kepler displaced it. It is significant, however, that Ptolemy at least laid no claims to truth. He had constructed a mathematical scheme which accounted for the motions of the celestial bodies, a theory which worked, but he did not profess that God had so designed the universe. Unfortunately people's confidence in the *truth* of a doctrine increases with the length of time it holds sway, and since Ptolemaic theory was accepted for about 1500 years, people came to regard it as an absolute and unchallengeable truth. No other product of the entire Greek era rivals the *Almagest* in the profound influence it exerted on conceptions of the universe and none, except Euclid's *Elements*, achieved such unquestioned authority.

The theory of Hipparchus and Ptolemy is the final Greek answer to Plato's problem of rationalizing the appearances in the heavens and is the first really great scientific synthesis. Whereas the Greeks of the classical period were convinced on philosophical and intuitive grounds that nature was rationally designed, Ptolemaic theory provided overwhelming, concrete evidence.

8-4 THE EVIDENCE FOR THE MATHEMATICAL DESIGN OF NATURE

Let us look back for a moment to see the total evidence which the Greeks could muster for their momentous doctrine that nature is mathematically designed. The astronomical theory of Hipparchus and Ptolemy was certainly the most impressive evidence not only because it dealt with the grandest natural spectacle but because it showed design in a maze of phenomena whose outward appearances scarcely suggested design. To this achievement we must add Euclidean geometry. We have already pointed out the larger significance of this body of knowledge; it demonstrated that the shapes and sizes of earthly figures conform to a reasoned system of doctrines. One might very well prove on the basis of self-evident axioms and of reasoning that satisfies the mind that the sum of the angles of a triangle is 180°. But when one constructs triangle after triangle for various purposes and finds in every case that the sum is indeed 180°, one cannot escape the implication that this and the other theorems of Euclidean geometry express essential principles of nature. Moreover, because these principles are all part of one reasoned body of knowledge, it seems clear that nature is designed in accordance with a reasoned plan.

In the domains of light and sound (music), the progress made by the Greeks was not nearly so impressive, but they had produced the law of reflection and they did know and use the properties of curved mirrors to concentrate light. The Greeks were sure that further investigation would reveal additional laws, and almost every Greek mathematician worked on light. Many, among them Euclid, Archimedes, Apollonius, Heron, and Ptolemy, wrote mathematical books on the subject. The development of a mathematical theory of musical sounds was initiated by the Pythagoreans and, as in the case of light, pursued by many later Greeks.

The Greeks also applied mathematics to various other classes of natural phenomena and found the mathematical laws applicable. Archimedes wrote a still famous book on the mathematical laws of the lever. Another of his works investigated the weight and stability of various shapes placed in water. It was primarily motivated by the experience that a ship whose shape is not well chosen may readily overturn in water. Still another study dealt with the centers of gravity of various shapes, an important bit of knowledge if bodies are to be balanced or remain upright.

Phenomena of motion were also studied by the Greeks. Here too they adopted what seemed to be self-evident principles and made deductions which fitted their limited experience. In the Aristotelian theory of matter all objects were composed of lightness, heaviness, wetness, and dryness. Those in which lightness dominated (for example, fire) always sought to rise. Those in which heaviness dominated (for example, metals) sought to fall. Every object had a natural place and, when not hindered, sought it. Thus the natural place of light objects was a region near the moon, whereas heavy objects tended to congregate at the center of the universe which was, of course, the center of the earth. Force is required to set an object in motion, and a measure of this force was the product of the weight and the velocity given to the body. Also, a force must constantly be applied to keep a body in motion or else the motion would cease. Forces are transmitted by material agents. Thus one body must strike another to transmit motion to the latter.* The Greeks made progress in other scientific fields such as geography and geodesy which we discussed somewhat in the preceding chapter.

In all of the fields discussed above mathematics was, at the very least, considerably involved. In fact, in the classical period mathematics *meant* arithmetic, geometry, astronomy, and music and, by the end of the Alexandrian period, it had come to mean, in addition, mechanics (motion, the lever, the hydrostatics of Archimedes), optics, geodesy, and logistics (practical arithmetic).

From these scientific investigations one major fact stood forth: the universe *is* mathematically designed. Mathematics is immanent in nature; it is the truth about its structure, or, as Plato would have it, the reality of the physical world. Moreover, human reason could penetrate the divine plan and reveal the mathematical structure of nature. Almost all of the mathematical and scientific research which has taken place since Greek times has been inspired by the conviction that there is law and order in the universe, and that mathematics is the key to this order.

The Greek miracle has not been rivaled, not even by our modern civilization. A relative handful of people produced in a few hundred years supreme works not only in mathematics and science but in literature, art, music, logic, and in many branches of philosophy.

8-5 THE DESTRUCTION OF THE GREEK WORLD

It is accurate to say of the Greeks that God proposed them but man disposed of them. We have already related in Chapter 2 that the Romans conquered the Greek lands and that Roman practicality affected adversely the theoretical studies in Alexandria. We have also mentioned the rise of Christianity and that the Christian reaction to Roman persecution was to condemn and forbid all pagan learning, though, of course, the new religion did absorb some Greek philosophic doctrines, notably Aristotle's. The destruction of what remained at Alexandria, Christian and pagan, was completed by the Mohammedans. The Arabs had been inspired by Mohammed to adopt a new religion. Mohammed died in 632 A.D., but his successors undertook to convert the world by the sword. They conquered Alexandria in 646 and burned the Museum on the ground that if the books there contained anything contrary to the teachings of Mohammed, they were wrong, and if in agreement, superfluous. With this stroke the dusk settled on Alexandria.

Although the Museum was destroyed and the scholars dispersed, Greek learning did ultimately become an integral part of European civilization and

* See also Section 13-5.

culture. Just how the Greek creations found a new home in Western Europe through one of the quirks of history has already been indicated briefly in Chapter 2, and we shall say more about it in later chapters.

EXERCISES

1. What essential differences can you find between the pre-Greek and the Ptolemaic view of the heavens?
2. What is the Pythagorean doctrine concerning the essence of reality?
3. What is the meaning of the statement that Ptolemaic theory is a geocentric theory?
4. Describe the basic idea in Ptolemaic theory.
5. Suppose a planet moves on an epicycle at twice the speed with which the center of the epicycle moves on the deferent. Suppose, further, that the radius of the deferent is three times the radius of the epicycle. Sketch the path of the planet around the earth.
6. What is meant by the rationality of nature?
7. How does Ptolemaic theory support the belief in the mathematical design of nature?
8. How does Euclidean geometry tend to establish the mathematical design of nature?

Topics for Further Investigation

1. The mathematical doctrines of the Pythagoreans. Use the references on the history of mathematics in Chapter 7.
2. The accomplishments of Greek physical science.
3. The astronomical theory of Eudoxus.
4. The astronomical theory of Aristarchus.
5. The astronomical theory of Ptolemy.
6. Pre-Greek views of the universe. Use Dreyer in the references below.

Recommended Reading

CLAGETT, MARSHALL: *Greek Science in Antiquity*, Abelard-Schuman, Inc., New York, 1955.

DAMPIER-WHETHAM, WM. C. D.: *A History of Science*, Chap. 1, Cambridge University Press, Cambridge, 1929.

DREYER, J. L. E.: *A History of Astronomy*, 2nd ed., Chaps. 1 through 9, Dover Publications, Inc., New York, 1953.

FARRINGTON, BENJAMIN: *Greek Science*, 2 vols., Penguin Books, Harmondsworth, England, 1944 and 1949.

JEANS, SIR JAMES: *The Growth of Physical Science*, 2nd ed., Chaps. 1 through 3, Cambridge University Press, Cambridge, 1951.

JEANS, SIR JAMES: *Science and Music*, pp. 160–190, Cambridge University Press, Cambridge, 1947.

KUHN, THOMAS S.: *The Copernican Revolution*, Chaps. 1 through 3, Harvard University Press, Cambridge, 1957.

SAMBURSKY, S.: *The Physical World of the Greeks*, Routledge and Kegan Paul, London, 1956.

SARTON, GEORGE: *A History of Science*, Vols. I and II, Harvard University Press, Cambridge, 1952 and 1959.

SINGER, CHARLES: *A Short History of Science*, Chaps. 1 through 4, Oxford University Press, London, 1953.

CHAPTER 9*

THE AWAKENING OF EUROPE

Solicit not thy thoughts with matters hid,
Leave them to God, Him serve and fear.
. . . . be lowly wise;
Think only what concerns thee and thy being.

JOHN MILTON

9-1 THE MEDIEVAL CIVILIZATION OF EUROPE

It is perhaps a comfort after reading about the destruction of the Greek civilization to turn to a new one—the civilization of western Europe. We know that Europe did acquire the Greek creations and built upon them a vast, scientifically oriented civilization. How did this come about? To answer this question and to understand the special nature of subsequent developments in Europe, we must note a few historical facts.

The Germanic tribes, who have occupied western and central Europe as far back as history goes and who are the forefathers of most Americans, were barbarians. We know very little about their early history because they had no writing and hence no records were kept. From Roman historians, notably Tacitus (first century A.D.), we know that the Germanic tribes possessed a very primitive civilization. Tacitus describes them as honest, hospitable, hard-drinking, hating peace, and proud of the loyalty of their wives. Their dwellings were huts of timber and straw located in woods and surrounded by crude fortifications. Animal skins and coarse linens served for clothes, while herds of cattle, hunting, and the cultivation of grain crops provided food. Industry was unknown; just enough iron was mined to provide crude weapons. Trade was effected through barter and supplemented by plundering other tribes and more civilized regions. There were no arts, no science, and no learning. The chief activities were eating, sleeping, carousing, and fighting other tribes. Since such activities are also characteristic of peoples we call civilized, we may say that to that extent the Germanic tribes were civilized.

Although the Romans won many battles with the Germanic tribes, the Empire grew weaker for a variety of reasons which we cannot survey here, and the barbarians finally conquered it. Barbarians became kings of Rome and what was left of the Empire. Only a small region around Constantinople, which we call the Eastern Roman or Byzantine Empire, managed to remain independent and isolated. The Eastern Roman Empire, incidentally, also withstood the Mohammedans, who in the seventh century conquered Egypt, the Near East, and the lands bordering the Mediterranean Sea.

By the time that the Roman Empire collapsed in the fifth century A.D., the Catholic Church had become a strong organization with good leadership. It gradually converted the heathens to Christianity, established schools in Europe, and taught reading, writing, and ethics. Moreover, it perpetuated and imposed the legal and political organization of Rome. The Christian influence was certainly beneficial in that it produced a more stable state of affairs and even induced the barbarians to remain at peace for longer periods of time, a restraining influence which the barbarians did not resent because they soon learned that civilization had its advantages. With a little thought they found that peaceful interludes permitted them to develop methods of mass destruction and so do as much killing at intervals as previously in constant warfare.

Cities and small states governed by powerful leaders were established in Europe. Trade between cities developed, producing the wealth necessary to support scholarship. But study was almost entirely confined to understanding the word of God as fostered, expounded, and dictated by the Fathers of the Church. Those Greek works which had survived destruction by Romans, Christians, and Mohammedans lay almost unnoticed in neglected public buildings, in private libraries, or in the isolated, beleaguered Eastern Roman Empire.

What little knowledge of nature was deemed necessary in the life prescribed by the Church was derivable, so the Christian leaders said, from the Bible. St. Augustine (354–430), a man learned in Greek and Christian thought, even declared that the authority of the Scriptures is greater than the capacity of the human mind. Unfortunately the Biblical statements about the nature and structure of the physical world are of Babylonian origin and hence decidedly inferior to the knowledge acquired by the Greeks.

Of course, some of the actual phenomena of nature were observed, and questions raised about them. The medieval intellectuals who pursued such matters offered a kind of explanation which is satisfying to some minds. They believed that natural processes were mainly means to an end, i.e., they adopted what is called a teleological viewpoint. Thus rain existed to nourish the crops. Crops and animals existed to provide food for man. Sickness was a punishment from God. Plagues and earthquakes were expressions of God's anger. In general, all explanations focused on the phenomenon's value to, or effect on, man. Man was the center of the universe not merely geographically but also in terms of the ultimate purposes served by nature.

Although nature existed to serve man, man himself existed on this earth only to serve an apprenticeship during which he prepared his soul for a life in heaven with God—or elsewhere. Life on earth was but an unimportant prelude, to be endured but not enjoyed. To prepare for the afterlife man had to wrest his soul from a stubborn flesh which was guilty of original sin. Participation in the bounty of nature, food, clothing, and sex, tainted the soul and so had to be severely restricted. Medieval man, certain of his sins and doubtful of salvation, had to bend all his efforts to attain redemption. By earning divine grace man could escape from this foul earth to the divine empyrean.*

9-2 MATHEMATICS IN THE MEDIEVAL PERIOD

We see that a new civilization did arise in Europe, but from the standpoint of the perpetuation of mathematical learning or the creation of mathematics, it was totally ineffective. Although this civilization did spread ethical teachings, fostered Gothic architecture and great religious paintings, no scientific, technical, or mathematical concept gained any foothold. In none of the civilizations which have contributed to the modern age was mathematical learning reduced to so low a level.

Superficially mathematics did seem to play an important role. In the medieval schools the standard curriculum consisted of seven subjects, the quadrivium and the trivium. The quadrivium comprised arithmetic, the science of pure numbers; music as an application of numbers; geometry, or the study of magnitudes such as length, area, and volume at rest; and astronomy, the study of magnitudes in motion. But the scope of these studies was terribly limited. Even the first universities of Europe, which began to function about 1100 A.D., offered merely a minimum of arithmetic and geometry. Arithmetic consisted of simple calculations mingled with complex superstitions. Geometry was confined to the first part of Euclid, far less than we learn in high-school courses today. The most advanced point reached in some of these institutions of learning was the very elementary theorem that the base angles of an isosceles triangle are equal.

The little mathematics kept alive in the schools served various purposes in the medieval period. Some astronomy was pursued to keep the calendar. Here a minimum of arithmetic and geometry sufficed for the accuracy needed, just as it did in ancient Egypt and Babylonia. This work was usually performed by monks because the clergy was the most learned class. Astronomy and therefore elementary mathematics played a larger role in medieval life in that they provided the factual information needed for astrology, which was regarded as a science.

One more medieval use of mathematics is worthy of mention. Plato's belief that the study of mathematics trains the mind for philosophy was taken over by the Church which, however, substituted theology for philosophy.

---

* A term of medieval cosmology referring to the "highest heaven" or paradise.

Clearly the interest here was not in mathematics as such but as a preparation for grasping the subtle reasoning which the Church employed to build and strengthen the foundations of religious doctrines.

9-3 REVOLUTIONARY INFLUENCES IN EUROPE

Whether or not the civilization of medieval Europe might in due time have given rise to mathematical activity will never be known. But dramatic changes, largely initiated by non-European forces, drastically altered the Christian world. The earliest influence tending to transform thought and life in medieval Europe may be credited to the Arabs. While the Church was gradually civilizing the European barbarians and establishing the Christian way of life, the Arabs, perhaps more ruthless in proselytizing and certainly more dynamic and aggressive, succeeded in establishing their own civilization and culture in southern Europe, North Africa, and the Near East. Though fanatic in the advancement of their own religion, once their empire was stabilized, the Arabs displayed great tolerance toward alien ideas and learning, readily absorbed the mathematics and science of the Greeks and Hindus, and built cultural centers in Spain and the Near East. They translated the Greek works into Arabic and added commentaries and contributions of their own to mathematics, astronomy, medicine, optics, meteorology, and science in general.

By about 1100 A.D. Europeans were trading freely with Arabs. The Crusades, which attempted to wrest Palestine from the Arabs, brought further contacts between Christians and Moslems. Through these channels the Europeans became aware of the Greek works and Arab additions. They were so fascinated by this material that they aroused themselves to acquire it. Wealthy merchants, princes, and popes sent agents to the Arab centers to purchase manuscripts. Many Europeans went to live in Spain and learned Arabic in order to read the works and translate them into Latin. Others were assisted by Jewish and Arab scholars in making the translations. Plato, Aristotle, Euclid, Ptolemy, and the Greek literary works were avidly grasped.

In the fifteenth century Italy made new contacts with the Greek heritage. Ambassadors from Constantinople, the capital of the Eastern Roman empire, which still possessed the largest collection of ancient manuscripts, came to Italy several times in the first half of the fifteenth century, largely to seek help against the Turks. The Italians learned about the Greek works and like the Europeans of three centuries earlier, sought eagerly to possess them. In addition, some Greek scholars discouraged by the poverty in Eastern Europe and Alexandria migrated to Italy. When the Turks finally captured Constantinople in 1453, a flood of these men bringing their manuscripts with them came to Italy.

By financing the geographical explorations of the fifteenth and sixteenth centuries, which were intended to discover new trade routes, the merchants affected the life of Europe. The discovery of America and of a route to China around Africa resulted in acquainting Europe with strange lands, beliefs, customs, religions, and ethical doctrines. Catholics met Mohammedans, Chinese, and the American Indian. To the broadening influence of trade itself was added knowledge which conflicted sharply with the doctrines and way of life hitherto accepted in Europe. Questioning of the accepted doctrines and values ensued.

The merchant class and the large classes of artisans and free laborers introduced new interests. Employers and employees sought material gain, and so looked for commodities, machinery, and natural phenomena which might be employed to advantage. The rulers of the Italian cities and states also spurred on these interests. They coveted power and magnificence and, to acquire the necessary wealth, favored trade, industries, and inventions. The cities competed to surpass one another in skills, devices, and quality of merchandise. These groups, though selfishly motivated, were nevertheless effective in orienting the civilization toward the physical world and in fostering the accumulation of empirical knowledge.

The Protestant Revolution, or the Reformation as it is called, also upset the old culture in Europe. We are not concerned here with justification of the break from the Church. But Luther fanned the fire of discontent which had spread throughout Europe. Disputes about the nature of the sacrament, the validity of the control of the Church by Rome, and the meaning of passages of the Scriptures raised doubts in many people, who were thus emboldened to turn to other sources of knowledge, notably the physical world itself.

Several discoveries and inventions of the late medieval period had effects far greater than one might at first expect. In the twelfth century the Europeans learned from the Chinese about the compass. The introduction of the compass was important because it was an immense aid to navigators on long sea voyages. The explorers who dared the Atlantic might not have been willing to do so without it.

The introduction of gunpowder in the thirteenth century produced as its most obvious effects changes in methods of warfare and the design of fortifications. It also introduced a new physical problem, the motion of projectiles. An indirect result was the granting of more power to the common man because with a musket he could be effective in warfare. Previously only those who could afford expensive armor, that is, the wealthy nobles, could wield military power.

The invention of printing (about 1450) was immensely important in helping to spread Greek knowledge across Europe. Another invention, paper made of cotton and later of rags, which replaced costly parchment, also helped to make books plentiful and cheap. Many editions and translations of Greek works were printed in the century following these inventions. They helped to bridge the gulf between the learned and the untutored just at the time when great numbers were seeking to obtain knowledge.

Advances in the subject of optics had a vast effect on future scientific activity. The first was the discovery made in the thirteenth century that lenses can be used to magnify objects and thus aid in the examination of materials and natural phenomena. Lens grinders began to produce spectacles. Early in the seventeenth century, two of them discovered that a pair of lenses held at some distance from each other could be used to make distant objects seem close. Thus the telescope became available and was immediately applied to astronomy with results we shall describe later. At about the same time, it was found that a combination of lenses would do even better than a single lens to magnify nearby objects, and the microscope was invented. The investigation of the biological world and the revelation of hitherto unsuspected small-scale phenomena soon followed.

9-4 NEW DOCTRINES OF THE RENAISSANCE

It was to be expected that the insular world of medieval Europe accustomed for centuries to one rigid, dogmatic system of thought would be shocked and aroused by the series of events we have just described. The European world was in revolt. As John Donne put it, "All in pieces, all coherence gone." Europe revolted against scholastic domination of thought, rigid authority, and restrictions on the physical life. It revolted against the Scriptures as the source of all knowledge and the authority for all assertions. It revolted against enforced conformity to the established canons of conduct.

A leading figure in the revolt from the old modes of thought is Leonardo da Vinci (1452–1519). Because he saw how most scholars accepted as authoritative all that they read, he distrusted the men who took their learning only from books and professed their knowledge so dogmatically. He describes them as puffed up and pompous, strutting about, and adorned only by the labors of others whom they merely repeated. These were only the reciters and trumpeters of other people's learning. Leonardo determined to learn for himself and made exhaustive studies of plants, animals, the human body, light, the principles of mechanical devices, rocks, the flight of birds, and hundreds of other subjects. Although he is most often remembered as one of the great masters of painting, he also was a psychologist, linguist, botanist, zoologist, anatomist, geologist, musician, sculptor, architect, and engineer.

Many scholars turned to exhaustive studies of the Greek authors, to translations, and to compilations. They gave to these works the same infinitely detailed and critical attention that they and others had formerly given to biblical documents. The writings of Luca Pacioli (1445–1514) show this tendency. He was a monk, who in 1499 published *Summa de Arithmetica, Geometrica, Proportione et Proportionalita*. As a full, almost encyclopedic, account of the mathematical knowledge available to Europe by 1500, it was enormously helpful.

More interesting as a transitional figure is Jerome Cardan whom we met in Chapter 5. He wrote a great number of works which exhibit a critical attitude only in the sense that he traced the origins of stories, miracles, and "facts" to the authorities. However, he accepted freely any number of medieval superstitions, legends, accounts of supernatural events, pseudo-sciences, and even magical medical treatments. He believed in the significance of dreams, ghosts, portents, palmistry, and astrology, which to him were sciences. He also wrote volumes on moral aphorisms and on the varieties of beings and bodies which fill the universe. Among these were spirits which took the form of sylphs, salamanders, gnomes, and ondines. Communion with these spirits was the highest aim in life.

Cardan's writings in the above fields were compilations; much of the material, incidentally, he stole from Leonardo da Vinci, who was a friend of Cardan's father. In his mathematical and scientific work, however, he shows the new influences. His still famous *Ars Magna* (1545), which contains a full account of the algebraic methods known to the Arabs, also contains results due to himself and his contemporaries. He is the first European mathematician of consequence. Some indication of what was new in his work was given in Chapter 5.

Pacioli and Cardan are mathematical figures in the movement commonly known as humanism. The humanists, and we speak now of those active in all fields, have been criticized because they idolized the past too much and looked backward rather than forward. They slavishly accepted the Greek works and pored over them, even undertaking extensive philological studies to determine the meanings of dubious words. To their credit may be noted that they prepared the atmosphere for the revival of reason, spread the Greek ideas through Europe, secularized education, and stressed the individual, experience, and the natural world.

The period devoted to the collection and study of the classics was followed by one in which intellectuals groped for positive doctrines and methods to replace or at least alter the medieval culture. We cannot trace in detail the oscillations of thought, the mixture of medieval fantasy and rational speculations, the commingling of fine observations with outmoded principles, all of which one finds especially in the sixteenth century. Many European thinkers finally broke away from the endless rationalizing on the basis of dogmatic principles which were vague in meaning and unrelated to experience, and chose human inquiry rather than divine authority.

It was from the Greek works that the leaders in this intellectual revitalization of Europe derived the principles of a new approach to man and the universe. They learned that man could enjoy a physical life and find pleasure in food, sports, and the development of his own body. Beauty was not a snare, and pleasure not a sin. Man, the unworthy creature, who had been commanded to regard himself as a sinner, to spend his life in abstinence, penance and abjectness, and to prepare for death, the only real event of life, could find dignity in his own being, and demand a full life on this earth as his birthright. In place of sin, death, and judgment, men should seek beauty, pleasure, and joy. The Renaissance world began to see man as the goal of God rather than God as the goal of man.

The human spirit was emancipated and inspired to refashion its ideals of existence. Perhaps the most important decision was to turn to nature herself as the source of knowledge. "Back to nature" became the new cry. Europeans turned to nature's laws instead of divine pronouncements gleaned from the Scriptures, to the universe of God instead of God. Man himself was included in the study of nature.

Leonardo is a representative figure in this shift to nature as the prime focus. He almost boasts that he is not a man of letters and that he chose to learn from experience. His observations and inventions recorded in his notebooks give evidence of his extensive and detailed physical studies. He says, "If you do not rest on the good foundation of nature, you will labor with little honor and less profit." Sciences which arise in thought and end in thought do not give truths because no experience enters into these purely mental reflections, and without experience no thing is sure.

A new school of biologists arose, of whom Andreas Vesalius (1514–64) was the leader. His *On the Structure of the Human Body* (1543) may be regarded as the beginning of modern anatomy. Although this work is based on Galen, he corrected many of Galen's errors and added new observations. Vesalius asserted that the true Bible is the human body, and he dissected corpses to learn the human structure. William Harvey (1578–1657), the famous seventeenth-century doctor, voices the spirit of Vesalius in the preface to his book *On the Movement of the Heart and the Blood:* "I profess to learn and to teach anatomy, not from books, but from dissections; not from the positions of philosophers, but from the fabric of nature." Harvey also followed Galen but, like Vesalius, added new material derived from his own observations and thought. Andrew Cesalpinus (1520–1603), the botanist, clearly advocated starting from observation and then proceeding through careful differentiation of the species observed to inductive truths.

We shall see in the next chapter how the artists, too, turned to the study of nature and to new goals in painting which obliged them to study anatomy, perspective, light, and mechanics. Regard for the primacy of observation forced Johannes Kepler to devise revolutionary doctrines in astronomy. Indeed, experience became the source of all basic scientific laws and, in this respect, usurped the role of mind.

The second guiding principle adopted by the Europeans of the Renaissance was to let reason be the judge of what to accept. Revelation, faith, and authority were to be subordinated as support for assertions about man and the universe, and reason was to be applied freely to all problems man sought to solve. Although the Church itself had used reason to erect its own theology, it had said that some matters were beyond reason. Moreover, the results obtained by reasoning were not put forth to be scrutinized rationally but rather to be accepted. In the Renaissance, mind replaced faith as the sovereign authority, and man was encouraged to apply it to the problems besetting his age.

The new impulse to study nature and the decision to apply reason instead of relying upon authority were forces which might in themselves have led to mathematical activity. But the Europeans also had the Greek works. From the Greeks the Europeans learned that nature is mathematically designed, and that this design is harmonious, aesthetically pleasing, and the inner truth about nature. Nature is not only rational, simple, and orderly but it acts in accordance with inexorable and immutable laws.

Almost from the beginning of the period in which Greek works began to be known in Europe, one finds leading thinkers impressed with the importance of the mathematical study of nature. In the thirteenth century, Roger Bacon believed that the laws of nature are but the laws of geometry. Mathematical truths are identical with things as they are in nature. Moreover mathematics is basic to the other sciences because it takes cognizance of quantity. Leonardo, too,—although his knowledge of Greek works was rather limited and his appreciation of what mathematical proof means almost nil—had caught the new spirit. He says that only by holding fast to mathematics can the mind safely penetrate to the essence of nature. "No human inquiry can be called true science unless it proceeds through mathematical demonstrations." He also says, "The man who discredits the supreme certainty of mathematics is feeding on confusion and can never silence the contradictions of sophistical sciences, which lead to eternal quackery." Leonardo was not a mathematician, and his understanding of the principles of mechanics, the study of bodies at rest and in motion, was intuitive and but a dim foreshadowing of the work of Galileo and Newton, but he had prophetic vision. He says in one of his notebooks, "Mechanics is the paradise of the mathematical sciences because in it we come to the fruits of mathematics." Leonardo does stress the role of theory in science and says, "Theory is the general; experiments are the soldiers." However he did not appreciate the precise role of theory or foresee what later became the true method of science. He, in fact, lacked methodology. Copernicus and Kepler, whom we shall study in more detail later, were also convinced that the world is mathematically and harmoniously designed, and this belief sustained them in their scientific endeavors.

Galileo speaks of mathematics as the language in which God wrote the great book—the universe—and unless one knows this language, it is impossible to comprehend a single word. René Descartes, father of coordinate geometry, was convinced that nature is but a vast geometrical system. He says that he "neither admits nor hopes for any principles in Physics other than those which are in Geometry or in abstract Mathematics, because thus all the phenomena of nature are explained, and some demonstrations of them can be given." Certainly by 1600 the conviction that mathematics is the key to nature's behavior had taken firm hold and stimulated the great scientific work which was to follow.

To the intellectuals of the Renaissance mathematics appealed for still another reason. The Renaissance, as we have seen, was a period in which medieval civilization and culture were challenged and new influences, information, and revolutionary movements were sweeping Europe. These men sought new and sound bases for the erection of knowledge, and mathematics offered such a foundation. Mathematics remained the one accepted body of truths amid crumbling philosophical systems, disputed theological beliefs, and changing ethical values. Mathematical knowledge was certain knowledge and offered a secure foothold in a morass. The search for truth was redirected toward mathematics.

9-5 THE RELIGIOUS MOTIVATION IN THE STUDY OF NATURE

The decisions to study nature, to apply reason, and to seek the mathematical design of nature led to a revival of mathematical activity and to the emergence of great mathematicians. But the thinking of these men took a turn which is of interest because it shows one of the strong motivations for mathematical activity over a couple of centuries and because it played a role in the subsequent cultural history.

The mathematicians and scientists of the Renaissance were brought up in a religious world which stressed the universe as the handiwork of God. The scientists whom we shall meet shortly, Copernicus, Brahe, Kepler, Pascal, Galileo, Descartes, Newton, and Leibniz, accepted this doctrine. These men were in fact orthodox Christians. Copernicus was a member of the Church. Kepler studied for the ministry although he did not take orders. Newton was deeply religious and, when late in life he felt too exhausted to pursue creative scientific work, turned to religious studies.

However, in the sixteenth century the new goal in the intellectual world became to study nature through mathematics and indeed to uncover the mathematical design of nature. Now Catholic teachings had by no means included this last principle, which is Greek. How then was the attempt to understand God's universe to be reconciled with the search for the mathematical laws of nature? The answer was to add a new doctrine, namely, that God had designed the universe mathematically. Thus the Catholic doctrine postulating the supreme importance of seeking to understand God and his creations took the form of a search for God's mathematical design of nature. Indeed the work of the sixteenth, seventeenth, and even some eighteenth-century mathematicians was a religious quest, motivated by religious beliefs, and justified in their minds because their work served this larger purpose. The search for the mathematical laws of nature was an act of devotion. It was the study of the ways and nature of God which would reveal the glory and grandeur of his handiwork. The Renaissance scientist was a theologian studying nature instead of the Bible. Copernicus, Kepler, and Descartes speak repeatedly of the harmony which God imparted to the universe through his *mathematical* design. Mathematical knowledge, being in itself truth about the universe, is as sacrosanct as any line of the Scriptures. Galileo says, "Nor does God less admirably discover Himself to us in Nature's actions than in the Scripture's sacred dictions." Man could not hope to perceive the divine plan as clearly as God himself understood it, but man could with humility and modesty seek to at least approach the mind of God.

One can go further and assert that these men were sure of the existence of mathematical laws underlying natural phenomena and persisted in the search for them because they were convinced *a priori* that God had incorporated them into the construction of the universe. Each discovery of a law of nature was hailed as evidence testifying more to God's brilliance than to the ingenuity of the investigator. Kepler in particular wrote paeans to God on the occasion of each discovery. The beliefs and attitudes of the mathematicians and scientists exemplify the larger cultural phenomenon which swept Renaissance Europe. The Greek works impinged on a deeply devout Christian world, and the intellectual leaders born in one and attracted by the other fused the doctrines of both.

EXERCISES

1. In view of what we know about Greek and medieval attitudes toward the physical world and mathematical activities in these two cultures, would you draw any conclusion about the connection between interest in the physical world and the pursuit of mathematics?
2. What events and influences led to a revival of interest in mathematics?
3. How did Renaissance scientists and mathematicians reconcile the Greek doctrine that the world is mathematically designed and the Christian doctrine that the universe is the creation of God?

Topics for Further Investigation

1. The rise of algebra in the sixteenth century.
2. Hindu and Arab mathematics.
3. The life and work of Roger Bacon.
4. The life and work of Jerome Cardan.
5. The life and work of Leonardo da Vinci.

Recommended Reading

BALL, W. W. ROUSE: *A Short Account of the History of Mathematics*, 4th ed., Chaps. 6 to 12, Dover Publications, Inc., New York, 1960.

CAJORI, FLORIAN: *A History of Mathematics*, 2nd ed., pp. 83–129, The Macmillan Co., New York, 1938.

CARDAN, JEROME: *The Book of My Life*, E. P. Dutton and Co., New York, 1930.

CROMBIE, A. C.: *Augustine to Galileo*, Chaps. 1 to 5, Falcon Press, London, 1952. Also published in paperback under the title *Medieval and Early Modern Science*, 2 vols., Doubleday and Co. Anchor Books, New York, 1959.

CROMBIE, A. C.: *Robert Grosseteste and the Origins of Experimental Science*, Oxford University Press, London, 1953.

DAMPIER-WHETHAM, WILLIAM C. D.: *A History of Science*, pp. 65–138, Cambridge University Press, London, 1929.

DA VINCI, LEONARDO: *Philosophical Diary*, Philosophical Library, Inc., New York, 1959.

EASTON, STEWART C.: *Roger Bacon and His Search for a Universal Science*, Columbia University Press, New York, 1952.

HOFMANN, JOSEPH E.: *The History of Mathematics*, Chaps. 3 and 4, The Philosophical Library, New York, 1957.

MACCURDY, EDWARD: *The Notebooks of Leonardo da Vinci*, George Braziller, New York, 1954.

ORE, OYSTEIN: *Cardano, The Gambling Scholar*, Princeton University Press, Princeton, 1953.

RANDALL, JOHN HERMAN, JR.: *The Making of the Modern Mind*, rev. ed., Chaps. 1 through 9, Houghton Mifflin Co., Boston, 1940.

RUSSELL, BERTRAND: *A History of Western Philosophy*, pp. 324–545, Simon and Schuster, New York, 1945.

SMITH, DAVID EUGENE: *History of Mathematics*, Vol. 1, Chaps. 5 through 8, Dover Publications, Inc., New York, 1958.

VALLENTIN, ANTONINA: *Leonardo da Vinci*, The Viking Press, New York, 1938.

MATHEMATICS AND PAINTING IN THE RENAISSANCE

Mighty is geometry; joined with art, resistless.

EURIPIDES

10-1 INTRODUCTION

The new currents of thought in the European Renaissance, the search for new truths to replace the discredited ones, the turn to the study of nature to obtain reliable facts, and the revived Greek conviction that the essence of nature's behavior should be sought in mathematical laws, bore fruit first in the field of art rather than science. While philosophers and scientists sought to unearth basic facts which might somehow be incorporated into their yet to be formulated new scientific method, and while mathematicians were still digesting the Greek works and awaiting inspiration for new themes, the artists, particularly the painters, reacted far more quickly and revolutionized the art of painting.

That the painters turned to mathematics to formulate their new style of painting is a little surprising, but the phenomenon has an explanation. The painters of the fourteenth, fifteenth, and sixteenth centuries were the architects and engineers of their time. They were also the sculptors, inventors, goldsmiths, and stonecutters. They designed and built churches, hospitals, palaces, cloisters, bridges, dams, fortresses, canals, town walls, and weapons. Thus Leonardo da Vinci, in offering his services to Lodovico Sforza, ruler of Milan, promises to serve as engineer, constructor of military works, and designer of war machines, as well as architect, sculptor, and painter. The artist was even expected to predict the motion of cannon balls, by no means a simple problem for the mathematics of those times. In view of these manifold activities the painter necessarily had to be something of a scientist.

Further, the Renaissance painter, unlike the builder of Gothic cathedrals, was influenced by the current doctrines which proclaimed that he learn truths from nature and that the essence of natural phenomena is best expressed through mathematics. Again, in comparison with his predecessors, he had the advantage of gleaning some mathematical knowledge from the newly recovered Greek works that were exciting the Europeans. The Renaissance painters went so far in assimilating this knowledge and in applying mathematics to painting that they produced the first really new mathematics in Europe. In the fifteenth century they were the most accomplished and also the most original mathematicians.

10-2 GROPINGS TOWARD A SCIENTIFIC SYSTEM OF PERSPECTIVE

Before we examine just how Renaissance painters employed mathematics and thereby revolutionized the art of painting, let us see what had been going on in this field. Rather early in the medieval period painting became an extensive activity. Kings, princes, and church leaders commissioned works of art to enhance buildings. The system which the medieval painters used until about 1300 was conceptual. Their objective was to portray and embellish the central themes in the Christian drama. Since the intent was to stir up religious feelings rather than to present real scenes, people and objects were drawn in accordance with conventions which had acquired symbolic meaning. Thus people were placed in unnatural, stylized positions; the general impression was one of flatness; and the entire painting had a two-dimensional effect. The backgrounds were usually solid gold to suggest that the action or people existed in some supra-earthly region.

Examples of this style of representation are abundant. A classic example of the late medieval period is found in Simone Martini's (1285–1344) "Majesty" (Fig. 10–1). Clearly this is no real scene. The background is blue. Despite the assemblage the scene looks flat; the throne especially lacks depth. There is hardly the suggestion of a floor on which the figures stand, and these appear lifeless and unrelated to one another. Moreover, sizes are not important. This painting also illustrates another conceptual device used in medieval painting, known as terraced perspective. To show a group of people arranged in depth, those farther back are placed somewhat above those in front.

Toward the end of the thirteenth century, the painters began to be influenced by the Renaissance. Since the preoccupation with religious themes still existed and paintings, in fact, continued to be commissioned mainly by church officials, the same subjects appear but in more realistic settings. The painters had turned to the observation of nature and saw a real world, physical beings, earth, sea, and air. Their paintings reveal this interest in natural scenes by reflecting their efforts to render space, depth, mass, volume, and other visual effects largely through the use of lines, surfaces, and other geometric forms. To achieve naturalism they also tried to render emotions and to depict drapery folding around parts of the body as drapery actually does. People began to look like real individuals instead of types. Mysticism gradually gave way to realism and art became more and more secular.

Cimabue (*ca.* 1300), Cavallini (*ca.* 1250–1330), Duccio (1255–1318), and Giotto (1266–1337) were the leaders of the new movement to inject realism

Fig. 10–1.
Simone Martini: *Majesty.* Pallazzo Communale, Siena.

into painting and to incorporate the beauty of nature. Giotto, in particular, is often called the father of modern painting. In the works of the men cited and in those of their immediate successors, we can readily observe the search for an optical system of perspective.

Duccio's "Last Supper" (Fig. 10–2) shows what could be a real scene and offers an ambitious attempt at depth. The receding wall and ceiling lines create this effect. Moreover, pairs of lines which are parallel in the actual scene and symmetrically placed with respect to the center are drawn so as to meet on a vertical line through the center of the painting. This scheme is referred to as vertical perspective and was developed further by other painters.

On the whole the picture is not too successful. The table seems to slant toward the front. The objects on the table are too much in the foreground and appear to be on the point of sliding off or toppling over. The table and the room are not seen from the same point of view. The various parts lack proportion. The failure of the painting to depict depth properly causes one to look from side to side instead of into the painting. An interesting feature

Fig. 10–2. Duccio: *Last Supper.* Opera del Duomo, Siena.

Fig. 10–3. Giotto: *Birth and Naming of St. John the Baptist.* Church of Santa Croce, Florence.

characteristic of the period is the setting in a partially boxed-in room. The artists were beginning to treat nature, but for the moment limited themselves to scenes which had both interior and exterior components. They were already looking into space and were about to venture into the wide world.

Giotto painted with the definite goal of reproducing visual perceptions and spatial relations, and his paintings tend to produce the effect of photographic copies. His figures possess mass, volume, and vitality, are grouped appealingly, and are interrelated. His "Birth and Naming of St. John the Baptist" (Fig. 10–3) is typical. The partially boxed-in interior is again evident as is the use of lines and surfaces. The side walls are drawn small or foreshortened to suggest depth. The ground plane is a clear surface. Although Giotto's paintings are not visually correct and although he introduced no new principles, his results are far better than those of his predecessors. He chose homelike scenes, gave human feelings to his figures, and distributed them in space. He catches shades of emotions and expresses them through the features and postures of the bodies. There is no mysticism nor ecstatic piety; "real" angels, Christ, and disciples stand before us. He was aware of the progress he had made and he delighted in showing his skill.

A step forward in the achievement of realism was made by Ambrogio Lorenzetti (fourteenth century). His outdoor panoramas are the best of this period. However, from the standpoint of the significant development which was to follow, his "Presentation in the Temple" (Fig. 10–4) is more worthy of attention. There is a definite foreground or horizontal plane as opposed to the background or vertical plane. The lines on the floor clearly recede and meet in one point. Other pairs of receding parallel lines meet in respective points of a vertical line. Also significant is the gradual decrease (foreshortening) in the size of the floor blocks to suggest distance. But the floor and the rest of the painting are not unified.

These few samples of fourteenth-century Renaissance painting show the increasing efforts to achieve naturalism, real scenes, and three-dimensionality. The innovators were groping for an effective technique but did not succeed. Visualization and sheer artistic skill were not enough.

10–3 REALISM LEADS TO MATHEMATICS

There was rather little progress in the second half of the fourteenth century because the Black Death seriously disturbed the life of Europe and decimated the population. The fifteenth century witnessed, as we noted in the preceding chapter, a new flood of Greek works to Italy, a new series of translations, and enormous support for artists. The Greek ideals became better known and were discussed enthusiastically in Italy. Secularization was hastened, and the artist acquired a heightened interest in humanity and in the study of nature, and a zeal for science.

Fig. 10–4.
Ambrogio Lorenzetti: *Presentation in the Temple.* Uffizi, Florence.

To achieve an accurate delineation of actual objects and a system of painting which would yield sound portraiture, painters studied nudes, the body in various postures, anatomy, expression, light, and color. The Madonna and Son were portrayed as human beings suffering human emotions, and Church history was enacted by real people. Religious themes became predominantly a conventional or habitual outlet for the depiction of the real world. Later, instead of humanizing religious themes, the artists turned to glorifying man and nature. The ascetic, mystical, and devotional attitudes were dropped entirely. Still later pagan subjects were adopted. The glory and gladness of nature, the delight in physical existence, the beauty of earth, sea, and air were the new values. Painting became entirely secular.

In their striving for realism the artists went one step further and decided that their function was to imitate nature, to depict what they saw as realistically as they could. Nature was to be the authority for what appeared on canvas, and painting was to be the science of reproducing nature accurately. The objective of painting, says Leonardo da Vinci, is to reproduce nature and the merit of a painting lies in the exactness of the reproduction. Even a purely imagined scene must appear to the spectator as if it existed exactly as pictured. Painting was to be a veridical reproduction of reality.

But how was the reproduction to be achieved? Here, too, the Renaissance artist adopted a Greek ideal. By the fifteenth century he had become thoroughly familiar and imbued with the Greek doctrine that mathematics is the essence of the real world. Hence to penetrate to the real substance of the theme he sought to display on canvas, the Renaissance artist believed that he must reduce it to its mathematical content. To capture the essence of forms, the organization of objects in space, and the structure of space the artist decided that he must find the underlying mathematical laws.

But realistic painting includes more than the mathematical properties of the objects being portrayed. The eye sees the painting, and this must create on the eye the same impression as the scene itself. Also, since vision and the light which carries the scene to the eye are involved, these too must be analyzed. But the study of light also led to mathematics. From Greek times on, as we have already seen in earlier chapters, light had been shown to be subject to mathematical laws. Indeed, the few mathematical laws of light were about the only precise knowledge about the phenomenon which the Greek and Renaissance worlds possessed, because the nature of light itself was a mystery. And so, to study the impress of scene and painting on the eye, the artists were once again led to mathematics.

Thus, although the artists made extensive and intensive physical studies of light and shade, color, the chemistry of pigments, the laws of movement and balance, the eye, anatomy, and the effect of distance on sight, they were chiefly dominated by the new thought that mathematics must be used to achieve realism in painting and that geometry is the key to the solution of this problem. Thereupon they created and perfected a totally new mathematical system of perspective which enabled them to "place reality on their canvases."

10–4 THE BASIC IDEA OF MATHEMATICAL PERSPECTIVE

The mathematical system of perspective which the Renaissance painters created and which is known as the system of focused perspective was founded about 1425 by the architect and sculptor Brunelleschi (1377–1446). His ideas were furthered and written down by the architect and painter Leone Battista

Alberti (1404–1472). It is not Alberti's artistic work which entitles him to fame but his technical knowledge. He studied architecture, painting, perspective, and sculpture, wrote several books explaining theoretical matters to artists, and exercised enormous influence. In his *Della Pittura* (1435) Alberti says that learning is essential to the artist. The arts are learned by reason and method; they are mastered by practice. He says further that the first necessity of a painter is to know geometry and that painting by incorporating and revealing the mathematical structure of nature can even improve on nature.

The mathematical scheme was developed and perfected by Paolo Uccello (1397–1475), Piero della Francesca (1416–1492), and Leonardo da Vinci (1452–1519). The system these men and others created and which Leonardo called the rudder and guide rope of painting has been used since the Renaissance by all artists who seek exact depiction of reality, and is taught in art schools today.

In their study of light, vision, and the representation of objects on canvas, these artists discovered the following facts. Suppose that a person looks at a real scene from a fixed position. Of course, he sees with both eyes, but each eye sees the same scene from a slightly different position. Although in ordinary vision we need both sensations to give us some perception and measure of depth, this perception is really not very good. Experience teaches us how to interpret the combined sensations, as Leonardo points out in his *Treatise on Painting*. The Renaissance artists decided to concentrate on what one eye sees and to compensate for the deficiency by shading, shadows where pertinent, and by what is known as aerial perspective, that is the gradual diminution of the intensity of colors with distance.*

Let us imagine that lines of light are drawn from one eye to various points on the objects in the scene. This collection of lines is called a *projection*. Let us imagine next, as did Alberti, Leonardo, and the German artist Albrecht Dürer (1471–1528), that a glass screen is interposed between the eye and the scene itself. Thus when one looks out of a window at a scene outside, the window serves as the glass screen. The lines of the projection will pierce the glass screen, and we may imagine a dot placed on the screen where each line pierces it. The figure formed by these dots on the screen is called a *section*. The most important fact which the Renaissance artists discovered is that *this section makes the same impression on the eye as does the scene itself*, for all that the eye sees is light traveling along a straight line from each point on the object to the eye, and if the light emanates from points on the glass screen but travels along the very same lines, it should still create the same impression.

* The difference between a drawing made according to the laws of perspective and a three-dimensional picture is clear when one views a stereoscopic drawing with both eyes through colored glasses.

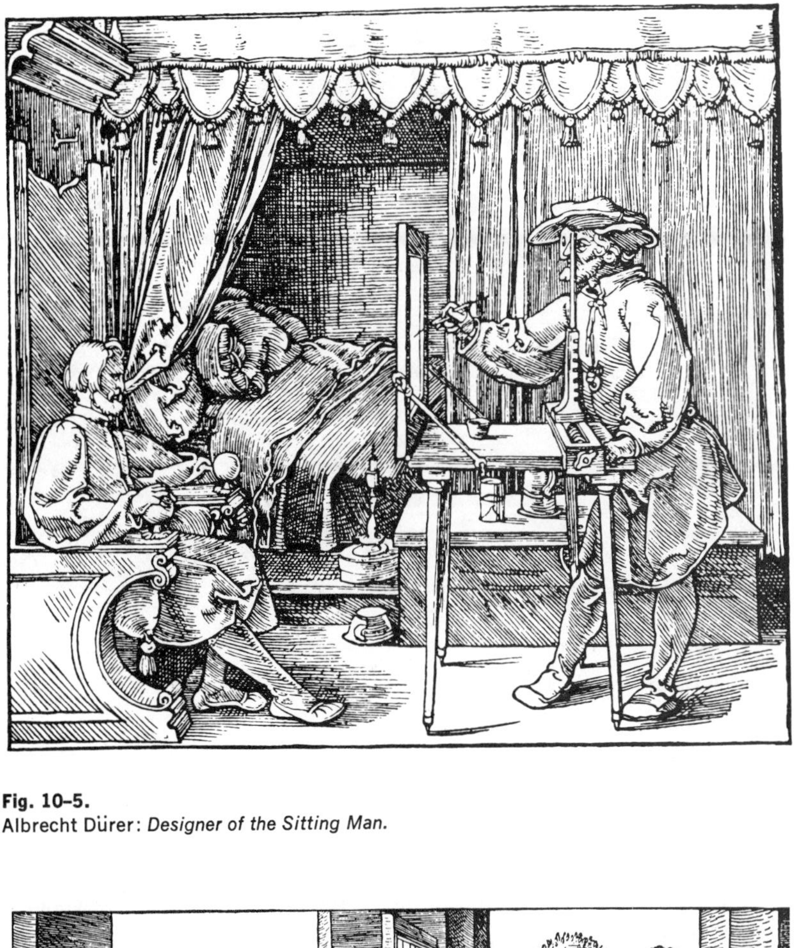

Fig. 10–5.
Albrecht Dürer: *Designer of the Sitting Man.*

Fig. 10–6.
Albrecht Dürer: *Designer of the Lying Woman.*

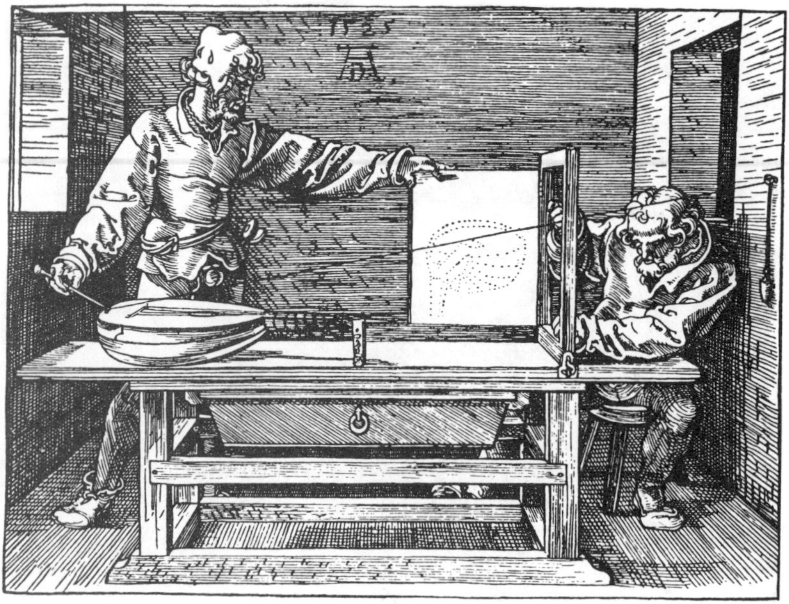

Fig. 10–7.
Albrecht Dürer: *Designer of the Lute.*

Hence this section, which is two-dimensional, is what the artist must place on the canvas to create the correct impression on the eye. Dürer used the word "perspective" because the Latin verb from which it is derived means "to see through."

Before we investigate just how the painter is to put this section on canvas, let us study the idea of projection and section. Fortunately some woodcuts made by Dürer, who learned the mathematical system of perspective in Italy and then returned to Germany to teach it to his countrymen, are very helpful. The woodcuts are in Dürer's text *Underweysung der Messung mit dem Zyrkel und Rychtscheyed* (1525). The first of these, "The Designer of the Sitting Man" (Fig. 10–5), shows an artist looking through a glass screen; he holds his eye at a fixed position, and marks on the screen the point at which a line of light from his eye to some point on the man's body pierces the screen.

The second woodcut, "The Designer of the Lying Woman" (Fig. 10–6), shows the artist again holding his eye at a fixed position and noting on paper the points where the lines of light from his eye to the woman pierce the screen. To facilitate the process of reproducing the correct location of the dots on the paper, he has divided screen and paper into little squares.

The third woodcut, "The Designer of the Lute" (Fig. 10–7), delineates on the screen the section which the eye would see if it viewed the lute from the point on the wall where the rope is attached.

These woodcuts, then, illustrate what the artists meant by a section on a glass screen. Of course, a section depends upon the position of the glass screen as well as on the position of the observer. But this implies no more than that there can be many different paintings of the same scene. Thus, for example, two paintings can be the same except for size, and size is determined by the distance between glass screen and eye. Two paintings may differ in that one shows a frontal view and the other represents the same scene viewed somewhat from the side. The difference is due to a change in the observer's position.

10–5 SOME MATHEMATICAL THEOREMS ON PERSPECTIVE DRAWING

Let us accept, then, the principle that the canvas must contain the same section that a glass screen placed between the eye of the painter and the actual scene would contain. Since the artist cannot look through his canvas at the actual scene and may even be painting an imaginary scene, he must have theorems which tell him how to place his objects on the canvas so that the *painting will, in effect, contain the section made by a glass screen.*

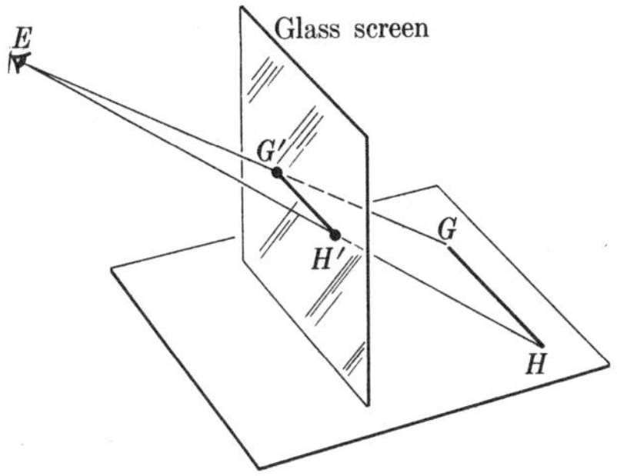

Fig. 10–8.
The image of a line horizontal and parallel to the screen is horizontal.

Suppose then that the eye at E (Fig. 10–8) looks at the horizontal line GH and that GH is parallel to a vertical glass screen. The lines from E to the points of GH lie in one plane, namely the plane determined by the point E and the line GH, for a point and a line determine a plane. This plane will cut the screen in a line, $G'H'$, because two planes which meet at all meet in a line. It is apparent that the line $G'H'$ must also be horizontal, but we can prove this fact and so be certain. We can imagine a vertical plane through GH. Since GH is parallel to the screen and the latter is also vertical, the two planes must be parallel. The plane determined by E and GH cuts these parallel planes, and a plane which intersects two parallel planes intersects them in parallel lines. Hence $G'H'$ is parallel to GH, and since GH is horizontal, so is $G'H'$. But GH was any horizontal line parallel to the screen. *Hence the image on the screen of any horizontal line parallel to the screen or picture plane must be horizontal.* Thus in a painting which is to contain what this glass screen contains, the line $G'H'$ must be drawn horizontally.

We can present practically the same argument to show that the image of any vertical line, which is automatically parallel to the vertical screen, must appear on the screen as a vertical line. *Thus all vertical lines must be drawn vertically.*

Now let us consider a somewhat more complicated situation. Suppose that AB and CD (Fig. 10–9) are two parallel, horizontal lines in an actual scene. Moreover, assume that these lines are *perpendicular* to the screen. The eye is at E. If we now imagine that lines go from E to each point of AB, these lines, that is the projection, will lie in one plane for the point E, and the line AB will determine this plane by virtue of the theorem of solid geometry already mentioned. Similarly, E and the line CD determine another plane. The screen cuts the two planes we have just described. The sections must lie on the screen, and our problem is to determine where they should lie.

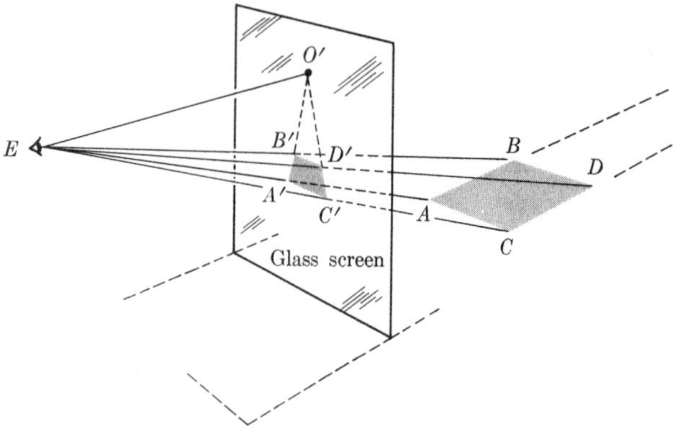

Fig. 10–9.
The images of two horizontal parallel lines which are perpendicular to the screen meet at a point on the screen.

Of course, the intersection of two planes is a line, and so the section corresponding to AB and that corresponding to CD will be lines, $A'B'$ and $C'D'$, respectively. Moreover, as the eye at E looks farther and farther out along the parallels AB and CD, the lines of sight will become more and more horizontal. As the eye follows AB and CD to infinity, so to speak, the lines from E tend to merge into one horizontal line which will be *parallel* to AB and CD. This line from E will pierce the screen at some point, say O', and this point corresponds to the imaginary point O where AB and CD seem to meet at infinity. Of course, AB and CD are parallel and do not meet, but it is convenient to think of them as meeting at a point at infinity. Indeed, the eye gets the impression that they do meet. Then the line EO' will be perpendicular to the screen because it is parallel to AB and CD and these two lines are perpendicular to the screen. The point O' corresponds to the imagined meeting point at infinity of AB and CD, but because this point does not actually exist, O' is called the *principal vanishing point.* It vanishes in the sense that it does not correspond to any actual point on AB or CD, whereas other points on $A'B'$ or $C'D'$ do correspond to actual points on AB or CD, respectively.

Now the lines AB and CD extend out to infinity to the hypothetical meeting point O; that is, ABO and CDO are lines in the real scene. The sections of these lines, $A'B'O'$ and $C'D'O'$, must therefore meet at O'. What we have shown then is that $A'B'$ and $C'D'$ must be placed on the screen so that they meet at O', and O' is the foot of the perpendicular extending from the eye to the screen. Let us now note that AB and CD are *any* horizontal lines perpendicular to the screen. *Hence all horizontal lines which are perpendicular to the screen must be drawn so as to go through* O', the principal vanishing point, which is the foot of the perpendicular from the eye to the screen.

We may draw another important conclusion from the preceding situation. The distances AC and BD are equal, for they are the distances between parallel lines. However, the corresponding images $A'C'$ and $B'D'$ are not equal because the lines $A'B'$ and $C'D'$ converge to O'. Moreover, $B'D'$ will be shorter than $A'C'$ because it is closer to O'. But $B'D'$ corresponds to the actual distance BD which is farther from the screen than AC is. Hence lengths which are farther from the screen must be drawn shorter than equal lengths closer to the screen. This fact is often described by the statement that, to obtain proper perspective in a painting, *lengths farther away from the observer must be foreshortened.*

We shall establish one more theorem about perspective drawing. Let us now suppose that JK (Fig. 10–10) is a horizontal line which makes an angle of 45° with the screen. Assume that the eye at E looks out along the line JK toward infinity. Then the line from the eye to the point at infinity on JK will be parallel to JK. Since JK is horizontal, the new line, EL in Fig. 10–10, will also be horizontal. It will pierce the screen at some point, say D_1, and will also make an angle of 45° with the screen. The triangle D_1EO' is a right triangle because EO' is perpendicular to the screen. In view of the acute angles of 45°, $O'D_1 = EO'$. Then the point D_1 is as far from O' as E is. The projection from E to the various points of JK cuts the screen in some line, $J'K'$, say. As the eye continues to follow JK toward infinity, the projection cuts the screen in points lying on an extension of $J'K'$, and we have already established that when the eye looks toward infinity on JK, the projection cuts the screen at D_1. Hence $J'K'$ must go through D_1. We now have another important result. The image of any horizontal line which makes an angle of 45°

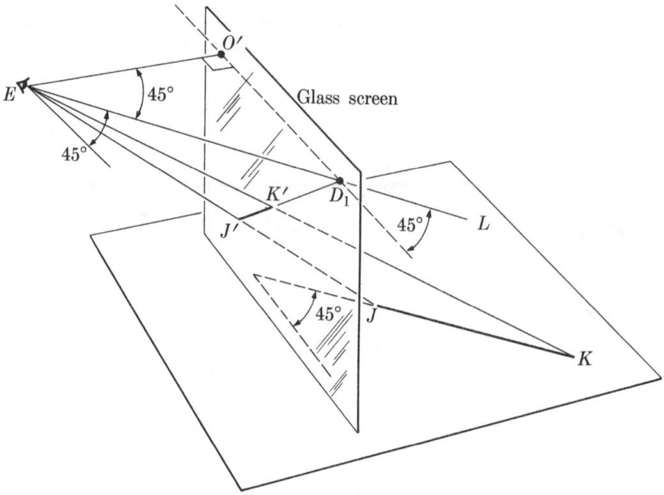

Fig. 10-10.
The image of a horizontal line which makes a 45°-angle with the screen goes through a diagonal vanishing point.

with the screen must go through the point D_1 which lies on the screen, on the same level as E, but is as far to the right of the principal vanishing point as E is from the principal vanishing point. The point D_1 is called a *diagonal vanishing point.*

Had we considered instead of JK lines which make an angle of 135° with the screen, we would have found that their images must go through a point D_2 which lies as far to the left of O' as E is from O'. The point D_2 is also called a diagonal vanishing point.

We see, then, that the points O', D_1, and D_2 correspond to points at infinity in the actual scene. As a matter of fact, all points on the horizontal line $D_2O'D_1$ correspond to points at infinity in the actual scene, and this line is called the *vanishing line.* It is the image of what one might call the horizon in the actual scene, that is, the points at infinity toward which the eye gazes when it looks in a horizontal direction.

The above theorems hardly begin to illustrate what one must know and apply to draw actual scenes realistically. The treatment of curves is especially difficult. For example, actual circles and spheres cannot, in general, be drawn as circles unless their centers happen to lie on the perpendicular from the eye to the screen. In all other cases, they must be drawn as ellipses or as arcs of parabolas or hyperbolas, depending upon their position relative to the observer. This fact becomes clear if one considers that the lines from the eye to each point on the edge of the circle or sphere, the projection, in other words, form a cone and that the section of this cone on the screen will be one of the conic sections discussed in Chapter 6. We shall not investigate the more complicated theorems because to do so would require a course in the subject and because the detailed theorems are of interest only for the specific purpose of learning to paint realistically. We may have seen enough of the basic principles to appreciate that the problem of painting realistically is handled by the application of a thoroughly mathematical system.

We know that the construction of a painting in accordance with the focused scheme presupposes a definite fixed position of the painter in relation to the scene. To view properly a painting so constructed, the observer should place himself in precisely the position the painter used in planning the painting. Otherwise the observer will get a distorted view. Strictly speaking, paintings in museums should be hung so that the observer can conveniently take that position.

10–6 RENAISSANCE PAINTINGS EMPLOYING MATHEMATICAL PERSPECTIVE

Renaissance painters achieved their goal of devising a mathematical system which permitted the realistic representation of actual scenes and joyously hastened to employ it. Realistic paintings constructed in accordance with the focused scheme of perspective begin to appear about 1430.

The artist who contributed key principles of mathematically determined perspective, including new methods of construction, and who was the best mathematician of his times is Piero della Francesca. This highly intellectual painter with a passion for geometry planned all his works mathematically to the last detail. Each scene to be painted was a mathematical problem. The placement of each figure was calculated to ensure its correctness in relation to other figures and to the painting as a whole. He loved geometrical forms so much that he used them for hats, parts of the body, and other details in his paintings. Piero practically identified painting and perspective. His *De prospettiva pingendi*, a treatise on painting and perspective in which he uses Euclid's deductive method, presents perspective as a science and provides sample constructions illustrating how perspective problems are to be handled. Though incidental to our purposes, it is worth noting that Piero painted the first Renaissance portraits of real people, the Duke and Duchess of Urbino, Federigo de Montefeltro and his wife Battista Sforza.

There are numerous examples which illustrate Piero's excellent perspective. His "Flagellation" (Fig. 10–11) is one of the best. As in all of his paintings a geometric framework underlies the design. The principal vanishing point is chosen to be near the figure of Christ. This device of placing the principal vanishing point within the most important area in the painting is deliberate because the eye tends to focus on that vanishing point. All objects are carefully foreshortened; this is especially noticeable in the marble blocks on

Fig. 10–11.
Piero della Francesca: *The Flagellation.* Ducal Palace, Urbino.

Fig. 10–12.
Piero della Francesca: *Architectural View of a City.* Kaiser Friedrich Museum, Berlin.

Fig. 10–13.
Leonardo da Vinci: *Study for the Adoration of the Magi.* Uffizi, Florence.

the floor and in the beams. The immense labor which went into the calculation of these sizes is indicated by a drawing in the book referred to above wherein he explains a similar construction.

Piero achieves unity of the various parts by means of the system of perspective. All parts are mathematically tied together to produce this synthesis. Indeed, it was somewhat because of this effect that the Renaissance painters valued the system and were excited about it. The example shown here should be compared with the fourteenth-century works (Section 10–2), where unity is lacking. The entire layout of Piero's painting is so carefully planned that movement is sacrificed to the unity of design.

To illustrate the power of perspective Piero painted several scenes of cities. His "Architectural View of a City" (Fig. 10–12) gives a striking illusion of depth. These examples of Piero's paintings show his obsession for perspective and his great technique.

Leonardo da Vinci's work provides excellent examples of paintings embodying mathematical perspective. Leonardo prepared for painting by deep and extensive studies in anatomy, perspective, geometry, physics, and chemistry. In his *Treatise on Painting*, a scientific treatise on painting and perspective, Leonardo gives his views. He opens with the statement, "Let no one who is not a mathematician read my works." Painting, he says, is a science which should be founded on the study of nature and, like all sciences, must also be based on mathematics. He scorns those who think they can ignore theory and by mere practice produce art: "Practice must be founded on sound theory."

Fig. 10–14.
Leonardo da Vinci: *Adoration of the Magi.* Uffizi, Florence.

Fig. 10–15.
Leonardo da Vinci: *Annunciation.* Uffizi, Florence.

Painting, which he regarded as superior to architecture, music, and poetry, is a science because it deals with the geometry of surfaces.

The detailed mathematical studies which Leonardo undertook in preparation for his paintings are illustrated by one of several sketches he made for his "Adoration of the Magi" (Fig. 10–13). The painting itself, which was never completed, is shown in Fig. 10–14. His "Last Supper" is another excellent example of mathematical perspective, but is so well known that we shall reproduce instead "The Annunciation" (Fig. 10–15). Although the action takes place in the foreground and the chief figures are far apart, they and the distant scene in the rear are all brought together by the perspective structure.

Raphael (1483–1520) supplies many superb paintings which exhibit excellent perspective. In his "School of Athens" (Fig. 10–16) he boldly tackles an enormous scene encompassing a vast number of people within a magnificent architectural setting. The portrayal of depth, the harmonious organization, coherence, and exactness of proportions achieved despite the difficulty of the undertaking are extraordinary. This picture, especially, shows how perspective unifies a composition and ties figures at the sides to the central theme.

The history of this painting is of interest. Pope Julius II (1443–1513) was impressed with ancient learning and regarded Christianity as the climax of

Fig. 10–16.
Raphael: *School of Athens.* Vatican.

Fig. 10–17.
Raphael: *Fire in the Borgo.* Vatican.

Jewish religious thought and Greek philosophy. He wished to have his idea embodied in paintings and commissioned both Michelangelo and Raphael to develop this theme. Michelangelo treated it in his frescoes on the ceiling of the Sistine Chapel, where he shows the human race led to Christ through a long line of Jewish prophets and pagan sibyls. Raphael executed the same theme in a somewhat different manner. In four frescoes which cover the walls of the Pope's principal official room, the Camera della Segnatura, he teaches that the human soul is to aspire to God through each of its faculties: reason, the artistic capacity, the sense of order and good government, and the religious spirit. "The School of Athens" glorifies reason and naturally exhibits the people who excelled in the intellectual sphere. Plato and Aristotle are the central figures. Plato points upward to the eternal ideas and Aristotle down to the earth as the field of experience. At Plato's left is Socrates. In the left foreground Pythagoras writes in a book. The right foreground shows the bald-headed Euclid; Archimedes stoops to demonstrate a theorem; Ptolemy holds up a sphere. All the way to the right is Raphael himself.

Raphael offers so many examples of excellent perspective that it is difficult to limit oneself to one or two representative samples. His "The Fire in the

Borgo" ((Fig. 10–17) shows exquisite depth, perfect handling of figures in various positions, the proper foreshortening, and again the unification of a scene in which many actions take place.

10–7 OTHER VALUES OF MATHEMATICAL PERSPECTIVE

We could offer countless examples illustrating the application of the mathematical system of perspective by the Renaissance masters.* All painters of this period in which western European art reached one of its pinnacles employed it and employed it well. Many, among them Uccello and Piero, were obsessed by it and painted scenes viewed from unusual positions just to solve the mathematical problems involved. The essential difference between the art of the Renaissance and that of the Middle Ages is the introduction of the third dimension, and Renaissance painting is characterized by the importance attached to realism, to the realistic rendering of space, distance, and forms, achieved by means of the mathematical system of perspective. Through it, the process of seeing was rationalized; the extended world was brought under control; and the rational interests of the painters were satisfied.

In the history of culture the accomplishments of the Renaissance artists have broad significance. Their apparent goal was to gaze at nature and to depict what they saw on canvas, but their true, more profound objective was to uncover the very secrets of nature. The Renaissance artist was a scientist, and painting was a science not merely in the sense that it had a highly technical and even mathematical content, but because it was inspired by the ultimate goal of science, understanding nature. Art and science are never separated in the thinking and work of Ghiberti, Alberti and Leonardo, for example. Leonardo's *Paragone*, *A Comparison of the Arts* (*Treatise on Painting*) contains a chapter on "Painting and Science" in which he asserts that painting seeks the truths of nature. The artist of that period regarded himself as the servant of science. These men who explored and represented nature with methods peculiar to their art were motivated precisely by the spirit and objectives of the scientists who studied astronomy, light, motion, and other phenomena. They were in fact the forerunners in spirit and goals of the great physical scientists of modern times and they revealed truth in a form which means more to many people than the deep and intricate analyses of modern mathematical physics. That mathematics proved to be the foundation of painting and thereby enabled painting to reveal the structure of nature was no more than fitting, for the Greeks had already shown that mathematics was the essence of design, and later scientists were to confirm this fact in ever more striking fashion.

---

* The same theme is treated in the author's *Mathematics in Western Culture*. Other examples can be found there.

The works of the Renaissance artists are hung in art museums. They could, with as much justification, be hung in science museums. The lover of Renaissance art is consciously or unconsciously a lover of science and mathematics.

EXERCISES

1. Relate the "back to nature" movement of the Renaissance to the development of a mathematical system of perspective.
2. Distinguish between conceptual and optical systems of perspective.
3. Which artists did most to create a mathematical system of perspective?
4. What is the principle of projection and section in the theory of perspective?
5. Draw the rear wall, and the visible portions of the side walls, ceiling, and floor of a room as seen by an observer in the room whose eye is looking directly at the rear wall.
6. Add to the drawing of the preceding exercise a square table, two of whose edges are parallel to the rear wall.
7. Draw a cube positioned in such a way that one edge is closest to you and that neighboring edges make angles of 45° and 135°, respectively, with the canvas. Go as far as you can with the theorems at your disposal.
8. State three theorems of the geometry of perspective drawing.

Topics for Further Investigation

1. The influence of mathematics on Renaissance painting.
2. Theories of the artists on human proportions. Use the reference to Panofsky's *Meaning in the Visual Arts.*
3. Vision and painting. Use the reference to Helmholtz.

Recommended Reading

BLUNT, ANTHONY: *Artistic Theory in Italy*, Oxford University Press, London, 1940.

BUNIM, MIRIAM: *Space in Medieval Painting and the Forerunners of Perspective*, Columbia University Press, New York, 1940.

CLARK, KENNETH: *Piero della Francesca*, Oxford University Press, New York, 1951.

COLE, REX V.: *Perspective*, Seeley, Service and Co., Ltd., London, 1927.

COOLIDGE, JULIAN L.: *Mathematics of Great Amateurs*, Dover Publications, Inc., New York, 1963.

DA VINCI, LEONARDO: *Treatise on Painting*, Princeton University Press, Princeton, 1956.

FRY, ROGER: *Vision and Design*, pp. 112–168, Penguin Books Ltd., Baltimore, 1937.

HELMHOLTZ, HERMAN VON: *Popular Lectures on Scientific Subjects*, pp. 250–286, Dover Publications, Inc., New York, 1962.

IVINS, WM. M., JR.: *Art and Geometry*, Dover Publications, Inc., New York, 1964.

JOHNSON, MARTIN: *Art and Scientific Thought*, Part Four, Faber and Faber, Ltd., London, 1944.

KLINE, MORRIS: *Mathematics in Western Culture*, Chap. 10, Oxford University Press, New York, 1953.

LAWSON, PHILIP J.: *Practical Perspective Drawing*, McGraw-Hill Book Co., Inc., New York, 1943.

PANOFSKY, ERWIN: "Dürer as a Mathematician," pp. 603–621 of James R. Newman: *The World of Mathematics*, Simon and Schuster, New York, 1956.

PANOFSKY, ERWIN: *Meaning in the Visual Arts*, Chap. 6, Doubleday Anchor Books, New York, 1955.

POPE-HENNESSY, JOHN: *The Complete Work of Paolo Uccello*, Phaidon Press, London, 1950.

PORTER, A. T.: *The Principles of Perspective*, University of London Press Ltd., London, 1927.

VASARI, GIORGIO: *Lives of the Most Famous Painters, Sculptors and Architects*, E. P. Dutton, New York, 1927, and many other editions.

CHAPTER 11

PROJECTIVE GEOMETRY

The gods did not reveal all things to men at the start; but as time goes on, by searching, they discover more and more.

XENOPHANES

11-1 THE PROBLEM SUGGESTED BY PROJECTION AND SECTION

The origins of mathematical ideas are far more novel and surprising than is commonly believed. Practical and scientific problems no doubt most frequently suggest new areas of exploration. However, not only are there other sources, but these often give rise to major branches of mathematics, some of which become valuable tools in scientific and practical endeavors. The questions raised by the painters during their work on the mathematics of perspective caused them and, later, professional mathematicians to develop the subject known as projective geometry. This subject, the most original creation of the seventeenth century, is now one of the principal branches of mathematics.

Let us see just what problems led to this development. The basic mathematical concepts in the system of focused perspective are projection and section. A projection we may recall, is a set of lines of light from the eye to the points of an object or scene; a section is the pattern formed by the intersection of these lines with a glass screen placed between the eye and the object viewed.

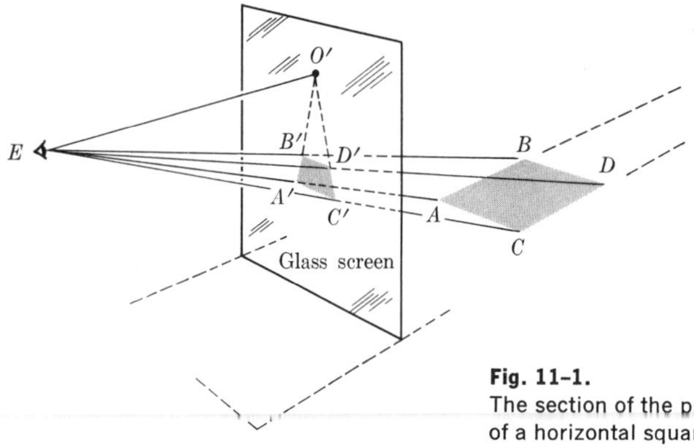

Fig. 11-1.
The section of the projection of a horizontal square.

Let us consider, as an example, the section of the projection of a square. Suppose that the square is horizontal (Fig. 11–1) and is viewed from a point somewhere above the level of the plane in which it lies. Furthermore, let us suppose that a vertical glass screen is placed parallel to the front and back edges of the square. We know from our work on perspective that the section on the screen of the two sides of the square which are perpendicular to the glass screen will consist of two line segments which tend to meet in the principal vanishing point. That is, the extensions of $A'B'$ and $C'D'$ meet at O'. Since the sides AC and BD of the square are parallel to the screen, they give rise to the parallel sections $A'C'$ and $B'D'$.

The section $A'B'D'C'$ is not a square because the sides $A'B'$ and $C'D'$ are not parallel. Nor is it a rectangle. The angles, then, of the section do not equal the corresponding angles of the original figure. The size of the section clearly depends upon where the glass screen is placed. Hence neither the lengths of the sides of the section nor the area equals the corresponding quantities in the original figure. In the language of Euclidean geometry, we may say that the section is neither congruent, similar, nor equivalent to the original figure. But the section does create the same impression on the eye as the original figure does. Hence it should possess some properties in common with the original. The question then becomes, What geometrical properties do the section and original figure have in common?

It is a natural step from this question to the next one. If two observers view the same scene from different positions, two different projections are formed (Fig. 11–2). If a section of each of these projections is made, then, in view of the fact that the sections are determined by and suggest the same scene, the sections should possess common geometrical properties. What are they?

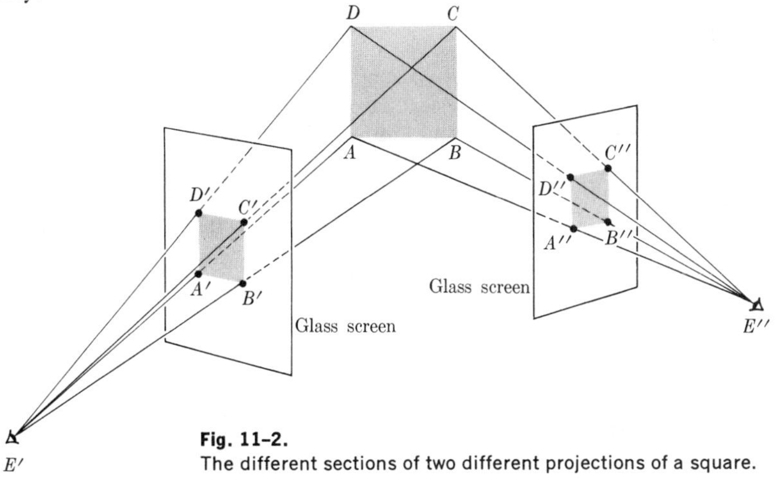

Fig. 11–2.
The different sections of two different projections of a square.

These questions, which were originally raised by the artists, could occur to anybody who thinks about what he sees. Suppose that an observer looks at a rectangular picture frame from various positions. The figure he actually perceives varies with his position. Thus he would be led to ask the same question that we raised above, What properties do these various shapes have in common? Again, as a man walks along a street and near a street lamp, his shadow changes in size and shape. The projection here consists of the lines of light from the street lamp, which takes the place of the eye, extending to the outlines of the man's body and then continuing to the ground. The section made by the plane of the ground is the shadow, although in this case the lamp and the section are on opposite sides of the actual object. The mathematical question raised by this example is, What properties do the various shadows have in common?

Another familiar, contemporary example illustrating projection and section is the photograph. In this case, the eye is the lens of the camera, and rays of light proceed from the scene through the lens to the film. The section is made by the film. We know that different sections can be obtained by placing the camera closer or farther from the scene being photographed or by tilting the camera. Yet all photographs of the same scene should contain some common geometrical properties.

The study of properties common to a figure and a section of a projection of that figure, or to two sections of the same projection, or to two sections of two different projections of the same figure has led to new concepts and theorems which today comprise a whole new branch of geometry, namely projective geometry. We shall attempt in this chapter to gain some understanding of the nature of this subject.

11-2 THE WORK OF DESARGUES

If we study any section of the projection of a figure and consider its relation to the original figure, then a few facts are readily noted. Mathematically the eye is a point, and this point and any line in the actual figure determine a plane, which is the projection determined by the line. The glass screen which cuts the projection is also a plane, and since two nonparallel planes meet in a line, the section is a line. Hence corresponding to a line in the actual scene there is a line in the section. We may therefore say that the property of linearity is common to an actual line and any section of a projection of that line. Similarly, it is easy to visualize that two intersecting lines of the actual figure will generate two intersecting lines in the section. This, then, is another minor mathematical property which is common to actual object and section. It follows that a triangle will give rise to a triangular section, although the shape of the triangle in the section will not necessarily be the same as that of the original triangle. Likewise a quadrilateral will correspond to a quadrilateral.

But the discovery of these few properties which a figure and its section or two sections of the same projection have in common hardly elates one nor does it give us any significant answers to the question of what properties figures and sections or various sections have in common. The first man to explore the problem and come up with nontrivial answers was Girard Desargues (1593–1662). Desargues was not a professional mathematician; he was a self-educated architect and engineer. His motive in tackling the subject was to help his colleagues. He believed that he could compile in compact form the many theorems on perspective that were useful to architects, engineers, painters, and stone-cutters. He even invented a special terminology which he thought would be more comprehensible to craftsmen and artists than the usual mathematical language. Of his motives Desargues writes:

> *I freely confess that I never had taste for study or research either in physics or geometry except in so far as they could serve as a means of arriving at some sort of knowledge of the proximate causes . . . for the good and convenience of life, in maintaining health, in the practice of some art, . . . having observed that a good part of the arts is based on geometry, among others the cutting of stone in architecture, that of sundials, that of perspective in particular.*

Desargues began by organizing numerous theorems and published one book on the construction of sundials and another on the application of his own geometric theories to stone-cutting and masonry. He lectured in Paris about 1626, and wrote a pamphlet on perspective ten years later. Desargues' chief contribution, a book on projective geometry, appeared in 1639.

The basic theorem of projective geometry, a theorem now fundamental in the entire field of mathematics, was formulated and proved by Desargues and is named after him. It illustrates how mathematicians responded to the questions raised by perspective.

Suppose the eye at point O (Fig. 11–3) looks at a triangle ABC. The lines from O to the various points on the sides of the triangle constitute, as we know, a projection. A section of this projection will then contain a triangle, $A'B'C'$, where A' corresponds to A, B' to B, and C' to C. Alternatively, we may regard both triangles as sections of a projection of a third triangle. The two triangles, ABC and $A'B'C'$, are said to be perspective from the point O. Desargues' theorem states an important geometrical fact which relates triangles ABC and $A'B'C'$. If we prolong the corresponding sides AC and $A'C'$, they will meet in a point P; the sides AB and $A'B'$ extended will meet in a point Q; and the extensions of BC and $B'C'$ will meet in a point R. Then, the theorem asserts, P, Q, and R will lie on a straight line. In more compact language, the theorem says:

> *If two triangles in different planes are perspective from a point, the three pairs of corresponding sides meet in three points which lie on one straight line.*

The proof of this theorem is simple. The lines AC and $A'C'$ lie in one plane because OAA' and OCC' are two intersecting lines and, as such, determine a plane. Then AC and $A'C'$ will meet in a point because any two lines in one plane meet in a point.* Let us denote this point by P. The same argument shows that AB and $A'B'$ meet in a point Q, and that BC and $B'C'$ meet in a point R.

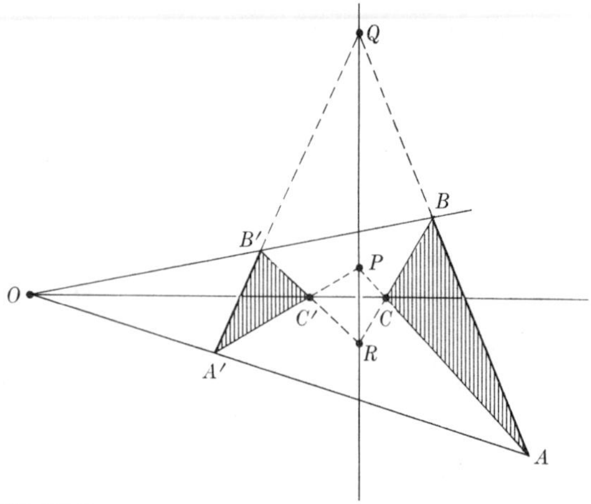

Fig. 11–3.
Desargues' theorem.

We now wish to show that P, Q, and R lie on one line. Now P, Q, and R lie in the plane of triangle ABC because P lies on AC, Q on AB, and R on BC. Likewise P, Q, and R lie in the plane of triangle $A'B'C'$. Now the points common to two planes must lie on one line, the line of intersection of the two planes. Hence, P, Q, and R lie on one line.

The reader may be troubled about the assertion in Desargues' theorem that each pair of corresponding sides of the two triangles must meet in a point. He may ask, If these sides happen to be parallel, does the theorem fail? Desargues had taken account of this possibility. We observed in the preceding chapter that a set of lines which are parallel in a particular scene being painted may have to be drawn on the canvas so as to meet in a point. In this case, there is a point in the section which does not correspond to any point in the scene itself. This breakdown of the correspondence between points in the scene and points in the section can be repaired by agreeing that any set of parallel lines is to be regarded as having a point in common. Where is this point? The answer is that it cannot be visualized, although the student is often advised to think of it as being at infinity, a bit of advice which es-

* We shall neglect for the moment the special case of two parallel lines.

sentially amounts to answering a question by not answering it. However, whether or not one can visualize the point common to parallel lines, a point distinct from the usual, finitely located points of the lines, it is convenient to say that they have a point in common. In addition, it is agreed that two or more parallel lines are to have just *one* point in common as do nonparallel lines. Hence we say of *any* two lines in projective geometry that they meet in one and only one point. This agreement is further recommended by the argument that projective geometry is concerned with problems which arise from the phenomenon of vision, and we never *see* parallel lines, as the familiar example of apparently converging railroad tracks illustrates.

One more agreement must be made about these new points introduced in projective geometry. We just agreed that any set of parallel lines has one point in common. Since there are many sets of parallel lines in one plane, each set having its own direction, there are many such new points in the plane of projective geometry. It is agreed that all these new points lie on one new line, sometimes called the line at infinity.

Let us now turn back to Desargues' theorem. If each of the three pairs of corresponding sides of the triangles presented in Desargues' theorem consisted of parallel lines, it would follow from our agreements that the two lines of each pair intersect in one point and that the three points of intersection lie on one line, the line at infinity. These conventions or agreements about points and a line at infinity obviate the necessity for making special statements when parallel lines happen to be involved in the theorems. This is as it should be, because the property of parallelism plays no role in projective geometry as opposed to Euclidean geometry. The reader who balks at accepting the agreements about parallel lines may nevertheless accept the theorems of projective geometry, with the mental reservation that they fail to state the truth when parallel lines are involved.

Desargues discovered another fundamental property which is common to a figure and to a section of a projection of that figure. Let us consider the figure consisting of the line l on which any four points are selected and designated A, B, C, and D (Fig. 11–4). We may form a projection of this figure from an arbitrary point O and cut this projection with the usual glass

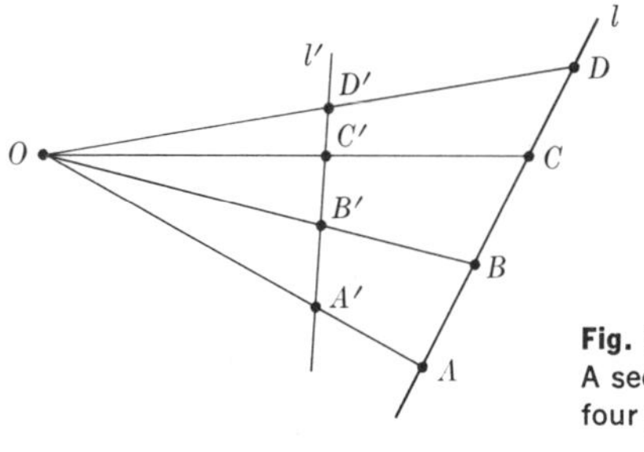

Fig. 11–4.
A section of a projection of four points on a line.

screen to obtain a section. The original figure and the point O determine a plane. Then the section consists of the line l' on which A' corresponds to A, B' to B, and so on. Since length in the figure differs from length in a section or, to use more technical phraseology, length is not invariant under projection and section, we should not expect that $A'B'$ would equal AB, or that any segment on l would equal the corresponding segment on l'. One might next consider the ratio CA/CB and venture that perhaps this ratio would equal the corresponding ratio $C'A'/C'B'$. This conjecture is not correct. However, one can prove the surprising fact, namely that

$$\frac{CA/CB}{DA/DB} = \frac{C'A'/C'B'}{D'A'/D'B'}.$$

Thus this ratio of ratios, or cross ratio as it is called, is a projective invariant. This is a very surprising fact. It does not matter where the points A, B, C, and D lie on the line l or which points are labeled A, B, C, or D. The cross ratio of the lengths they determine and the cross ratio of the corresponding lengths in the section will be the same.

Incidentally the fact that the cross ratio of four points on a line is the same in the section as in the original figure permits us to check the correctness of a painting executed in accordance with the system of focused perspective. If four points A', B', C', D' in the painting correspond to four points A, B, C, D which lie on one line in the original scene, then the cross ratio of the first set must equal the cross ratio of the second set. This fact, however, is not too useful in constructing the painting itself.

EXERCISES

1. What fact or facts may you assert about:
   a) a section of a projection of an equilateral triangle;
   b) a section of a projection of a square?
2. Why should one expect that a figure and a section of a projection of that figure should possess some geometrical properties in common?

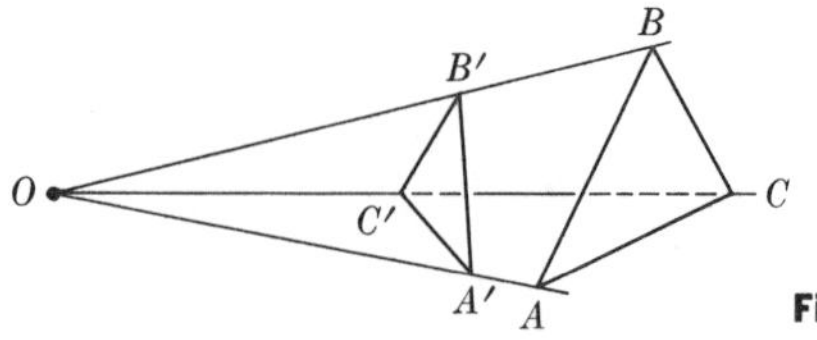

Fig. 11–5

3. Figure 11–5 shows two triangles lying in the same plane, the plane of the paper. Moreover, these two triangles are perspective from the point O.
   a) What is the difference between this figure and the one considered in the text?

b) Verify by actually drawing lines that the pairs of corresponding sides of the two triangles in Fig. 11–5 meet in three points which lie on one straight line.
c) What generalization of Desargues' theorem is suggested by the result in (b)?

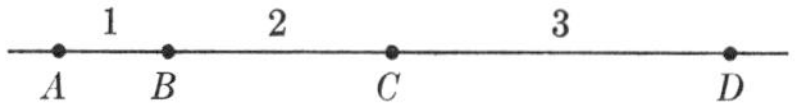

Fig. 11–6

4. Given the points and lengths in Fig. 11–6, what is the cross ratio of the lengths determined by A, B, C, and D?
5. What is the meaning of the statement that a geometrical property is invariant under projection and section?

Fig. 11–7

6. In projective geometry, lengths are sometimes regarded as directed. For the positions shown in Fig. 11–7, CA would be taken to be negative, whereas CB is positive because it is in the opposite direction. Then the cross ratio

$$\frac{CA/CB}{DA/DB}$$

in the present case is negative. When the cross ratio is -1, the four points A, B, C, D are said to form a *harmonic* set. Suppose now that D is moved indefinitely far to the right and C is moved so as to keep the cross ratio -1. Can you discover any special property of the point C in relation to A and B?

11–3 THE WORK OF PASCAL

In the domain of projective geometry, Desargues' ideas were further advanced by Blaise Pascal (1623–1662). This man of many contradictory qualities, beset by deep emotional conflicts, is commended to us also by his superb original work in other branches of mathematics, physics, literature, and theology. His father, a judge and tax commissioner, recognized that his son was bright and guided his education. He decided that Blaise should not tackle mathematics until he was 16 years old, but somehow the boy got started on his own and learned a good deal quickly.

When Pascal was still a youth in Paris his father took him to the weekly sessions of a group of noted intellectuals, Roberval, Mersenne, Mydorge, and others. There Blaise met Desargues and as a result became interested in studying properties of geometric figures which remain invariant under projection and section. At the age of 16 he proved a famous theorem, still called Pascal's theorem, which we shall examine shortly. He then wrote the *Essay on Conic Sections*, which contains many original results. Mathematics became his great passion. To aid his father he conceived the idea of having a machine perform arithmetic operations, and he constructed the first successful computing machine. He also was one of the notable precursors of Newton and Leibniz in the creation of the calculus and, together with Pierre de Fermat, another great French mathematician whom we shall meet later, founded the theory of probability.*

Pascal made some famous physical experiments which confirmed the discovery by Evangelista Torricelli, a pupil of Galileo, that the air presses down upon us, or that the air has weight, and he also clarified and furthered the study of pressure in liquids, a study which, in technical parlance, is known as hydrostatics.

It was clear to Pascal that the data of experience are the starting points of knowledge and he respected and superbly exercised the power of reason. His *Spirit of Geometry*, an essay on method and rules of thought, may well be classed with Descartes' *Discourse on Method*, another landmark on the role of reason. But as he grew older, he became more and more dissatisfied with the limited results attained by reason. About ten years before he died, he began to find emptiness in the knowledge of nature and acquired some distaste for it. "Don't overrate science," he cautioned. He became convinced that the truths of mathematics were not broad enough to encompass all of man's world. He would frequently say that all the sciences could not comfort one in days of affliction, but that the doctrines of Christian truth would comfort one at all times both in affliction and in one's ignorance of these sciences. Famous are his epigrams, "The heart has its reasons which the reason knows nothing of" and "Nothing that has to do with faith can be the concern of reason." More and more he turned to religion. He had been brought up as a Catholic, but he would not accept the strict dogmatic theology of the powerful Jesuits. He became a Jansenist and his *Provincial Letters*, one of his famous literary works, is filled with anti-Jesuitical polemics. In his *Pensées*, another literary classic, he penned many more of his thoughts on religion. The conflict between science and faith ended with a victory for religion. Ironically, Pascal, the defender of faith, helped immensely to found the ensuing Age of Reason.

Typical of theorems in projective geometry is the one conceived and proved by Pascal. This theorem, like Desargues', states the property of a geometrical figure which is common to the figure and to any section of any projection of that figure; that is, it states a property of a geometrical figure which is invariant under projection and section.

Pascal has this to say: Draw any six-sided polygon (hexagon) inscribed in a circle and letter the vertices A, B, C, D, E, F (Fig. 11–8). Prolong a pair of opposite sides, AB and DE say, until they meet in a point, P. Extend another pair of opposite sides, AF and CD say, until they meet in a point, Q.

---

* See Chapter 23.

Finally, prolong the third pair until they meet in a point, R. Then, Pascal asserts, P, Q, and R will always lie on one straight line. The mathematician, with his usual passion for brevity, says:

> *If a hexagon is inscribed in a circle, the pairs of opposite sides intersect in three points which lie on one straight line.*

We shall not give the proof of this theorem because it would require more time than we should devote to the subject. The statement, however, offers another illustration of the type of theorem investigated in projective geometry.

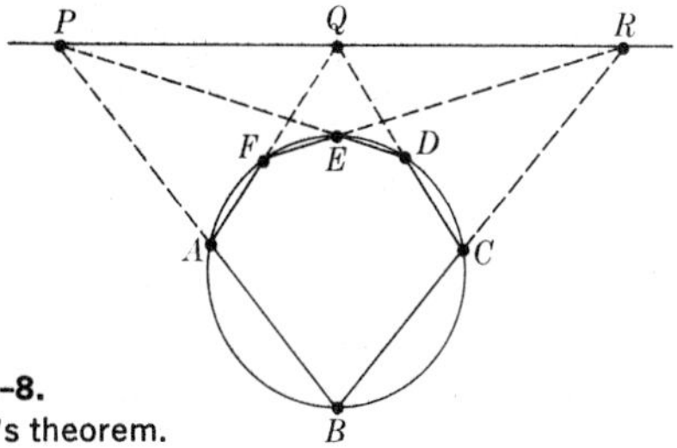

Fig. 11–8.
Pascal's theorem.

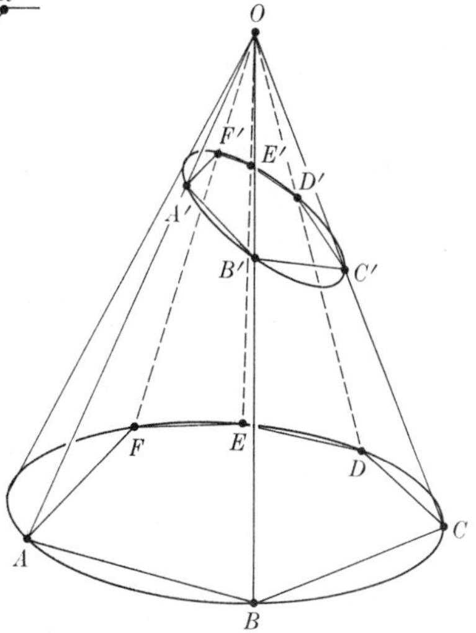

Fig. 11–9.
A section of a projection of a hexagon inscribed in a circle.

As stated, Pascal's theorem seems to have little to do with properties common to all sections of a projection. However, let us visualize (Fig. 11–9) a projection of the figure involved in Pascal's theorem and a section of this projection. The projection of the circle is a cone, and a section of this cone will not necessarily be a circle, but, as we know from the work of the Greeks, it may be an ellipse, a parabola, or a hyperbola, that is, a conic section. To each side of the hexagon inscribed in the circle there corresponds a side of the hexagon inscribed in this conic section, and to each intersection of lines in the original figure there belongs an intersection of the corresponding lines in the section. Finally, since the points P, Q, R lie on one line in the original figure, the corresponding points will be on one line in the section. Hence Pascal's theorem states a property of a circle which continues to hold in any section of any projection of that circle.

EXERCISES

1. Draw a circle, choose any six points on it as the vertices of a hexagon, find the points of intersection of the three pairs of *opposite* sides, and see whether they lie on a straight line.
2. Draw any two straight lines. Choose three points on one and label them A, B, C. Choose three points on the second line and label them A', B', C'. Find the point of intersection of AB' and $A'B$; do the same for AC' and $A'C$, and for BC' and $B'C$. Do you observe any interesting fact about these three points of intersection?

11-4 THE PRINCIPLE OF DUALITY

It would be pleasant to relate that the innovations of Desargues and Pascal were immediately appreciated by their fellow mathematicians and that the potentialities in their methods and ideas were eagerly seized upon and further developed. Actually this pleasure is denied to us. Perhaps Desargues' novel terminology baffled his contemporaries just as many people today are baffled and repulsed by the language of mathematics. At any rate, except for Descartes, Pascal, and Fermat, Desargues' colleagues exhibited the usual reactions to radical ideas: they dismissed them, called Desargues crazy, and forgot projective geometry. Desargues himself became discouraged and returned to the practice of architecture and engineering. Every printed copy of Desargues' book, originally published in 1639, was lost. Pascal's work on conics and his other studies on projective geometry, though published in 1640, also remained unknown until almost 1800. Fortunately, a pupil of Desargues, Philippe de La Hire, made a manuscript copy of Desargues' book which was accidentally picked up in a bookshop by the nineteenth-century geometer Michel Chasles, and thus the world finally learned the full extent of Desargues' major work. Apart from some results which La Hire used and which were incorrectly credited to him for 150 years, Desargues' and Pascal's discoveries had to be remade one by one by the nineteenth-century geometers.

Another reason for the neglect of projective geometry during the seventeenth and eighteenth centuries was that analytic geometry, created by Desargues' contemporaries, Descartes and Fermat (see Chapter 12), and the calculus, developed chiefly by Newton and Leibniz during the latter half of the seventeenth century (see Chapters 16 and 17), proved to be so useful in the rapidly expanding branches of physical science that mathematicians concentrated on these subjects.

The study of projective geometry was revived through a series of accidents and events almost as striking as those which had first stimulated interest in this discipline. The problem of designing fortifications attracted the geometrical talents of Gaspard Monge (1746–1818), the inventor of descriptive geometry. It is relevant that this subject, though distinct from projective geometry, uses projection and section. Monge was a most inspiring teacher, and there gathered about him a host of bright pupils, among them Charles J. Brianchon (1785–1864), L. N. M. Carnot (1753–1823), and Jean Victor Poncelet (1788–1867). These men were so impressed by Monge's geometry that they sought to show that geometric methods could accomplish as much and more than the algebraic or analytical methods of treating geometry introduced by Descartes. Carnot in particular wished "to free geometry from the hieroglyphics of analysis." As if to take revenge on Descartes whose creation had caused the abandonment of pure geometry, the early nineteenth-century geometers made it their objective to outdo Descartes.

The revival of projective geometry was launched dramatically by Poncelet. While serving as an officer in Napoleon's army, he was captured during the campaign in Russia and spent the year 1813–1814 in a Russian prison. There Poncelet reconstructed without the aid of any books all he had learned from Monge and Carnot and then proceeded to create new results in projective geometry. He was the first mathematician to appreciate fully that this subject was indeed a totally new branch of mathematics, and he consciously sought properties of geometrical figures which were common to all sections of any projection of a given figure. A group of French and, later, a group of German mathematicians continued Poncelet's work and developed intensively the subject of projective geometry.

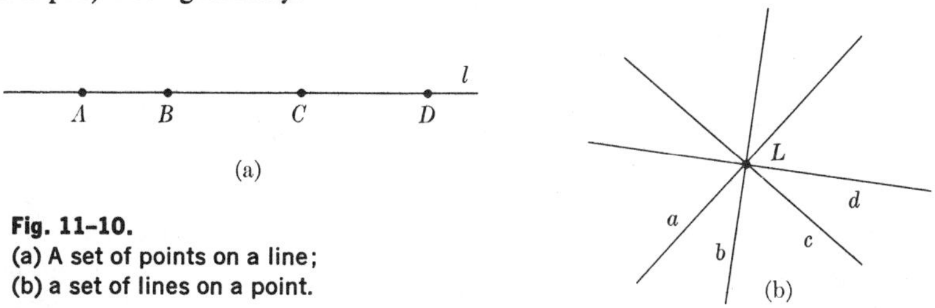

Fig. 11–10.
(a) A set of points on a line;
(b) a set of lines on a point.

The many accomplishments of this period were capped by the discovery of one of the most beautiful principles in all mathematics—the principle of duality. It is true in projective geometry, as in Euclidean geometry, that any two points determine one line or, as we now prefer to put it, *any two points lie on one line*. But in projective geometry it is also true that *any two lines* determine, or *lie on*, *one point*. (The reader who has refused to accept the convention that parallel lines in Euclid's sense are also to be regarded as having a point in common will have to forego the next few paragraphs and pay for his stubbornness.) It will be noted that the second italicized statement can be obtained from the first one by merely interchanging the words point and line. We say in projective geometry that we have dualized the original statement or that one is the dual of the other. If we are discussing a *set of points on a line* and interchange "point" and "line," we obtain the phrase a *set of lines on a point*. Figure 11–10 illustrates the two dual statements.

A triangle consists of *three points not all on the same line and the lines joining them.* We could speak of *three lines not all on the same point and the points joining them.* We usually do not speak of a point as joining two lines; rather, we refer to such a point as the point of intersection of the lines. But the meaning is clear either way. The figure described by the rephrased or dualized statement is again a triangle. Because the dual figure of the triangle is a triangle, the triangle is called self-dual.

In projective geometry the quadrilateral is defined as a figure consistng of *four lines and the six points in which the lines join in pairs.* This use of the term "quadrilateral" differs slightly from the one common in Euclidean geometry. A picture of a quadrilateral is shown in Fig. 11–11(a). We can equally well speak of a figure consisting of *four points and the six lines which join the points in pairs* (Fig. 11–11b). This new figure is called a quadrangle. Hence quadrilateral and quadrangle are dual figures.

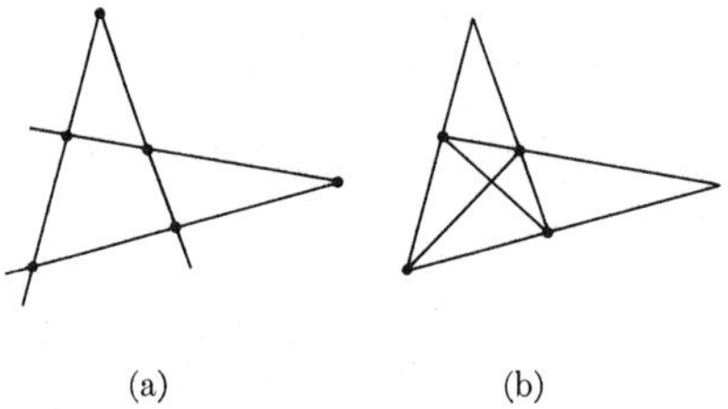

Fig. 11–11.
(a) A quadrilateral;
(b) a quadrangle.

We seem to be able to take the statement describing any figure and by dualizing the statement obtain a new figure. Let us try next something more ambitious. We shall dualize Desargues' theorem. We shall consider the case where the two triangles and the point O from which the triangles are perspective all lie in one plane, and see what results when we interchange point and line. We shall use the fact already noted that the dual of a triangle is a triangle.

Desargues' Theorem	*Dual of Desargues' Theorem*
If we have two triangles such that lines joining corresponding vertices lie on one point, O, then corresponding sides join in three points which lie on one straight line.	If we have two triangles such that points which join corresponding sides lie on one line, o, then corresponding vertices are joined by three lines which lie on one point.

If we examine the new statement we see that it is actually the converse of Desargues' theorem; that is, the hypothesis and the conclusion in Desargues' theorem are now interchanged. Hence by interchanging point and line we have discovered a possible theorem. It would be too much to ask that the proof of the new statement should be obtainable from the proof of the original theorem by interchanging point and line. Although it is too much to ask, the gods have been generous beyond our merits, for the new proof can indeed be obtained in this way.

The principle of duality, as thus far described, tells us how to obtain a new statement or theorem from a given one involving points and lines. But projective geometry also deals with curves. How should one dualize statements describing curves? The clue lies in the fact that a curve is after all but a collection of *points* satisfying some condition. For example, the circle is the set of all points at a fixed distance from a given point. The principle of duality suggests, then, that the figure dual to a given curve might be a collection of *lines* satisfying the condition dual to the condition defining the given curve. (However the definition given for the circle is not in the form which can be dualized.) This collection of lines may also be called a curve, for a collection of lines suggests a curve as well as does a collection of points (Fig. 11–12). It is called a line curve.

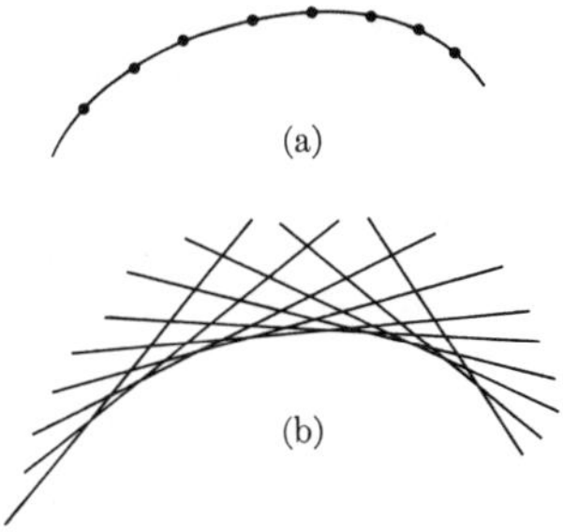

Fig. 11–12.
(a) A point curve; (b) a line curve.

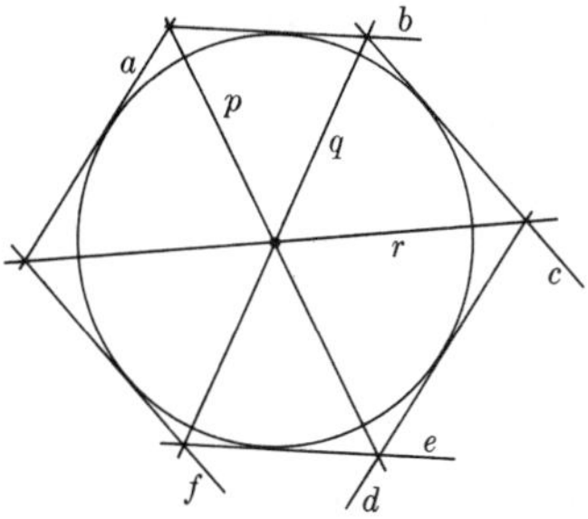

Fig. 11–13.
Brianchon's theorem, the dual of Pascal's theorem.

For conic sections, the figure dual to a point conic, that is, a conic regarded as a collection of points, turns out to be the collection of tangents to that point conic. Thus if the conic section is a circle, the dual figure is the collection of tangents to that circle. This collection of tangents suggests the circle as well as does the usual collection of points, and we shall call this collection of tangents the line circle.

We have dualized statements about simple figures with suggestive results. Let us now see whether the application of the principle of duality to theorems on curves is equally productive. As a test we shall dualize Pascal's theorem. Figure 11–13 illustrates the content of the dual statement.

Pascal's theorem	*Dual of Pascal's theorem*
If we take six points, A, B, C, D, E, and F on the point circle, then the lines which join A and B and D and E join in the point P; the lines which join B and C and E and F join in the point Q; the lines which join C and D and F and A join in the point R. The three points P, Q, and R lie on one line, l.	If we take six lines a, b, c, d, e, and f on the line circle, then the points which join a and b and d and e are joined by the line p; the points which join b and c and e and f are joined by the line q; the points which join c and d and f and a are joined by the line r. The three lines p, q, and r lie on one point, L.

Geometrically the dual statement has the following meaning: Since the line circle is the collection of tangents to the point circle, the six lines on the line circle are any six tangents to the point circle, and these six tangents form a hexagon circumscribed about the point circle. Hence the dual statement tells us that if we circumscribe a hexagon about a point circle, the lines joining opposite vertices of the hexagon, lines p, q, and r in the dual statement, meet in one point. This dual statement is indeed a theorem of projective geometry. It is called Brianchon's theorem after the man who discovered it by applying the principle of duality to Pascal's theorem.

The *principle of duality* in projective geometry says that we can interchange point and line in a theorem about figures lying in one plane and obtain a meaningful statement. Moreover—although nothing said so far justifies the assertion—the new, or dual, statement will itself be a theorem; that is, it can be proved. However, it is possible to show by one proof that every rephrasing of a theorem of projective geometry in accordance with the principle of duality must lead to a theorem. The principle of duality is a remarkable property of projective geometry. It reveals the symmetry in the roles which point and line play in the structure of that geometry, and this symmetry in turn reveals that line and point are equally fundamental concepts.

The principle of duality also gives us some insight into the process of creating mathematics. Whereas the discovery of this principle as well as of theorems such as Desargues' and Pascal's calls for imagination and genius, the discovery of new theorems by means of the principle is an almost mechanical procedure.

EXERCISES

1. Given the figure consisting of four points no three of which are on the same line, what is the dual figure?
2. State the principle of duality.
3. Given the figure consisting of four points all on one line, what is the dual figure?

4. Given the figure consisting of three points on one line, a fourth point not on that line, and the lines joining any two of these points, what is the dual figure?
5. In what way is the principle of duality a means of discovering new theorems?

11-5 THE RELATIONSHIP BETWEEN PROJECTIVE AND EUCLIDEAN GEOMETRIES

Projective geometry offers many more exciting concepts than we can hope to survey. Let us see rather what the broader features of the subject are. The basic concept is projection and section and the main goal is to find properties of geometric figures which hold for any section of any projection of those figures. A careful examination of the properties which prove to be invariant under projection and section shows that these properties deal with the collinearity of points, that is, with points lying on the same line; with the concurrence of lines, that is with a set of lines meeting in one point; with cross ratio; and with the fundamental roles of point and line as exhibited by the principle of duality. On the other hand, Euclidean geometry, which, of course, was well known to the nineteenth-century projective geometers, deals, for example, with the equality of lengths, angles, and areas. A comparison of these two classes of properties suggests that the projective properties are simpler than those treated in Euclidean geometry. One might say that projective geometry deals with the very formation of the geometrical figures whose congruence, similarity, and equivalence (equal area) are discussed in Euclidean geometry.

With hindsight to aid us, we can see that there should be a geometry more fundamental than Euclidean geometry. Anybody first perceives the position in space of trees, houses, roads, and other objects and only then thinks of distances and sizes. When traveling, one must first choose a particular road to follow before being concerned with how far to move along the road. That is, position and relative position precede distance in importance both practically and logically.

Hence one might suspect that logically projective geometry is the more fundamental and encompassing subject and that Euclidean geometry is in some sense a specialization. This conjecture is correct. The clue to the relationship between the two geometries may be obtained by again examining projection and section. Consider a geometric figure, a rectangle say (Fig. 11–14). Let a projection of this figure be formed from an arbitrary point O, and then let a section of this projection be made by a plane *parallel* to the plane of the rectangle. By applying some theorems of Euclidean geometry one can prove that the section will be a rectangle similar to the original one. Hence the relationship of similarity is a special type of projective relationship in which the plane of the section and the plane of the original figure are parallel.

If, now, the point O moves indefinitely far to the left, the lines of the projection come closer and closer to being parallel to each other. When the center of the projection is the "point at infinity"(!), these lines are parallel;

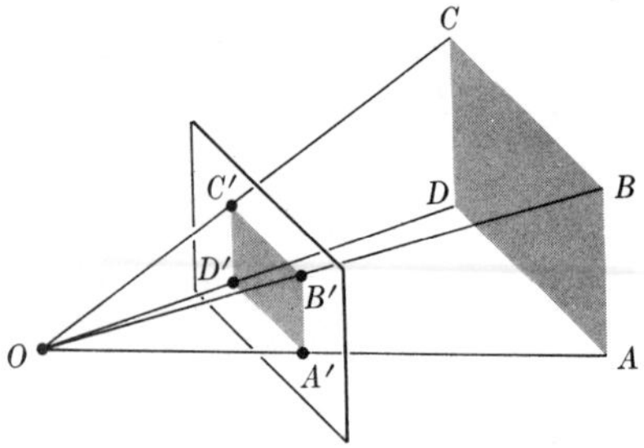

Fig. 11–14.
Similar figures related by projection and section.

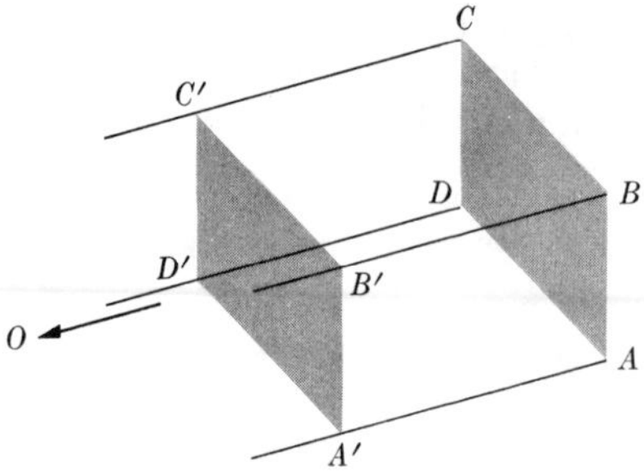

Fig. 11–15.
Congruent figures related by projection and section.

then the section made by a plane parallel to the rectangle is a rectangle congruent to the original one (Fig. 11–15). This last type of projection, called parallel projection, thus yields congruent sections. In other words, from the standpoint of projective geometry, the relationships of congruence and similarity which are so intensively studied in Euclidean geometry can be studied through projection and section for special projections. We see therefore that Euclidean geometry is not only a logical subdivision of projective geometry, but we can now look upon it in a new light, namely, as a study of properties of geometric figures which are invariant under special projections.

Although projective geometry was initiated by Desargues for the very practical purpose of extending and systematizing theorems which might help the artists, it is not highly significant from the standpoint of applications to art or science. The subject has been developed and cultivated by mathematicians who sought and found pleasure in its ideas. The Renaissance artists and geometers opened up new themes which the Greek world did not grasp, the investigation of the properties of intersections of lines, cross ratio, duality, projection and section, and, above all, the theme of properties invariant under projection and section. Projective geometry is now a vast branch of mathematics because it does offer latitude to intuition, new methods of proof, elegant results, and aesthetically satisfying ideas. This subject born of art makes its primary contribution to mathematics as an art.

EXERCISES

1. What major mathematical problem was suggested by the artists' use of projection and section?
2. How would you distinguish projective geometry from Euclidean geometry with respect to the properties of geometrical figures?
3. Write a short essay on how the rise of realistic painting in the Renaissance stimulated a new mathematical development.

Recommended Reading

BELL, E. T.: *Men of Mathematics*, Chaps. 5 and 13, Simon and Schuster, New York, 1937.

IVINS, WM. M., JR.: *Art and Geometry*, Dover Publications, Inc., New York, 1964.

KLINE, MORRIS: "Projective Geometry," an article in James R. Newman: *The World of Mathematics*, pp. 622–641, Simon and Schuster, New York, 1956.

MORTIMER, ERNEST: *Blaise Pascal: The Life and Work of a Realist*, Harper and Bros., New York, 1959.

SAWYER, W. W.: *Prelude to Mathematics*, Chap. 10, Pelican Books Ltd., England, 1955.

YOUNG, JACOB W. A.: *Monographs on Topics of Modern Mathematics*, Chap. 2, Dover Publications, Inc., New York, 1955.

YOUNG, JOHN W.: *Projective Geometry*, The Open Court Publishing Co., Chicago, 1930.

CHAPTER 12

COORDINATE GEOMETRY

In order to seek truth it is necessary once in the course of our life to doubt as far as possible all things.

DESCARTES

12-1 DESCARTES AND FERMAT

Doubts as to the soundness of the knowledge and outlook possessed by medieval Europe had already been raised during the Renaissance. The revived Greek knowledge, great explorations, new inventions, the rise of an artisan class with problems of its own which could not be answered by purposive or teleological explanations of natural phenomena, and the advocacy of experience as the source of all knowledge, all tended to undermine the old foundations. No one appreciated more the need for the reconstruction of knowledge than did René Descartes (1596–1650).

Born to moderately wealthy parents in La Haye, France, Descartes received an excellent formal and traditional education at the Jesuit College of La Flèche. But while still at school, he had already become critical of the truths which so many of his contemporaries and teachers professed so confidently and he began to question the kind of knowledge that was being imparted to him. Of the traditional studies, he said, eloquence has incomparable force and beauty, and poetry has its ravishing graces and delights. However these attainments he judged to be gifts of nature rather than the fruits of study. He respected theology because it pointed out the path to heaven and he, too, aspired to heaven, but "being given assuredly to understand that the way is not less open to the most ignorant than to the most learned, and that the revealed truths which lead to heaven are above our comprehension," he did not presume to subject these truths to his impotent reason. Philosophy, he agreed, "affords the means of discoursing with an appearance of truth on all matters, and commands the admiration of the more simple," but, though cultivated for ages by the most distinguished men, it had not produced doctrines which were beyond dispute. Law, medicine, and other professions secure riches and honor for their practitioners, but since these subjects borrow principles from philosophy, they could not be solid structures, and fortunately he was not obliged to pursue them to better his fortune. Logic he also deprecated because its syllogisms and the majority of its other rules are of use only in the communication of what one already knows or in speaking without judgment about things of which one is ignorant. It does not in itself proffer knowledge. Treatises on morals contain useful precepts and exhortations to be virtuous but no evidence that these are founded on truths.

Because he had a critical mind and because he lived at a time when the world outlook which had dominated Europe for a thousand years was being vigorously challenged, Descartes could not be satisfied with the tenets so forcibly and dogmatically pronounced by his teachers and other leaders. He felt all the more justified in his doubts when he realized that he was in one of the most celebrated schools of Europe and that he was not an inferior student. At the end of his course of study he concluded that all his education had advanced him only to the point of discovering man's ignorance.

At the age of 20, after having graduated from the University of Poitiers, where he studied law, Descartes decided to learn some things that were not in books. He began by living a gay life in Paris, after which he retired for a period of reflection in a quiet corner of that city. To see the world he joined an army, participated in military campaigns, and traveled. Finally he decided to settle down.

Because Descartes thought he could more easily find peace and seclusion in the saner atmosphere of Holland, he secured in 1628 a house in Amsterdam. There he devoted himself over a period of twenty years to critical and profound thinking about the nature of truth, the existence of God, and the physical structure of the universe. There he created his best works. As he continued to write, he and his audience became more and more impressed with the greatness of his work. Lucid thoughts set forth in literary classics which revealed the clarity, precision, and effectiveness of the French language made Descartes famous and his philosophy popular.

His retirement from the world was broken by an invitation to serve as a tutor to Queen Christina of Sweden. Reluctant as he was to leave his comfortable home, he could not resist the attraction of royalty and so moved to Stockholm. The queen preferred to begin her day at 5 a.m. by studying in an icy library, and her tutor was obliged to meet her at that hour. This regimen was too much for the frail Descartes. His flesh was weak, and his spirit unwilling. He caught cold and died in the year 1650.

From the profound thinking and writings carried on during the years in Holland there emerged new foundations of knowledge. Descartes is the acknowledged father of modern mathematics and philosophy, and he founded a new cosmology which dominated the seventeenth century until it was ultimately displaced by the work of Galileo and Newton. Convinced that the knowledge he had acquired in school was either unreliable or worthless,

Descartes swept away all opinions, prejudices, dogmas, pronouncements of authorities, and, so he believed, preconceived notions. He began reconstruction by seeking a new method of obtaining sure and reliable knowledge. The answer, he says, came to him in a dream while he was on one of his campaigns.

The "long chains of simple and easy reasonings by which geometers are accustomed to reach the conclusions of their most difficult demonstrations" led him to the conclusion that "all things to the knowledge of which man is competent are naturally connected in the same way." He decided, then, that a sound body of philosophy could be deduced only by the methods of the geometers, for only they had been able to reason clearly and unimpeachably and to arrive at universally accepted truths. Having concluded that mathematics "is a more powerful instrument of knowledge than any other that has been bequeathed to us by human agency," he sought to distill from a study of the subject some general principles which would provide a method of obtaining exact knowledge in all fields, a method which he called a "universal mathematic."

Following the pattern of mathematics which builds on axioms, he decided that he would accept nothing as true which was not so clear and distinct to his mind as to exclude all doubt. He would begin, in other words, with unquestionable, self-evident truths. The next principle of his method was to break down larger problems into smaller ones; he would proceed from the simple to the complex. Then he would write out the steps of his reasoning and review them so thoroughly that nothing would be inadvertently assumed or necessary arguments omitted. These four principles are the core of his method.

However, he first had to find those simple, clear, and distinct truths which would play the part in his philosophy that axioms play in mathematics proper. And here Descartes took a backward step. Whereas his age was turning to experience as the reliable source of knowledge, Descartes looked into his mind. After much critical reflection he decided that he was sure of the following truths: (a) I think, therefore I am. (b) Each phenomenon must have a cause. (c) An effect cannot be greater than the cause. (d) The mind has within it the ideas of perfection, space, time, and motion.

He then proceeded to reason on the basis of these axioms. The full story of his search for method and of the application of the method to actual problems of philosophy is presented in his famous *Discourse on Method* (1637). In later writings he continued to follow the procedure outlined in this work and thereby founded the first great modern system of philosophy. What is relevant at the moment is that the truths of mathematics and mathematical method served as a beacon to a great thinker lost in the intellectual storms of the seventeenth century and enabled him to develop a philosophy which was more rational, less mystical, and less bound to theology than the systems produced by all his European predecessors.

To show what his new method could accomplish in fields outside of philosophy, Descartes applied it to geometry and published these results in his *Geometry*, an appendix to his *Discourse on Method*. But before we examine how Descartes revolutionized method in geometry, we must note another great seventeenth-century thinker, Pierre de Fermat, who, equally concerned with improving geometrical methods, independently arrived at the same broad idea.

In contrast to Descartes' adventurous, romantic, and purposive life Fermat's was highly conventional. He was born in 1601 to a French leather merchant. After studying law at Toulouse, he earned his living as a lawyer and served as King's councillor for the parliament of Toulouse, a position much like that of a modern district attorney. Fermat's home life was also quite ordinary. He married and brought up five children. His evenings he devoted to study. Whereas Descartes cared little for knowledge as such or for the beauty and harmony in mathematics and the arts but sought truths and useful knowledge, Fermat was faithful to the Greek ideals of speculative knowledge and intellectual pleasures. He was a student of Greek literature, wrote poetry, joined in the solution of the scientific problems of his day, and above all regaled himself in all branches of mathematics. Despite the brief amount of time he could devote to study, he made fundamental contributions to algebra, the calculus, the mathematical theory of probability, coordinate geometry, and the theory of numbers. Fermat's mathematical achievements entitle him to the honor of being considered one of the best mathematicians the world has had.

12-2 THE NEED FOR NEW METHODS IN GEOMETRY

Descartes and Fermat developed a new approach to geometry, specifically, an algebraic method of representing and analyzing curves. Why were the mathematicians of the seventeenth century so much concerned with ways of working with curves? The general reason is that the rise of science and the vast expansion of commercial and industrial activities had raised problems involving curves. Let us see what these problems were.

The heliocentric theory of Copernicus and Kepler, which we shall examine in Chapter 15, won gradual acceptance during the seventeenth century, and scientists and mathematicians, at least, began to apply it intensively to the purely scientific problem of understanding the heavenly motions and to the more practical concerns, such as navigation, wherein astronomical knowledge was essential. Now the heliocentric theory called for the use of ellipses and, to some extent, of parabolas and hyperbolas. Many new facts about these curves were needed.

In the seventeenth century the idea that a ship's longitude could be determined most easily and accurately by means of a clock was actively pursued. The precise details of this method of determining longitude do not matter at the moment, but it is relevant that there were no clocks at that time which could be carried conveniently aboard ships. The adaptation of a spring and a pendulum to a clock was initiated by several scientists, notably Galileo Galilei, Robert Hooke, and Christian Huygens. The motion of the bob of a pendulum and of objects suspended from springs (see Chapter 18) is studied by the use of curves.

The very motion of ships at sea raised problems involving curves. Although the ideal path of a ship on a spherical surface is a great circle, the actual path cannot always be that since ships must obviously detour around land. Moreover, on maps the ideal and actual paths have to be represented by even more complicated curves whose shapes depend on the method of projection used to draw the flat map.

The increased interest in light raised numerous problems involving curves. When light travels long distances through the earth's atmosphere, it is gradually bent or refracted and hence follows a curved path (see Chapter 1 and Section 6 of Chapter 7). Since observations of the positions of heavenly bodies may be in error because the path of light is curved, it is obviously necessary to know something about these curved paths to correct our observations. A knowledge of curves was needed to design lenses used in telescopes and microscopes. Both of these instruments had been invented in the early part of the seventeenth century and attracted considerable attention. Lenses for spectacles had by this time been in use for 300 years, but improvement in design was a constant problem. Both Descartes and Fermat were very much interested in optics, and Descartes in particular did a great deal of work on the design of lenses. A good part of his findings are contained in the essay *Dioptrics* which, incidentally, appeared with the *Geometry* as one of three appendices to his *Discourse on Method*. Fermat, too, was a notable contributor to optics; his *principle of least time*, which we described earlier (Chapter 7), still stands as a basic postulate in that subject.

Another class of problems calling for the study of curves was presented by the increasing use of cannons. The balls or shells shot from cannons are called projectiles, and a number of questions were raised concerning their motion. What paths or curves do projectiles follow? How do the paths depend upon the angle at which the cannon is inclined? What is the range or horizontal distance traveled by the projectile? And how is the path affected by the initial velocity imparted to the ball?

The motion of projectiles was but one of a wider class of problems involving motion. As we shall note more fully in a later chapter, the motion of objects on and near the surface of the earth became an active concern in the early seventeenth century because the heliocentric theory raised basic problems in this domain. Since all such motions take place along straight-line or curved paths, these more fundamental problems of motion also led to the study of curves.

Of course, this study was not new to mathematics. The Greeks had studied extensively the line, circle, and conic sections and had deduced hundreds of theorems about these curves. These contributions were known to seventeenth-century Europe. Why, then, did Fermat and Descartes decide that mathematics needed new methods of working with curves? The reason was stated by Descartes. He complained that Greek geometry was so much tied to figures "that it can exercise the understanding only on condition of greatly fatiguing the imagination." Descartes also deplored that the methods of Euclidean geometry were exceedingly diverse and specialized and did not allow for general applicability. Each theorem required a new type of proof, and much imagination, effort, and ingenuity had to be expended to find such proofs. The Greeks, with ample time at their disposal and no concern for immediate application, were not troubled by the lack of general procedures. However, these conditions no longer obtained in the seventeenth century. Moreover, new curves were needed for the applications which were of importance to the seventeenth century, and the Greek geometrical methods did not seem to be effective in these cases.

There is another limitation inherent in Greek geometry which the seventeenth century could no longer tolerate. One might indeed determine by some geometrical argument what type of curve a projectile shot from a cannon follows and one might prove some geometrical facts about this curve, but geometry could never answer such questions as how high the projectile would go or how far from the starting point it would land. The seventeenth century sought quantitative or numerical information because such data are paramount in practical applications.

It was clear to Descartes and Fermat that entirely new methods of working with curves were needed. Descartes, impatient with the methods of the Greeks and their disinterest in application, says,

> *I have resolved to quit only abstract geometry, that is to say, the consideration of questions which serve only to exercise the mind, and this, in order to study another kind of geometry, which has for its object the explanation of the phenomena of nature.*

Both Descartes and Fermat, working independently of each other, saw clearly the potentialities of algebra for the representation and study of curves. This realization was not entirely a bolt from the blue. A great deal of progress had been made in algebra during the latter half of the sixteenth and the early part of the seventeenth century, much of it contributed by Descartes and Fermat themselves. Cardan, Tartaglia, Vieta, Descartes, and Fermat had extended the theory of the solution of equations (cf. Chapter 5), had introduced symbolism, and had established a number of algebraic theorems and methods. What impressed Descartes especially was that algebra enables man to reason efficiently. It mechanizes thought, and hence produces almost automatically results that may otherwise be difficult to establish. This value of algebra has already been pointed out earlier in this book, but historically it was Descartes who clearly perceived and called attention to this feature. Whereas geometry contained the truth about the universe, algebra offered the science of method. It is, incidentally, somewhat paradoxical that great thinkers should be enamored of ideas which mechanize thought. Of course, their goal is to get at more difficult problems, as indeed they do.

EXERCISES

1. How did mathematics help Descartes build his philosophy?
2. What were the four steps of mathematical method emphasized by Descartes?
3. Why did Descartes and Fermat seek a new method of deriving properties of curves?
4. What scientific problems of the seventeenth century required more knowledge about curves?

12-3 THE CONCEPTS OF EQUATION AND CURVE

We can understand more readily what Descartes and Fermat accomplished if we examine a somewhat modernized version of their idea. In his general study of methodology, Descartes had decided to solve problems by proceeding from the simple to the complex. Now the simplest figure in geometry is the straight line, and so Descartes sought to analyze curves by working with straight lines. He observed, first of all, that if one introduces (Fig. 12–1) a horizontal line, OX, then the shape of a curve C could be studied by observing how vertical line segments such as Q_1P_1, Q_2P_2, Q_3P_3, . . . changed in length.

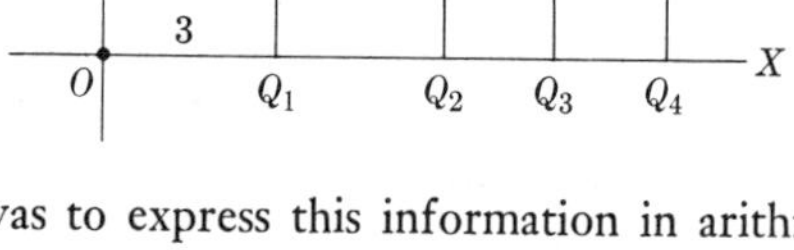

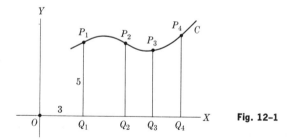

Fig. 12–1

The next step was to express this information in arithmetical terms. The position of Q_1, for example, could be specified by stating its distance from a fixed point O, called the origin, on the horizontal line. The length Q_1P_1 could certainly be specified by a number. Thus the position of P_1 would be determined by two numbers, the length OQ_1 and the length Q_1P_1. The length OQ_1, which is 3 in Fig. 12–1, is called the *abscissa* of P_1, and the length P_1Q_1, which is 5 in Fig. 12–1, is called the *ordinate*. The line OX is called the X-axis, and the line OY, which is perpendicular to OX and which shows the direction in which Q_1P_1 is taken, is called the Y-axis.

Stated in more general terms, Descartes' and Fermat's first step was to describe the position of any point P on a curve by two numbers, an abscissa and an ordinate. The first expresses the distance or length from O along the X-axis to the point Q directly below P, and the second denotes the distance or length from Q to P along a line perpendicular to the X-axis or parallel to the Y-axis. The pair of numbers is called the coordinates of P and is written thus: (3, 5).

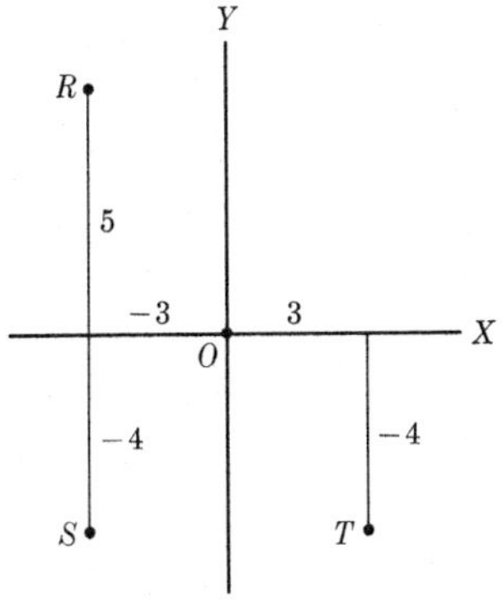

Fig. 12–2.
Plotting points on a rectangular Cartesian coordinate system.

To distinguish points which are reached by proceeding along the X-axis to the right of O from those which are reached by proceeding to the left of O, distances to the left are represented by negative numbers. Thus to arrive at the point R of Fig. 12–2 one proceeds 3 units to the left along the X-axis and 5 units upward in the direction of the Y-axis. The coordinates of R are therefore -3 and 5 and are represented as $(-3, 5)$. The distinction between upward and downward is also made by using positive and negative numbers, and as a consequence the coordinates of S in Fig. 12–2 are -3 and -4, and the coordinates of T are 3 and -4.

Thus far, then, Descartes and Fermat had a simple scheme for representing the position of any point in the plane by means of numbers, these numbers being distances from two arbitrarily chosen but fixed axes. To each point there corresponds a pair of numbers, and to each pair of numbers a unique point. The system using axes and coordinates to represent points is called a *rectangular Cartesian coordinate system.*

To represent a curve such as C of Fig. 12–1, one could list the coordinates of the many points on the curve, that is, the coordinates of P_1, P_2, P_3, . . . But such a representation would hardly be convenient, for each curve consists of an infinite number of points. Descartes and Fermat had a better idea. First of all they introduced the letters x and y to stand for the coordinates of *any* one of the points on the curve. When x and y have specific numerical values, for example, 2 and 3, respectively, then they refer, of course, to a definite point. Otherwise they represent an arbitrary point. The use of x and y here is analogous to using the words "man" and "woman" to represent any man or woman in the United States, whereas John and Mary would describe a particular couple.

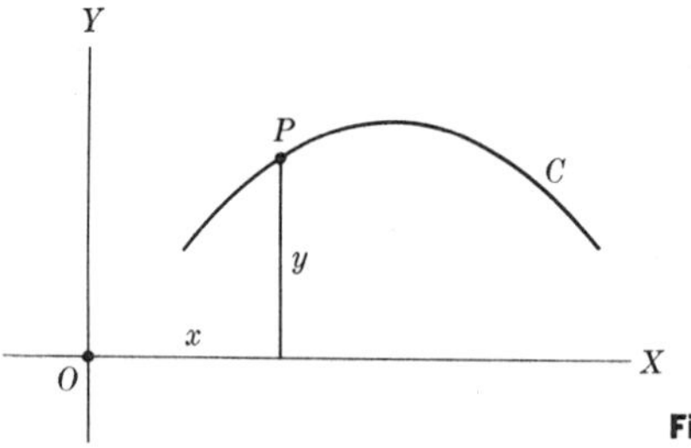

Fig. 12–3

Now, if one looks at the curve C of Fig. 12–3 and observes the abscissas and ordinates of the points on this curve, he notices that as the abscissas increase (as one looks from left to right), the corresponding ordinates at first increase and then decrease. The behavior of the ordinates changes as the abscissas change. Might it not be possible to describe the relationship between these abscissas and ordinates by specifying how large any ordinate is when the corresponding abscissa is named? This description should hold for all points on the curve and apply to that curve and no other. The answer is yes, and the general description turns out to be an algebraic equation involving x and y, the coordinates of an arbitrary point. This statement is a bit vague; let us see, therefore what it means in concrete examples.

Suppose, first, that the curve happens to be a straight line inclined 45° to the horizontal.* To describe the line algebraically, we introduce a horizontal line which passes through any point O of the given line (Fig. 12–4) and consider this horizontal line as the X-axis of our coordinate system. The Y-axis is then a vertical line through O. Consider *any* point P on the line. The coordinates of this general point P are the x and y shown in the figure. Now Euclidean geometry tells us that the triangle OQP is an isosceles right triangle. Hence $OQ = QP$ or $x = y$. Therefore the line OP appears to be characterized by the algebraic equation

$$y = x \tag{1}$$

because for *any* point P on the line, the coordinates are such that $y = x$.

* In coordinate geometry and in higher mathematics, in general, the word "curve" includes straight lines.

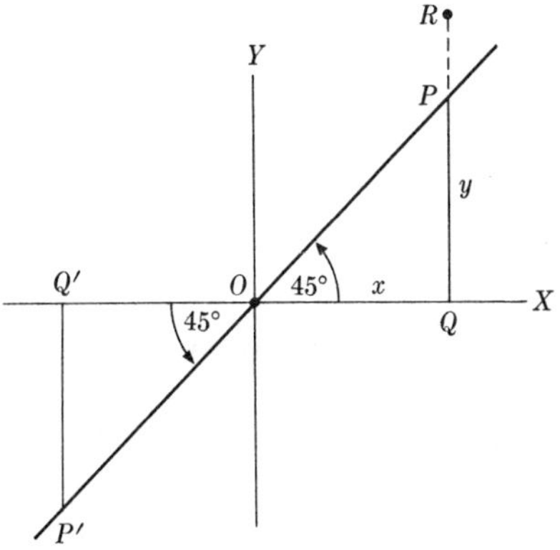

Fig. 12–4.
A straight line on a rectangular Cartesian coordinate system.

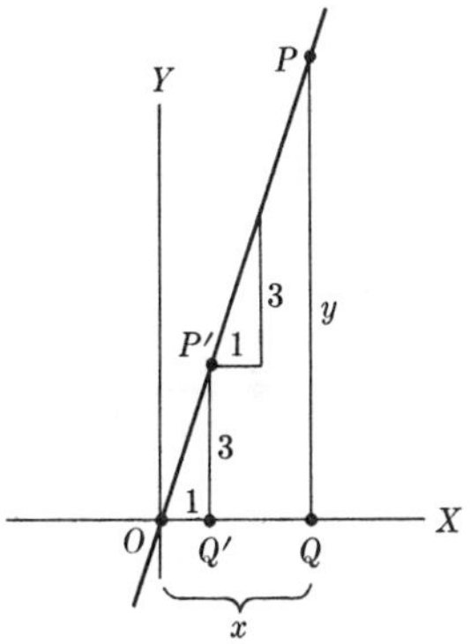

Fig. 12–5.
A straight line of slope 3 on a rectangular Cartesian coordinate system.

We should note that this equation also describes points, such as P', which lie to the left of the vertical axis. Thus, for example, the abscissa of P′ may be -4; now angle $Q'OP'$ is also 45°, and hence the ordinate of P' must also be -4. Since abscissa and ordinate are equal, $x = y$ also holds for P'. Then the all-important fact about the line $P'P$ is that it may be described algebraically by the equation $y = x$. In other words, we may say that the coordinates of any point on the line satisfy the equation $y = x$. On the other hand, points not on that line, such as R, will have ordinates unequal to the abscissas because, while R has the same abscissa as P, the ordinate of R is larger than the ordinate of P, and hence for R, y does not equal x.

Let us consider a second example. One may describe the line of Fig. 12–4 by saying that it rises one unit for each unit of horizontal distance or, in the customary expression, that it has slope 1. We now consider a line which rises more steeply, for example, one which rises 3 units for each horizontal distance of 1 unit (Fig. 12–5). Again let P be *any* point on the line. The coordinates of this general point P are then (x, y). Then from the similar triangles $OQ'P'$ and OQP we may argue that

$$\frac{y}{x} = \frac{3}{1}.$$

Hence

$$y = 3x \tag{2}$$

is the equation of the line.

The nature of equations such as $y = x$ and $y = 3x$ requires some attention. In elementary algebra we also treat equations, for example, $x^2 - 5x + 6 = 0$

or $2x + 3 = 7$. In these latter equations, however, x represents some definite but unknown quantity and our aim is to find the value or values of x. On the other hand, when we represent a curve by an equation such as $y = 3x$, we are not seeking to determine unknowns. In fact, x and y are not unknown. They represent the coordinates of *any* point on the line. Thus $x = 3$ and $y = 9$ are one pair of values satisfying the equation; $x = 4$ and $y = 12$ constitute another such pair; there are millions of others. The end product of the process of finding the equation of a curve is then an equation involving x and y which states the relationship between x and y peculiar to all points on the curve. Of course, if one wishes to find the coordinates of a particular point on the curve, he can, provided he knows the abscissa of the point, substitute this number for x in the equation and now solve for the ordinate. Thus, if the given line has the equation $y = 3x$ and we wish to determine the ordinate of the point whose abscissa is $2\frac{1}{2}$, we substitute $2\frac{1}{2}$ for x and immediately find that the ordinate of this point is $7\frac{1}{2}$.

Let us consider next another example which will enlarge our understanding of these new equations. Suppose we are given two straight lines as shown in Fig. 12–6. The line OP is the one we have just discussed and its equation is $y = 3x$. The line $O'P'$ is supposed to be parallel to OP and 2 units above it; that is, PP' is 2. What is the equation of the line $O'P'$?

To answer this question, we again seek the relation between x and y which holds for any point on this line. Now P and P' have the same x-value or abscissa, namely OQ. But the ordinate of P' is larger than that of P by the amount PP'. The distance PP' is given to be 2. Since the two lines are parallel, the vertical distance between a point on OP and a point on $O'P'$ will always be 2. Hence, whereas the ordinate of each point on the line OP is always 3 times the abscissa, the ordinate of each point on $O'P'$ will be 3 times the abscissa plus 2. That is, the equation of O′P′ is

$$y = 3x + 2. \tag{3}$$

We should note that although the straight line $O'P'$ is identical except for position with the straight line OP, its equation is different. The difference results from the fact that OP passes through the origin O of the system of coordinates, whereas $O'P'$ does not. *Hence the very same curve may be represented by a different equation if its position with respect to the coordinate axes is changed.*

Why do we bother with different equations for the same curve? If a straight line having a slope of 3 can always be placed on a coordinate system so that its equation is $y = 3x$, why do we have to consider the more complicated form $y = 3x + 2$? The answer is that if one wishes to work with two straight lines simultaneously and keep these lines in the same relative positions which they may happen to have in some physical application, one cannot assign to them an identical position on the set of axes.

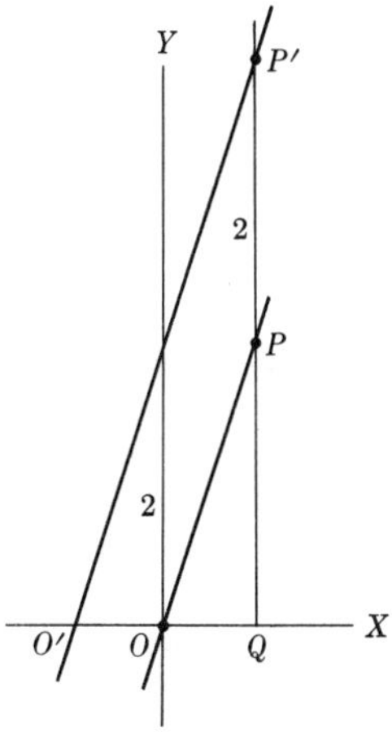

Fig. 12-6.
Two parallel lines of slope 3 on a rectangular Cartesian coordinate system.

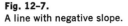

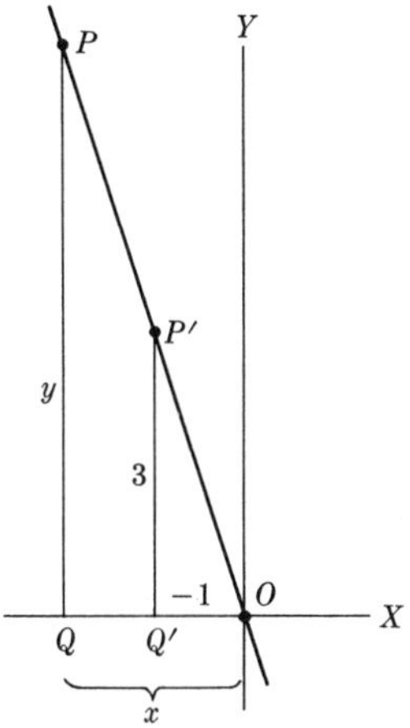

Fig. 12-7.
A line with negative slope.

While we are discussing equations of straight lines, we wish to note one more case which will be of interest later. Let us compare the line *OP* of Fig. 12–7 with the line *OP* of Fig. 12–5. The line in Fig. 12–7 may be said to "fall" as one views it from left to right, whereas that in Fig. 12–5 rises. Alternatively, we may say that the ordinates in Fig. 12–7 *decrease* as the abscissas increase. What is the equation of the line *OP* of Fig. 12–7? Suppose that the coordinates of some point P' on the line are $(-1, 3)$. The point P is an arbitrary point of the line *OP*, so let us denote its coordinates by x and y. Again we have the similar triangles OQP and $OQ'P'$. Insofar as mere lengths, without regard to sign, are concerned, we can say that

$$\frac{y}{x} = \frac{3}{1}.$$

However, we know that for the line *OP*, the ordinate of any point is always opposite in sign to the abscissa of that point. That is, when y is a positive number, x is a negative one and conversely. Hence, in the present case, not $y = 3x$, but

$$y = -3x \tag{4}$$

is the equation of the line *OP*. Thus the fact that *OP* falls to the right is reflected in the negative sign of the coefficient of x. To distinguish, with respect to slope, lines which fall to the right from those which rise to the right, we say that the former have negative slope. Thus the slope of *OP* is -3.

To illustrate Descartes' and Fermat's idea once more we shall seek the equation of a circle. Suppose that we choose our axes so that the origin is at the center of the circle (Fig. 12–8). Now the circle is defined to be the collection of all points which are at the same distance from one point, called the center. Let us assume that this distance is 5 units. (The quantity 5 is, of course, the length of the radius.) Since each point on the circle is described by a pair of coordinates, the problem of finding the algebraic equation of this circle can be solved by answering the following question. What property or relationship do the coordinates of points on the circle possess which distinguishes them from those of other points? If we consider a typical point *P* on the circle and let x and y represent its two coordinates, then we see that the lengths x, y, and 5 form a right triangle. According to the Pythagorean theorem of Euclidean geometry, the square of *OP* must equal the square of *OQ* plus the square of *QP*; hence

$$x^2 + y^2 = 5^2. \tag{5}$$

This same statement also holds for points on the circle such as P', for even though the coordinates of P' are negative, their squares are positive and so satisfy equation (5). Equation (5) is the algebraic representation of the circle. It says in words that the square of the abscissa of any point on the curve plus the square of the ordinate equals the square of the radius.

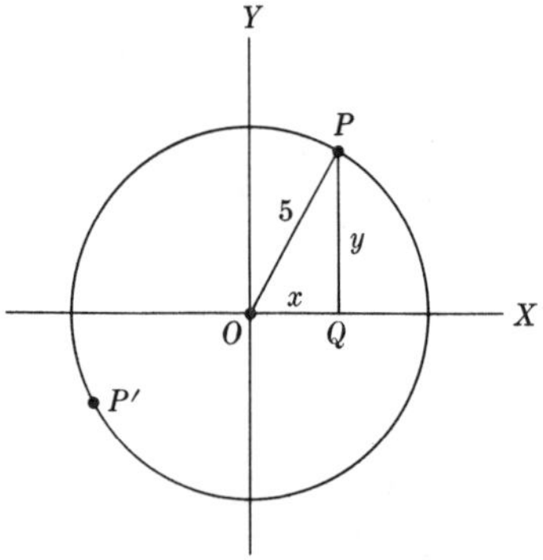

Fig. 12–8.
A circle on a rectangular Cartesian coordinate system.

To decide algebraically whether any point belongs to the circle represented by equation (5), we have to test whether the coordinates of the point satisfy the equation. Thus the point whose abscissa is 3 and whose ordinate is 4 belongs to the circle under discussion because, substituting 3 for x and 4 for y in equation (5), we see that the resulting left side equals the right side, that is, $3^2 + 4^2 = 5^2$. As another example, let us consider the point whose abscissa is 2 and whose ordinate is $\sqrt{21}$. Again, substituting 2 for x and $\sqrt{21}$ for y in equation (5), we find that

$$2^2 + (\sqrt{21})^2 = 25,$$

because the square of $\sqrt{21}$ is 21. Thus the point whose coordinates are $(2, \sqrt{21})$ lies on the circle.

EXERCISES

1. What is meant by the coordinates of a point?
2. State in your own words what the equation of a curve is.
3. Find the equation of the straight line which
   a) rises 2 units for each horizontal distance of one unit and passes through the origin of the coordinate system chosen;
   b) makes an angle of 30° with the X-axis and passes through the origin [*Suggestion:* $\tan 30° = 1/\sqrt{3}$.];
   c) falls 4 units for each unit of horizontal distance traversed and passes through the origin;
   d) passes through the origin and has slope 4;
   e) passes through the origin and has slope -4.
4. Find the coordinates of one point which lies on the curve whose equation is

   a) $x + 2y = 71$; b) $x^2 + y^2 = 36$.

5. Does the point whose coordinates are $(-3, 5)$ lie on the curve whose equation is $x^2 + 2y^2 = 59$?
6. Determine whether the point whose coordinates are $(3, -2)$ lies on the curve whose equation is $x^2 + y^2 = 4x + 1$.
7. Describe the curve whose equation is

   a) $y = 3x + 7$, b) $x^2 + y^2 = 49$, c) $x^2 + y^2 = 20$,

   d) $x + 2y = 6$, e) $y^2 = 20 - x^2$.

8. Would latitude and longitude serve as a coordinate system for points on the surface of the earth?
9. Can a curve have more than one equation? If so, how is this possible?
10. Can you say anything about the slope of the line whose equation is $y = mx + 2$?
11. What are the coordinates of the point of intersection of the line $y = 3x + 7$ and the Y-axis?
12. What are the coordinates of the point of intersection of the line $y = 3x + b$ and the Y-axis?
13. What is the slope of the line whose equation is $y = mx + b$, and what are the coordinates of the point where the line cuts the Y-axis?

14. The equation of a circle with center at the origin and radius 1 is

$$x^2 + y^2 = 1. \tag{a}$$

The equation of a circle with center at the origin and radius 2 is

$$x^2 + y^2 = 4. \tag{b}$$

If we subtract equation (a) from equation (b), we obtain the result $0 = 3$. What is wrong?

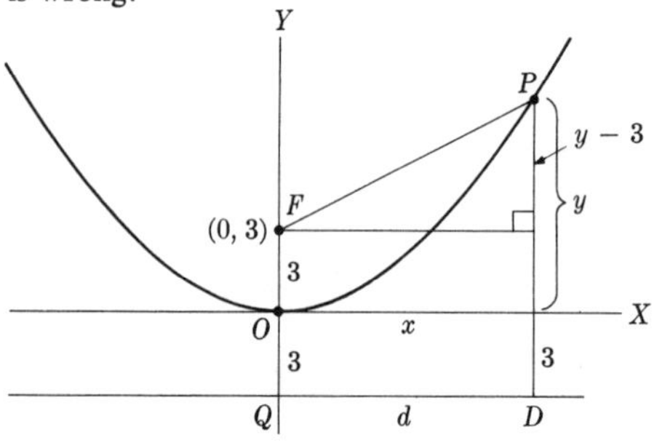

Fig. 12–9.
A parabola on a rectangular Cartesian coordinate system.

12–4 THE PARABOLA

The curve most widely used, next to the straight line and circle, is the parabola. Let us see how this curve is represented algebraically. Recalling the definition of the parabola given in Chapter 6, we start with a fixed line d, called the directrix, and a fixed point F (Fig. 12–9), called the focus. We now consider all points each of which is equidistant from d and F. This set of points is called a parabola. Thus if P is a typical point on the parabola determined by d and F, then the distance from P to F must equal the distance from P to d, that is

$$PF = PD. \tag{6}$$

To obtain the equation of this curve, we first introduce a set of coordinate axes. We know from the discussion of the straight line that the same curve may have different equations, depending upon how one chooses the axes in relation to the curve. Mathematicians have learned by experience that a simple equation results if the axes are chosen in the following way (see Fig. 12–9). Let the Y-axis be the line through F and perpendicular to d. The X-axis, which is, of course, perpendicular to the Y-axis, is drawn halfway between F and d.

Since the point F and the line d are fixed, the distance from F to d, namely FQ, is fixed. Let us suppose that this distance is 6 units. Then the distance OF is 3 units, and the distance OQ is also 3 units because the X-axis is halfway between F and d. Then the coordinates of F are $(0, 3)$. We now wish to express equation (6) in algebraic terms. Let P be any point on the parabola. Then its coordinates are (x, y). We see that PF is the hypotenuse of a right triangle whose sides are x and $y - 3$. Hence, by the Pythagorean theorem,

$$PF = \sqrt{x^2 + (y - 3)^2}.$$

The perpendicular distance from P to d is $y + 3$. Thus, in algebraic terms, equation (6) states that

$$\sqrt{x^2 + (y - 3)^2} = y + 3. \tag{7}$$

Equation (7) is the equation of the parabola.

Next we shall perform some algebraic manipulations to simplify the form of this equation. We begin by squaring both sides of the equation, an operation which amounts to multiplying equals by equals. Squaring the left side removes the radical, and squaring the right side yields $y^2 + 6y + 9$. Hence we now have

$$x^2 + (y - 3)^2 = y^2 + 6y + 9.$$

We now write out in full the square called for on the left side and obtain

$$x^2 + y^2 - 6y + 9 = y^2 + 6y + 9.$$

Subtracting y^2 and 9 from both sides yields

$$x^2 - 6y = 6y.$$

We add $6y$ to both sides and obtain $x^2 = 12y$ or, as we prefer to write it,

$$12y = x^2.$$

Dividing both sides by 12, we obtain

$$y = \tfrac{1}{12}x^2. \tag{8}$$

This equation is much simpler than (7) and yet expresses the same fact.

We might note incidentally that the number 12 in the denominator is twice the distance from F to d. Had we called this distance a, the resulting equation would have read

$$y = \frac{1}{2a}x^2. \tag{9}$$

In Fig. 12–9 we drew a curve which resembles a parabola, but we did not know at the time its position in relation to the axes we chose. To obtain a

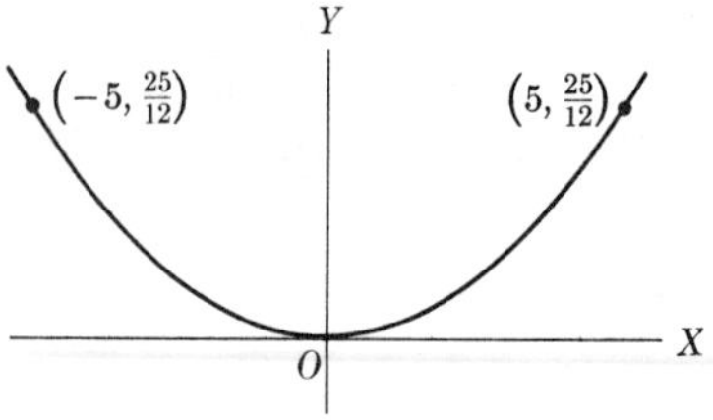

Fig. 12-10

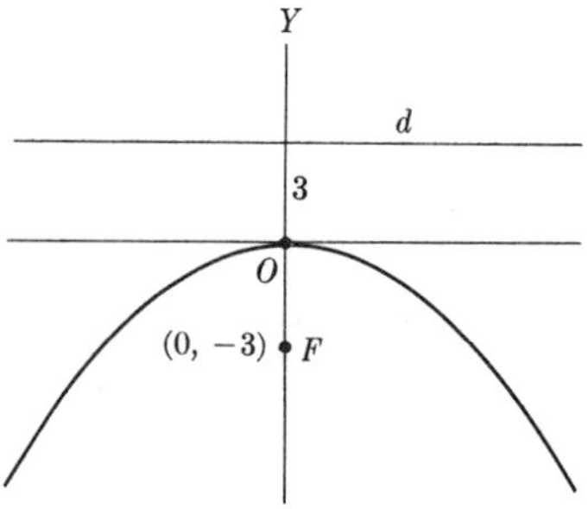

Fig. 12-11.
A parabola with focus below the directrix.

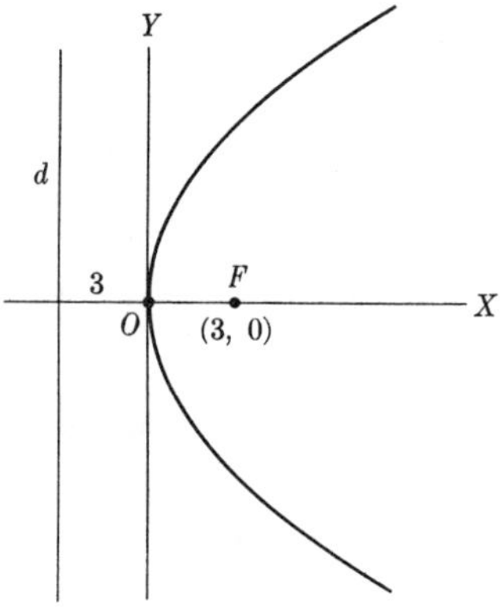

Fig. 12-12.
A parabola with focus to the right of the directrix.

rough idea of this position, let us substitute into equation (8) any positive value of x, say 5. Then $y = \frac{25}{12}$. This tells us that $(5, \frac{25}{12})$ is a point on the curve (Fig. 12–10). But if we substitute -5 for x, we obtain the same result for y. Thus $(-5, \frac{25}{12})$ is another point on the curve. These two points are symmetrically situated with respect to the Y-axis. Moreover, no matter which abscissa and ordinate we calculate, the negative of that abscissa will always yield the same ordinate because the abscissa has to be squared. This means that to each point on the curve to the right of the Y-axis there corresponds a point symmetrically situated to the left of the Y-axis. Hence, once we have determined the shape of the curve to the right of the Y-axis, we automatically know what the curve looks like to the left.

If we are interested only in the shape of the curve, it is now sufficient to observe that as x increases from zero to any value, x^2 increases, and $x^2/12$ also increases. This means that the points on the curve move out and up from the origin. Of course, there are many curves that meet this description. To obtain a more precise picture of the curve we need to calculate a few sets of coordinates. Thus, when $x = 3$, $y = \frac{9}{12}$, so that $(3, \frac{3}{4})$ are the coordinates of a point on the curve. We should note, too, that the curve lies entirely above the X-axis except, of course, for the point at the origin, because for every value of x, y is zero or positive.

In work involving parabolas and their equations it is sometimes convenient to consider one whose focus lies below the directrix. If we now choose the axes as shown in Fig. 12–11, what is the equation of the parabola? We could, of course, obtain the answer by going through steps analogous to those contained in equations (6) through (8). However, there is no need to do so since Figs. 12–9 and 12–11 differ only in the sign of the y-values: whereas the y-values in the former are positive, the y-values in the latter must be negative. Hence the equation of the parabola in Fig. 12–11 is

$$y = -\tfrac{1}{12}x^2. \tag{10}$$

Let us consider Fig. 12–11, but suppose now that distances *downward* on the Y-axis are chosen to be positive. How does this change affect the equation of the parabola? The suggested situation does not really differ from that shown in Fig. 12–9. If we imagine this whole figure rotated about the X-axis through 180°, that is through half a rotation, we obtain the situation we have just proposed. Since the position of the curve in relation to the X- and Y-axes is exactly the same in Fig. 12–11 as in Fig. 12–9, the parabola has the equation

$$y = \tfrac{1}{12}x^2. \tag{11}$$

Let us consider one more variation. Suppose that directrix and focus happen to lie as in Fig. 12–12. We choose the X- and Y-axes as shown, with the Y-axis halfway between focus and directrix. The upward direction on the Y-axis is positive. What is the equation of the parabola determined by this choice of focus, directrix, and axes? To answer this question we have but to compare Figs. 12–9 and 12–12. The X-axis in Fig. 12–9 plays the role of the Y-axis in Fig. 12–12, and the Y-axis in Fig. 12–9 plays the role of the X-axis in Fig. 12–12; in other words, the roles of abscissa and ordinate are exchanged. Hence the equation of the parabola in Fig. 12–12 is

$$x = \tfrac{1}{12}y^2. \tag{12}$$

These last few equations illustrate some of the various forms which the equation of the parabola can take. Which of these one should use is a matter of convenience in application, as we shall see in later chapters.

EXERCISES

1. Determine the shapes of the following parabolas by choosing a number of values of x, calculating the corresponding values of y, and plotting the points whose coordinates are thereby determined.

a) $y = 3x^2$ b) $y = \tfrac{1}{10}x^2$ c) $y = -3x^2$ d) $x = 2y^2$

e) $x = \tfrac{1}{2}y^2$ f) $2x = y^2$

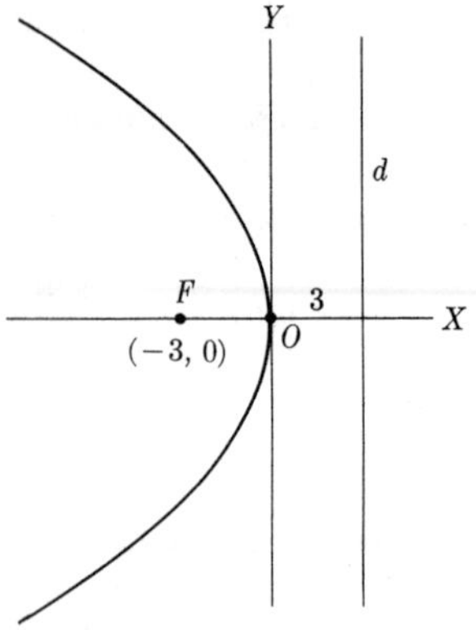

Fig. 12–13

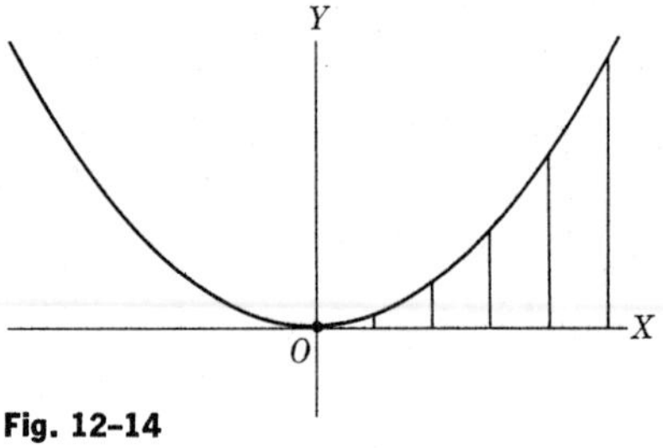

Fig. 12–14

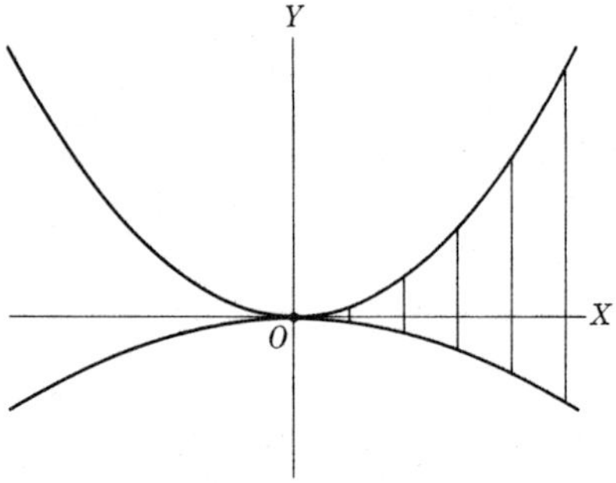

Fig. 12–15

2. Compare the curves of Exercises 1(a) and 1(b).
3. For the parabola shown in Fig. 12–13, the distance from focus to directrix is 6. What is the equation of the parabola?
4. If the equation of a parabola is $y=\frac{1}{8}x^2$, what are the coordinates of the focus? Describe the position of the directrix.
5. The following exercises specify the position of focus and directrix of a parabola in relation to a coordinate system. Find the equation of the parabola.
   a) Focus (0, 4); directrix parallel to, and 4 units below, the X-axis.
   b) Focus (0, 6); directrix parallel to, and 6 units below, the X-axis.
   c) Focus (0, −5); directrix parallel to, and 5 units above, the X-axis.
   d) Focus (4, 0); directrix parallel to, and 4 units to the left of, the Y-axis.
6. Suppose that the designer of a bridge has decided to use the parabolic cable $y=x^2$ for the range $x=-5$ to $x=5$ (Fig. 12–14). The roadbed of the bridge is to be the X-axis. Compute the lengths of straight wire needed to suspend the roadbed from the cable at $x=1$, 2, 3, 4, and 5.
7. Suppose that the designer of a bridge has decided to use the parabolic cable $y=x^2$ for the range $x=-5$ to $x=5$ (Fig. 12–15). The roadbed is to have the shape $y=-\frac{1}{10}x^2$. Compute the length of straight wire needed to suspend the roadbed from the cable at $x=1$, 2, 3, 4, and 5.

12-5 FINDING A CURVE FROM ITS EQUATION

The great merit of the idea conceived by Fermat and Descartes is that it permits us to represent a curve algebraically and, as we shall see later, learn much about the curve by working with the equation. But another value of their idea, hardly secondary in importance, is that any equation in x and y determines a curve. Hence by merely writing down any equation we please and by finding out what curve belongs to the equation we can discover many new curves. Although we shall not, at this time, make any sensational discoveries, let us see how the process of determining the curve of an equation can be carried out.

Suppose we consider the equation

$$y = x^2 - 6x \tag{13}$$

and try to determine the curve which has this equation. A direct method would be to calculate coordinates of points on the curve and then plot these points. Thus, when $x=2$, $y=-8$, and so $(2, -8)$ are the coordinates of a point on the curve. By calculating many such sets of coordinates and by plotting the corresponding points one can determine the shape of the curve. However, one often learns more by applying a little algebra.

If equation (13) had been $y=x^2$, we would know at once by comparing with equation (9) that it is a parabola with distance from focus to directrix equal to $\frac{1}{2}$. Let us see whether a little algebraic juggling with equation (13) might bring it into a form which will permit us to identify the curve. By adding 9 to both sides of equation (13) we obtain

$$y + 9 = x^2 - 6x + 9.$$

Now our knowledge of algebra tells us that the right side is $(x - 3)^2$. Hence we have

$$y + 9 = (x - 3)^2. \tag{14}$$

Suppose we introduce new letters x' and y' such that

$$x' = x - 3 \quad \text{and} \quad y' = y + 9. \tag{15}$$

Then substitution in equation (14) yields

$$y' = x'^2. \tag{16}$$

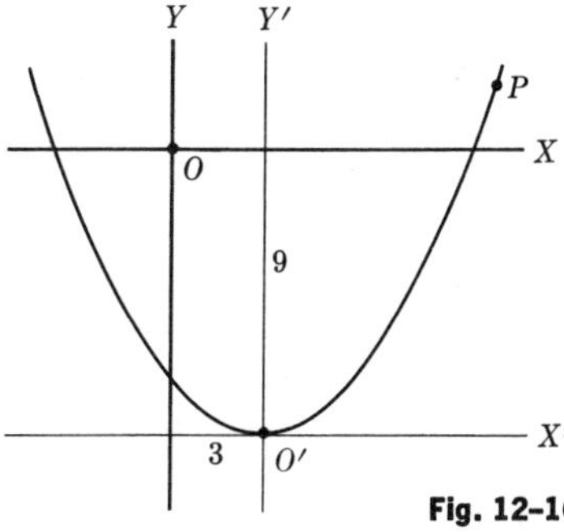

Fig. 12–16

The curve corresponding to this equation is the parabola shown in Fig. 12–16, where it is graphed with respect to the X'- and Y'-axes. Of course, we wish to find the curve corresponding to equation (13), not that described by (16). However, equations (15) provide the necessary connection. The equation $x' = x - 3$, or $x = 3 + x'$, tells us that the abscissas x of points belonging to (13) should be 3 more than the abscissas x' of points belonging to (16). How can we increase by 3 the abscissas of each point in Fig. 12–16? The answer is simple. We draw a new Y-axis 3 units to the left of the Y'-axis. Then the x-value of a typical point such as P is 3 units more than its x'-value. We now use the second equation in (15), that is, $y' = y + 9$, or $y = y' - 9$. This equation says that the y-values of points should be 9 units less than the y'-values. How can we reduce by 9 the ordinates of the curve in Fig. 12–16? We have already indicated the essential trick. We introduce a new X-axis 9 units above the X'-axis. Now the y-value of P is 9 units less than the y'-value. Hence with respect to the X- and Y-axes, the points of the parabola $y' = x'^2$ have the correct x- and y-values called for by equation (13). Since we did not change the curve in any way by introducing the X- and Y-axes, but merely changed the axes, the curve of equation (13) is a parabola placed with respect to the X- and Y-axes as shown in Fig. 12–16.

We were a bit lucky in analyzing equation (13) because the introduction of new coordinates in equation (15) reduced equation (13) to equation (16) whose curve we already knew. If this change to a familiar form is not possible, then the initial equation may indeed represent some new curve, and by analyzing the equation we might get to know the properties of this new curve. The results of such studies would be a further addition to the stock of knowledge about equations and their corresponding curves, a type of knowledge which the professional mathematician builds up for his work just as a writer may build up a bigger and bigger vocabulary. Thus through the notion of equation and curve, Fermat and Descartes opened up to mathematicians a vast variety of new curves.

EXERCISES

1. For equation (13) of the text calculate the coordinates of a number of points on the curve and plot the points. Then sketch in the curve. Does your graph look like the one in Fig. 12–16?
2. Determine the curve whose equation is $y = x^2 - 10x$.
3. Determine the curve whose equation is $y = -x^2 + 6x$. [*Suggestion:* Note that the given equation is the same as $-y = x^2 - 6x$ and use the results obtained for equation (13) of the text.]
4. Sketch the curve whose equation is $y = -x^2 + 6x$ by finding and plotting the coordinates of a number of points on the curve.
5. Knowing the curve which corresponds to $y = x^2 - 6x$, can you determine the curve which corresponds to $y = x^2 - 6x + 9$?
6. What does one mean by the statement that a curve can be associated with any equation in x and y?

7. Sketch the curves whose equations are given below.

a) $y = x^3$ b) $y = x^3 + 9$ c) $y = \frac{1}{x}$

Does the sketch of part (c) suggest one of the conic sections?

12-6 THE ELLIPSE

Another very widely used curve is the ellipse. Let us review the definition given in Chapter 6. We start with two fixed points, F and F', called foci, and a constant quantity which is greater than the distance FF'. We now consider all points, the sum of whose distances from F and F' is the constant quantity. This set of points is called an ellipse. To be more concrete, suppose that the distance FF' is 6 and that the constant quantity is 10. If P is a point such that $PF + PF'$ is 10, then P is a point on the ellipse.

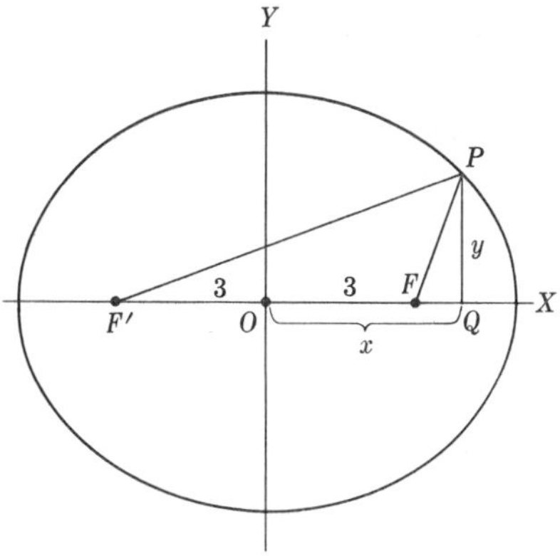

Fig. 12-17. An ellipse on a rectangular Cartesian coordinate system.

From the standpoint of coordinate geometry, the first thing of interest about the ellipse is its equation. Let us see whether we can find it. As in the case of the parabola, experience has taught mathematicians that the resulting equation will be simplest if the line FF' (Fig. 12-17) is chosen as the X-axis and the Y-axis is chosen to be the line perpendicular to the X-axis and halfway between F and F'. Let us consider the ellipse for which the length FF' is 6 units. Then the coordinates of F are (3, 0) and those of F' are (−3, 0). Now let P be any point on the curve and let us denote its coordinates by (x, y). If the constant quantity which determines the ellipse is 10, then the condition which any point P on the ellipse satisfies is

$$PF + PF' = 10. \tag{17}$$

We wish to express this condition algebraically. The procedure is straightforward. The distance PF is the hypotenuse of the right triangle PQF whose arms are $x - 3$ and y. Hence $PF = \sqrt{(x-3)^2 + y^2}$. The distance PF' is the hypotenuse of the right triangle PQF' whose arms are $x + 3$ and y. Hence $PF' = \sqrt{(x+3)^2 + y^2}$. Thus equation (17) amounts to

$$\sqrt{(x-3)^2 + y^2} + \sqrt{(x+3)^2 + y^2} = 10. \tag{18}$$

We are now in the same position that we were in when we arrived at equation (7) for the parabola. We could maintain that (18) is the equation of the ellipse, for it is indeed the condition which the coordinates (x, y) of any point on the ellipse satisfy. However, as for the parabola, a little algebra applied to (18) will simplify the equation. We shall not carry out the algebraic steps explicitly because they are uninteresting, and it is not important for us to acquire great facility in algebra. The result is

$$16x^2 + 25y^2 = 400. \tag{19}$$

We know what an ellipse looks like. But we do not know how our ellipse lies in relation to the axes chosen in Fig. 12–17. An analysis of equation (19) will supply the answer. First of all let us note that if (a, b) should happen to be the coordinates of a point which satisfy equation (19), that is, if

$$16a^2 + 25b^2 = 400, \tag{20}$$

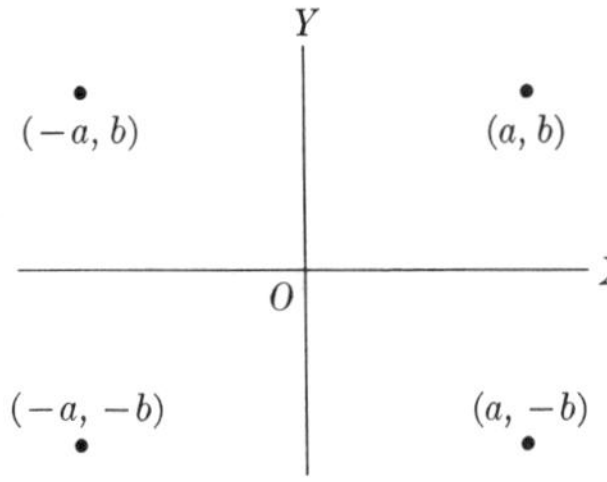

Fig. 12–18

then the sets of coordinates $(-a, b)$, $(a, -b)$, and $(-a, -b)$ will also satisfy the equation, because the substitution of any one of these latter three pairs of coordinates will yield the same equation as (20). Figure 12–18 shows where the various points (a, b), $(-a, b)$, $(a, -b)$, and $(-a, -b)$ lie in relation to the axes. We see, for example, that (a, b) and $(-a, b)$ are symmetrically placed with respect to the Y-axis. What we have learned so far is that if the ellipse contains a point (a, b) which lies in the first quadrant, it contains the point $(-a, b)$ which is symmetrically situated with respect to the Y-axis; it contains the point $(a, -b)$ which is symmetrically situated with respect to the X-axis; and it contains the point $(-a, -b)$ which is symmetric to $(-a, b)$ with respect to the X-axis. Hence, if we can determine which points lie on the ellipse in the first quadrant, we can, by symmetry, decide what the ellipse looks like in the other three quadrants.

The shape of the ellipse in the first quadrant is easily determined. We have but to calculate the coordinates of a number of points in the first quadrant and plot the points carefully with respect to the coordinate axes. By symmetry we obtain the shape of the curve in the other three quadrants. The final graph is that shown in Fig. 12–17.

We could investigate the equations of other curves and the curves corresponding to other equations. But what we have done should make the primary idea clear. To each curve there corresponds an equation which describes that curve. The equation depends upon how we choose the axes, but once this choice is made, the equation is unique. Conversely, given an equation involving x and y, we can find the curve which this equation describes, namely the collection of points whose coordinates satisfy the equation.

EXERCISES

1. For equation (19) of the text, calculate the coordinates of the points whose abscissas are 0, 1, 2, 3, 4, 5. Plot the points.
2. Sketch the ellipse whose equation is $9x^2 + 16y^2 = 144$.
3. Calculate the length of the X-axis which is contained within the ellipse represented by equation (19). Does this length have any relation to any of the quantities which determine the ellipse? This length is called the major axis of the ellipse.
4. Suppose that the constant quantity which defines an ellipse, 10 in the example discussed in the text, is retained, but the distance between the foci F and F' is 0. What changes must one make in equation (18)? Can you now simplify the equation and recognize the curve that it represents?
5. Kepler's first law of planetary motion says that the path of each planet is an ellipse with the sun at one focus. Let us suppose that equation (19) of the text is the equation of some planet's path and that the sun is at F. What is the planet's distance from the sun when it crosses the positive X-axis and what is the distance when it crosses the negative X-axis?

* 12–7 THE EQUATIONS OF SURFACES

The mathematician has but to get hold of an idea, and he will develop it for all that it is worth. It had already occurred to Descartes that the idea of equations for curves might be extended to finding equations for surfaces. This possibility was soon explored.

A curve can lie in one plane, but surfaces such as a sphere or the ellipsoid (of which the surface of the earth and the surface of a football are examples), do not lie in one plane. They exist in three-dimensional space. To pursue the idea of finding equations for surfaces, we must first introduce coordinates for points in space. This is readily done. One introduces three mutually perpendicular lines (Fig. 12–19) as axes instead of the two lines used for points in the plane. These are called the coordinate axes. The X- and Y-axes determine a plane called the XY-plane. Similarly, the X- and Z-axes determine the XZ-plane, and the Y- and Z-axes define the YZ-plane.

The location of a point P in space is described by three numbers. For example, the point P of Fig. 12–19 is described by (3, 4, 5). The number 5

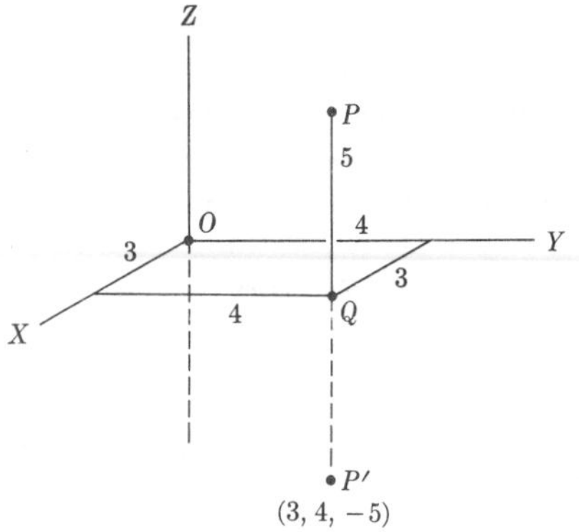

Fig. 12-19.
A three-dimensional rectangular Cartesian coordinate system

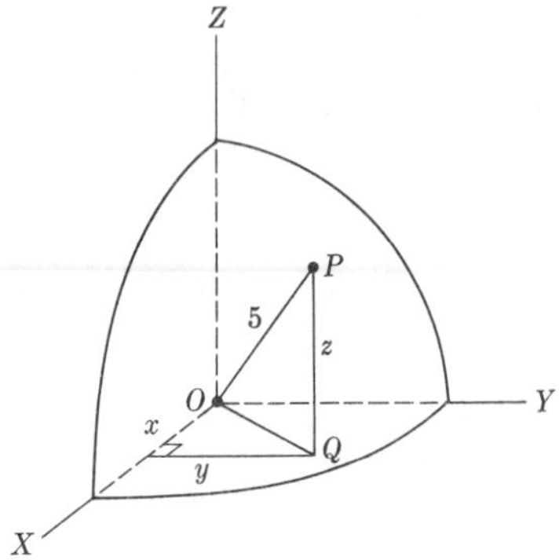

Fig. 12-20.
A sphere on a three-dimensional rectangular Cartesian coordinate system.

describes the perpendicular distance of P above the XY-plane, while 3 and 4 are the x- and y-coordinates of Q, the foot of the perpendicular from P to this plane. Alternatively one can say that if one proceeds a distance of 3 units along the X-axis, then a distance of 4 units along a parallel to the Y-axis, and finally travels upward a distance of 5 units along the perpendicular to the XY-plane, he will arrive at the point P. To represent all points in space, we must, as in the two-dimensional system, use negative numbers also. Thus points below the XY-plane have negative third coordinates.

Let us now consider the problem of finding the equation of a surface. We shall use the sphere as an example. A surface, like a curve, has some defining property which states just which points belong to it. By definition, the sphere is the set of all points in space at a given distance, the radius, from a fixed point called the center. To be concrete let us suppose that all points of our sphere are 5 units from the center and that the sphere is located so that its center is at the origin of our three-dimensional coordinate system (Fig. 12–20). The general point on a surface is represented by three letters, x, y, z. Thus the coordinates of the general point P are (x, y, z). Let us now express algebraically the fact that the distance of any point (x, y, z) on the sphere is 5 units from the origin. The lengths x, y, and z are shown in Fig. 12–20. Now x and y are the arms of a right triangle (lying in the XY-plane) whose hypotenuse is OQ. Then by the Pythagorean theorem,

$$x^2 + y^2 = OQ^2. \tag{21}$$

Further OQ and z are the arms of the right triangle OQP, whose hypotenuse is OP or 5 units. Hence

$$OQ^2 + z^2 = 25. \tag{22}$$

But OQ^2 has the value given by equation (21). If we substitute this value in equation (22), we obtain

$$x^2 + y^2 + z^2 = 25. \tag{23}$$

This is the equation of a sphere in the sense that the left side equals the right side when and only when the coordinates of a point on the sphere are substituted for x, y, and z.

We could readily obtain the equations of a few other surfaces, such as plane, paraboloid, and ellipsoid. But we shall not do so because we shall not utilize three-dimensional coordinate geometry and the procedure involved is a more or less apparent extension of a familiar concept.

EXERCISES

1. Plot the points whose coordinates are given below.

   a) (1, 2, 3) b) (1, 2, −3) c) (−1, 2, 3) d) (1, −2, 3)

2. What geometrical figure does an equation in x, y, and z represent?
3. Describe the surface whose equation is $x^2 + y^2 + z^2 = 49$.
4. We know that an equation such as $x + y = 5$ represents a straight line. What does the equation $x + y + z = 5$ represent?

* 12-8 FOUR-DIMENSIONAL GEOMETRY

Our experience is limited to figures lying in a plane and in space. But our intellects are not. The idea of a four-dimensional world and of figures in it had tantalized mathematicians such as Pascal at a time when coordinate geometry was still being fashioned. During the next 200 years the subject was mentioned occasionally, but it was not taken seriously until some startling developments (which we shall consider in Chapter 20) caused a number of mathematicians, notably Bernhard Riemann, to investigate it. Four-dimensional geometry proved to be more than a speculation, for some of the deepest developments in modern science, notably the theory of relativity, use this concept. Let us see how coordinate geometry can be employed to portray a four-dimensional world.

We have seen that the equation of a circle of radius 5 in a two-dimensional coordinate system is

$$x^2 + y^2 = 25, \tag{24}$$

and we have just seen that the equation of a sphere of radius 5 in a three-dimensional coordinate system is

$$x^2 + y^2 + z^2 = 25. \tag{25}$$

Even idle speculation would suggest that we at least write down the equation

$$x^2 + y^2 + z^2 + w^2 = 25 \tag{26}$$

and consider what meaning it might have. It would seem reasonable to interpret x, y, z, and w as the coordinates of a point in four-dimensional space and, by analogy with equations (24) and (25), to interpret equation (26) as the equation of a hypersphere in four-dimensional space. This is exactly what mathematics does. However, mathematics does *not* suppose that there is anywhere a real four-dimensional space in which four mutually perpendicular axes can be set up. Nor do mathematicians claim that they, wise and farsighted as they think they are, can visualize figures in a four-dimensional space. It follows that no one else can either.

Four-dimensional geometry is entirely a creation of the mind; it is a geometry without pictures. One speaks of the coordinates (x, y, z, w) as representing a point, and one uses the term hypersphere as though it were a real geometrical figure corresponding to equation (26), but these geometrical terms are merely a convenience and a carry-over from two- and three-dimensional geometry. The words are suggestive but not descriptive of actual figures.

Suppose it is agreed then that four-dimensional geometry is indeed a mental creation. Is it of any value? There is excellent reason to study this "geometry", and there is excellent use for it. We can understand these facts better if we backtrack a bit. Consider the equation $x^2 + y^2 = 25$. It represents, as we know, a circle. But where is this curve that knows no end, this "arc unbroken", the cherished figure of the Greeks? Every geometrical fact we know or can establish about the geometrical circle has its algebraic equivalent and can be derived algebraically from the equation of the circle. Hence the equations of the circle and of other curves in plane, or two-dimensional, geometry are a complete substitute for their geometrical counterparts, and we could, if we wished to, eliminate geometry altogether. In four-dimensional "geometry" we have only the equations, but we talk about them as though they represented figures in a four-dimensional world. The properties of these figures are completely specified by the equations. What is lacking is the possibility of actually constructing such figures.

So far then we have tried to see what meaning this four-dimensional geometry has. Now we wish to know how it is used. One application is made in studying physical events wherein time plays a role. Consider for example, the motion of a planet. The location of a planet is described by three coordinates x, y, and z. But the instant at which the planet occupies that location is also important. An eclipse of the sun, for example, occurs because planet and sun are in certain positions at the same instant. Hence the full description of the position of a heavenly body requires four coordinates, the fourth being a value of time. The path of a celestial object is described by equations involving four letters, usually x, y, z, and t.

But is there any value to thinking geometrically about equations involving four letters? There is. Let us consider the usual sphere for a moment. Some curves on this sphere, a circle of latitude, for example, lie in one plane, and hence only two-dimensional geometry is needed to visualize some curves which lie on a three-dimensional sphere. Similarly, the path on which a planet moves in the four dimensions of space and time may be a curve which can exist in three-dimensional space. This curve may be part of a "geometrical structure" which lies in four-dimensional space, just as the circle of latitude is part of a sphere and yet the curve can be visualized. This visualization aids the understanding. This same visualization might be possible for the paths of other planets and one can consequently better understand their motions. Yet the proper interrelationship of these several paths can be represented only in four-dimensional space just as the relationship of the circles of latitude to one another can be represented only in three-dimensional space. We see therefore that it is helpful to think in terms of geometrical figures lying in a four-dimensional space.

This brief presentation of four-dimensional geometry may give some further indication of the direction in which scientific thought has been moving with the aid of mathematics. Copernicus asked the world to accept a theory of planetary motions which violated some sense impressions for the sake of a better mathematical account. The utilization of a four-dimensional geometry which has no sensuous or visual content means complete reliance upon the mind.

12-9 SUMMARY

From the purely mathematical standpoint coordinate geomery offers a brand-new thought, the representation of geometrical figures by equations. It also offers, as Descartes and Fermat had expected, a new mathematical method of deriving properties of figures from equations. For example, the fact that a curve is symmetric with respect to some line is readily seen from the equation, as we observed in the case of parabola and ellipse.

But Descartes' and Fermat's union of algebra and geometry means far more than a new mathematical method of working with curves. The forms of all physical objects which are studied for any reason whatsoever are, at least when idealized, curves and surfaces. The fusilage of an airplane, the wings of an airplane, the hull of a boat, and the shape of a projectile are surfaces. The paths of all moving objects, a ball thrown by a child, an electron expelled from an atom, a ship on the ocean, a plane in the air, the planets in the heavens, and the tracks of light, are curves. These surfaces and curves can be represented by equations, and the shapes or motions studied by applying algebra to these equations. In other words, Descartes and Fermat made possible the algebraic representation and the study by algebraic means of the various objects and paths of interest to scientists. In addition, algebra supplies quantitative knowledge. This method of working with curves and surfaces is so basic in science that Descartes and Fermat may very well be called the founders of mathematical physics. Part of Descartes' greatness and perhaps the largest part of his contribution was his vision of what his method accomplished; he said he had "reduced physics to mathematics." The investigation of nature which Renaissance Europe had determined to undertake was enormously expedited as we shall soon see. The story of coordinate geometry illustrates how an interest in geometric method became immensely valuable for science and engineering.

EXERCISES

1. In what sense does a four-dimensional geometry exist?
2. What geometrical language would be appropriate to describe the figure whose equation is $x + y + z + w = 5$?
3. Did Descartes and Fermat introduce a new method for working with curves? If so, describe it.
4. Does coordinate geometry replace Euclidean geometry?

REVIEW EXERCISES

1. What is the x-coordinate of any point on the Y-axis?
2. What is the y-coordinate of any point on the X-axis?
3. If two points lie on a line parallel to the Y-axis, what can you say about their x-coordinates?
4. Write the equation of a line whose slope is 4 and which
   a) passes through the origin,
   b) cuts the Y-axis at the point (0, 2).
5. Write the equation of a line whose slope is -4 and which
   a) passes through the origin,
   b) cuts the Y-axis at the point (0, 2),
   c) cuts the Y-axis at the point (0, -2).
6. Show that the point whose coordinates are $(2, \sqrt{21})$ lies on the circle

$$x^2 + y^2 = 25.$$

7. Sketch the curves whose equations are given below.

a) $y = \frac{1}{8}x^2$ b) $y = 8x^2$ c) $y = -\frac{1}{8}x^2$

8. Sketch the curves whose equations are given below.

 a) $y = x^2 + 6x$ b) $y = -x^2 - 6x$
 c) $y = x^2 + 6x + 9$, by relating the curve to the one in part (a).

9. Plot the graph of $16x^2 + 25y^2 = 400$ in the first quadrant by solving for y and then making a table of values.
10. Plot the entire graph of the equation $x^2 - y^2 = 4$. Can you identify the curve?
11. One can regard the quadratic equation $x^2 - 6x = 0$ as a special case of

$$y = x^2 - 6x,$$

the special nature being that the values of x which satisfy the first equation are those which correspond to $y = 0$ in the second one. Can you then suggest a graphical method of solving the quadratic equation?

Topics for Further Investigation

1. The life and work of René Descartes.
2. The life and work of Pierre de Fermat.
3. Four-dimensional geometry.

Recommended Reading

ABBOTT, E. A.: *Flatland, A Romance of Many Dimensions*, Dover Publications, Inc., New York, 1952.

DESCARTES, RENÉ: *Discourse on Method*, Penguin Books Ltd., Harmondsworth, England, 1960 (also many other editions).

DESCARTES, RENÉ: *La Géométrie* (the original French and an English translation), Dover Publications, Inc., New York, 1954.

HALDANE, ELIZABETH S.: *Descartes, His Life and Times*, J. Murray, London, 1905.

MANNING, H. A.: *The Fourth Dimension Simply Explained*, Dover Publications, Inc., New York, 1960.

SAWYER, W. W.: *Mathematicians' Delight*, Chap. 9, Penguin Books Ltd., Harmondsworth, England, 1943.

SCOTT, J. F.: *The Scientific Work of René Descartes*, Taylor and Francis, Ltd., London, 1952.

WHITEHEAD, ALFRED N.: *An Introduction to Mathematics*, Chaps. 9 and 10, Holt, Rinehart and Winston, Inc., New York, 1939 (also in paperback).

CHAPTER 13

THE SIMPLEST FORMULAS IN ACTION

When you can measure what you are talking about and express it in numbers, you know something about it.

LORD KELVIN

13-1 MASTERY OF NATURE

We have already mentioned not only a revival of interest in the study of nature but a decided effort on the part of manufacturers, artisans, and engineers to utilize materials effectively and to lighten the burden of labor. This more practical interest in exploiting knowledge of materials and of natural phenomena in behalf of economic and social needs was an incentive for scientific activity which was adjoined to the older goal so strongly pursued by the Greeks, namely, the understanding of nature. The new motivation for scientific work, mastery of nature for the welfare of man, was proclaimed and advocated by such prominent and respected thinkers as Francis Bacon (1562–1626) and René Descartes. Bacon criticizes the Greeks. He says that the interrogation of nature should be pursued not to delight scholars but to serve man. It is to relieve suffering, to better the mode of life, and to increase happiness. Let us put nature to use. Knowledge should bear fruit in works; science should be applied to industry. In Bacon's words, let us ascend to knowledge and descend to work. Man should reconstitute his knowledge to apply it to the relief of man's estate. "The true and lawful goal of science is to endow human life with new powers and inventions." Bacon foresaw that science could provide man with "infinite commodities" and minister to the conveniences and comfort of man.

Descartes, too, is explicit about employing science for practical ends. He says, "It is possible to attain knowledge which is very useful in life, and instead of that speculative philosophy which is taught in the schools, we may find a practical philosophy by means of which, knowing the force and action of fire, water, air, the stars, heavens and all other bodies that environ us, as distinctly as we know the different crafts of our artisans, we can in the same way employ them in all those uses to which they are adapted, and thus render ourselves the masters and possessors of nature."

The founder of modern chemistry, Robert Boyle, expressed the same thought: "The good of mankind may be much increased by the naturalist's insight into the trades." The mathematician and philosopher Leibniz, about whom we shall learn more later, proposed in 1669 the organization of a society devoted to making inventions in mechanics and discoveries in chemistry and physiology which would be useful to people. He, too, wanted to put knowledge to use. He called the universities monkish and said that they were absorbed in trifles. They possessed learning but no judgment. Instead he urged the pursuit of real knowledge, mathematics, physics, geography, chemistry, anatomy, botany, zoology, and history. To Leibniz the skills of the artisan and the practical man were more valuable than the learned subtleties of professional scholars.

One should not conclude that science and mathematics became concerned exclusively with the solution of problems facing society. It is true that the scientists of the seventeenth century worked on many specific practical problems, the invention of a clock, the improvement in methods of determining longitude, the design of better lenses, and so on. And they focused even their more general theoretical effort on those fields of pure science—astronomy, motion, and optics—in which practical problems predominated or whose investigation gave promise of solving practical problems. But the desire to understand nature's ways was by no means lost; it remained the outstanding motivation for the truly great scientists and mathematicians.

13-2 THE SEARCH FOR SCIENTIFIC METHOD

We have tried to point out thus far that the scientific needs and interests of seventeenth-century Europe were great. But need and interest do not in themselves produce results. A need for money and an interest in money do not provide money. The question still remains, How did the European scientists go about solving scientific problems? How does one come to grips with nature either to understand or subjugate her? One might be tempted to guess that the Europeans found the proper scientific method in the revived Greek literature. But this was not the case.

We shall review a few principles of Greek science and late medieval science, such as it was, to appreciate the changes made in the seventeenth century. First of all most Greeks and medieval thinkers believed that the basic truths exist within the human mind. They are already implanted at birth and are called upon when desired, or they are so clearly truths that when proposed, the mind immediately recognizes them as such. Thus the axioms of Euclidean geometry were accepted by the Greeks as self-evident truths. The medievalists added revelation from God as another source of truth, but again a source communicated to man's mind. The task of science, then, was to determine the implications of these principles by reasoning.

To this source of knowledge, Aristotle and his followers added observation and induction on the basis of observations. Although Aristotle, Galen, the celebrated physician, and astronomers such as Hipparchus and Ptolemy certainly made observations, inductive conclusions did not play a great role. Also, observational results were more likely to be forced to fit a preconceived notion than allowed to suggest some new conclusion. For example, the principle that heavenly motions must somehow consist of circular paths because only circular motion is complete and perfect dominated all Greek and medieval astronomy.

Another methodological principle employed by the Greeks was classification, an approach stressed by Aristotle and taken over by his medieval followers. Thus one observed varieties of animals, flowers, fruits, and humans and classified them according to genus and species. This method is, of course, still used in biology and has some general applicability. It at least reduces the variety of organisms to a few major types and permits systematic study of whole classes in one swoop. It is relevant that Aristotle himself was a physician.

The Aristotelians pursued another scientific doctrine which is best described by the key words "qualitative study of nature." They believed that all phenomena could be explained in terms of the acquisition or loss of basic substances. Thus they and the Platonists believed that heat, coldness, wetness, and dryness were basic substances, and these substances, combined in different proportions, produced other substances. Heat and dryness produced fire; heat and wetness produced air; coldness and wetness produced water; and coldness and dryness produced earth. The hardness or softness, coarseness or fineness of various substances was accordingly determined by the relative abundance of the four basic elements in them. Solids, fluids, and gases were also distinguished by the possession of special substances. Thus a fluid such as mercury possessed some quality, fluidity, which gold did not have. To change mercury into gold meant that one had to take away the fluidity and substitute a new quality which supplied rigidity. Today we recognize that solidity, fluidity, and gaseousness are states of the same matter. However, explanation in terms of special substances was employed right up to modern times. Early chemists, for example Robert Boyle, ascribed the fact that substances such as sulphur were easily set afire to the presence of a special substance called phlogiston. Until the nineteenth century heat was considered to be a substance called caloric which bodies lost or gained as they lost or gained heat. Electricity in the eighteenth century was conceived of as a fluid which flowed through metals.

Aristotelian and medieval science also emphasized another objective for science, namely explanation. To explain meant to give the cause of a phenomenon. However, there were four distinct types of causes, each important in its own way. Suppose that an architect builds a church. The material cause of the church is the brick, stone, and mortar of which the church is constructed. The formal cause is the design which the architect has in mind. Then there is the effective cause, that is, the actual building process. The fourth type, called final cause, is the purpose which the entire project serves. In the present example, the purpose might be to provide a house of worship or to glorify God. Of these four types the final cause was considered most important because it supplied the meaning people usually seek. Thus when we ask why some one was killed and are told that the killer sought revenge, we are satisfied. An entirely different explanation might be furnished in terms of the physical and physiological processes which took place. But such an explanation is usually not of as much interest.

In medieval thought, the final cause dominated. Rain falls to water the crops and supply drinking water. Plants grow to supply food for man. Balls fall to earth because all objects seek their natural place, and the natural place of heavy objects is the center of the universe which is the center of the earth.

By the sixteenth century many scholars realized that science could not be advanced by such means. They recognized that new principles and entirely new methods were needed, but did not have a clear conception of what these should be. Prior to Galileo's work, one idea emerged distinctly from the writings of Aristotle's critics, namely the need for systematic experimentation. Francis Bacon issued the manifesto for the experimental method. He attacked preconceived philosophical systems, barren speculations, and idle displays of learning. Scientific work, he said, should not become entangled in a search for final causes which belonged to philosophy. In his *Advancement of Learning* (1605) and in his *Novum Organum* (1620) he points out the feebleness of efforts and the paucity of results in past studies of nature. Man, he observes, has put very little thought and labor into science. Let us come to grips with nature. Let us not have desultory and haphazard experimentation, but let it be thorough and directed. He then makes the acute and most important statement that the only hope for progress lies in a change of method for science. All knowledge begins with observations. But *it must proceed by gradual and successive inductions rather than by hasty generalizations.* He contrasts the anticipation of nature with the interpretation of nature. The one skims; the other is orderly. We gain our ends only if we start with correct laws of nature. He criticizes the then current notions of substance, quality, action, being, heaviness, lightness, density, rareness, moistness, dryness, generation, corruption, attraction, and repulsion. The Aristotelian emphasis on form, he says, is fantastic and ill defined. Man masters nature by understanding her.

Bacon's stress on experimentation and induction from experimental results did reflect what was beginning to take place in Europe. The work of the biologists Vesalius, Cesalpinus, and Harvey was mentioned in Chapter 9. Famous for his systematic experimentation is William Gilbert (1540–1603), physician to Queen Elizabeth. In his *De Magnete* (1600) he presents the details and results of his clear and fruitful work on magnetism, a phenomenon about which practically nothing was known. Gilbert states explicitly that we must start from experiments. Kepler's regard for observational facts has already been mentioned. Galileo, too, performed some key experiments on motion; about his results in this area we shall say more later. Moreover, he and his pupils, notably Evangelista Torricelli (1608–1647), having convinced themselves that air has weight, proceeded to carry out relevant experiments. Torricelli also investigated the flow of water through nozzles. Blaise Pascal and Robert Boyle (1627–1691) worked on the pressure of fluids. Boyle and the French priest Edmé Mariotte (1620–1684) studied gases such as air. Otto von Guericke (1602–1686) invented the air pump and used it to demonstrate the pressure of air. René Descartes experimented in chemistry, biology, and optics. Robert Hooke was a famous experimenter, whose work on springs we shall discuss later, and Christian Huygens (1629–1695) obtained distinguished results from his experiments with the pendulum. Newton's work on light was one of the greatest experimental achievements of the seventeenth century.

It is also true that the artists, engineers, and craftsmen, concerned with the practical problems of their trades or professions, did not wait for new scientific methods to gain further knowledge of nature. They investigated mechanical forces, the design of lenses, the chemistry of paints, the motion of cannon balls, and other phenomena and discovered new facts. To this class belongs the self-educated sixteenth-century mathematician Tartaglia, who worked on projectile motion and arrived at results which contradicted Aristotelian physics. The Dutch engineer Simon Stevin (1548–1620) learned about the pressure exerted by water on the walls of canals, and made precise observations of the nature of stable and unstable equilibrium of bodies. He also studied the motion of bodies on slopes. It was spectacle makers who, without discovering a single law of optics, nevertheless invented the telescope and microscope. Many of these men sought not ultimate meanings but common useful knowledge.

There is no doubt that experimentation and practical investigations by technicians and engineers did produce new facts and even opened up new lines of inquiry, but the rise of experimentation was not the reason that science suddenly blossomed in the seventeenth century. The value and import of seventeenth-century experimentation has been vastly overrated. Modern science owes its origins and present flourishing state to a new scientific method which was fashioned almost entirely by Galileo Galilei. Galileo's method is doubly important to us because, as we shall see, it assigned a major role to mathematics.

13-3 THE SCIENTIFIC METHOD OF GALILEO

Galileo, born in Pisa in 1564, entered the university of his native city to study medicine. He also took private lessons in mathematics and was so strongly attracted to the subject that he decided to make mathematics his profession. At the age of 23 when his application for a teaching position at the University of Bologna was rejected because he did not seem worthy of an appointment, he accepted a professorship of mathematics at Pisa. Galileo was one of the men who attacked Aristotelian science, and he did not hesitate to express his views even though these criticisms alienated his colleagues. He had also begun to write important mathematical papers which aroused jealousy in the less competent. Galileo was made to feel uncomfortable, and left in 1592 to accept the position of professor of mathematics at the University of Padua. After 18 years at Padua he was invited to Florence by the Grand Duke Cosimo II. He appointed Galileo "Chief Mathematician" of his court, gave him a home and handsome salary, and protected him from the Jesuits who had gained domination of the Papacy and who had already threatened Galileo because of his defense of Copernican theory. In Florence Galileo had leisure to pursue his studies and to write. There he spent 23 years. In gratitude Galileo named the satellites of Jupiter, which he discovered in the first year of his service under Cosimo de'Medici, the Medicean stars.

After his condemnation by the Roman Inquisition in 1633 he was forbidden to publish any more. But he undertook to write up his years of thought and work on phenomena of motion and the strength of materials. The manuscript, entitled *Discourses and Mathematical Demonstrations Concerning Two New Sciences* (also referred to as *Dialogues Concerning Two New Sciences*), was secretly transported to Holland and published there in 1638. Galileo defended his actions with the words that he had never "declined in piety and reverence for the Church and my own conscience." He died in 1642.

Galileo began his investigation of the methodology of science by asking what is fundamental about the world of phenomena perceived by the senses, a question also considered by Descartes. Both agreed, as some philosophers had asserted earlier, that color, tastes, smells, sounds, and the various sensations of heat, hardness, and softness of objects are not distinct physical substances, but are effects which physically existing properties produce in human beings. What then does exist outside of man and is independent of man? The extension of objects, their shapes and sizes, and their motion are real and external to human perception. Galileo says,

> *If ears, tongues, and noses were removed, I am of the opinion that shape, quantity* [*size*], *and motion would remain, but there would be an end of smells, tastes, and sounds, which, abstractedly from the living creature, I take to be mere words.*

Descartes' famous words in this connection are, "Give me extension and motion and I will construct the universe." The idea advocated by these two men is known as the doctrine of primary and secondary qualities. The primary qualities exist in the physical world, and their effects on the sense organs of human beings produce the secondary qualities.

Thus in one sweeping blow Descartes and Galileo stripped away a thousand phenomena and qualities to concentrate on matter and motion. But this was only the first step in the new approach to nature which Galileo was fashioning. His next thought, one also voiced by Descartes and even by Aristotle, was that any branch of science should be patterned on the model of mathematics. This implies two essential steps. Mathematics starts with axioms, that is, clear, self-evident truths. From these it proceeds by deductive reasoning to establish new truths. So any branch of science should start with axioms or first principles and proceed deductively

Galileo departs radically from the Greeks, medievalists, and even Descartes in the method of obtaining these first principles. As noted earlier, the pre-Galileans believed that the mind supplies the basic principles. These men, we might say, first decided how the world should function and then fitted what they saw into their preconceived principles. Galileo decided that in physics as opposed to mathematics basic principles must come from experience and experimentation; they will be correct if attention is paid to what nature says rather than what the mind prefers. He openly criticized scientists and philosophers who accepted principles which conformed to their preconceived ideas of how nature should and must behave. He said that nature did not first make men's brains and then arrange the world so that it would be acceptable to human intellects. To the medievalists who kept repeating Aristotle and debating the meaning of his works, Galileo addressed the criticism that knowledge comes from observation and not from books. It was useless to debate about Aristotle. Those who did he called paper scientists who fancied that science was to be studied like the *Aeneid* or the *Odyssey* or by collation of texts. "When we have the decrees of nature, authority goes for nothing; . . ." Of course some Renaissance thinkers and Galileo's contemporary, Francis Bacon, had also arrived at the conclusion that experimentation was necessary. With respect to this particular plank of his new method Galileo was not ahead of all others. Yet even a modernist as great as Descartes did not grant the wisdom of Galileo's reliance upon experimentation. The facts supplied by the senses, he said, can only lead to delusion. Reason penetrates such delusions. Particular phenomena of nature can be deduced from, and understood in terms of, the innate general principles. In much of his scientific work, Descartes did experiment and require that theory fit facts, but in his philosophy he was still tied to truths of the mind.

The phenomena one observes are so numerous, so varied, so unlike each other that one can well despair of finding any principles at all in nature. Galileo decided that he must penetrate to the core of a phenomenon and begin there. He says in his *Two New Sciences* that it is impossible to treat the infinite variety of weights, shapes, and velocities. But he had observed that different objects fall with more nearly equal speeds in air than in water. Hence the thinner the medium, the smaller the difference in speed of fall among bodies. "Having observed this I came to the conclusion that in a medium totally devoid of resistance all bodies would fall with the same speed." What Galileo was doing here was to strip away the incidental or minor effects in the effort to get at the essential or major one. He then imagined what would happen if all resistance were removed, that is, if bodies *fell in a vacuum*, and he obtained the principle that in a vacuum all bodies fall according to the same law. Thus Galileo did not just experiment and infer from experiments. He tried to discard the relatively unimportant and nonessential, and here he showed genius, for, as any card player knows, to recognize what to discard is wisdom. In other words, he *idealized*. He did just what the mathematician does in studying real figures. The mathematician strips away molecular structure, color, and thickness of lines, to get at some basic properties, and he concentrates on these. So did Galileo penetrate to basic physical principles.

Of course, actual bodies do fall in resisting media. What could Galileo say about such motions? His answer was

> *. . . hence, in order to handle this matter in a scientific way, it is necessary to cut loose from these difficulties (air resistance, friction, etc.) and having discovered and demonstrated the theorems, in the case of no resistance, to use them and apply them with such limitations as experience will teach.*

Thus far Galileo had formulated a number of methodological principles, many of which were suggested by the pattern mathematics employed in algebra and in geometry. His next principle was to apply mathematics itself. Galileo proposed to seek for science axioms and theorems of a special kind. Unlike the Aristotelians and the medieval scientists, who had fastened upon the notion of fundamental qualities, studied the acquisition and loss of these qualities, or debated their meaning, Galileo proposed to seek *quantitative* axioms. The change is most important, and we shall see its full significance in several succeeding chapters. But an elementary example may help at the moment to demonstrate some of its implications. The Aristotelians said that a ball falls because it has weight and it falls to the earth because it, like every object, seeks its natural place, and the natural place of heavy bodies is the center of the earth. The natural place of a light body, such as fire, is in the heavens, and hence fire rises. These principles are qualitative. By contrast let us consider the statement that the speed (in feet per second) with which a ball falls is 32 times the number of seconds it has been falling. This statement can be expressed more briefly in symbols. If we denote by v the speed of the body and by t the number of seconds it has been falling, then the above assertion amounts to $v = 32t$. This simple statement illustrates many important ideas. But the relevant one at the moment is that it is primarily quantitative. It tells us the speed that a ball will acquire in a given number of seconds. In two seconds its speed will be 64 feet per second; in 3 seconds, 96 feet per second; and so on. In the expression $v = 32t$, the letters v and t stand for many values. We can substitute for t any number we please and calculate the corresponding value of v. Technically, v and t are called variables, and the relation $v = 32t$ is called a *formula*.

Galileo intended to adopt such formulas as his axioms, and he expected, by mathematical means, to deduce from them new formulas which would serve as theorems. Since formulas give quantitative knowledge, we can perhaps begin to comprehend the meaning of the statement that Galileo sought quantitative knowledge. Moreover we see that mathematics was to be the essential medium in his scientific reasoning.

The decision to seek quantitative knowledge expressed in formulas engendered another decision which was also radical, although at first contact it hardly reveals its full significance. As pointed out earlier in this chapter, the Aristotelians believed that one of the tasks of science was to explain why things happened, and explanation meant unearthing the causes of a phenomenon. The statement that a body falls because it has weight gives the effective cause of the fall, and the statement that it seeks its natural place gives the final cause. But the quantitative statement $v = 32t$, for whatever it may be worth, does not explain why a ball falls. It tells only how speed changes with time. In other words, formulas do not explain; they describe. And the knowledge of nature Galileo sought was descriptive. He says, for example, in his *Two New Sciences* that he will investigate and demonstrate some of the properties of motion without regard to what the causes might be. Positive scientific inquiries were to be separated from questions of ultimate causation.

First reactions to this thought of Galileo are likely to be negative. Descriptions of phenomena in terms of formulas hardly seem to be more than a first step. It would appear that the Aristotelians had really grasped the true function of science, namely, to explain why phenomena happened. Even Descartes protested Galileo's decision to seek descriptive formulas. He said, "Everything that Galileo says about bodies falling in empty space is built without foundation: he ought first to have determined the nature of weight." Further, said Descartes, Galileo should reflect about ultimate reasons. But we shall see more clearly in the space of a few chapters that Galileo's decision to aim for description was the most profound and the most fruitful thought that anyone has had about scientific methodology. We merely wish to recapitulate here that the scientific knowledge which Galileo envisioned was to consist of a series of mathematical formulas deduced from a few fundamental ones.

Since the laws Galileo proposed to find were to be quantitative, they obviously had to relate measures, sizes, or amounts of some physical quantities, just as $v = 32t$ relates measures of speed and time. Here, too, Galileo made a fundamental contribution. Whereas the Aristotelians had talked in terms of qualities such as earthiness, fluidity, rigidity, essences, natural places, natural and violent motion, potentiality, actuality, and purpose, Galileo not only introduced an entirely new set of concepts but chose concepts which were measurable so that their measures could be related by formulas. Some of his concepts, such as distance, time, speed, acceleration, force, mass, and weight, are, of course, familiar to us and so the choice does not surprise us. But to Galileo's contemporaries these choices and in particular their adoption as *funda-*

mental concepts, were startling. However, these very ones did prove to be most instrumental in the task of understanding and mastering nature.

We have described the essential features of Galileo's program. Some of his ideas had been espoused by others. Some were entirely original with him. But what establishes Galileo's greatness in the invention of this methodology is that he saw clearly what was wrong or deficient in the scientific efforts of his age, completely shed the older ways, and formulated the new steps—almost in so many words. Moreover, he applied his method to problems of motion and in this work not only managed to provide a lucid example of the procedure but succeeded in obtaining brilliant results. He showed, in other words, that it worked. Galileo was fully conscious of what he had accomplished. He says toward the end of his *Two New Sciences*, "So that we may say the door is now opened, for the first time, to a new method fraught with numerous and wonderful results which in future years will command the attention of other minds." But others were also aware of Galileo's greatness. The seventeenth-century philosopher Thomas Hobbes said of Galileo, "He has been the first to open to us the door to the whole realm of Physics. . ."

Since we are interested in the role of mathematics in the modern world, it may be worth while to emphasize one point. The scholars who fashioned modern science, Descartes, Galileo, and Newton, approached the study of nature as mathematicians. They proposed to find broad, profound, but also simple and clear mathematical principles either through intuition or through crucial observation and experiments and then expected to deduce new laws from these principles, entirely in the manner in which mathematics proper had constructed its geometry and algebra. Mathematical deduction was to take up the major share of scientific activity. Galileo says he valued a scientific principle, whether or not obtained by experimentation, far more because of the abundance of theorems which he could deduce from it than because of the knowledge afforded by the principle itself.

What these great thinkers envisioned did in fact prove to be the profitable course. For the next two centuries, scientists formulated precise and sweeping mathematical laws of nature on the basis of slim, almost trivial, observations and experiments. The greatest progress in the seventeenth and eighteenth centuries occurred in mechanics and in astronomy, and in both these fields experimental results were hardly startling and certainly not decisive. The significant contribution, as we shall see, was the creation of vast branches of mathematical theory.

The expectations of these scientists, seemingly rash, can be explained. These men were convinced that nature is mathematically designed and therefore saw no reason why they could not proceed in scientific matters as mathematics had proceeded in the study of numbers and geometric figures. As Randall says in his *Making of the Modern Mind*, "Science was born of a faith in the mathematical interpretation of Nature, held long before it had been empirically verified."

EXERCISES

1. What properties of physical objects did Descartes and Galileo regard as fundamental and real?
2. What is the distinction between a qualitative and a quantitative study of nature?
3. Describe the essential principles in Galileo's plan of scientific activity and contrast them with those of his predecessors.
4. Contrast the Greek objectives in the study of nature with those advocated by Bacon and Descartes.
5. How does mathematics enter into Galileo's scientific method?

13-4 FUNCTIONS AND FORMULAS

We intend to pursue the seventeenth-century developments initiated by Galileo and to pay particular attention to the role of mathematics in Galileo's method. Let us recall just what Galileo set out to do. He proposed to find fundamental quantitative physical principles or laws and to apply mathematical reasoning to these quantitative statements in order to deduce new physical laws. These physical laws would then provide the answers to a variety of scientific and practical problems. To express the physical principles in the manner he regarded as significant, Galileo introduced a new mathematical concept, the extremely important concept of a function. For the next two centuries mathematicians devoted themselves to the construction of functions and to the study of their properties. But the purely mathematical aspect of these creations is in itself rather barren. It is merely the sketch of a picture. And the picture in the present case is precisely the physical world which Galileo set out to investigate. Hence, as we study functions, we shall also study the situations which gave rise to them and the good that was accomplished with them. In fact, it is artificial to separate the physical thinking from the accompanying mathematics, for the two were developed as one. The leading mathematicians of the seventeenth and eighteenth centuries were also the leading scientists. And the accomplishments of these two centuries were a triumph of mathematics and science conjoined.

Before proceeding with Galileo's work, we shall familiarize ourselves with the notion of a function. Let us consider the situation in which a ball is dropped from some point above the ground and let us suppose that we wish to describe the distance the ball falls with increasing time. (Why we should seek such a description and what we can do with it are questions we shall answer later.) It is understood that the distance is measured downward from the point at which the ball begins to drop, and the time of fall is measured from the instant the ball begins to fall. Then the correct description, which we can accept for the moment as a fact, says that *the distance the ball falls, measured in feet,* is *16 times the square of the number of seconds it falls.* The italicized statement is an example of a *function.* As such it is important in two respects. First of all, it deals with varying quantities or *variables*. The number of seconds that the ball falls increases from zero to larger and larger values. The distance that the ball falls also increases from zero to larger and larger values. Secondly, the statement specifies exactly the relationship between the variables time and distance. What is characteristic of functions, then, is that they are precise statements of relationships among variables.

We know that verbal statements are clumsy to work with. Our experience with algebra teaches us that we can be more effective by introducing symbols. Let us then introduce the symbol t to stand for any number of seconds that the ball has been falling and the symbol d to stand for the distance that the ball falls in t seconds. In these symbols the italicized statement above says that

$$d = 16t^2.$$

The algebraic expression of a functional relationship is called a *formula*. Several facts about formulas are important for their proper understanding and use. In the present case, for example, we must be sure to note that the letters d and t represent not just one particular value of distance and time but whole ranges of values. Thus, if the ball falls for 5 seconds, the variable t can represent any number from 0 to 5. The variable d can represent any distance which the ball may have fallen during the 5 seconds. Of course, the values of d are not independent of the values of t. In fact, the whole point of the formula is to tell us precisely what d is for a given t. Thus when t is 2, for example, d is $16 \cdot 2^2$, or 64. That is, by substituting a particular value of t in the formula, we can calculate the distance d that an object has fallen in that number of seconds chosen for t. The values of d depend upon the values of t and, for this reason, t is called the *independent variable* and d, the *dependent variable*. One also says that the formula expresses d as a function of t. Since we can calculate d for millions of values of t, the formula is indeed a compact representation of millions of bits of information.

Suppose that the dropped ball falls for just 5 seconds. It then hits the ground and remains at rest. However, the formula $d = 16t^2$ does not "stop" at the end of 5 seconds. We could substitute 6 for t and find that d is $16 \cdot 36$, or 576. Likewise, we could substitute 7, or $9\frac{1}{2}$, or even -2 for t and in each case calculate the correpending value of d. Thus, the mathematical formula has meaning for all positive and negative values of t. However, if the ball falls for only 5 seconds, the formula represents the physical situation only for values of t from 0 to 5. In other words, the mathematical formula is more extensive than the physical situation.

We used the letters d and t to represent the variables distance and time. We could have used y and x, in which case the very same formula would read

$$y = 16x^2.$$

The letters d and t happen to be better because they suggest the physical meaning. But nothing would be altered mathematically if we used y and x.

Discussion of a formula and of its physical significance is often aided by utilizing the ideas of coordinate geometry. We can think of $d = 16t^2$ as the equation of a curve. Since the choice of particular letters does not have any mathematical significance, we can introduce axes d and t (Fig. 13–1). The curve corresponding to $d = 16t^2$ consists of those points whose abscissa and ordinate or whose t- and d-coordinates satisfy the equation. Thus, since for $t = 1$, we have $d = 16$, the point whose abscissa is 1 and whose ordinate is 16 lies on the curve. (For convenience we use a smaller unit on the d-axis.)

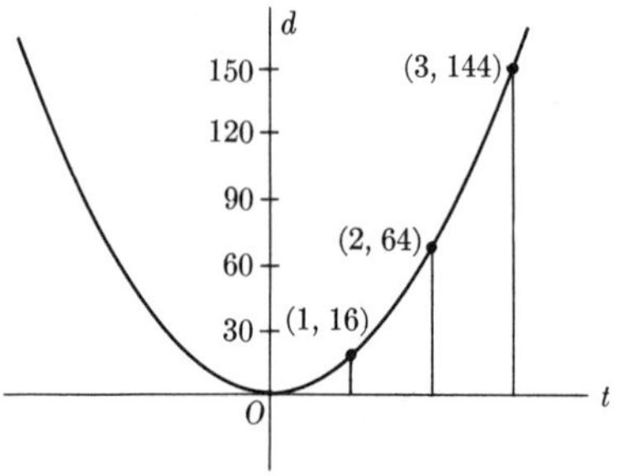

Fig. 13–1.
The graph of $d = 16t^2$.

The curve need not be and is not here a picture of the physical motion. Nevertheless, it does show that compared with t, d increases very rapidly as t increases beyond the value 0. Moreover, the curve reveals that the formula has mathematical meaning for all positive and negative values of t, whereas only that part of the curve which extends from $t = 0$ to $t = 5$ represents the physical situation.

EXERCISES

1. What is a function?
2. Distinguish between a function and a formula.
3. "Cook up" some mathematical formulas of your own, whether or not they have any physical significance.
4. For the formula $v = 32t$, calculate the value of v when t is 0, 3, 7, $4\frac{1}{2}$, and -6.
5. In the formula $A = \pi r^2$, which quantity would you regard as the independent variable and which as the dependent one?
6. For the formula $d = 16t^2$, calculate the value of d when $t = 2\frac{1}{2}$, -4, and 7.
7. For the formula $v = 32t$, calculate the value of t when $v = 64$, 80, 128.
8. For the formula $d = 16t^2$, calculate the value of t when $d = 144$. Are both answers physically significant?

9. For any given temperature, the relationship between readings in the fahrenheit scale and the centigrade scale is

$$F = \frac{9}{5}C + 32.$$

What is F when $C = 0$? What is F when $C = 100$? Do these values of F and C have special physical significance?

13-5 THE FORMULAS DESCRIBING THE MOTION OF DROPPED OBJECTS

Galileo not only formulated the general program for science, but he put it into effect. And here, too, he showed immense wisdom. He did not try, as had scientists and philosophers before him, to tackle the whole universe or to embrace man and nature in one theory. He decided to concentrate on a few classes of phenomena, and principally on motions near the surface of the earth. Galileo possessed the restraint which proves the master.

As we noted earlier in this chapter, Galileo thought as a mathematician, and he began his work by *idealizing* the problem he set out to solve. He, too, considered the motion of a ball, say rolling along the ground, and he asked, What if air resistance, friction between ball and ground, and any other hindering forces were not present? What would the ball do once it were set into motion? He concluded that the ball will continue indefinitely to move at a constant speed in a straight line. In more general terms, *if no force acts on a body and the body is at rest, it will remain at rest; if no force acts and the body is in motion, it will continue to move at a constant speed in a straight line.* This fundamental principle of motion or axiom of physics is now known as Newton's first law of motion. We should note that it contains two important assertions. The first, that a body in motion will continue to move in a straight line, is no innovation. It says that straight-line motion is the natural motion of bodies, the motion they will pursue unless they are forced to deviate from such a path. But the second assertion, that the body will continue to move at a constant speed indefinitely, is a radical departure from Aristotle, for Galileo was saying that no force is needed to keep the body going once it is set into motion.

But what if a force is applied to a body? Galileo answered that if the body is at rest, the force will set it in motion and change its speed from zero to some nonzero quantity. If the body is already in motion, the *force will change the speed, the direction of the motion, or both.* Thus, an object set into motion along a rough surface encounters friction. Friction is a force, and its effect is to reduce the speed of the object. Galileo's second principle, then, states that force produces *change* in speed or direction. In other words, Galileo said that force causes acceleration.

Let us consider an object which moves along a straight line, but which is being accelerated. The acceleration is a gain or loss in speed per unit time. Thus, if an object has been moving at a speed of 30 feet per second and if in one second its speed increases to 40 feet per second, then the acceleration is 10 feet per second for that one second, or 10 feet per second per second or, in scientific shorthand, 10 ft/sec^2. If the increase in speed had been 10 feet per second over two seconds, then the acceleration would have been 5 ft/sec^2.

Now, as the Greeks had observed, a body which is dropped falls with increasing speed; its motion is accelerated. Since falling bodies possess acceleration, that is, they do not move at a constant speed, it follows that some force must be causing the change in speed. By Galileo's time the concept of gravity had become more or less accepted. The earth exerts a force on any object and this force, if not offset by some other force, gives the object an acceleration. The surprising fact which Galileo discovered is that if one neglects air resistance, then an object falls to earth with a constant acceleration, and, moreover, this constant is the same for *all* bodies, namely 32 ft/sec^2. Thus if we let a stand for acceleration, the third fundamental law of motion, an axiom of physics, states that for all bodies falling to earth*

$$a = 32. \tag{1}$$

We now have a few fundamental principles about motion. Let us see next whether, in accordance with Galileo's plan, mathematical reasoning can lead to new information. Let us consider the motion of an object which is dropped, that is, whose initial speed is zero. Galileo's third principle says that the object gains speed each second at the rate of 32 ft/sec. Hence, at the end of one second its speed is 32 ft/sec. At the end of two seconds its speed is 2 times 32 ft/sec or 64 ft/sec. At the end of t seconds its speed is t times 32 ft/sec or $32t$ ft/sec. If we let v denote the speed at the end of t seconds, then

$$v = 32t. \tag{2}$$

We now have a formula which tells us the precise speed which a dropped body acquires in t seconds. It can, of course, be used to calculate v for any given value of t. Thus at the end of 6 seconds the speed of the body is 192 ft/sec.

Formula (2) is of some interest but hardly a surprise. Let us see whether we can obtain more significant conclusions by the further application of mathematics. We wish to determine the distance which a dropped body falls in t seconds. To be specific, let us consider for the moment that $t=6$. Now the speed at the end of 6 seconds is 192 ft/sec. To obtain the distance traveled in 6 seconds, one is tempted to multiply 192 by 6, that is, the speed by the time. However, the object did not travel at 192 ft/sec throughout the 6 seconds. In fact, it started with zero speed and only gradually increased its

* This axiom applies only to objects near the surface of the earth. We shall say more about it in Chapter 15.

speed to 192. Which speed should we use to compute the distance traveled? Presumably, the average speed.

A reasonable guess would be that the average speed is the arithmetic average of the initial and final speeds, that is, $(0 + 192)/2$ or 96 ft/sec. Let us see whether we can establish the correctness of this guess.

Suppose we calculate the speeds at the instants $t = 0$, 1, 2, 3, 4, 5, and 6. If we substitute these values of t in (2) we obtain the speeds

$$0, 32, 64, 96, 128, 160, 192.$$

If we take the average of these seven speeds, that is,

$$\frac{0 + 32 + 64 + 96 + 128 + 160 + 192}{7},$$

we obtain 96 ft/sec. Of course, this calculation does not prove that the average speed is 96 because the object falls with varying speed even during the first second, the second second, and so forth. We might therefore average the speeds attained by the object after each half-second of fall, that is, the speeds at $t = 0, \frac{1}{2}, 1, \frac{1}{2}, 2, \ldots, 6$. These speeds are

$$0, 16, 32, 48, 64, 80, 96, 112, 128, 144, 160, 176, 192. \tag{3}$$

If we average these speeds we again get 96 ft/sec. This calculation is no more of a proof that 96 is correct than the preceding one because even in each half-second the object falls with a varying speed. Yet we continue to get the average of 96. Let us see whether we can find the reason that 96 results each time.

We note that 96 is the speed the object attains when $t = 3$, because if we substitute 3 for t in (2) we get $v = 96$. Now it seems significant that the average speed occurs halfway through the time interval from 0 to 6. This probably is the case because the speed at each instant before $t = 3$ and the speed at some corresponding instant after $t = 3$ average to 96. Indeed, if we examine the speeds in (3) we see that 0 and 192, 16 and 176, 32 and 160, and so on, each averages to 96. In other words, if for each instant before $t = 3$, for example, $t = 1\frac{1}{2}$, and for the instant as far beyond $t = 3$, in our example $t = 4\frac{1}{2}$, the average speed is 96, then the average for all the instants will be 96. Just to check once more, at $t = 1\frac{1}{2}$ the speed is 48 and at $t = 4\frac{1}{2}$ the speed is 144. The average of 48 and 144 is 96.

This argument can be generalized. Let h be any interval of time. Then $3 - h$ is some instant before 3 seconds, and, by formula (2), the speed of the falling body at the instant $t = 3 - h$ is $32(3 - h) = 96 - 32h$. At the instant $t = 3 + h$, which is h seconds after 3 seconds, the speed is $32(3 + h)$ or $96 + 32h$. We see that the speed at $t = 3 - h$ is $32h$ less than 96, and the speed at $t = 3 + h$ is $32h$ more than 96. Hence the average speed for these two instants is 96 because

$$\frac{96 - 32h + 96 + 32h}{2} = 96.$$

Since the object falls for as many instants during the interval from $t = 0$ to $t = 3$ as it does from $t = 3$ to $t = 6$, and since the pairing of instants equally far from $t = 3$ produces the average of 96, the average speed over the entire interval of 6 seconds is 96 ft/sec. It is important to note that this average speed is attained after one-half of the time of travel.

If, instead of 6 seconds, we had used the general value of t seconds, then our conclusion would read that the average speed is attained after $t/2$ seconds. Since $v = 32t$, the average speed in the interval from 0 to t is given by

$$\text{average speed} = 16t.$$

The argument used to derive the average speed utilizes formula (2) and this is correct only when (1) holds. One should not, then, use the conclusion that the average speed is the speed attained at $t/2$ seconds in other kinds of motion. However, the argument does hold when the acceleration is constant, even if that constant is not 32.

Now that we have the average speed of an object which is dropped and falls for t seconds, we can calculate the distance fallen. The average speed is that constant speed with which the object could have fallen to cover the same distance in t seconds. Since this average is a constant speed, we have but to multiply it by the time of travel to obtain the distance fallen. Then since $16t \cdot t = 16t^2$, if we let d represent the distance fallen in t seconds, we have the new result:

$$d = 16t^2. \tag{4}$$

Formula (4) says, for example, that in 3 seconds the object falls $16 \cdot 3^2$, or 144 feet. With a little mathematics we have derived an important law of falling bodies. It tells us the distance which any body that is dropped and freely falling travels in t seconds.

We can derive a few significant consequences of formulas (2) and (4) by the application of simple algebra. Dividing both sides of formula (4) by 16 and taking the square root of both sides of the resulting equation, we obtain

$$t = \pm\sqrt{\frac{d}{16}}\,.$$

This result tells us the time required for a dropped body to fall d feet. Of course, of the two roots (one positive and one negative), only the positive value possesses physical significance because we are dealing with a physical situation in which time is positive and measured from the instant the body begins to fall. Hence we shall forget about the negative root and consider that

$$t = \sqrt{\frac{d}{16}}. \tag{5}$$

If we now wished to calculate how long it takes an object to fall 1000 feet, we would substitute this value in formula (5) and calculate t.

From formula (5) one can draw a most significant conclusion. The formula does not tell us the name of the President of the United States, but this is not so surprising. However, it is surprising that the formula does not involve the weight or any other property of the falling body. This means that *all* bodies take the *same* time to fall a given distance, provided, of course that air resistance is neglected. A feather and a piece of lead take the same time to fall a given distance in a vacuum. This is the lesson which Galileo is supposed to have learned by dropping various objects from the leaning tower of Pisa. Many people still hesitate to accept this conclusion because they observe bodies falling in air, and the resistance encountered by feathers is quite different from that offered to lead. Undoubtedly it was this difference gleaned from actual observations which led the Aristotelians to the conclusion that heavier bodies fall faster.

Formula (5) was derived by merely rearranging, so to speak, formula (4). But Galileo's plan envisaged also combining existing formulas to obtain new knowledge. To illustrate this process, suppose one takes the value of t given by (5) and substitutes it in the formula $v = 32t$. This yields

$$v = 32\sqrt{\frac{d}{16}}.$$

Now the square root of a fraction is equal to the square root of the numerator divided by the square root of the denominator. Hence

$$v = 32\,\frac{\sqrt{d}}{4} = 8\sqrt{d}. \tag{6}$$

The new formula enables us to calculate the speed which a dropped body will acquire in falling d feet. While this information is implicit in formulas (4) and (5), which yield (6), we now see clearly something we might not have appreciated before. We should note that formula (6) says that the speed increases as the square root of d. The predecessors of Galileo believed that the speed increased directly with distance.

EXERCISES

1. Was Aristotle wrong in asserting that, to keep an object moving at a constant speed in a real medium, a force must constantly be applied?
2. Suppose that gravity does not exist and a man steps off the roof of a building. What would his subsequent motion be? What would it be in the presence of gravity?
3. An automobile travels at the speed of 10 mi/hr for 59 min and at a speed of 50 mi/hr for 1 min. What is its average speed?
4. What is the speed of an object 4 sec after it is dropped? What is its average speed during the 4 sec? At what instant does the object actually possess this average speed?
5. Distinguish between speed and acceleration.
6. We may regard the formula $v = 32t$ as an equation in v and t, and we may therefore plot t-values as abscissas and v-values as ordinates. Draw the curve of the formula $v = 32t$.
7. Using the formula $d = 16t^2$, calculate how far a body will drop in 5 sec, $6\frac{1}{2}$ sec, 10 sec. Is the drop the same from one second to another?
8. Apply the instructions of Exercise 6 to the formula $d = 16t^2$. What is the name of the resulting curve?
9. Graph the curve of $d = 16t^2$, but let the downward direction of the vertical axis, that is the d-axis, be positive.
10. A window washer at the 50th floor of a skyscraper (500 ft above the street) steps back to observe the results of his work. Describe mathematically his subsequent behavior.
11. Using the formula $v = 8\sqrt{d}$, calculate the speed with which an object dropped from the top of the Empire State Building (about 1000 ft above street level) hits the ground.
12. If the relation between speed and distance were $v = 8d$ instead of $v = 8\sqrt{d}$, what difference would there be in the behavior of falling bodies?
13. Suppose that we are considering the motion of an object which is dropped from a point near the surface of the moon. On the moon all objects also fall to the surface with a constant acceleration, and the value of this acceleration is 5.3 ft/sec². What change would you make in formulas (2) and (4) to have them represent speed acquired and distance traveled for objects falling to the moon's surface? Incidentally, the moon has no atmosphere, and hence air resistance can surely be neglected.
14. Suppose that an object is dropped and falls with a constant acceleration a. What would you propose as formulas for speed and distance fallen in time t?
15. Show first that formula (6) implies $d = v^2/64$. Now suppose a dropped object acquires a speed of 88 ft/sec. What distance must it fall to acquire this speed?
16. Suppose an object is traveling along a straight line with a speed of 88 ft/sec and then starts to lose speed, that is decelerates, at the constant rate of 32 ft/sec². What distance must it travel for its speed to become zero? [*Suggestion:* The

distance it must travel to reach zero speed is the same as the distance it would travel if it started with zero speed and accelerated at 32 ft/sec^2 to attain a speed of 88 ft/sec.]

17. As a direct generalization of the thought in Exercise 16, we may state that if an object is traveling in a straight line at a speed of v ft/sec and then loses speed at the rate of 32 ft/sec^2, the distance d it travels before attaining zero speed is $d = v^2/64$. Suppose the deceleration is 11 ft/sec^2. What formula gives the distance the object travels before attaining zero speed?
18. Using the result of Exercise 17, answer the following question. An automobile is traveling at 60 mi/hr (or 88 ft/sec), and the brakes are applied. The action of the brakes decelerates the automobile at the rate of 11 ft/sec^2. How far will the automobile travel before stopping? The answer gives the minimum distance in which one can, even under most favorable road conditions, stop a car traveling at 60 mi/hr. However, it takes about 1 sec before a person who decides to apply the brakes actually does so. What distance will the automobile travel in that time?
19. A man drops a stone into a well and listens for the sound of the splash. He finds that $6\frac{1}{2}$ sec elapse from the instant the stone is dropped until he hears the sound. How far below is the surface of the water? Assume that sound travels at 1152 ft/sec.

13-6 THE FORMULAS DESCRIBING THE MOTION OF OBJECTS THROWN DOWNWARD

Thus far we have seen how simple formulas describe the motion of a body which is dropped. By employing slightly more complicated formulas Galileo was able to tackle further phenomena of motion. Suppose that instead of being dropped a ball is thrown downward. Now the ball does not start its motion with zero speed but with whatever speed the hand imparts to it. The problem we shall look into is, What is the subsequent motion of the ball? To be specific, suppose the hand imparts to the ball a speed of 96 ft/sec. Neglecting for the moment the action of the force of gravity, we can say that the ball will continue to travel downward in a straight line with a speed of 96 ft/sec. The basis for this assertion is, of course, the first law of motion. We know, however, that gravity will also act on the ball and give it a speed of $32t$ ft/sec in t seconds. Since both speeds will operate simultaneously to make the ball move downward, the total speed, v, of the ball is represented by the formula:

$$v = 96 + 32t. \qquad (7)$$

Let us compare this formula with $v = 32t$. We see that the term 96 in formula (7) represents the speed given to the ball by the hand. Both formulas are said to be of the first degree in t because the independent variable, t, appears only to the first power. That is, the formulas contain $32t$ as opposed to $32t^2$, or $32t^3$, or some other power of t. First-degree formulas are often called *linear functions* because the curve representing each is a straight line.

We can also obtain the formula for the distance, d, which the ball will fall in t seconds. If there were no gravity, the ball would fall a distance of $96t$ feet in t seconds because it would have the constant speed of 96 ft/sec imparted by the hand. But during the same t seconds, the force of gravity will exert an additional downward pull which, according to formula (4), will cause the ball to fall $16t^2$ feet. Since both forces, the hand and gravity, cause the ball to fall downward, the total distance, d, traveled in t seconds is

$$d = 96t + 16t^2. \tag{8}$$

Comparison of formula (8) with formula (4) shows that formula (8) contains a new term, $96t$, which represents the contribution made by the action of the hand to the distance the ball falls. Formulas (4) and (8), incidentally, are of the second degree in t because the independent variable t occurs to the second power. Second-degree functions are also called *quadratic functions.*

EXERCISES

1. If a ball, instead of being merely dropped, is thrown downward with a speed of 128 ft/sec, will its speed be greater in t seconds? Will the distance fallen in t seconds be greater?
2. Write the formula representing the speed acquired and distance traveled in t seconds by a ball which is thrown downward with a speed of 128 ft/sec.
3. Suppose a ball is thrown downward with a speed of 96 ft/sec. What are the speed and distance traveled after 3 sec? after $4\frac{1}{2}$ sec?
4. Graph formula (8) by plotting points whose coordinates satisfy the equation. What is the name of the resulting curve?
5. Graph formula (8) by applying the method of change of coordinates presented in Chapter 12.

13-7 FORMULAS FOR THE MOTION OF BODIES PROJECTED UPWARD

A more interesting phenomenon both physically and mathematically is the motion of a ball thrown straight up into the air. Suppose, for example, that the ball is thrown *upward* with a speed of 96 ft/sec, and let us again consider the questions, What are the speed and distance traveled after t seconds of motion? If gravity is neglected, then the action of the hand will cause the ball to start upward with a speed of 96 ft/sec and, according to the first law of motion, it should continue to travel upward at that speed indefinitely. However, we know that the downward pull of gravity causes the ball to acquire in t seconds a downward speed of $32t$ ft/sec. Since the hand gives the ball an upward speed of 96 and gravity gives it a downward speed of $32t$, the net speed, v, of the ball at the end of t seconds is

$$v = 96 - 32t. \tag{9}$$

The minus sign in formula (9) takes care of the fact that the speed resulting from the action of gravity reduces the speed imparted by the hand. Formula (9) should be compared with formula (7).

Let us turn to the second question: How far does the ball travel? Since the ball travels upward to some maximum height and then falls down, we shall instead ask the more pertinent question, What height above the ground does the ball possess at any time t? If there were no gravity, the ball would move upward at the constant velocity of 96 ft/sec. Hence in t seconds it would travel upward $96t$ feet. However, we know that a ball moving above the surface of the earth for t seconds will experience a downward pull of gravity amounting to a distance of $16t^2$ feet in t seconds. Hence the net height, d, reached by the ball is

$$d = 96t - 16t^2. \tag{10}$$

As in formula (9), the minus sign here represents the fact that the action of gravity offsets the action of the hand.

We can now answer some questions about the motion of the ball. We know from experience that the ball will rise to some height and then fall back to the ground. How high will it go? We would expect the ball to continue to rise until its upward speed, which is continually decreasing, becomes zero. This fact can be put to use through formula (9). We now ask the question, What is t when $v = 0$? Suppose we denote by t_1 this particular unknown value of t. Then we may say on the basis of formula (9) that

$$0 = 96 - 32t_1.$$

To find t_1 we have but to solve this simple equation. Clearly $t_1 = 3$.

We have determined the time it takes the ball to reach its maximum height but not the height itself. However, formula (10) gives us the height at any time t. Suppose then that we substitute the value $t_1 = 3$ in (10) and calculate d_1, the height of the ball above the ground at this instant. Substitution of the quantity 3 for t in (10) yields

$$d_1 = 96 \cdot 3 - 16 \cdot 3^2 = 144.$$

Thus the maximum height above the ground to which the ball will rise is 144 feet.

We know that the ball will fall to the ground after reaching this maximum height. Will it take as much time to reach the ground from its maximum height as it did to travel from the ground to the maximum height? Those people who have lots of confidence in their intuition should answer this question before we settle it by mathematical reasoning.

To obtain an answer we must proceed somewhat indirectly. Our information about the motion is contained in formulas (9) and (10). Of these two formulas, (10) offers some prospect of being useful because it relates time and height reached by the ball. There is one bit of information which might be used in connection with formula (10), namely, that when the ball reaches the ground, the height of the ball above the ground is zero. Let us therefore find the time at which the ball reaches the ground and then see what it suggests.

We denote by t_2 the value of t at which the ball reaches the ground. Hence, by formula (10),

$$0 = 96t_2 - 16t_2^2. \tag{11}$$

Our problem now is to determine the value of t_2 which, according to (11), satisfies a second-degree equation. This equation is easily solved. We apply the distributive axiom to justify writing

$$0 = 16t_2(6 - t_2). \tag{12}$$

We are seeking the value or values of t_2 for which the right side of (12) equals the left side. Clearly when $t_2 = 0$, one factor on the right side is zero, and hence the product is zero. Likewise, when $t_2 = 6$, the product is zero. Hence, there are two values of t, namely 0 and 6, when d, the height above the ground, is zero.

Why two values? Mathematically the two values result from the fact that we are solving a quadratic or second-degree equation. Physically the two values are readily understandable. The value $t_2 = 0$ is the value of t at the instant at which the ball is about to start out; $t_2 = 6$ is the value of t at the instant at which the ball hits the ground after traveling up and down. With respect to the problem in hand, the second value is the interesting one because it tells us that 6 seconds elapse during the upward and downward travel of the ball. Since we found earlier that the ball requires 3 seconds to reach its highest position, it is evident that only 3 seconds are required for the ball to return to the ground. Hence it takes exactly the same time for the ball to go up as it does to come down.

We can now ask mathematics to answer another question for us. What speed does the ball possess when it strikes the ground? Is this speed the same, more, or less than the 96 ft/sec with which it was thrown up? The answer can be obtained at once. Formula (9) gives the speed of the ball at any instant of its flight. The ball strikes the ground at the instant $t_2 = 6$. If we substitute 6 for t in (9), we find that v_2, the speed at the instant the ball strikes the ground, is

$$v_2 = 96 - 32 \cdot 6 = -96.$$

Thus mathematics tells us that the speed is 96 ft/sec, the very same speed with which it was projected upward. Obligingly, mathematics also tells us, through the minus sign, that the speed is in the opposite direction to that of the upward throw.

EXERCISES

1. For a ball thrown upward with a speed of 128 ft/sec, what formula describes the relationship between the subsequent height above the ground and the time of travel?
2. If a ball is thrown upward with a speed of 160 ft/sec, then the formula relating its subsequent height above the ground and the time of travel is $d = 160t - 16t^2$, and the speed of the ball is given by the formula

$$v = 160 - 32t.$$

   a) How high is the ball after 4 sec?
   b) What is its speed at the end of the fourth second?
   c) How high will the ball go?
3. If the height of a ball thrown upward is given by the formula $d = 144t - 16t^2$, what is d when $t = 9$? Interpret the result physically.
4. If the height above the ground of a ball is representable by the formula $d = 192t - 16t^2$, then its height after 4 sec is 512 ft, and its height after 8 sec is also 512 ft. Verify these heights and account for the fact that the height is the same after 4 additional seconds.
5. If a gun capable of firing a bullet at the speed of 1000 ft/sec is fired straight upward, how high will the bullet go?
6. Suppose that a ball is dropped from the top of a building 100 ft high. Let d represent the *height of the ball above the ground* and t the time of travel measured from the instant the ball is dropped. Write a formula representing the motion in terms of d and t.
7. Since man seems to be preparing for experiences on the moon, it may be well to consider the following question. Suppose that a ball is thrown up from the surface of the moon with a speed of 96 ft/sec. How high will it go and how long will it take to reach its maximum height? Remember that on the moon the value of 5.3 ft/sec^2 corresponds to the acceleration of 32 ft/sec^2 which holds on the earth.
8. Suppose that a bullet shot straight up into the air returns to the ground 60 sec later. What was the initial speed? [*Suggestion:* Use formula (10). However, the initial speed, which was 96 in formula (10), is now unknown.]
9. A rocket is shot straight up into the air to a height of 50 mi at which point its velocity is 300 mi/hr. Its fuel is now exhausted, and hence the rocket receives no further acceleration from this source. Write a formula which describes the subsequent motion of the rocket. [*Suggestion:* You can choose the 50-mi point as the origin for height and zero time as the instant when the rocket is at that height.]
10. A ball is thrown up into the air from the roof of a building with a speed of 96 ft/sec. Write a formula for its subsequent height above the roof as a function of time.
11. Using the data of Exercise 10 and the additional fact that the roof is 112 ft above the ground, find the time when the ball reaches the ground.

REVIEW EXERCISES

In all of the exercises below the units are feet and seconds.

1. An object which is dropped near the surface of the earth acquires a speed in ft/sec of $v = 32t$ in t sec. Calculate v when t is

   a) 7 b) $2\frac{1}{2}$ c) $3\frac{1}{2}$ d) $4\frac{3}{4}$ e) 9.

2. An object falls and acquires the speed $v = 32t$ in t sec. How long does it take to acquire a speed of

   a) 128 b) 160 c) 400 d) 16?

3. If an object acquires speed according to the formula $v = 32t$, what is its average speed during
   a) the first 5 sec of fall,
   b) the first 8 sec of fall?

4. Suppose an object acquires speed according to the formula $v = 32t$, and we wished to compute the average speed during the eighth second of fall. Could we use the argument in the text (properly modified) to average the speed at $t = 7$ and the speed at $t = 8$?

5. Suppose an object is dropped and falls with a constant acceleration of g ft/sec^2 instead of 32 ft/sec^2. What formula relates the speed and time of fall?

6. Suppose an object is dropped and falls with a constant acceleration of g ft/sec^2. Then in t sec it acquires a speed of $v = gt$ ft/sec. One might convince himself that the argument given in the text where the acceleration is 32 ft/sec^2 and the average speed proves to be $16t$ carries over to the case where the acceleration is g. Suppose the argument does carry over.
   a) What is the average speed for t sec of fall?
   b) What is the distance fallen in t sec?

7. If an object is dropped near the surface of the earth, it falls $d = 16t^2$ feet in t sec. Calculate d when t is

   a) 4 b) 7 c) $3\frac{1}{2}$ d) $3\frac{3}{4}$ e) $5\frac{1}{4}$.

8. If an object falls according to the formula $d = 16t^2$, how much is t when d is

   a) 64 b) 96 c) 144 d) 200 e) 169?

9. If an object is thrown downward from a point near the surface of the earth with an initial speed of 64 ft/sec, the speed it acquires in t sec is given by the formula $v = 64 + 32t$. Calculate the speed when t is

   a) 3 b) $3\frac{1}{2}$ c) 5 d) $5\frac{1}{4}$ e) 7.

10. If an object acquires speed according to the formula $v = 64 + 32t$, how long does it take to acquire a speed in ft/sec of

    a) 96 b) 100 c) 300 d) 150?

11. Suppose the constant acceleration of a falling object is g ft/sec^2 and the object is thrown downward at speed of 100 ft/sec. Guess the formulas which represent
    a) the speed acquired in t sec,
    b) the distance fallen in t sec.

12. Suppose an object falls d ft in t sec where $d = 128t + 16t^2$. How long does it take to fall

    a) 320 ft b) 768 ft c) 304 ft d) 156 ft?

13. Suppose the constant acceleration of a falling object is g ft/sec^2 and the object is thrown upward with an initial speed of 128 ft/sec. Guess the formulas which represent
    a) the speed acquired in t sec,
    b) the height above the ground in t sec.

14. If the height above the ground of an object is given by the formula $d = 96t - 16t^2$, how high is the object when t is

    a) 3 b) $2\frac{1}{2}$ c) 5 d) $5\frac{1}{2}$?

15. The height above the ground of an object is given by $d = 96t - 16t^2$. Calculate the height when $t = 7$. What is the physical meaning of the result?

16. If an object is thrown up from the roof of a building 200 ft high and the height of the object above the roof is given by $d = 96t - 16t^2$, what is the height of the object when $t = 7$? What is the physical meaning of the result?

17. If we had reason to believe that the acceleration of any object which is dropped near the surface of any planet, the sun, or the moon is a constant, could we carry over the mathematics of this chapter to motions of objects near the surface of these bodies?

Topics for Further Investigation

1. Galileo's scientific work.
2. Huygens' scientific work.
3. The importance of experimental work versus that of mathematical deduction from basic principles in seventeenth-century science.
4. The scientific ideas espoused by Francis Bacon.

Recommended Reading

BELL, A. E.: *Christian Huygens and the Development of Science in the Seventeenth Century,* Edward Arnold and Co., London, 1947.

BONNER, FRANCIS T. and MELBA PHILLIPS: *Principles of Physical Science,* pp. 37–65, Addison-Wesley Publishing Co., Inc., Reading, Mass., 1957.

BURTT, E. A.: *The Metaphysical Foundations of Modern Physical Science,* 2nd ed., Chaps. 1 through 6, Routledge and Kegan Paul Ltd., London, 1932.

BUTTERFIELD, HERBERT: *The Origins of Modern Science,* Chaps. 4 through 7, The Macmillan Co., New York, 1951.

COHEN, I. BERNARD: *The Birth of a New Physics,* Chap. 5, Doubleday and Co., Anchor Books, New York, 1960.

CROMBIE, A. C.: *Augustine to Galileo*, Chap. 6, Falcon Press Ltd., London, 1952. Also in paperback under the title: *Medieval and Early Modern Science*, 2 vols., Doubleday and Co., Anchor Books, New York, 1959.

DAMPIER-WHETHAM, WM. C. D.: *A History of Science and Its Relations with Philosophy and Religion*, Chap. 3, Cambridge University Press, London, 1929.

FARRINGTON, BENJAMIN: *Francis Bacon: Philosopher of Industrial Science*, Henry Schuman, Inc., New York, 1949.

GALILEI, GALILEO: *Dialogues Concerning Two New Sciences*, pp. 147–233, Dover Publications, Inc., New York, 1952.

HOLTON, GERALD and DUANE H. D. ROLLER: *Foundations of Modern Physical Science*, Chaps. 1, 2, and 13 through 15. Addison-Wesley Publishing Co., Inc., Reading, Mass., 1958.

KLINE, MORRIS: *Mathematics and the Physical World*, Chaps. 12 and 13, T. Y. Crowell Co., New York, 1959. Also in paperback, Doubleday and Co., New York, 1963.

MOODY, ERNEST A.: "Galileo and Avempace: Dynamics of the Leaning Tower Experiment," an essay in PHILIP P. WIENER and AARON NOLAND: *Roots of Scientific Thought*, Basic Books, Inc., New York, 1957.

RANDALL, JOHN HERMAN, JR.: *Making of the Modern Mind*, rev. ed., Chaps. 9 and 10, Houghton Mifflin Co., Boston, 1940.

SAWYER, W. W.: *Mathematician's Delight*, Chaps. 8 and 9, Penguin Books Ltd., Harmondsworth, England, 1943.

SMITH, PRESERVED: *A History of Modern Culture*, Vol. I, Chap. 6, Henry Holt & Co., New York, 1930.

STRONG, EDWARD W.: *Procedures and Metaphysics*, University of California Press, Berkeley, 1936.

TAYLOR, HENRY OSBORN: *Thought and Expression in the Sixteenth Century*, 2nd ed., Vol. II, Chaps. 30 through 35, The Macmillan Co., New York, 1930.

TAYLOR, LLOYD WM.: *Physics, The Pioneer Science*, Chaps. 3 through 7, Dover Publications, Inc., New York, 1959.

WHITEHEAD, ALFRED N.: *Introduction to Mathematics*, Chaps. 2 through 4, Holt, Rinehart and Winston, Inc., New York, 1939.

WHITEHEAD, ALFRED N.: *Science and the Modern World*, Chap. 3, Cambridge University Press, London, 1926.

WOLF, ABRAHAM: *A History of Science, Technology and Philosophy in the 16th and 17th Centuries*, 2nd ed., Chap. 3, George Allen and Unwin Ltd., London, 1950. Also in paperback.

CHAPTER 14

PARAMETRIC EQUATIONS AND CURVILINEAR MOTION

I now propose to set forth those properties which belong to a body whose motion is compounded of two other motions, namely, one uniform and one naturally accelerated; these properties, well worth knowing, I propose to demonstrate in a rigorous manner.

GALILEO

14-1 INTRODUCTION

We saw in the preceding chapter that simple functions can be used to express physical principles and that by applying algebra to the formulas which express the functions symbolically, we can obtain new physical knowledge. To some extent, then, we have come to recognize the broader significance and usefulness of functions and mathematical processes for science in general. However, we have hardly penetrated as yet the mathematical domain of functions nor have we learned enough applications to sense its real power.

In this chapter we shall extend slightly the use of functions. In Chapter 13, we represented the acceleration and speed attained and distance traveled by a falling body by using one formula for each physical quantity. We were enabled thereby to study motion along straight-line paths. We shall now examine motion along curved paths, for example, the motion of an object dropped from a moving plane, or the motion of a projectile shot out from a cannon. It was again Galileo who perceived the basic principle underlying the phenomenon of *curvilinear* motion. He presented the concept and its mathematical treatment to the world in the *Dialogues Concerning Two New Sciences*, the very same book in which he treated motion in a straight line. Galileo's purpose in investigating curvilinear motion was to study the behavior of cannon balls, or projectiles in general. The cannon, introduced in the fourteenth century, had undergone such improvement by Galileo's time that it could fire a projectile over several miles. However, the theory of projectile motion was not well understood before Galileo's work because mathematicians and physicists had attempted to apply Aristotle's laws of motion, and these were not correct.

The problems that Galileo treated, e.g., the motion of cannon balls, unfortunately did not lose their importance in the succeeding centuries. In fact, they have become even more common and more complicated in our times, since such phenomena as the motion of bombs dropped from moving airplanes, the trajectories of death-dealing projectiles capable of traveling thousands of miles, and similar problems of modern "civilization," also fall within the puissance of Galileo's method. However, the value of this phase of Galileo's work is not limited to meting out death and destruction. Aside from using his results as an illustration of the power of mathematics, we shall see in the space of one chapter how an extension of Galileo's ideas on projectile motion led, in the hands of Newton, to the greatest advance in science which our civilization has achieved.

14–2 THE CONCEPT OF PARAMETRIC EQUATIONS

Let us suppose that a stone is thrown out horizontally from the top O of a cliff (Fig. 14–1). We know from physical experience that the stone will travel out and down and follow the curved path OAB. If we introduce a set of coordinate axes on which the positive direction of the Y-axis is downward, then we know from our work in coordinate geometry that this curve can be represented by an equation. Let us suppose, for definiteness, that this equation is $y = x^2$. Then it follows from equation (9) of Chapter 12 that the path is part of a parabola opening downward. We shall call $y = x^2$, which is, of course, also the formula that tells us how y changes when x changes, the *direct* relationship between x and y.

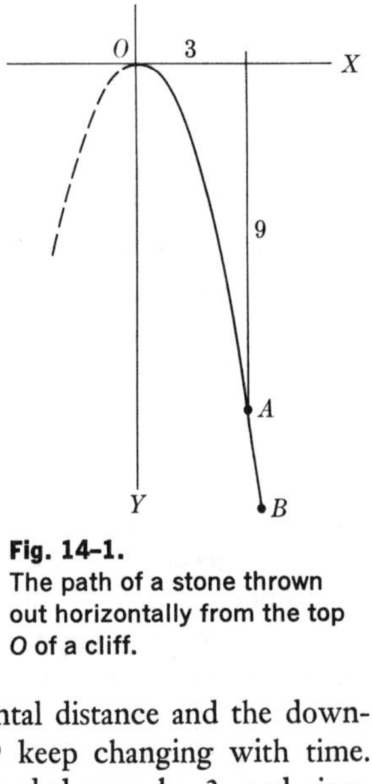

Fig. 14–1. The path of a stone thrown out horizontally from the top O of a cliff.

As the stone travels out and down, the horizontal distance and the downward distance which it travels from the point O keep changing with time. Thus, at the point A, the horizontal distance traveled may be 3, and since $y = x^2$, the vertical distance must be 9. At the point B, the horizontal distance may be 4, in which case the vertical distance must be 16.

The direct relationship between x and y is frequently useful, but it does not involve the time that the object is in motion. We may wish instead to utilize the equation which gives the relationship between horizontal distance traveled and time and the equation which relates vertical distance and time. Let us suppose for the moment that the stone travels *straight out* at the rate of 3 ft/sec. Then the relationship between horizontal distance and time is $x = 3t$. Since $y = x^2$, then in terms of t, $y = (3t)^2$ or $y = 9t^2$.

The two equations:

$$x = 3t, \qquad y = 9t^2, \tag{1}$$

are called the *parametric equations* of the curve OAB. They describe the curve OAB just as well as does the single equation $y = x^2$, provided that we understand how parametric equations are to be used. For each value of t, equations (1) yield a value of x and a value of y. These values of x and y which belong to the same value of t are the coordinates of one point on the curve OAB. Thus for $t = 1$, $x = 3$, and $y = 9$. Then (3, 9) are the coordinates of a point on the curve, namely, the point A, which we discussed earlier. For $t = \frac{4}{3}$, $x = 4$ and $y = 16$, and (4, 16) are the coordinates of the point B.

We may also say that the two formulas $x = 3t$ and $y = 9t^2$ are equivalent to the single formula $y = x^2$. Whether we speak of equations of curves or formulas is really immaterial. The word formula emphasizes the idea of change because formulas are relationships among variables, and we often like to think of what happens to one variable as another, related variable changes. On the other hand, when a curve is given in its entirety, the concept of change may not be relevant, and then we speak of the equation of the curve.

If the two formulas in (1) are entirely equivalent to the single formula $y = x^2$, why do we bother with two formulas instead of one? There are two reasons: (1) When one argues from physical principles, it is often easier to arrive at the parametric representation of a given phenomenon, and (2) it is easier to study the phenomenon by working with parametric equations. We shall recognize the utility of parametric representations as we study the next few sections.

There is one more mathematical detail. Suppose that we find the parametric formulas describing a motion and we wish to determine the direct relationship between x and y. Can we do this? Yes indeed. For example, if $x = 3t$ and $y = 4t^2$ are the parametric formulas, we can solve the first one for t and obtain $t = x/3$. We substitute this value of t in $y = 4t^2$ and obtain

$$y = 4\left(\frac{x}{3}\right)^2, \qquad \text{or} \qquad y = 4\left(\frac{x^2}{9}\right), \qquad \text{or} \qquad y = \frac{4x^2}{9}.$$

This is the direct relationship between x and y.

EXERCISES

1. If the parametric formulas representing a phenomenon are $x = 2t$ and $y = 3t$, what is the direct relationship between x and y? What curve represents the parametric formulas or the direct relationship?
2. If the parametric formulas are $x = 4t$ and $y = 5t^2$, what is the direct relationship between x and y? What curve represents the direct relationship?

3. Suppose the parametric formulas are $x = 2t$ and $y = 10t + 4t^2$. What is the direct relationship between x and y and what curve describes it?
4. Suppose $x = 3t$ and $y = (\frac{4}{3})t$ are the parametric equations of a curve. Sketch the the curve by using the parametric equations only.

14-3 THE MOTION OF A PROJECTILE DROPPED FROM AN AIRPLANE

Let us see now how parametric formulas arise in the study of physical phenomena and how they can be useful in deducing new information about the phenomena. Suppose a bomb is released from an airplane which is flying horizontally at 60 miles per hour (an unrealistic figure used for computational convenience). If there were no gravity, the bomb would continue to move forward alongside the airplane at the rate of 60 miles per hour. This fact seems surprising, but it is a consequence of the first law of motion, which states that if an object is in motion and no force is applied to alter that motion, then the object will continue to move indefinitely at the speed it already has. Since the bomb has been moving with the airplane, it already possesses a horizontal speed of 60 miles per hour. We have assumed that no forces are acting on the bomb and hence it will continue to move forward at that speed. There are more familiar analogous situations which may make the truth of what was just said a little more acceptable. Suppose that a person rides in an automobile which is moving at the rate of 60 miles per hour and the driver suddenly applies the brakes. The automobile's motion is then checked, but the passenger's motion is not, and he continues to move forward at 60 miles per hour, at least until he hits the windshield.

Let us return to the motion of the bomb released from the plane. We had assumed that gravity was not acting. But it does act and it pulls the bomb downward at the same time as the bomb moves forward so that the bomb follows a curved path. Here Galileo made a discovery applying to projectile motion, namely, that one could study its horizontal and vertical motions as though they were occurring separately, and that the position of the bomb at any time could be determined by finding how far it had traveled horizontally and vertically. This idea was new and radical in Galileo's time. Aristotle had argued that one motion would interfere with the other, and that only one could operate at any given time. Thus he would have said that the violent motion imparted to the bomb by the airplane would prevail until the acting force was used up, and then the natural motion downward would take over and cause the bomb to fall straight down.

Let us apply Galileo's way of analyzing the motion. The bomb moves horizontally at the constant speed of 60 miles per hour, or 88 feet per second. If we measure time from the instant the bomb is released from the plane, and if we measure horizontal distance from the point at which it is released, then the horizontal distance x covered by the bomb in t seconds is given by the formula

$$x = 88t. \tag{2}$$

This formula describes the horizontal motion.

According to Galileo, the vertical motion downward takes place as though it were independent of the horizontal motion. But the vertical motion is due to gravity only, and we know that an object which falls straight down under the action of gravity and starts with zero speed falls $16t^2$ feet in t seconds. Hence, if we let y represent the distance *downward* from the point at which the bomb is released, then

$$y = 16t^2. \tag{3}$$

Formulas (2) and (3) together describe the entire motion. We observe that x and y, the horizontal and vertical distances traveled, are given in terms of a third variable, t. In fact, they are the parametric formulas for the motion in question. To draw the graph of the path described by the bomb, we may adopt either of two methods. We can choose various values of t, say $t=0, 1, 2, 3$, and so on, and calculate the values of x and y for each value of t. Thus when $t=1$, $x=88$ and $y=16$; then (88, 16) are the coordinates of one point on the curve. Calculating many such sets of coordinates will give us some idea of the shape of the curve (Fig. 14–2).

Or we may proceed by the second method, that is, determine the direct relationship between x and y. Solving equation (2) for t, we have $t=x/88$. Substituting this expression for t in (3) yields

$$y = 16\left(\frac{x}{88}\right)^2$$

or

$$y = \frac{x^2}{484}. \tag{4}$$

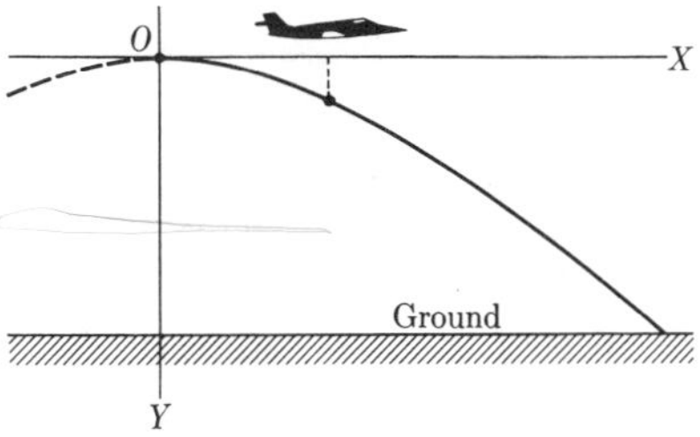

Fig. 14–2.
Path of a bomb released from an airplane flying horizontally.

From formula (9) of Chapter 12 we know that the curve is a parabola. We have thus called upon our knowledge of curve and equation to determine that the curve is a parabola. If we had not been familiar with the curve of equation (4), we would have had to analyze the equation or plot points whose coordinates satisfy (4) and thereby determine the curve. In other words, we would have been faced with a problem of coordinate geometry.

We should note that only part of the parabola is of physical interest. The full parabola extends to the right and left of the Y-axis. However, only the part to the right, that is, the half corresponding to positive x-values, represents the motion of the bomb. And of this right-hand half, which mathematically extends downward indefinitely, only an arc is of physical interest, namely the arc from O to the ground.

We have learned so far that the path of the bomb released from an airplane traveling horizontally is an arc of a parabola. Let us now see whether we can use mathematics to derive more information about the motion of bombs or, in general, about objects which move outward and downward. Suppose that two airplanes flying horizontally at speeds of 60 and 120 miles per hour, respectively, release bombs from the same point at the same instant of time. Which of these bombs would reach the ground sooner? The reader might try to answer this question by using his intuition before resorting to the use of mathematics.

Both bombs must fall the same vertical distance to reach the ground. The vertical motion is independent of the horizontal motion and is governed by formula (3). Hence this formula applies to both bombs. When they reach the ground, the value of y will be the same for both bombs. It follows that the value of t will also be the same for both. That is, both bombs will reach the ground at the same time.

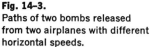

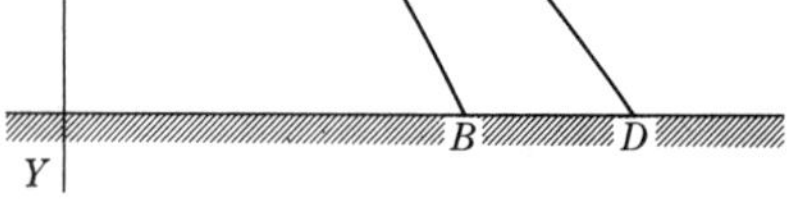

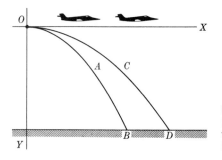

Fig. 14–3.
Paths of two bombs released from two airplanes with different horizontal speeds.

How does the difference in the speeds of the two airplanes affect the motion? The plane flying at 60 miles per hour gives its bomb a horizontal speed of 60 miles per hour, or 88 feet per second, and the plane flying at 120 miles per hour imparts to its bomb a speed of 120 miles per hour, or 176 feet per second. Hence the bombs move with different horizontal speeds, and in the same time, t, the second one will travel farther horizontally. Thus OCD, the path of the second bomb, will be a wider parabola than OAB, the path traveled by the first bomb (Fig. 14–3).

EXERCISES

1. Suppose that there were no force of gravity, and that an object is released from an airplane flying horizontally at the rate of 100 mi/hr. Describe the subsequent motion of the object.
2. One object is dropped from an airplane flying horizontally at the rate of 100 mi/hr and another from a plane flying horizontally at 200 mi/hr. Both planes are at the same altitude. Compare the times required for the two objects to reach the ground. What principle is involved?
3. From a cliff 500 ft high a stone is thrown horizontally with a speed of 100 ft/sec. How long does it take the stone to reach the ground below? What horizontal distance has the stone traveled by the time it strikes the ground?
4. A gun installed in a plane which is flying in a horizontal line at a speed of 2000 ft/sec fires a bullet in the direction of the plane's motion at the initial speed of 1000 ft/sec. What is the horizontal speed of the bullet relative to the ground?
5. A bullet fired horizontally hits a point on a wall 300 ft away. The point is 1 ft below the level at which the bullet is fired. What is the horizontal speed of the bullet?
6. A plane is traveling in a horizontal line at a speed of 300 ft/sec and at an altitude of 1 mi. Where (at what horizontal distance from the target) should the gunner release a bomb to hit a given point on the ground?
7. Suppose that a plane flying in a horizontal line at the rate of 200 ft/sec releases a bomb and continues to fly horizontally at the same rate. Where is the plane in relation to the bomb when the bomb strikes the ground?

14-4 THE MOTION OF PROJECTILES LAUNCHED BY CANNONS

A slight extension of the mathematics just introduced to treat the motion of bombs dropped from airplanes will enable us to handle the motion of projectiles shot out from cannons inclined at some angle to the ground. It was this latter problem which Galileo investigated in the seventeenth century. We shall see how neatly mathematics answers a variety of problems raised by such motions.

Suppose that a cannon inclined at an angle of 30° to the ground fires a shell with a velocity* of 1000 ft/sec (Fig. 14-4). What is the subsequent motion of the shell? We know from intuition or experience with balls thrown at a similar angle of elevation that the shell will travel out and up along some curved path and will then return to the ground. This qualitative knowledge is not, of course, sufficient to answer significant questions about the motion.

The initial velocity of the shell is in the direction which makes an angle of 30° to the ground. To treat the motion of the shell, it is mathematically

---

* The terms speed and velocity are often used interchangeably. However, the word *velocity* implies that direction as well as magnitude of the speed are under discussion.

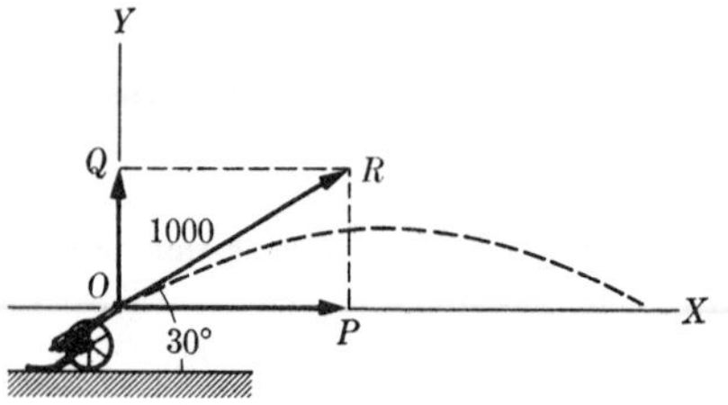

Fig. 14–4.
Shell fired with an initial velocity of 1000 ft/sec from a cannon inclined at an angle of 30° to the ground.

simpler to consider its horizontal and vertical motions separately, that is, to obtain the parametric formulas. For this purpose we must know the horizontal and vertical velocities of the shell.

Suppose that the shell travels for one second in the direction OR in which it is fired. How far will it travel horizontally and vertically in that second? Let us drop a perpendicular from R onto the X-axis and from R onto the Y-axis. Thus we determine the lengths OP and OQ, respectively. The length OP is the horizontal distance which the shell travels in one second and the length OQ is the corresponding vertical distance. Since OP and OQ are distances traveled in one second, they also represent the horizontal and vertical velocities.

The horizontal and vertical velocities are then OP and OQ. What are their magnitudes? We see from Fig. 14–4 that

$$\cos 30° = \frac{OP}{1000}$$

or

$$OP = 1000 \cos 30° = 1000(0.8660) = 866 \text{ ft/sec.}$$

Similarly,

$$\sin 30° = \frac{PR}{1000}$$

or

$$PR = 1000 \sin 30° = 1000(0.5000) = 500 \text{ ft/sec.}$$

Since $PR = OQ$, the horizontal and vertical velocities of the shell are 866 ft/sec and 500 ft/sec, respectively.

We now utilize the physical fact that the horizontal and vertical motions can be treated independently. Let us begin with the horizontal motion. The shell has an initial horizontal velocity of 866 ft/sec, and no force acts to accelerate or decelerate the horizontal motion. Hence the shell will continue to move horizontally at a constant speed of 866 ft/sec, and the horizontal distance x traveled in time t is given by

$$x = 866t. \qquad (5)$$

Next we consider the vertical motion of the shell. Gravity gives the shell a constant acceleration downward of 32 ft/sec². Since the upward direction has been chosen to be positive, the downward acceleration must be written

$$a = -32. \tag{6}$$

The downward velocity acquired in time t is $-32t$. However, the shell has an initial upward velocity of 500 ft/sec which, by the first law of motion, would continue indefinitely, were it not affected by gravity. The net velocity v is then (compare Section 13–7)

$$v = -32t + 500. \tag{7}$$

To obtain the distance traveled upward in any time t, we use the same reasoning as in the preceding chapter. If only the velocity of 500 ft/sec were acting, the distance traveled upward in t seconds would be $500t$. But in that time gravity pulls the shell downward a distance of $16t^2$. Hence y, the net height above the ground, is

$$y = -16t^2 + 500t. \tag{8}$$

Formulas (5) and (8) give the horizontal and vertical distances from the starting point O. We note that once more motion is represented by parametric formulas.

Several questions about the motion arise in practice. The first is, What path does the shell take? We may save ourselves the work of plotting the curve by determining the direct relationship between x and y, provided we recognize the curve of the resulting equation. Let us try this. Solving (5) for t yields $t = x/866$. We substitute this value of t in (8) and obtain

$$y = -16\left(\frac{x}{866}\right)^2 + 500\left(\frac{x}{866}\right)$$

or

$$y = -\frac{x^2}{46{,}872} + \frac{250}{433}x. \tag{9}$$

In Section 12–5 we discussed equations of the form (9)—albeit by means of numerically simpler examples. We could have proved the quite general statement that an equation of the form

$$y = -ax^2 + bx, \tag{10}$$

where a and b are any positive numbers, represents a parabola which opens downward and passes through the origin. (See Exercise 3 of Section 12–5, which is a special case.) Hence with just a little more work in coordinate geometry, we could have proved what we shall now state without proof, namely, that equation (9) describes a parabola. Thus the parabola, as Galileo readily established, appears once more as the path of a projectile.

What is the *range* of the shell? That is, how far from the starting point will the projectile strike the ground again? The answer is important because it tells us whether a given target on the ground can be reached. Unfortunately neither formula (5) nor formula (8) answers this question directly. However, when the shell reaches the ground, the value of y in (8) should be zero. Let us determine, then, the value of t, say t_1, when $y=0$. From (8)

$$0 = -16t_1^2 + 500t_1. \tag{11}$$

Equation (11), which is of the second degree in t_1, is rather easy to solve. Applying the distributive axiom, we may write

$$0 = t_1(-16t_1 + 500). \tag{12}$$

The right side of (12) equals zero when either factor is zero, that is, when $t_1=0$ and when

$$-16t_1 + 500 = 0.$$

The second alternative leads to

$$t_1 = \tfrac{125}{4}. \tag{13}$$

The first value, $t_1=0$, corresponds to the instant when the shell first starts its flight. Then the second value, 125/4, must be the time when the shell returns to the ground.

To determine the range, one more step is necessary. Formula (5) tells us how far the shell travels horizontally in any time t. Since the shell travels 125/4 seconds by the time it reaches the ground, we have but to substitute this value of t in (5) to get the range. If x_1 denotes the range, then (see Fig. 14–5)

$$x_1 = 866\,\tfrac{125}{4} = 27{,}063 \text{ feet}. \tag{14}$$

We might also like to know how high the shell will go in its flight and how long it takes to reach that height. These questions are readily answered. At the highest point in its flight, the *vertical* velocity is zero, else the shell would continue to rise. Formula (7) gives us the vertical velocity at any time t. Let us ask for the value of t, say t_2, when $v=0$. Then

$$0 = -32t_2 + 500$$

or

$$t_2 = \tfrac{500}{32} = \tfrac{125}{8}. \tag{15}$$

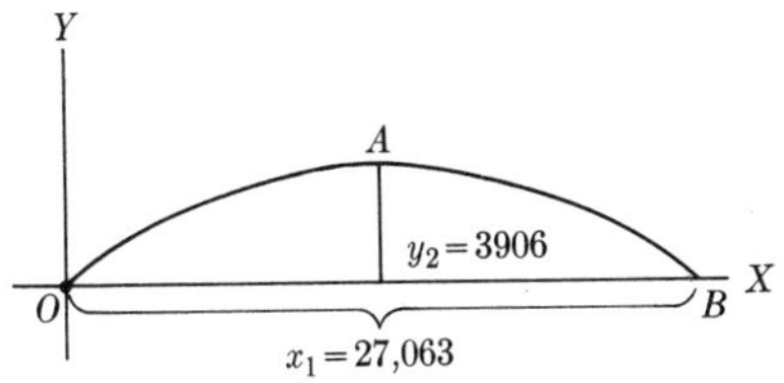

Fig. 14–5.
Path of a shell shot from a cannon.

Hence it takes 125/8 seconds for the shell to reach the highest point. Now formula (8) tells us how high the shell is at any time t. Let us therefore find the height, y_2, when $t = 125/8$. We substitute 125/8 for t in (8) and obtain

$$y_2 = -16\left(\tfrac{125}{8}\right)^2 + 500\left(\tfrac{125}{8}\right)$$

or

$$y_2 = 3906 \text{ feet.} \tag{16}$$

Hence the shell reaches a maximum height of 3906 feet above the ground.

Another interesting question is whether the shell takes as long to travel from the cannon to its maximum height as it does to return from the latter position to the ground. Or, to fit the situation shown in Fig. 14–5, we may restate the question and ask, Does it take as long for the shell to travel from O to A as from A to B? We considered an analogous problem in the preceding chapter while discussing the motion of an object thrown straight up into the air and found that the two time intervals were equal. What does intuition suggest as the answer in the present case?

We can show at once that the time required to get from O to A is the same as the time required to travel from A to B. Equation (13) supplies the time it takes the shell to reach B, that is, to travel the path OAB. Equation (15) gives the time required to travel the path OA. We see at once that the value of t_1 is twice the value t_2. Hence the time of travel along path AB must equal the time of travel along path OA.

EXERCISES

1. Suppose that a shell is fired in a direction making an angle of 40° with the ground and with a velocity of 300 ft/sec. What are the horizontal and vertical velocities of the shell? What are the parametric equations describing the motion?
2. Suppose that the parametric formulas for the motion of a projectile are $x = 20t$ and $y = -16t^2 + 30t$. What is the direct relationship between x and y? What is the nature of the curve represented by the direct relationship between x and y?
3. Suppose that the parametric formulas for the motion of a projectile are $x = 3t$ and $y = -16t^2 + 5t$. Working with these formulas, plot a few points of the path.
4. Find the range of the projectile whose motion is described in Exercise 2.

5. Find the maximum height of the projectile whose motion is described in Exercise 2.
6. What velocity does the shell whose motion is treated in Section 14–4 have on striking the ground? How does this terminal velocity compare with the initial velocity?

* 14–5 THE MOTION OF PROJECTILES FIRED AT AN ARBITRARY ANGLE

In the preceding section we saw how we could study the motion of a shell fired from a cannon which is inclined at an angle of 30° to the ground. The initial velocity of the shell in this direction was 1000 ft/sec. Since the angle of fire and initial velocity given in our example are representative values, the example teaches a great deal about the phenomenon of projectile motion. However, suppose that we sought to answer such questions as: What is the effect of the initial velocity on the range and on the maximum height attained by the projectile? What is the effect of the angle of fire on the range and on the maximum height of the projectile? At what angle should one fire a projectile to hit a given target? One could repeat the procedures pursued in the preceding section, using different initial velocities and angles of fire, and thus perhaps obtain answers to some of these questions. But the work would be considerable and still leave us with the problem of trying to infer a general conclusion from a number of special cases. The mathematician would not proceed in this way. He would suppose that the initial velocity is an arbitrary value, V, and that the angle of fire is an arbitrary angle, A, and then study the motion with these arbitrary values V and A. He might thereby obtain conclusions about all such motions because his results would hold for *any* initial velocity and *any* angle of fire.

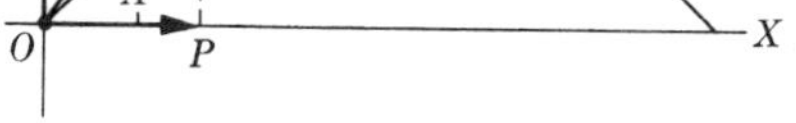

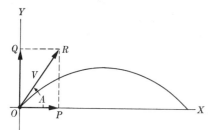

Fig. 14–6.
Shell fired with an initial velocity of V ft/sec from a cannon inclined at an angle A to the ground.

Let us pursue this *general* investigation of projectile motion. Suppose that a shell is fired from a cannon which is inclined at an angle A to the ground (Fig. 14–6), and that the initial velocity of the shell is V ft/sec. What is the subsequent motion of the shell?

We shall follow the procedure of the preceding section. The first major point to remember is the physical principle that the horizontal and vertical motions of the projectile can be studied as though the motions were taking place independently. Hence let us find the initial horizontal and vertical velocities of the shell. By the very same argument that we used in Section 14–4, we know that the components of the velocity OR in the horizontal and vertical direction (Fig. 14–6) are obtained by dropping a perpendicular from R onto the X- and Y-axes, respectively. Thus OP is the horizontal velocity, and OQ is the vertical velocity. Now

$$\cos A = \frac{OP}{OR}.$$

Hence

$$OP = OR \cos A$$

or

$$OP = V \cos A. \tag{17}$$

Since

$$\sin A = \frac{PR}{OR},$$

and since $OQ = PR$, we have

$$OQ = OR \sin A$$

or

$$OQ = V \sin A. \tag{18}$$

Formulas (17) and (18) give us the initial horizontal and vertical velocities, respectively. We must now see what happens when the shell is in motion. The horizontal motion is uniform; i.e., no force acts to speed it up or slow it down. The shell will therefore continue to travel indefinitely in the horizontal direction at the velocity given by (17). Let us use v_x to indicate the velocity in the X-direction. Then at any time t,

$$v_x = V \cos A. \tag{19}$$

Since the velocity in the horizontal direction is constant, the distance traveled is velocity multiplied by time. Then

$$x = (V \cos A)t.$$

This expression is best written as

$$x = Vt \cos A, \tag{20}$$

so that there is no confusion about the fact that the quantity whose cosine is to be taken is A, whereas if we had written $V \cos At$, one might think that the quantity is At. Formula (20) is a generalization of formula (5), for the 866 in (5) is just 1000 cos 30°.

To obtain the vertical velocity of the shell we must take into account the fact that gravity does produce a vertical acceleration which affects the vertical velocity. This vertical acceleration is 32 ft/sec² and is downward. Since we have chosen the upward direction as positive, we have

$$a = -32.$$

Because the acceleration is constant, the downward velocity gained by the shell in t seconds is $-32t$. However the shell has an upward initial velocity of $V \sin A$. Hence the net vertical velocity v_y is

$$v_y = -32t + V \sin A. \tag{21}$$

Next we use an old argument to determine the vertical height attained by the shell in t seconds. If only the velocity $V \sin A$ were acting, then in t seconds the shell would reach the height $(V \sin A)t$ or $Vt \sin A$. However, in these t seconds gravity pulls the shell downward a distance of $16t^2$. The net height, y, of the shell therefore is

$$y = -16t^2 + Vt \sin A. \tag{22}$$

Formulas (19) through (22) supply the general equations of projectile motion. We are now in a position to answer with respect to the arbitrary values V and A the same questions that were discussed in the preceding section for a specific numerical example. Thus, by solving (20) for t and substituting this result in (22), we could find the direct relationship between x and y. We could then see that for any fixed value of V and any fixed value of A, the path is a parabola. Similarly, we could obtain general expressions for the maximum height reached by the shell, the time required to reach that height, and, say, the time required for the shell to return to the ground. In other words, we could reproduce for any V and A the results derived in the preceding section for special values of V and A.

Let us turn instead to answering questions which we could not treat before. Let us study the effect of the initial velocity, V, and angle of fire, A, on the range of the cannon. To do this we must first determine the general expression for the range. The method is the same as the one used in the preceding section.

We begin with the physical fact that when the shell strikes the ground, the y-value of its position is zero. Hence let t_1 be the value of t when $y=0$. Setting $y=0$ in (22), we have

$$0 = -16t_1^2 + Vt_1 \sin A.$$

We may apply the distributive axiom to write

$$0 = t_1(-16t_1 + V \sin A).$$

Now the right side is zero when $t_1 = 0$ and when

$$-16t_1 + V \sin A = 0. \tag{23}$$

Since $t_1 = 0$ corresponds physically to the instant when the shell starts its motion, it follows that the value given by (23) is the value of t at the instant the shell strikes the ground. If we solve (23) for t_1, we find

$$t_1 = \frac{V \sin A}{16}. \tag{24}$$

To determine the shell's range, that is, its horizontal distance from the starting point at the instant when t_1 has the value just found, we use formula (20). Thus, if x_1 is the value of x when t has the value t_1, then

$$x_1 = V \cos A \, \frac{V \sin A}{16},$$

or the range is

$$x_1 = \frac{V^2}{16} \sin A \cos A. \tag{25}$$

Formula (25) answers one question immediately. If the angle A is held fixed, then the range depends upon V^2. If V is increased, then x_1 increases, and indeed x_1 increases rapidly because it depends upon V^2 rather than upon just V. Also, if we wished to attain a given range with a given angle of fire, that is, if x_1 and A were specified, we could use (25) to calculate the necessary initial velocity, V.

The more practical problem is to study the dependence of range upon angle of fire, for it is easier to change the angle of fire of a cannon than it is to change the initial velocity of the shells. Let us suppose, then, that the initial velocity V is fixed and ask the question, What is the maximum range that can be obtained by varying angle A? Since V is fixed, our question, in view of (25), amounts to: For what value of A is the product $\sin A \cos A$ a maximum? A little mathematics provides the answer.

We know from our work on the trigonometric ratios that $\sin A$ and $\cos A$ are certain ratios of sides of a right triangle. The size of the right triangle used to determine $\sin A$ and $\cos A$ does not matter because all possible right triangles containing a definite angle A are similar and therefore the ratio of two particular sides in any one of these triangles is always the same.* Hence

* The reader can review this point in Chapter 7, where it was first mad

let us choose a right triangle whose hypotenuse AB (Fig. 14–7) is the diameter of a definite circle. The vertex of the right angle of this right triangle must lie on the circle because a theorem of plane geometry states that the vertices of all right triangles with vertex C and hypotenuse AB must lie on a circle with AB as diameter. Let us denote the sides of the right triangle ABC by a, b, and c. We draw the perpendicular CD and denote it by h. Then, using the right triangle ADC, we obtain

$$\sin A = \frac{h}{b}\cdot$$

From the right triangle ABC we have

$$\cos A = \frac{b}{c}.$$

So far, then, we have found that

$$\sin A \cos A = \frac{h}{b}\cdot\frac{b}{c} = \frac{h}{c}\cdot \qquad (26)$$

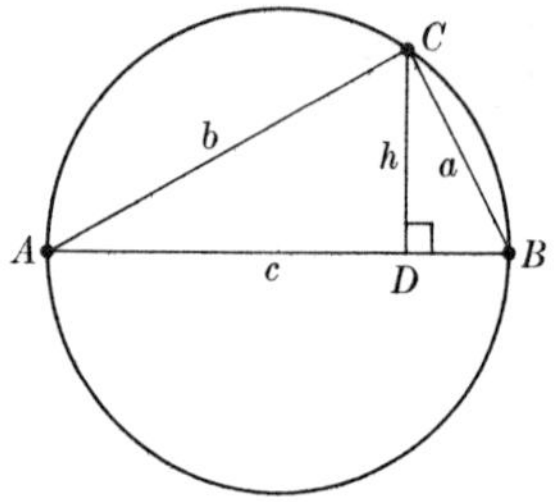

Fig. 14-7

We may vary angle A and continue to regard it as an angle of a right triangle with hypotenuse AB because, as we noted above, the size of the right triangle is immaterial. Of course, variations in the angle A will produce changes in the position of C, but for the reason given earlier, C will continue to be a point on the circle with diameter AB. Hence we can consider all possible acute angles A by considering all triangles ABC with AB fixed and C varying on the circle. Let us now look at (26) again. Since the quantity c is fixed, $\sin A \cos A$ will be a maximum when h is a maximum, and h is greatest when it is the radius of the circle. But when h is the radius, C is directly above the center of the circle. Then $AC = BC$. In this case angle $A = 45°$ because the right triangle ABC is isosceles. Hence

$$\sin A \cos A$$

is a maximum when angle $A = 45°$.

We may now return to formula (25). We have found that when V is fixed, the maximum range, that is, the maximum possible value of x_1, is obtained when $A = 45°$. Since for $A = 45°$, $\sin A = \cos A = \sqrt{2}/2$, the maximum range is given by the formula

$$\text{maximum } x_1 = \frac{V^2}{16}\cdot\frac{\sqrt{2}}{2}\cdot\frac{\sqrt{2}}{2} = \frac{V^2}{32}\cdot$$

This famous result was first proved by Galileo.

EXERCISES

1. Formulas (20) and (22) give the parametric representation of the motion of a projectile. Find the direct relationship between x and y.
2. Derive the formula in terms of V and A for the time it takes a projectile to reach its maximum height.
3. What is the formula in terms of V and A for the maximum height reached by a projectile?
4. Show generally that it takes as long for a projectile to return from its highest position to the ground as it does to travel from the cannon to the highest position.
5. What is the range of a projectile which is fired with an initial velocity of 2000 ft/sec and at an angle of 40° to the ground?
6. What is the maximum range of a projectile fired with an initial velocity of 800 ft/sec?
7. How long does it take for a projectile fired with an initial velocity of 2000 ft/sec to reach a target located at maximum range?
8. Show that the highest point which a gun can reach for all possible angles of fire but with the same initial velocity is attained by firing straight up.

14-6 SUMMARY

In this chapter we have seen how the simultaneous application of two simple formulas, the parametric formulas, makes it possible to represent an entire class of curves which physically happen to be the paths of projectiles. The value of the parametric formulas is, then, that they enable us to answer readily a variety of questions about projectile motion. To appreciate how much mathematics accomplishes in this area, one might consider how he would proceed experimentally to find, for example, the dependence of the range of a projectile on the angle of fire. One would have to fire at least dozens of projectiles at different angles, making certain that other factors, such as the velocity with which the projectiles are fired, the shape of the projectiles, and the state of the atmosphere, are constant, and accurately measure the range and angle each time. With all these precautions taken and the information secured, the experimenter might obtain some limited results. He might, for example, learn that as the angle of fire increases in 1°-degree steps from 1° to 40°, the range increases steadily, but he might miss the all-important fact that above 45° the range decreases. The dependence of range upon velocity would still be entirely unknown and require further experimentation. But we have seen how a little mathematics costing only pencil and paper can supply the full story of the dependence of range upon angle of fire and initial velocity.

Thus the present chapter, too, illustrates how the combination of mathematics and simple physical axioms, such as the fact that the acceleration of bodies near the surface of the earth is 32 ft/sec^2, the first law of motion, and the independence of the horizontal and vertical motions, permits us to deduce a vast amount of knowledge about our physical world. The knowledge referred to at the moment concerns projectile motion under idealized conditions; that is, the resistance of air is neglected; the earth is assumed to be flat over the short distances which the projectiles cover; and the projectiles are limited to travel near the surface of the earth. One might regard the whole story as of minor interest because it deals with just one phenomenon and one which seems limited to bombs and guns. However, the study of this phenomenon has proved to be of immeasurable scientific importance. First of all, the deductions made from the physical axioms mentioned above can be checked experimentally. If the deductions agree with experience we have some reason to believe that the axioms are correct. We must remember in this connection that physical axioms are generalizations from limited experience and that our confidence in them depends upon how well they continue to lead to new physical facts. Secondly, the study of motions near the surface of the earth, projectile motion in particular, led to the most important advance in science since 1600, namely Newtonian mechanics. The step to the broad science of mathematical mechanics will be taken in the next chapter.

REVIEW EXERCISES

1. Sketch the curves whose parametric equations are given below.

   a) $x = 3t, \quad y = 7t$
   b) $x = 3t, \quad y = 5t^2$
   c) $x = 3t^2, \quad y = 5t$
   d) $x = 3t + 7, \quad y = 5t + 9$
   e) $x = 5 \cos \theta, \quad y = 5 \sin \theta$
   f) $x = 2t, \quad y = 5t^2 + 3t$

2. Find the direct equations in parts (a), (b), (c), and (d) of Exercise 1.
3. Suppose a bomb is released from an airplane which is flying horizontally at a speed of 240 mph.
   a) Write the parametric equations of motion of the bomb.
   b) If the airplane is one mile above the ground, how long will it take the bomb to strike the ground?
   c) Suppose the airplane flies at 300 mph instead of 240. How long will it take the bomb to strike the ground?
   d) Suppose the airplane is 2 mi above the ground and flies at 240 mph. How long will it take the bomb to reach the ground?
   e) How far from the point on the ground directly below the point at which the bomb is released does the bomb strike the ground?
4. Suppose a shell is shot from a cannon inclined at an angle of 45° with the ground and the shell is given an initial velocity of 2000 ft/sec.
   a) What are the horizontal and vertical velocities of the shell?
   b) Write the parametric equations of the motion of the shell.
   c) Find the range of the shell.
   d) What is the maximum height above the ground attained by the shell during its flight?

Recommended Reading

GALILEI, GALILEO: *Dialogues Concerning Two New Sciences*, pp. 234 through 282, Dover Publications, Inc., New York, 1952.

HOLTON, GERALD and DUANE H. D. ROLLER: *Foundations of Modern Physical Science*, Chap. 3, Addison-Wesley Publishing Co., Inc., Reading, Mass., 1958.

KLINE, MORRIS: *Mathematics and the Physical World*, Chap. 14, T. Y. Crowell Co., New York, 1959. Also in paperback, Doubleday and Co., New York, 1963.

CHAPTER 15

THE APPLICATION OF FORMULAS TO GRAVITATION

. . . from motion's simple laws
Could trace the secret hand of Providence
Wide-working through this universal frame.

JAMES THOMSON in his
Memorial Poem to Newton

15-1 THE REVOLUTION IN ASTRONOMY

While Galileo was fashioning the new science of motion, Johannes Kepler was making dramatic contributions to one of the most far-reaching developments in the history of Western civilization. This development was begun by Nicolaus Copernicus and its essence was a radically new mathematical theory of planetary motions.

Up to the sixteenth century the only sound and useful astronomical theory was the geocentric system of Hipparchus and Ptolemy which we examined in Chapter 8. This was the theory accepted by professional astronomers and applied to calendar reckoning and navigation. It was, however, a rather sophisticated creation in that its strength lay entirely in the mathematical effectiveness of the scheme. The deferents and epicycles had no physical significance in themselves nor did the theory give any physical or intuitive reasons that the planets should move on epicycles attached to deferents.

The author of the next great celestial drama was Nicolaus Copernicus, who lived about 1400 years after Ptolemy. Copernicus was born in Poland in 1473 and, after studying mathematics and science at the University of Cracow, decided to go to Italy, the center of the revived Greek learning. At the University of Bologna, which he entered in 1497, he studied astronomy. Then for ten years he studied medicine and law and secured a doctor's degree in both fields. He also became learned in Greek and mathematics. In 1500 Copernicus was appointed a canon of the Cathedral of Frauenberg in East Prussia, but he did not assume his duties until 1512 when he had finished his studies in Italy. The job, which entailed mainly the management of estates owned by the Cathedral, left Copernicus with plenty of time to make astronomical observations and to think about the relevant theory. After years of reflection and observation Copernicus finally evolved a new theory of planetary motions which he incorporated in a classic work, *On the Revolutions of the Heavenly Spheres.* This appeared in 1543, the year in which Copernicus died.

As we have already noted, when Copernicus began to think about astronomy, the Ptolemaic theory was the only sound and effective system in existence. This theory had become somewhat more complicated during the intervening centuries in that more epicycles had been added to those introduced by Ptolemy, to make the theory fit the increased amount of observational data gathered largely by the Arabs. In Copernicus' time the theory required a total of 77 circles to describe the motion of the sun, moon, and the five planets known then.

Copernicus had studied the Greek works and had become convinced that the universe was mathematically and harmoniously designed. Harmony demanded a more pleasing theory than the complicated extensions of Ptolemaic theory. Copernicus read that some Greek authors, notably Aristarchus, had suggested the possibility that the sun might be stationary and that the earth revolved about the sun and rotated on its axis at the same time. He decided to explore this possibility. He was in a sense overimpressed with Greek thought, for he, too, believed that the motions of heavenly bodies must be circular or, at worst, a combination of circular motions since circular motion was natural motion. Moreover, he also accepted the belief that each planet must move at a constant speed on its epicycle, and that the center of each epicycle must move at a constant speed on the circle which carried it. Such principles were axiomatic for him. Copernicus even adds an argument which shows the somewhat mystic character of sixteenth-century thinking. He says that a variable speed could be caused only by a variable power; but God, the cause of all motions, is constant.

The upshot of such reasoning was that Copernicus used the scheme of deferent and epicycles to describe the motions of the heavenly bodies, with, however, the all important difference that the sun was at the center of each deferent, while the earth itself became a planet moving about the sun and rotating on its axis. Nevertheless, he achieved considerable simplification. He was able to reduce the total number of circles, deferents and epicycles, to 34 instead of the 77 required under the geocentric view.

However, the remarkable simplification was achieved by Johannes Kepler, one of the most intriguing figures in the history of science. In a life beset by many personal misfortunes and hardships occasioned by social and political events, Kepler had the good fortune to become in 1600 an assistant to the famous astronomer Tycho Brahe. Brahe was then engaged in making extensive new observations, the first such major undertaking since Greek times. These observations, together with others which Kepler made himself, were invaluable to him in his later work. When Brahe died in 1601 Kepler succeeded him as Imperial Mathematician to the Emperor Rudolph II of Austria, King of Bohemia.

Kepler's scientific reasoning is fascinating. Like Copernicus, he was a mystic and, like Copernicus, he believed that the world was designed by God in accordance with some simple and beautiful mathematical plan. This belief dominated all his thinking. But Kepler also had qualities which we now associate with scientists. He could be coldly rational. His fertile imagination triggered the conception of new theoretical systems. But he knew that theories must fit observations and, in his later years, saw even more clearly that empirical data may indeed suggest the fundamental principles of science. Copernicus, too, wanted his theory to fit observational data; yet he held to the heliocentric view, although the differences between theoretical predictions and astronomical data were greater than might be accounted for by experimental errors alone. Kepler, on the other hand, sacrificed his most beloved theories when he saw that they did not fit observational data, and it was precisely this incredible persistence in refusing to tolerate discrepancies which any other scientist of his day would have disregarded that led him to espouse radical ideas. He also had the humility, patience, and energy to perform extraordinary labor which mark great men.

In his book *On the Motion of the Planet Mars*, published in 1609, Kepler announced the first two of his three famous laws of planetary motion. The first of these is especially remarkable, for Kepler broke with the tradition held for 2000 years that circles or spheres must be used to describe heavenly motions. Instead of resorting to deferent and several epicycles, which both Ptolemy and Copernicus had used to describe the motion of any one planet, Kepler found that a single ellipse would do. His first law states that each planet moves on an ellipse and that the sun is at one (common) focus of each of these elliptical paths (Fig. 15–1). The other focus of each ellipse is merely a mathematical point at which nothing physical exists.

Kepler's first law utilizes a geometrical figure which had been introduced and studied by the Greeks of the classical period. Had the ellipse and its properties not yet been known, and had Kepler been faced with the double

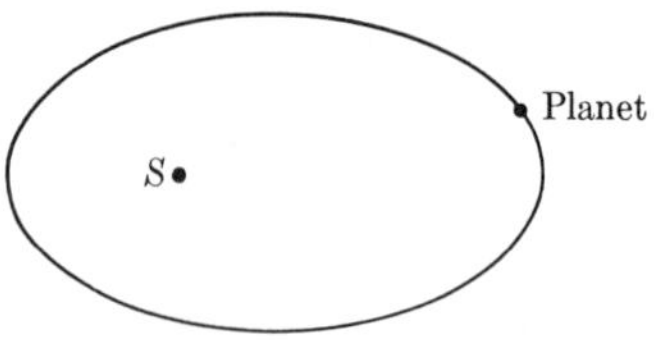

Fig. 15–1.
Each planet moves in an ellipse about the sun.

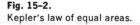

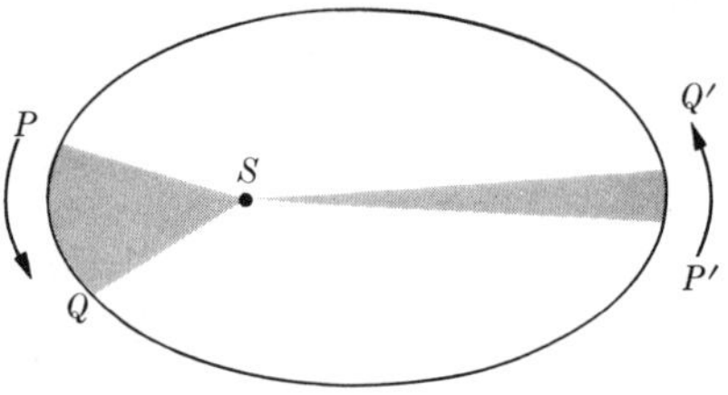

Fig. 15–2.
Kepler's law of equal areas.

problem of abstracting the proper path from a multitude of data *and* conceiving the ellipse, he might possibly have ended up in an impasse. By working out the properties of this curve Euclid, Apollonius, and Archimedes determined the course of our civilization just as decisively as if they had stood at Kepler's side.

Kepler's first law is of immense value in comprehending readily the paths of the planets. But astronomy must go much further if it is to be interesting in itself and useful. It must tell us how to predict the positions of the planets. If one finds by observation that a planet is at a particular position, P say in Fig. 15–2, he might like to know when it might be at some other position, a solstice or an equinox, for example. What is needed is the velocity with which the planets move along their respective paths.

Here, too, Kepler made a radical step. Copernicus, as we noted earlier, and the Greeks had always used constant velocities. A planet moved along its epicycle so as to cover equal arcs in equal times, and the center of each epicycle also moved at a constant velocity on another epicycle or on a deferent. But Kepler's observations told him that a planet moving on its ellipse does not move at a constant speed. Kepler searched hard and long for the correct law of velocities and found it. What he discovered was that if a planet moves from P to Q (Fig. 15–2) in, say one month, then it will also move from P' to Q' in one month, provided that the *area* PSQ equals the *area* $P'SQ'$. Since P is nearer the sun than P' is, the arc PQ must be larger than the arc $P'Q'$ if the areas PSQ and $P'SQ'$ are equal. Hence the planets do not move at a constant velocity. In fact, they move faster when closer to the sun.

Kepler was overjoyed to discover this second law. Although it is not so simple to apply as a law of constant velocity, it nonetheless confirmed his fundamental belief that God had used mathematical principles to design the universe. God had chosen to be just a little more subtle, but a mathematical law clearly determined how fast the planets moved.

Another major problem remained open. What law described the distances of the planets from the sun? The problem was now complicated by the fact that a planet's distance from the sun was not constant but varied from a least to a greatest value (see Fig. 15–2). Hence Kepler searched for a new principle which would take this fact into account. Now he believed that nature was not only mathematically but harmoniously designed and he took this word "harmony" very literally. Thus he believed that there was a music of the spheres which produced a harmonious tonal effect, not one given off in actual sounds but discernible by some translation of the facts about planetary motions into musical notes. He followed this lead and after an amazing combination of mathematical and musical arguments, arrived at the law that if T is the period of revolution of any planet and D is its mean distance from the sun, then

$$T^2 = kD^3,$$

where k is a constant which is the same for all the planets. This statement is Kepler's third law of planetary motion and the one which he triumphantly announced in his book *The Harmony of the World* (1619).

15-2 THE OBJECTIONS TO A HELIOCENTRIC THEORY

The work of Copernicus and Kepler is by far the most dramatic, startling, and influential development in the formation of modern culture. The first surprising feature, and one which in itself makes their work astounding, is the sharp break from existing thought. Copernicus and Kepler were educated in a milieu which accepted the geocentric theory of Ptolemy as almost unquestionable truth. Moreover, both were scientifically cautious. Nor did either really have at his disposal any unusual observations which conflicted sharply with Ptolemy's theory. Copernicus, as a matter of fact, was not a great observer and did not seem to mind leaving his work somewhat at odds with observations. Kepler did have access to more numerous and more reliable data and showed greater tenacity in making the theory fit the data, but there was nothing in these observations which suggested that an entirely new theory must be introduced.

While casting aside the weight of centuries, Copernicus and Kepler could give only token rebuttals to the numerous scientific arguments against a moving earth which Ptolemy had advanced against Aristarchus. How could the heavy earth be put into motion? That the other planets were in motion even according to Ptolemaic theory was explained by the doctrine that these were made of special light matter and therefore easily moved. About the best answer Copernicus could give was that it was natural for a sphere to move. Another scientific argument against the earth's rotation maintained that rotation would cause objects to fly off into space just as an object on a rotating platform will fly off. Copernicus had no answer to this argument. To the further objection that a rotating earth should itself fly apart, Copernicus replied weakly that since the earth's motion was natural, it could not destroy the body. Then he countered by asking why the sky did not fall apart under the very rapid daily motion which the geocentric theory called for. Yet another objection declared that if the earth rotated from west to east, an object thrown up into the air should fall back to the west of its original position because the earth moved on while the object was in the air. If, moreover, the earth revolved about the sun, then, since the velocity of an object is proportional to its weight, or so at least Greek and Renaissance physics maintained, lighter objects on the earth should be left behind. Even the air should be left behind. To the last argument Copernicus replied that air is earthy and so moves in sympathy with the earth.

These scientific objections to a moving earth were weighty ones and could not be dismissed as the stubbornness of doubters who refused to see the truth

The substance of the matter is that a rotating and revolving earth did not fit in with the physical theory of motion due to Aristotle and common in Copernicus' and Kepler's time.

Another class of scientific arguments against a heliocentric theory came from astronomy proper. The most serious one stemmed from the fact that the heliocentric theory regarded the stars as fixed. In six months the earth changes its position in space by 186 million miles. Hence if one notes the direction of a particular star at one time and again six months later, he should observe a difference in direction. But this difference was not observed in Copernicus' and Kepler's time. Copernicus answered that the stars are so far away that the difference in direction was too small to be observed. However, his explanation did not satisfy the critics, who countered that if the stars were that distant, then they should not be clearly observable. In this instance, Copernicus' answer was correct. The change in direction over a six-month period for the nearest star is an angle of 0.76″, and this was first detected by the mathematician Bessel in 1838 who, of course, by that time had a good telescope at his disposal.

A further, powerful argument against a moving earth contended that we do not feel any motion despite the fact that the earth is presumably moving around the sun at 18 miles per second and that a person on the equator is rotating at the rate of about 0.3 miles per second. Our senses, on the contrary, tell us that the sun is moving in the sky. Of course, there are counterarguments which would mean more to us today because we have the experience of traveling at high speeds. A person traveling in an airplane at 400 miles per hour does not feel the motion. But to the people of Copernicus' time the argument that we do not feel ourselves moving at the very high speeds called for by the new astronomy was convincing.

15-3 THE ARGUMENTS FOR THE HELIOCENTRIC THEORY

In view of the numerous and sound arguments against the heliocentric theory and the challenge it posed to the prevailing religious thinking of the times, what made Copernicus and Kepler take up this long-discarded thought and pursue it so courageously? For what most other men would call a mess of pottage, they broke with established physics, philosophy, religion, and common sense.

Both were convinced that the universe is mathematically designed and hence that a true pattern of the motions was inherent. Moreover, this design was instituted by God, and God would surely have used a simple and harmonious pattern. But Ptolemaic theory had become so encumbered in the sixteenth century that it was no longer simple or beautiful. Hence Copernicus and Kepler believed, when each found a more harmonious and simpler theory, that their work was indeed a description of the divine order of things.

There are many passages in Copernicus' *On the Revolutions of the Heavenly Spheres* and in Kepler's writings which bear unmistakable testimony to this motivation as the central force in the search for a new theory and to their conviction that they had found the right one when it proved to be simpler. Copernicus says of his theory:

> *We find therefore, under this orderly arrangement, a wonderful symmetry in the universe, and a definite relation of harmony in the motion and magnitude of the orbs, of a kind that it is not possible to obtain in any other way.*

Kepler remarks of his later work wherein he had already introduced the elliptical theory of motion, "I have attested it as true in my deepest soul and I contemplate its beauty with incredible and ravishing delight." The work of Copernicus and Kepler is the work of men searching the universe for the harmony which their religious convictions assured them must exist and which must be describable mathematically and simply because God had so designed the universe. What distinguishes their religious convictions from those of their contemporaries is that they did not tie themselves to literal interpretations of the Holy Writings. They searched for the word of God in the heavens.

The core of the argument which Copernicus and Kepler presented for the heliocentric theory was its mathematical simplicity. Their philosophical and religious convictions assured them that the world is mathematically and simply designed; accordingly the fact that a heliocentric view was mathematically simpler than the geocentric one determined their position. The mathematical simplicity of the new view was, in fact, the sole argument they could advance. Only persons possessed of the unshakable conviction that mathematics is the essence of the design of the universe and that the omnipotent mathematician would necessarily prefer simplicity would dare to advance such a radical theory and would have had the courage to defend it against the opposition it was sure to and did encounter in those times.

The new theory appealed to astronomers, geographers, and navigators because it simplified their theoretical and arithmetical work. Hence many of these men adopted the new view just as a mathematical convenience, even though they were not convinced of its truth. While this feature of the new theory carried little weight with Copernicus and Kepler, it nonetheless had the effect of making more and more people think in terms of a heliocentric view, and, since one tends to accept as true what is familiar, there is no doubt that this practical aspect did, in the long run, help to gain adherents for the theory.

Support for the new theory came from an unexpected development. Early in the seventeenth century the telescope was invented, and Galileo, upon hearing of this invention, built one himself. He then proceeded to make observations of the heavens which startled his age. He detected four moons of Jupiter (we now can observe twelve), and this discovery showed that a moving planet can have satellites. Hence it was likely that the earth, too, could be in motion and yet have a satellite, our moon. Galileo saw irregular surfaces and mountains on the moon, spots on the sun, and a bulge around the equator of Saturn (which we now call the rings of Saturn). Here was further evidence that the planets were like the earth and certainly not perfect bodies composed of some special ethereal substance, as Greek and medieval thinkers had believed. The Milky Way, which had hitherto appeared to be just a broad band of light, could be seen with the telescope to be composed of thousands of separate stars, each of which gave off light. Thus, there were other suns and presumably other planetary systems suspended in the heavens. Moreover, the heavens clearly contained more than seven moving bodies, a number which had been accepted as sacrosanct. Copernicus had predicted that if human sight could be enhanced, then man would be able to observe phases of Venus and Mercury, that is, to observe that more or less of each planet's hemisphere facing the earth is lit up by the sun, just as the naked eye can discern the phases of the moon. Galileo did discover the phases of Venus.

All of Galileo's observations were made with a telescope of such limited power that, as has been said, it is remarkable he could find Jupiter, let alone the moons of Jupiter. Many of his discoveries were in direct support of a heliocentric theory; others served primarily to challenge current beliefs and to at least prepare some minds for a more objective examination of the new theory. Galileo, himself, though he lectured on Ptolemaic theory until 1605, had been converted to Copernicanism by a work of Kepler. In 1611 he openly declared for Copernicanism. His own observations convinced him that the Copernican system was correct, and in the classic *Dialogue on the Great World Systems* he defended it strongly. By the middle of the seventeenth century the scientific world was willing to proceed on a heliocentric basis.

One word of caution regarding the work of Copernicus and Kepler: these men believed that the heliocentric theory was true for the reasons already cited. This is not the view we hold today. If the criterion is truth, then heliocentric theory is not to be preferred to Ptolemaic theory. Scientific theories, we now believe, are the work of man. The mind supplies the patterns which organize observations. We may indeed prefer the heliocentric theory because it is simpler and agrees better with observations, but we do not regard it as the last word. Another theory, which still may not be the truth, may be conceived and produce even better results. As a matter of fact, one was—the theory of relativity. We shall not anticipate too much. The evolution of the concept of truth as it applies to mathematics and mathematical theories of science will be a continuing concern.

EXERCISES

1. What is a geocentric astronomical system? a heliocentric astronomical system?
2. Did Copernicus break completely with Greek astronomy?
3. Is the sun at the center of the Keplerian system?

4. What innovations did Kepler introduce into the Copernican system?
5. To reconstruct Kepler's improvement on Copernicus, suppose that a planet P moves once around its epicycle while the center of the epicycle moves once completely around the sun. What path does the planet seem to follow in relation to the sun? Would it be simpler to accept this single path as opposed to the combination of epicycle and deferent?
6. What scientific objections were there to an earth in motion?
7. Why did Copernicus and Kepler advocate the new heliocentric theory?
8. Why do you accept the heliocentric theory?
9. State Kepler's first law of planetary motion.
10. State Kepler's second law of planetary motion.
11. If we take the earth's average distance from the sun, 93,000,000 mi, as the *unit of distance* and the earth's period of revolution, 1 year, as the *unit of time*, then Kepler's third law says that $T^2 = D^3$. If the average distance of Neptune from the sun is 2,797,000,000 mi, how long does it take the planet to complete one revolution around the sun?

15-4 THE PROBLEM OF RELATING EARTHLY AND HEAVENLY MOTIONS

In view of the fact that Galileo had discovered the laws which underlie terrestrial motions and Kepler had discovered the basic laws of planetary motion one would expect that the scientists of the seventeenth century would have regarded the theory of motion to be complete. But to scientists who seek the ultimate design of our universe, the two accomplishments we have just described immediately suggested more profound problems. A comparison of these two classes of laws, namely Galileo's for terrestrial motions and Kepler's for heavenly motions, revealed several basic differences. In the first place, Galileo had started with clear physical principles, such as the first law of motion and the constant downward acceleration of objects moving near the surface of the earth, and had *deduced* the formulas which describe straight-line and curvilinear motions. Kepler's three laws, though they fitted observations within the limits of observational errors, did not rest on physical principles. They were merely accurate mathematical descriptions of collections of data. Moreover, the three laws were logically independent of one another. Secondly, for terrestrial motions, the parabola was found to be the basic path of curvilinear motion, whereas for planetary motion, the ellipse was the basic path.

This comparison raised several questions. Could one establish any logical relationship among the Keplerian laws or were they really independent? What physical principles determined planetary motions? The mathematical laws, accurate and succinct as they were, presented after all only a rather bleak account, without giving any insight into, or rationale for, the motions. And why should parabolic paths prevail on earth and elliptical paths in the heavens?

The overriding question, however, which bothered the leading scientists of the latter half of the seventeenth century was: Could one establish a connection between the laws of terrestrial motion and the laws of planetary motion? Perhaps the very same physical principles which Galileo had used to deduce the paths of objects moving near the earth could lead to the laws describing the motion of the planets. In this event, the two classes of laws would be united; the Keplerian laws would be related to each other by being deduced from a common basis; and the physical reasons for planetary motion would be revealed.

The thought that *all* the phenomena of motion should follow from one set of physical principles might seem grandiose and inordinate to reasonable people, but it occurred very naturally to the religious mathematicians of the seventeenth century. God had designed the universe, and it was to be expected that all phenomena of nature would follow one master plan. One mind designing a universe would almost surely have employed one set of basic principles to govern as many related phenomena as possible. Since the scientists of the seventeenth century were engaged in the quest for God's design of nature, it seemed very reasonable to them that they should seek the unity underlying the diverse earthly and heavenly motions. As phrased by Newton, this goal was

> *to derive two or three general principles of motion from phenomena, and afterwards to tell us how the properties and actions of all corporeal things follow from these manifest principles. . .*

A less cogent but to mathematicians nonetheless significant indication of the existence of some unity was furnished by the fact that parabola and ellipse were both conic sections. The common mathematical origin of these curves warranted some belief that parabolic and elliptical motions were but special cases of some fundamental principle of motion.

In the seventeenth century, there were other less weighty but perhaps more pressing reasons to pursue the study of motion beyond the stage reached by Galileo and Kepler. Another open question was how to relate heavenly and earthly motions in a more limited but practical connection. This was the problem described earlier of determining the longitude of a ship at sea. Although navigators had used the stars, sun, and moon to determine the locations of their ships, the positions of these celestial bodies at various times of the year had yet to be related more precisely to the longitudes of points on the earth. In the seventeenth century it seemed that the moon would be most suitable for the determination of longitude because its closeness permitted accurate observation of its position from points on the earth. Hence, more precise information about the motion of the moon around the earth was needed. This became a major scientific problem of the age.

15-5 A SKETCH OF NEWTON'S LIFE

Any great advance in mathematics and science is almost always the work of many men contributed bit by bit over hundreds of years. Then one man smart enough to distinguish the worthy ideas of his predecessors from the welter of suggestions and results and imaginative and audacious enough to fit the significant ideas into a master plan makes the culminating and definitive step. In the problem of unifying all the phenomena of motion, the decisive step was made by Isaac Newton.

He was born in 1642, premature and weak. His mother was already widowed and so preoccupied with running the family farm that she could pay no attention to the boy. The elementary education Newton received in local schools of a small English town could hardly have given him much of a start, and in his youth Newton showed no promise. His family sent him to Cambridge University, where he entered Trinity College in 1661. Here, at last, Newton got the opportunity to study the works of Copernicus, Kepler, and Galileo, and here he had at least one good teacher, the distinguished mathematician Isaac Barrow. His university work was not outstanding and he had, in fact, such difficulties with geometry that he almost changed his course of study from science to law. However, Barrow did recognize that Newton had ability.

Newton finished his undergraduate work; at that point an outbreak of the plague in the area around London led to the closing of the university. He, therefore, spent the years 1665 and 1666 in the quiet of the family home at Woolsthorpe. During this period Newton initiated his great work in mechanics, mathematics, and optics. He realized that the law of gravitation, which we shall examine shortly, was the key to an embracing science of mechanics; he obtained a general method for treating the problems of the calculus (see Chapters 16 and 17); and through experiments he made the epochal discovery that white light such as sunlight is really composed of all colors from violet to red. "All this," Newton said later in life, "was in the two plague years of 1665 and 1666, for in those days I was in the prime of my age for invention, and minded mathematics and philosophy [science] more than at any other time since."

Newton returned to Cambridge in 1667 and was elected a Fellow of Trinity College. In 1669, Isaac Barrow resigned his professorship of mathematics to devote himself to theology, and Newton was appointed in Barrow's place. Newton apparently was not a successful teacher, for few students attended his lectures; nor did anyone comment on the originality of the material he presented.

In 1684 his friend Edmond Halley, the astronomer of Halley's comet fame, urged him to publish his work on gravitation and even assisted him editorially and financially. Thus in 1687 the classic of science, the *Mathematical Principles of Natural Philosophy*, often briefly referred to as the *Principia* or the *Prin-*

ciples, appeared. This book received much acclaim and, aside from three Latin editions, appeared in many languages. One popularization was entitled *Newtonianism for Ladies*. The *Principia* is written in the deductive manner of Euclid; that is, it contains definitions, axioms, and hundreds of theorems and corollaries. Its conciseness makes it difficult reading. To excuse this aspect, Newton told a friend that he had made the *Principia* difficult on purpose "to avoid being baited by little smatterers in mathematics." He thereby hoped to avoid the criticisms heaped on his earlier papers on light.

After about thirty years of creative activity which included some work in chemistry, Newton became depressed and suffered a nervous breakdown. He left Cambridge University to become Warden of the British Mint in 1696 and thereafter confined his scientific activities to the investigation of an occasional problem. He did, however, devote himself to theological studies, which he regarded as more fundamental than science and mathematics because the latter disciplines concerned only the physical world. In fact, had Newton been born two hundred years earlier he would almost surely have become a theologian. An example of his theological writing is *The Chronology of the Ancient Kings Amended*, in which he sought to determine the dates of Biblical events by utilizing astronomical facts mentioned in connection with these events.

During his last years and posthumously he was honored in many ways. He was President of the Royal Society of London from 1703 to his death; he was knighted in 1705; and he was buried in Westminster Abbey.

15-6 NEWTON'S KEY IDEA

In his philosophy and method of science, Newton followed Galileo. He, too, believed that the universe was mathematically designed by God and that mathematics and science should strive to uncover that glorious design. Like Galileo, he was convinced that fundamental physical principles should be quantitative statements about the real qualities of the world, space, time, mass, weight, and force. From these principles and with the axioms and theorems of mathematics, it should be possible to deduce the laws of nature. Newton expressed this philosophy in the preface to his *Principles:*

> . . . *for the whole burden of philosophy* [*science*] *seems to consist in this—from the phenomena of motion to investigate the forces of nature, and from these forces to demonstrate the other phenomena* . . .

By investigating the forces of nature, he meant to arrive at the basic laws governing the operation of these forces and to deduce the consequences.

The first problem, then, in executing such a program is to discover the fundamental principles. Like Galileo, Newton insisted on obtaining these by direct study of the physical world rather than by searching one's mind for hypotheses that seemed to be reasonable or by accepting Biblical passages.

Here, too, Newton is explicit. In another of his famous books, *Opticks*, first published in 1704, he says:

> *Thus analysis consists in making observations and experiments and in drawing general conclusions by induction, and admitting of no objections against the conclusions, but such as are taken from experiments or other certain truths.*

What Newton sought to emphasize and what required emphasis in his time is that generalizations must be based on some experimental or observational grounds, and that no hypothesis can be tolerated which is contrary to a single bit of physical evidence. Further, deductions made from the basic principles must also be in accord with physical evidence, for only by continued agreement between deductively established conclusions and experimental tests can one acquire confidence that the original generalizations are correct.

With such principles of scientific method clearly in mind, Newton turned to the problem of finding the physical principles which would lead to a unifying theory of earthly and celestial motions. He was, of course, familiar with the principles unearthed by Galileo. But these were presumably not enough. It was clear from the first law of motion that the planets must be acted on by a force which pulls them toward the sun, for if no force were acting, each planet would move in a straight line. The idea of a force which constantly pulls each planet toward the sun had occurred to many men, Kepler, the famous experimental physicist, Robert Hooke, the physicist and renowned architect, Christopher Wren, Halley and others, even before Newton set to work. It had also been conjectured that this force exerted on a distant planet must be weaker than that exerted on a nearer one and, in fact, that this force must decrease as the square of the distance between sun and planet increased. But, prior to Newton's work, none of these thoughts about a gravitational force advanced beyond speculation.

Newton adopted these ideas. However, in his attempt to tie in the action of the gravitational force with motions on the earth, a line of thinking occurred to him which was highly imaginative and certainly original in his time but which is now an almost daily experience. He considered the problem of what happens when a projectile is shot out horizontally from the top of a mountain. As Newton knew and as we know from our study of Galileo's work, the projectile follows a parabolic path to earth (see Chapter 14, Fig. 14–3). If the horizontal speed of the projectile is increased, then the path is wider but remains parabolic. However, Galileo had assumed that the earth was flat and that the projectiles were given moderate initial horizontal speeds such as a cannon might impart to shells. Newton then asked himself what would happen if the sphericity of the earth were taken into account, and if the horizontal speeds of the projectiles were gradually increased. If the sphericity of the earth is taken into account, then projectiles with small horizontal speeds will follow the paths *PA* and *PB* of Fig. 15–3. As the speed is increased somewhat, the projectile might take a path such as *PC*. Suppose now that the speed is increased still more. Would the projectile fall off into space? Not necessarily. As the projectile travels into space, it is pulled toward the earth. But the pull of a spherical earth is directed toward the center, and hence the projectile, subjected to this continuous pull toward the center, need not fall off into space. It might, in fact, continue to circle the earth indefinitely if the earth pulled it in just enough so that it would not wander out into space and yet not fall to earth.

And so Newton concluded in his *Principles:*

> *And after the same manner that a projectile, by the force of gravity, may be made to revolve in an orbit, and go round the whole earth, the moon also, either by the force of gravity, if it is endowed with gravity, or by any other force, that impels it toward the earth, may be continually drawn aside towards the earth, out of the rectilinear [straight-line] way which by its innate force [inertia] it would pursue; and would be made to revolve in the orbit which it now describes; nor could the moon without some such force be retained in its orbit. If this force were too small, it would not sufficiently turn the moon out of a rectilinear course; if it were too great, it would turn it too much, and draw the moon from its orbit toward the earth. It is necessary that the force be of a just quantity, and it belongs to the mathematicians to find the force that may serve exactly to retain a body in a given orbit with a given velocity; . . .*

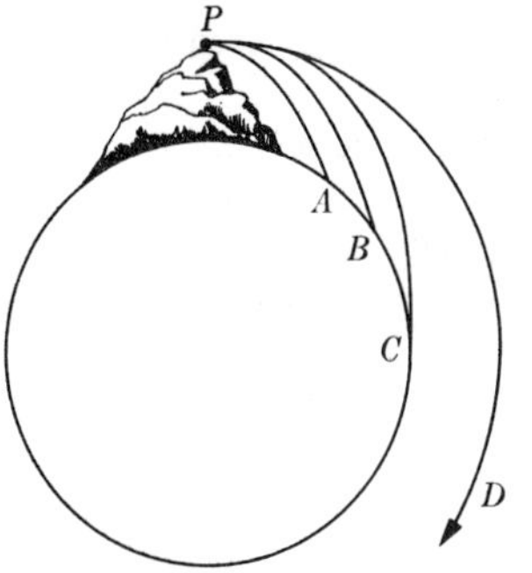

Fig. 15-3.
Projectiles shot out horizontally from the top of a mountain with increasing horizontal velocities.

This argument which showed how the motion of the moon around the earth could be related to motions occurring on earth was immediately extended to the motions of the planets about the sun. The planets, set into motion somehow, are attracted by the sun and are presumably pulled in just enough to keep them from flying off into space or from crashing into the sun.

Thus Newton had some reason to suppose that the same force which pulled projectiles to earth caused the moon to revolve around the earth and the planets to revolve around the sun. He had now to determine precisely how strong the force of gravitation is, that is, how it depends upon the bodies involved and upon the distances between the bodies.

15-7 MASS AND WEIGHT

Before we can understand Newton's law of gravitation we must distinguish two properties of matter, mass and weight. Newton's first law of motion says that if no force is applied, bodies continue at the speed they already have. Stated otherwise, the law says that bodies have inertia; they persist in the motion they already have unless compelled to do otherwise by the application of force. This inertia or resistance of matter to change in speed is called *inertial mass* or just *mass*.

Do all objects have the same mass? Not at all. Since mass exhibits itself in an object's resistance to change in speed, we can appeal to experience to see that different objects may possess different masses. Suppose, for example, that a small and a large ball of lead are at rest on the ground and one wishes to start them moving. Experience tells us that we must exert more force to get the larger ball rolling than to get the smaller one rolling with the same speed. Since more force is required in the first case, the larger ball must possess more mass. Or we can imagine the force that might be required to stop these balls if they were rolling toward us at the same speed. Again, more force would be required to stop the larger one. Thus the masses of objects are not the same.

We shall not present physical methods of measuring mass. It suffices to know that by adopting a unit of mass, just as we adopt a unit of length, we can compare all other masses with this unit and so determine exactly how much mass there is in any individual piece of matter. Mass is measured in pounds or grams, the pound being approximately 454 grams.

Bodies falling to earth possess acceleration. Hence some force must be acting to produce this change in speed. The force, as Galileo and others realized, is the pull of the earth or the force of gravity. We feel this pull when we hold an object in our hands. This particular force applied to an object is called the *weight* of the object. Hence weight and mass are by no means the same. Mass is inertia or resistance to change in speed, and weight is a force exerted by the earth.

However, there is a remarkable, experimentally determined relationship between the mass and weight of an object, namely, near the surface of the earth the weight is always 32 times the mass; in symbols:

$$w = 32m. \qquad (1)$$

The quantity 32 is precisely the acceleration which all bodies falling to earth possess. Thus equation (1) says that the force, w, which the earth exerts on a mass, m, is the acceleration with which it causes the mass to fall times the mass. When the mass m is measured in pounds, the weight w is measured in poundals. Thus a mass of one pound has a weight of 32 poundals. For convenience, 32 poundals are called one *pound of weight*. Of course, a pound of mass and a pound of weight are not the same physical quantities; and it is confusing at times to use the same unit for both mass and weight. Yet, as we shall see in a moment, this confusion is not too serious. (If the mass is measured in grams, the quantity which replaces 32 is 980, and the units are centimeters per sec². Then instead of (1) we have

$$w = 980m,$$

and the weight w is measured in dynes.)

Because weight and mass are so intimately related, we do not trouble in ordinary life to distinguish between the two properties. Large masses have large weights, and so, even in those instances where we are actually concerned with the mass of an object, we often tend to think of weight. For example, if one were to try to start an automobile rolling by pushing it, he would have to exert considerable force. The average person relates this fact to the great weight of the automobile. However, weight plays no role here because the force of gravity acts downward and has no effect on motion along the ground. The forceful push is required because the mass resists change in speed. Hence, it is the mass of the automobile rather than the weight which calls for the exertion of great force.

EXERCISES

1. Why do people usually fail to distinguish between mass and weight?
2. Let us assume that the relationship between weight and mass on the moon has the same mathematical form as on the earth; that is, the weight is a constant times the mass. The acceleration which the moon gives to all bodies falling toward its surface is 5.3 ft/sec². If a man weighs 160 lb on the earth, what will he weigh on the moon?
3. If the acceleration which the sun imparts to bodies falling toward its surface is 27 times that imparted by the earth, what would a man whose weight on the earth is 160 lb weigh on the sun?

15-8 THE LAW OF GRAVITATION

Newton adopted the conjecture already made by his contemporaries, namely, that the force of attraction, F, between *any* two bodies of masses m and M, respectively, separated by a distance r is given by the formula

$$F = G\frac{mM}{r^2}. \tag{2}$$

In this formula, G is a constant; i.e., it is the same number, no matter what m, M, and r may be. The numerical value of this constant, which we shall determine later, depends upon the units used for mass, force, and distance.

From the mathematical standpoint, formula (2) represents a new type of functional relationship. The quantity F is the dependent variable which depends upon three independent variables, m, M, and r, the quantity G being a constant. If we give values to m, M, and r, then the value of F is determined. Such a function is, of course, more complicated than, say the formula $d = 96t - 16t^2$, which contains just one independent variable, t, and one dependent variable, d. When one is working in a situation in which m and M are fixed and only r can vary, then F is a function of just one independent variable. For example, if G were 1, m were 2, and M were 3, then the relationship between F and r would be $F = 6/r^2$. This formula expresses the dependence of the gravitational force between two fixed masses on the distance between them.

Newton had yet to show that formula (2) was the correct quantitative expression for this force. To apply formula (2) and to work with forces in general, Newton adopted a second quantitative physical principle which proved to be just as important as his law of gravitation. As we noted in the preceding section, near the surface of the earth the force of gravity of the earth gives to objects an acceleration of 32 ft/sec², and the force is 32 times the mass of the object. Newton generalized this relationship and affirmed that whenever any force acts on an object it gives that object an acceleration. Moreover, the relationship of force, mass of the object, and acceleration imparted to the object, is

$$F = ma. \tag{3}$$

In this formula F is the amount of force applied to the object of mass m, and a is the amount of acceleration imparted to the object. In the special case where F is the weight w, the value of a is 32 ft/sec². Formula (3) is known as Newton's second law of motion. It applies to any force, whether or not it be the force of gravity. As in the case of formula (1), if m is measured in pounds and a in feet and seconds, then F is measured in poundals. Thus a force of 32 poundals gives a mass of 1 pound an acceleration of 32 ft/sec². The unit, pound, is also used for forces with the understanding that one pound of force equals 32 poundals.

Let us see how Newton tested his law of gravitation. We shall write (2) in the slightly different form

$$F = m\frac{GM}{r^2}\cdot \tag{4}$$

If we compare (3) and (4), we observe that the quantity GM/r^2 in (4) plays the role of a in (3); that is, the law of gravitation can be viewed as stating that the gravitational force F gives a mass m the acceleration GM/r^2. In symbols,

$$a = \frac{GM}{r^2}\cdot \tag{5}$$

Now let M be the mass of the earth and let m be the mass of a small body near the surface of the earth. Then there is the question of what r in (5) represents. It is supposed to represent the distance between the two masses. Shall we take it then to be the distance from the mass m to the surface of the earth or to some point in the interior? If the two masses were separated by millions of miles, as are the earth and sun, one might idealize each mass and regard it as concentrated at one point because the size of each mass is small compared with the distance between them. But for objects near the surface of the earth, the value of r depends heavily upon what point in or on the surface of the earth is chosen as the position of the earth. Newton conjectured (and later proved) that for purposes of gravitational attraction the mass of the earth could be regarded as though it were concentrated at the earth's center. Hence, with respect to the earth's gravitational acceleration acting on a mass m near the surface of the earth, r in (5) can be taken to be 4000 miles or 21,120,000 feet. This value of r is essentially the same for all objects *near* the surface of the earth. Moreover, the mass M of the earth is constant, and so is G. Hence, for all objects near the surface of the earth, the entire right side of (5) is constant. Consequently, the acceleration which gravity imparts to all objects near the surface of the earth is constant. This is precisely what Galileo had found and, in fact, he had determined that the constant is 32 ft/sec^2. Thus Newton's law of gravitation met its first test, for it yielded as a special case a well established fact.

EXERCISES

1. Suppose that the gravitational force varies with the distance between two definite masses according to the formula $F = 6/r^2$. Show graphically how F varies with r.
2. Knowing that the acceleration of objects near the surface of the earth is 32 ft/sec^2, use formula (5) to calculate the acceleration which the earth exerts on objects 1000 mi above the surface of the earth.
3. Suppose that an object falls to earth from a point 1000 mi above the surface. May we use the formula $d = 16t^2$ to compute the time it takes to fall this distance?
4. What is the mass of an object which weighs 150 lb? (One pound of *weight* is 32 poundals.)
5. How much force is required to give an automobile weighing 3000 lb an acceleration of 12 ft/sec^2?

15-9 FURTHER DISCUSSION OF MASS AND WEIGHT

With some support for the law of gravitation we can now, following Newton's example, adopt it as an axiom of physics and see what conclusions we may draw from this axiom and the other axioms of physics and mathematics. The law itself states that the force of gravitation F between any two masses m and

M is given by the formula

$$F = G\frac{Mm}{r^2}, \tag{6}$$

where r is the distance between the masses. Formula (6) leads immediately to a better understanding of the relationship between weight and mass and to an extension of the concept of weight. Let M be the mass of the earth and let m be the mass of some other object. Since F is the force with which the earth attracts this object, we can regard F as the weight of the object, for this attractive force is what we have meant by weight. However, we now see that the force or weight depends upon the distance r between the two masses. Hence, the weight of an object is not really a fixed number but varies with the distance of the object from the earth or, more precisely, from the center of the earth (see Section 15–8). If an object of mass m is at the surface of the earth, its weight is given by

$$F_1 = G\frac{Mm}{(4000 \cdot 5280)^2}, \tag{7}$$

but the same object taken 1000 miles above the surface of the earth will have the weight:

$$F_2 = G\frac{Mm}{(5000 \cdot 5280)^2}. \tag{8}$$

The value F_2 is considerably less than F_1 because the denominator in the second expression is much larger. We see, then, that the farther an object of mass m is from the surface of the earth, the less is its weight. On the other hand, the mass of the object, that is its resistance to change in speed, is the same at all locations. Thus we can see more clearly that the weight and mass of an object are quite different properties.

The concept of weight can, and in the present scientific era must, be extended still further. So far we have considered the weight of an object to be the force with which the earth attracts the object. But now let us imagine that the object were taken to the moon and, for simplicity, let us suppose that no matter other than the moon and the object exist in space. May we speak of the weight of the object on the moon? The law of gravitation applies to moon and object, and so the moon will attract the object. This attractive force will be the weight of the object on the moon. To calculate this weight, we have but to let M in (6) be the mass of the moon, m, the mass of the object, and r, the radius of the moon. We know that the radius of the moon is 1080 miles or $1080 \cdot 5280$ feet. The mass of the moon can be determined by methods similar to those used later (Section 15–10) to compute the masses of the earth and sun. The result of the calculation, which we may accept for present purposes, is that the weight of an object on the moon is $\frac{1}{6}$ its weight on earth.

We can extend the notion of weight still further. Suppose that an object is in space somewhere between earth and moon. According to the law of gravitation, the earth attracts the object, and so does the moon. Since these attractions oppose each other, we may regard the weight of the object as the net attraction. If we now think of the object as moving from the earth to the moon, then the attractive force of the earth decreases while that of the moon increases. At the outset, the earth's force is stronger, but at some point in the path to the moon the two forces will be equal and oppositely directed so that the net weight of the object will be zero. This point is located at a distance of about 24,000 miles from the moon along the line from the earth to the moon. All of the above considerations about weight are now no longer purely academic flights of fancy but are important factors in the process of determining the paths of rockets which are sent out to strike the moon.

EXERCISES

1. Suppose that a person weighs 150 lb at the surface of the earth where, of course, his distance from the center of the earth is 4000 mi. What would this person weigh at a point 4000 mi above the surface of the earth?
2. How does the law of gravitation enable us to further differentiate between the mass and weight of an object?
3. Suppose that of two objects on the earth, one has twice the mass of the other. Show that the force with which the earth attracts the first one is twice the force with which the earth attracts the second.
4. What would a man whose present weight is 150 lb weigh if the earth's mass were one-tenth of what it is?
5. Suppose that the earth's mass were twice as large as it is. What change would there be in the acceleration of falling bodies? Would a body which is dropped from a height of 1000 ft reach the ground sooner than it now does?
6. It is stated in the text that all bodies near the surface of the earth fall with the same acceleration. Suppose that an object is several thousand miles from the surface of the earth. How would the acceleration of its fall to the earth compare with the acceleration of a body near the surface?
7. Suppose that the mass of the moon were the same as the mass of the earth. The radius of the moon is about $\frac{1}{4}$ the radius of the earth. What would a man who weighs 150 lb on the earth weigh on the moon?
8. The earth's attractive force acts quite differently on objects in the interior of the earth than on objects outside the earth. In the former case the force is given by the formula $F = GmMr/R^3$, where m is the mass of the object, M the mass of the earth, R the radius of the earth, and r the distance of the object from the center of the earth. Compare the variation of this attractive force as r varies, with the force given by formula (6).

9. Suppose that the law of gravitation were $F = GmM/r$ instead of formula (6). Compare the variation of weight with distance from the center of the earth according to this formula with the variation of weight according to (6).
10. Consider all objects at a distance of 5000 mi from the center of the earth. Is the ratio of weight to mass the same for all these objects?

15-10 SOME DEDUCTIONS FROM THE LAW OF GRAVITATION

The essence of the scientific method created by Galileo and Newton is to establish basic quantitative physical principles and to apply mathematical reasoning to these principles. The law of gravitation and the first and second laws of motion are such physical principles. We shall see now that Newton was able to make some remarkable deductions from these principles.

The law of gravitation contains the constant G. Many calculations based on the law of gravitation require that one know G. In principle, this quantity is easily measured. One has but to take two known masses, place them a measured distance apart, and measure the force with which the two masses attract each other. Then, since

$$F = G\frac{mM}{r^2}, \tag{9}$$

we see that every quantity in (9) is given except G, so that we have a simple algebraic equation for G. The actual experiments which have been made to measure G are a little more complicated because the force F is small for ordinary masses. However, the experiments have been performed, and the value of G turns out to be $1.07/10^9$. The notation 10^9 is scientific shorthand for the product in which 10 occurs as a factor 9 times, i.e., one billion. This value of G presupposes that masses are measured in pounds, distances in feet, and forces in poundals (practical English system). (In the centimer-gram-second (cgs) system of units, G is $6.67/10^8$.)

With the value of G known, it is a simple matter to calculate the mass of the earth. We may recall that formula (5), which is an immediate consequence of the law of gravitation and the second law of motion, states that the acceleration which the earth imparts to any other mass is

$$a = G\frac{M}{r^2}, \tag{10}$$

where r is the distance between the two masses. We have also learned that when r is 4000 miles or 21,120,000 feet, then $a = 32$. Let us substitute these values and the value of G in (10). Then

$$32 = \frac{1.07}{10^9}\cdot\frac{M}{(21{,}120{,}000)^2}, \tag{11}$$

and we have obtained a simple equation for the unknown M. To shorten the somewhat complicated arithmetic, let us approximate and write 21,120,000 as 21,000,000 or as $21 \cdot 10^6$. Then by a theorem on exponents (see Section 5–3)

$$(21 \cdot 10^6)^2 = (21)^2 \cdot (10^6)^2 = 441 \cdot 10^{12}.$$

Substituting the value just obtained in (11) yields

$$32 = \frac{1.07}{10^9} \cdot \frac{M}{441 \cdot 10^{12}}. \tag{12}$$

The factors 10^9 and 10^{12} can be combined, for the first factor means $10 \cdot 10 \cdot 10 \cdots$, wherein 10 occurs 9 times, and the second factor means that 10 occurs 12 times. Then in the product of these two factors 10 occurs 21 times. Hence

$$32 = \frac{1.07M}{441 \cdot 10^{21}}. \tag{13}$$

Multiplying both sides of this equation by $441 \cdot 10^{21}$ and dividing both sides by 1.07, we obtain

$$M = \frac{32 \cdot 441 \cdot 10^{21}}{1.07}$$

or

$$M = 13.1 \cdot 10^{24} \text{ pounds}. \tag{14}$$

Since there are 2000 pounds in one ton, we may divide the right side by 2000 and write

$$M = 6.5 \cdot 10^{21} \text{ tons}. \tag{15}$$

Hence some simple algebra applied to formula (10) was all that was needed to calculate the mass of the earth. Let us note clearly that this quantity is *not* the weight of the earth. Technically the earth has no weight since weight is, by definition, the force which the earth exerts on *other* masses. However, a mass of $6.5 \cdot 10^{21}$ tons would weigh the same amount of tons, and so one can get some idea of the earth's mass.

From the knowledge just obtained we can deduce some information about the interior of the earth. The earth is approximately spherical in shape, and since the volume, V, of a sphere is $4\pi r^3/3$, where r is the radius, we can compute the volume of the earth. Thus

$$V = \tfrac{4}{3}\pi(4000 \cdot 5280)^3 = \tfrac{4}{3}\pi(4 \cdot 528)^3(10^4)^3 = \tfrac{4}{3}\pi(2112)^3 \cdot 10^{12}.$$

We shall approximate 2112 by $21 \cdot 10^2$ and use the value of 3.14 for π. Then

$$V = \tfrac{4}{3}(3.14)(21 \cdot 10^2)^3 \cdot 10^{12} = \tfrac{4}{3}(3.14)(21)^3 \cdot 10^6 \cdot 10^{12}.$$

Since $(\tfrac{4}{3})(3.14)(21)^3$ is about 39,000, we have

$$V = 39{,}000 \cdot 10^6 \cdot 10^{12} = 39 \cdot 10^{21} = 3.9 \cdot 10^{22} \text{ cubic feet.}$$

We next divide the mass of the earth, in pounds, by the volume to find the mass per cubic foot. Thus

$$\frac{M}{V} = \frac{13.1 \cdot 10^{24}}{3.9 \cdot 10^{22}} = \frac{1310 \cdot 10^{22}}{3.9 \cdot 10^{22}} = \frac{1310}{3.9} = 336. \tag{16}$$

The mass per cubic foot of water is 62.5 pounds. We see then that the mass per cubic foot of earth is about 5.5 times the mass per cubic foot of water. This figure of 5.5, incidentally, is the *density* of the earth.

Examination of the earth's surface shows that it consists mostly of water and sand. Since the quantity of rock visible on the surface does not account for the ratio 5.5, the conclusion follows that the interior of the earth must contain heavy minerals.

Only a little more work is required to compute the mass of the sun. We shall again begin with the law of gravitation and the second law of motion. The two masses involved now are the mass of the sun, S, and the mass of the earth, E. Then the law of gravitation states that the force with which the sun attracts the earth is

$$F = G\frac{SE}{r^2}, \tag{17}$$

where r is the distance from the earth to the sun. According to Newton's second law of motion, the force which the sun exerts on the earth gives the earth an acceleration a such that

$$F = Ea. \tag{18}$$

Since the forces in (17) and (18) are the same, we may equate the right sides. Then

$$Ea = G\frac{SE}{r^2}. \tag{19}$$

We may next divide both sides of (19) by E and obtain

$$a = \frac{GS}{r^2}. \tag{20}$$

In this last equation we know G and r. If we knew a, the acceleration of the earth, we could calculate S. Let us see what we can do about calculating a.

The acceleration which the sun imparts to the earth causes the earth to depart from a straight-line path, which it might otherwise pursue, and "fall" toward the sun just enough to keep it on its elliptical path. (The acceleration which the earth imparts to the moon has the same effect on the lunar orbit.)

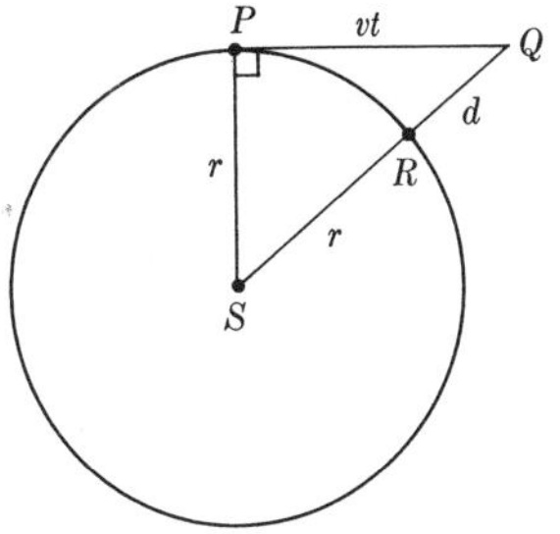

Fig. 15–4.
The sun's pull on the earth causes the earth to "fall" the distance QR in t seconds.

We shall suppose, for the sake of simplicity, that the path of the earth is circular. Let us imagine that the earth is at the point P (Fig. 15–4) in its path around the sun. If there were no gravitational force, the earth would shoot straight out along the tangent at P into space in accordance with the first law of motion. Let us suppose that in time t, the earth would have reached the point Q. The distance traveled would be the velocity of the earth in its path around the sun, v say, multiplied by t. Hence $PQ = vt$. However, during that time t, the sun pulls the earth in a distance QR or d. Since SPQ is a right triangle,

$$(r + d)^2 = r^2 + (vt)^2.$$

Squaring $r + d$ and substituting the result, we obtain

$$r^2 + 2dr + d^2 = r^2 + v^2t^2.$$

We subtract r^2 from both sides of this equation and find that

$$2dr + d^2 = v^2t^2.$$

Applying the distributive axiom on the left side permits us to write

$$2d\left(r + \frac{d}{2}\right) = v^2t^2. \tag{21}$$

Now d is the distance that the earth falls in time t. Let us suppose that it falls with constant acceleration. (We shall soon let t become very small so that the acceleration can well be taken as constant.) If a body falls a distance

d with constant acceleration a, then we know from our work in Chapter 13 (Section 13–5, Exercise 14) that

$$d = \tfrac{1}{2}at^2$$

or

$$2d = at^2. \qquad (22)$$

Let us substitute this value of $2d$ in (21). Then

$$at^2\left(r + \frac{d}{2}\right) = v^2t^2.$$

We now divide both sides of this equation by t^2 and obtain

$$a\left(r + \frac{d}{2}\right) = v^2. \qquad (23)$$

Thus far t was arbitrarily chosen, and d was the distance the earth fell toward the sun in time t. Our result so far, then, is valid for any value of t. If we now let t become smaller and smaller, d will also decrease. When $t=0$, it follows from (22) that $d=0$. In this case, (23) becomes

$$ar = v^2$$

or

$$a = \frac{v^2}{r}. \qquad (24)$$

This result states that the acceleration which the sun imparts to the earth at each point P of the earth's path is the square of the earth's velocity divided by the distance of the earth from the sun. This acceleration is called *centripetal* (i.e., center-seeking) acceleration, because it causes the earth to move toward the center of its path.

We now have the quantity a which we needed in (20). Substitution of (24) in (20) yields

$$\frac{v^2}{r} = \frac{GS}{r^2}.$$

We may multiply both sides by r and divide both sides by G to obtain

$$S = \frac{v^2r}{G}. \qquad (25)$$

Every term on the right side of this equation is known. The distance r is 93,000,000 miles or $4.9 \cdot 10^{11}$ feet. The velocity, v, of the earth is the circumference of the earth's path divided by the number of seconds in one year:

$$v = \frac{2\pi 4.9 \cdot 10^{11}}{365 \cdot 24 \cdot 60 \cdot 60} = \frac{30.8 \cdot 10^{11}}{3.15 \cdot 10^{7}} = 9.8 \cdot 10^{4} \text{ ft/sec.}$$

Hence

$$v^2 = (9.8 \cdot 10^4)^2 = (9.8)^2 \cdot 10^8 = 96 \cdot 10^8. \tag{26}$$

In Section 15–10 we learned that $G = 1.07/10^9$. Thus, using these values of r, v^2, and G in (25), we have

$$S = \frac{96 \cdot 10^8 \cdot 4.9 \cdot 10^{11}}{1.07/10^9} = \frac{96 \cdot 10^8 \cdot 4.9 \cdot 10^{11} \cdot 10^9}{1.07}$$

or

$$S = 440 \cdot 10^{28} = 4.40 \cdot 10^{30}. \tag{27}$$

Hence the mass of the sun is $4.40 \cdot 10^{30}$ pounds. Since the earth's mass was previously found to be $1.31 \cdot 10^{25}$ pounds, we see that the mass of the sun is $3.36 \cdot 10^5$ or 336,000 times the mass of the earth.

We can determine the mass per cubic foot of the sun in the same manner as we calculated the mass per cubic foot of the earth. The mass of the sun is now known, and the radius, computed in Chapter 7, is 432,000 miles, or $2.28 \cdot 10^9$ feet. We shall not reproduce the calculations, but state the result: the mass per cubic foot proves to be 90 pounds. Since a cubic foot of water has a mass of 62.5 pounds, we see that the mass per cubic foot of the sun is about $1\frac{1}{2}$ that of water; that is, the density of the sun is about $1\frac{1}{2}$.

The examples given in this section further illustrate how mathematical reasoning can be applied to physical laws (in our case, to the second law of motion and the law of gravitation) in order to deduce fundamental knowledge about the universe. We did, of course, also use some experimentally obtained facts such as the value of G and the acceleration of bodies near the earth's surface. However, mathematics has been the main tool, and it obtains for us such remarkable information as the mass of the earth and the mass of the sun.

EXERCISES

1. Suppose an object moves in a circle at a constant speed. Is the motion subject to an acceleration?
2. Does the formula for the acceleration of the earth given in (24) depend upon the law of gravitation?
3. If you whirl an object on a string of radius 5 ft, at the rate of 50 ft/sec, what is the centripetal acceleration acting on the object? What force exerts this centripetal acceleration?

4. Use formula (24) with the understanding that v is the velocity of the moon and r is the distance of the moon from the earth, to calculate the acceleration of the moon. (The period of the moon's path around the earth is $27\frac{1}{3}$ days, and the distance of the moon from the earth is 240,000 mi.)
5. Using the figures in the text for the mass and radius of the sun, calculate the ratio of the mass to the volume of the sun.

* 15-11 THE ROTATION OF THE EARTH

We have repeatedly used the quantity 32 ft/sec² as the acceleration which the earth gives to objects near its surface. This figure is perfectly satisfactory for most purposes, but it is not strictly accurate even for motions near the earth's surface. Actually, the acceleration of falling bodies decreases from 32.257 ft/sec² at either pole to 32.089 ft/sec² at the equator.* The discovery of this decrease was at first not surprising to the seventeenth-century scientists. Newton had already proved that the earth is not strictly spherical, but has the shape of a somewhat flattened sphere (Fig. 15–5); that is, for example, the lengths OA, OB, OC, and OD are not equal, but are successively larger. Since the general formula for the acceleration due to gravity [see (5)] is GM/r^2, where G and M are fixed and r is the distance from the center of the earth, this acceleration is less at C, say, than at B because r is larger at C than at B. Hence we should expect the acceleration due to gravity to decrease as the location varies from A to D. Now the values of G and M were known. Moreover, Newton and Huygens had computed lengths such as OA, OB, and so forth, and therefore were able to determine what the acceleration should be at points such as A, B, C, and D. These calculations based on the expression GM/r^2, call for only a small percentage of the decrease actually measured. Thus precise measurement revealed a discrepancy between the acceleration predicted by the law of gravitation and the actual acceleration of falling bodies. This discrepancy required explanation.

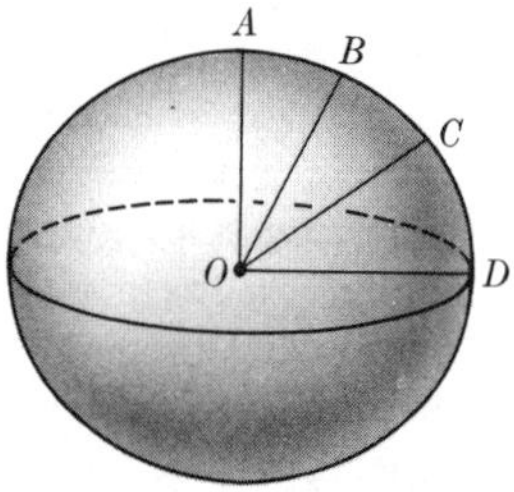

Fig. 15–5.
The spheroidal shape of the earth.

The problem was solved by Huygens. Objects on the surface of the earth would fly off into space if the earth did not pull them toward the center, just as an object whirled at the end of a string would fly off into space if the hand

* The numerical values can, in principle, be obtained by measuring the accelerations with which bodies near the surface fall to earth. However, a more accurate method utilizes the formula for the period of a pendulum.

at the center did not exert an inward pull. Thus the earth's gravitational force has two effects. Even if the earth did not rotate, it would pull all objects toward the center, simply because the earth's mass attracts the object. But since the earth does rotate, it must also exert an inward pull so that objects do not fly off into space but remain on or near the surface of the earth. This latter effect is a centripetal force. In a sense, the two effects of the earth's gravitational force, that is, the force which causes objects to fall to the surface, or weight, and the centripetal force, are of the same nature. The centripetal force also pulls objects toward the earth's center, but pulls them in just enough to keep them on a circular course. The weight, on the other hand, pulls objects toward the earth from the circular course to which they are kept by the centripetal force.

Let us express quantitatively what we have just described. By Newton's second law, the centripetal force must produce an acceleration (centripetal acceleration) on the object. Now formula (24) gives the centripetal acceleration which the sun exerts on the earth. However, this formula is really quite general, that is, if we replace sun and earth in the argument which led to (24) by the earth and an object on or near the earth's surface, then the argument still holds, provided that v is the velocity of the object and r is its distance from the center of the circle in which the object rotates. Newton's second law then tells us that the centripetal force must be the mass of the object times the centripetal acceleration, that is mv^2/r.

The force with which the earth pulls an object straight down, i.e., the weight, equals the mass of the object times the acceleration of its fall. It is this acceleration which we measure when we observe the fall of objects and which varies from pole to equator. We shall now denote it by g. Then the weight is mg.

According to Huygens the gravitational force which the earth exerts on objects supplies both the centripetal force and the weight. The centripetal force must be directed toward the center of the circle of latitude on which the object rotates. However, the weight is directed toward the center of the earth. Hence we cannot write a simple formula which expresses precisely how the earth's gravitational force is apportioned to provide the centripetal force and the weight at any latitude. However at the extreme cases of latitude, that is, at the equator, and at the poles, the apportionment is simple. At the equator the centripetal force must be directed toward the center of the earth. If we denote the radius of the earth by R, then at the equator

$$\frac{GMm}{R^2} = \frac{mv^2}{R} + mg. \tag{28}$$

At the North Pole, for example, an object does not travel in a circle as the earth rotates, and so no centripetal force is required to keep it rotating with the earth. Hence at that location,

$$\frac{GMm}{R^2} = mg. \qquad (29)$$

Clearly g is larger in (29) than in (28).

What can we say about the apportionment at intermediate latitudes? Formula (28) is no longer correct because the circle on which an object rotates does not have radius R, but has a smaller radius. Also the velocity, v, of the object depends on its latitude. Moreover, as we have already noted, the direction of the centripetal force required to keep the object rotating with the earth must be directed toward the center of the latitude circle. The effect of all these factors is to decrease the centripetal force required to keep the object rotating with the earth as the latitude increases, and this force is zero at the poles. Hence more and more of the gravitational force is applied to the weight of the object, the quantity mg, and since m is constant, g increases from the equator to the poles. Almost the full increase in g is due to the rotation of the earth, the balance being due to the shape of the earth. We can turn our argument around. We observe that g increases from the equator to the poles. This increase can be explained by assuming that the earth rotates. Hence we have reason to believe that the earth rotates.

The numerical value of g, that is, the acceleration of falling bodies, has, of course, been of importance for centuries. But it has additional importance today. Let us consider a satellite which circles the earth once every hour. The circular paths of satellites do have the center of the earth as their center. Although the satellite moves at a height of a few hundred miles above the surface of the earth, we shall ignore this distance and suppose that it travels very near the surface. What is significant is that the satellite covers 25,000 miles per hour. Hence the centripetal force required to keep it in its path is considerably greater than that required to keep an object which travels 25,000 miles in 24 hours from flying off into space. We see this fact from the middle term in (28) which tells us that the centripetal force increases with the square of the velocity. Thus a great deal of the earth's gravitational force must be expended in centripetal force. In fact, since the satellite does not fall to earth, the value of g, that is, the acceleration with which it should fall to earth, must be zero. In other words, the full gravitational force of the earth is expended in keeping the satellite on its circular path around the earth, and the satellite neither flies off into space nor falls to earth.

But the weight of any object is the product of its mass and the acceleration, g, with which gravity makes it fall to earth. Since for the satellite, $g=0$, it follows that the satellite has no weight. Objects contained in the satellite would also be weightless and so would not experience any earthward pull.

In view of the importance which satellites are likely to have in future scientific investigations, it is desirable to know the velocity which a satellite must possess if it is to stay in orbit at some desired distance from the center of the earth. This velocity is readily calculated from (28). Since the satellite does not fall to earth, the value of g for it must be zero. Then

$$\frac{GMm}{r^2} = \frac{mv^2}{r}.$$

If we divide both sides by m and multiply both sides by r, we obtain

$$v^2 = \frac{GM}{r}. \tag{30}$$

We know G and M, the mass of the earth. When r, the distance of the satellite from the center of the earth, is chosen, then we know all the quantities on the right side of (30). The quantity GM can be calculated once and for all. Thus

$$GM = \frac{1.07}{10^9}(13.1)10^{24} = 14 \cdot 10^{15}.$$

The value of r must be in feet. We can now calculate v.

EXERCISES

1. Since the weight of an object is mg, how does a person's weight change as he travels from the North Pole to the equator?
2. Suppose that a satellite stays close to the earth's surface. How fast would it have to travel to stay on its circular path and not fall to earth? [*Suggestion:* Use formula (30).]
3. The moon is a satellite of the earth. Since the moon stays on its path and does not fall to the earth, we may conclude that the earth's entire gravitational force acts as centripetal force on the moon. Using the assumption that the moon's path is a circle and that it is 240,000 mi from the earth, calculate the velocity of the moon. [*Suggestion:* Use (30).]
4. Using the result of Exercise 3, calculate the time it takes the moon to make one complete revolution around the earth.
5. Calculate the speed required to maintain a satellite in an orbit 500 mi above the surface of the earth.

* 15-12 GRAVITATION AND THE KEPLERIAN LAWS

Thus far in this chapter we examined the evidence which convinced Newton that the law of gravitation was correct, and we have seen how it can be applied to answer a variety of questions about objects and motions on the earth and in the heavens. We should now recall that one of the major problems challenging seventeenth-century scientists was the question whether the same physical principles could account for terrestrial and celestial motions. Since the law of gravitation when applied to bodies falling near the earth's surface reduces to fall with constant acceleration (see Section 15–8), Newton's principles certainly encompassed earthly motions. As to heavenly motions, the three famous laws of Kepler, which he had inferred from observations, were seemingly independent of the law of gravitation. The truly great triumph of Newton was his demonstration that all three Keplerian laws were mathematical consequences of the law of gravitation and the two laws of motion.

We shall illustrate what Newton did by showing how the third Keplerian law can be deduced from the basic laws just mentioned. However, we shall simplify Newton's work and suppose that the path of a planet around the sun is circular, whereas the true path, as Kepler proved, is an ellipse.

Let m be the mass of any planet, M the mass of the sun, and r the distance between them. Then the law of gravitation says that the force F exerted by the sun on the planet is

$$F = \frac{GmM}{r^2} \cdot \qquad (31)$$

We also know that the sun's force causes any planet to depart from straight-line motion and "fall" toward the sun with some acceleration. This acceleration, a, is none other than the centripetal acceleration given by formula (24), that is v^2/r. The derivation of (24) dealt with the sun and earth, but it applies to any planet, provided that v is the velocity of the planet and r is its distance from the sun. We may also assert, by the second law of motion, that the centripetal force F with which the sun attracts that planet is

$$F = m\frac{v^2}{r} \cdot \qquad (32)$$

The velocity v of any planet is the circumference of its path divided by the time T of revolution around the sun; that is, $v = 2\pi r/T$. Hence, from (32),

$$F = \frac{m}{r} \cdot \frac{4\pi^2 r^2}{T^2} = \frac{4\pi^2 rm}{T^2} \cdot \qquad (33)$$

Now formulas (31) and (33) yield two different expressions for the force with which the sun attracts any one planet.* Hence we may equate these

* In the light of Section 15–11 we can say that the gravitational force equals the centripetal force because the sun does not cause any planet to fall toward the sun *from the circular path.*

two expressions and obtain

$$\frac{GmM}{r^2} = \frac{4\pi^2 rm}{T^2}.$$

Dividing both sides of this equation by m eliminates that quantity. Multiplying both sides by T^2, we obtain

$$\frac{GMT^2}{r^2} = 4\pi^2 r.$$

If we now multiply both sides of this last equation by r^2/GM, we find

$$T^2 = \frac{4\pi^2}{GM} r^3. \tag{34}$$

The quantity $4\pi^2/GM$ is the same no matter what planet is being considered, because G is a constant, M is the mass of the sun, and $4\pi^2$ is a constant. Hence formula (34) says that T^2 is the product of some constant, say K, and r^3; in symbols,

$$T^2 = Kr^3. \tag{35}$$

Thus the square of the time of revolution of any planet is a constant (i.e., the same for all planets) times the cube of that planet's distance from the sun. Formula (35) is, then, Kepler's third law of planetary motion. We have derived it from the two laws of motion and the law of gravitation by a purely mathematical argument.

As we remarked earlier, Newton demonstrated that all three of Kepler's laws, which the latter had obtained only after years of observation and trial and error, were mathematical consequences of the laws of motion and gravitation. Hence the laws of planetary motion, which prior to Newton's work seemed to have no relationship to earthly motions, were shown to follow from the same basic principles as did the laws of earthly motions. In this sense, Newton "explained" the laws of planetary motion. These facts were as much a consequence of basic physical laws as the straight-line motion of objects falling to earth from rest or of projectiles following parabolic paths. Newton's original conjecture that the parabolic motion of projectiles should be intimately related to the elliptical motion of the planets was gloriously established. Further, since the Keplerian laws agree with observations, their derivation from the law of gravitation constituted superb evidence for the correctness of that law.

The few deductions from the laws of motion and gravitation which we have presented are just a sample of what Newton and his colleagues were able to accomplish. Newton applied the law of gravitation to explain a phenomenon which heretofore had not been understood, namely the tides in the oceans. He showed that these were due to the gravitational forces exerted by the moon and, to a lesser extent, the sun on large bodies of water. From data collected on the height of lunar tides, that is, tides due to the moon, Newton calculated the moon's mass. Newton and Huygens calculated the bulge of the earth around the equator. Newton and others showed that the paths of comets are in conformity with the law of gravitation. Hence the comets, too, were recognized as lawful members of our solar system and ceased to be viewed as accidental occurrences or visitations from God intended to wreak destruction upon us. Newton then showed that the attraction of the moon and the sun on the earth's equatorial bulge cause the axis of the earth to describe a cone over a period of 26,000 years instead of always pointing to the same star in the sky. This motion of the earth's axis causes a slight change each year in the time of the spring and fall equinoxes, a fact which had been observed by Hipparchus 1800 years earlier. Thus Newton explained the precession of the equinoxes.

Finally Newton solved a number of problems involving the motion of the moon. The plane in which the moon moves is inclined somewhat to the plane in which the earth moves. He was able to show that this phenomenon follows from the interaction of the sun, earth, and moon under the law of gravitation. As the moon travels around the earth, it cuts the plane of the earth's motion around the sun. The points in which it intersects are called the nodes. The nodes change in position, and this variation (regression of the nodes) also proved to be a consequence of the gravitational effect of the sun and earth on the moon. As the moon moves around the earth in an almost elliptical path, the point farthest from the earth, called the apogee, shifts about 2° per revolution. This effect, Newton showed, was due to the sun's attraction. Newton and his immediate successors deduced so many and such weighty consequences about the motions of the planets, the comets, the moon, and the sea, that their accomplishments were viewed as "the explication of the System of the World."

Today we have almost daily evidence that Newton had found sound physical principles which govern the operation of the universe. By applying just those principles man can now create satellites which circle the earth. In fact, Newton's suggestion that projectiles shot out horizontally and with large velocities from the top of a mountain would circle the earth is, in essence, the one used to launch satellites. Strictly speaking, scientists do not operate from mountain tops because accessible peaks are not high enough to ensure that the satellite will clear other mountains, and because the air resistance at such altitudes is still considerable. Instead rockets project the satellite upward to a high altitude where the air resistance is negligible; there a mechanism turns the satellite to a horizontal direction and another rocket gives it a horizontal velocity. Then the satellite follows an elliptical path.

Newton went further in his speculations and conjectured that the planets must have been shot from the sun at some angle and, upon reaching their present distances, must have retained enough "horizontal" velocity to start moving in their elliptical paths around the sun. This conjecture is still the accepted theory of the origin of our solar system.

EXERCISES

1. What reason would there be for calling Newton's law of gravitation a universal law?
2. In what sense did Newton incorporate the Keplerian laws in his science of motion?
3. What support did the heliocentric theory receive from Newton's work on gravitation?
4. What support did Newton's principles derive from the heliocentric theory?

* 15-13 IMPLICATIONS OF THE THEORY OF GRAVITATION

The work on gravitation presented mankind with a new world order, a universe controlled throughout by a few universal mathematical laws which in turn were derived from a common set of mathematically expressible physical principles. Here was a majestic scheme which embraced the fall of a stone, the tides of the oceans, the moon, the planets, the comets which seemed to sweep defiantly through the orderly system of planets, and the most distant stars. This view of the universe came to a world seeking to secure a new approach to truth and a body of sound truths which were to replace the already discredited doctrines of medieval culture. Thus it was bound to give rise to revolutionary systems of thought in almost all intellectual spheres. And it did. But, for the moment, we wish to confine ourselves to the implications and consequences of the theory of gravitation for mathematics proper.

Newton's work followed and considerably broadened the plan laid down by Galileo, who proposed to find basic quantitative physical principles and to deduce from them the description of physical phenomena. Galileo had discovered and utilized such axioms as the first law of motion, the constant acceleration of bodies moving near the surface of the earth, and the independence of the horizontal and vertical motions of projectiles. His results were confined to terrestrial motions. Newton added to the axioms the second law of motion and replaced the principle of constant acceleration of falling bodies by the more general law of gravitation. He then found that the resulting set of principles enabled him to deduce the description of all motions of matter on earth and in the heavens. Thus the scientific method of Galileo and Newton involves mathematics not only in the expression of axioms and the laws which are deduced but also in the deductive process itself. Indeed, mathematics offered not merely the vehicle for scientific expression but the most powerful tool for the real work of science, that is the acquisition of knowledge about the physical world and the organization of that knowledge in coherent systems. From the time of Newton, these roles of mathematics have been unquestionably accepted and utilized. Hence, as the success of Newtonian mechanics spurred efforts in other physical domains, mathematics was confronted with new challenges and received new suggestions for the creation of concepts and methods which in turn gave greater power to science. This interaction of mathematics and science has grown immensely since its beginning in the seventeenth century and has become the outstanding feature of the intellectual life of our own century.

The most surprising development of the theory of gravitation and one which established a new and unanticipated role for mathematics took place after Newton had deduced a number of conclusions about our solar system. Galileo and Newton had set about finding quantitative laws that related matter, space, time, forces, and other physical properties, but had wisely decided not to look into causal relationships; that is, they had deliberately avoided such questions as why bodies fall to earth or why planets move around the sun. In other words, they had concentrated on description. Nevertheless, they did utilize the force of gravitation, a concept which had been vaguely suggested even before Galileo's time—for example, by Copernicus and Kepler. Since the force of gravitation now assumed central importance, it was natural to ask, What is the mechanism that enables the earth to attract objects and the sun to attract planets? The heightened emphasis on this universal force could not but push such questions to the fore. The properties ascribed to the force of gravitation were indeed remarkable. It acted over distances of inches and millions of miles. It acted instantaneously and through empty space. Nor could the action of the force be suspended or blocked. Even when the moon was between the earth and the sun, the sun continued to attract the earth.

Although he tried to provide some physical explanation for the action of gravity, Newton did not succeed, and he concluded, "I have not been able to deduce from phenomena the cause of the properties of gravity and I frame no hypotheses." In spite of his ignorance of the workings of gravitation, Newton insisted on adopting the laws of motion and gravitation. He says,

> *But to derive two or three general principles of motion from phenomena, and afterwards to tell us how the properties and actions of all corporeal things follow from those manifest principles, would be a very great thing though the causes of those principles were not yet discovered: and therefore I scruple not to propose the principles of motion above mentioned, they being of very general extent, and leave their causes to be found out.*

Concerning his work in his *Principles*, he says,

> *But our purpose is only to trace out the quantity and properties of this force [gravitation] from the phenomena, and to apply what we discover in some simple cases as principles, by which, in a mathematical way, we may estimate*

> *the effects thereof in more involved cases; for it would be endless and impossible to bring every particular to direct and immediate observation. We said*, in a mathematical way [*note Newton's emphasis on the mathematics*], *to avoid all questions about the nature or quality of this force, which we would not be understood to determine by any hypothesis; . . .*

Newton was indeed troubled that he could give no explanation. But all he could do to justify the introduction of this force is summed up at the end of his *Principles*,

> *And to us it is enough that gravity does really exist, and act according to the laws we have explained, and abundantly serves to account for all the motions of the celestial bodies, and of our sea.*

Contrary to popular belief, no one ever discovered gravitation, for the physical reality of this force has never been demonstrated. However, the mathematical deductions from the quantitative law proved so effective that the phenomenon has been accepted as an integral part of physical science. What science has done, then, in effect is to sacrifice physical intelligibility for the sake of mathematical description and mathematical prediction. This basic concept of physical science is a complete mystery, and all we know about it is a mathematical law describing the action of a force *as though it were real.* We see therefore that the best knowledge we have of a fundamental and universal phenomenon is a mathematical law and its consequences. And it has become more and more true since Newton's days that our best knowledge of the physical world is mathematical knowledge.

REVIEW EXERCISES

1. Write as a decimal:

a) $\frac{1}{10^3}$ b) $\frac{1}{10^6}$ c) $\frac{1}{10^8}$ d) $\frac{2}{10^5}$ e) $\frac{10^2}{10^6}$

2. Express each of the following numbers as a number between 1 and 10 multiplied or divided by a power of 10:

a) 58,000 b) 58,790 c) $63.4 \cdot 10^3$ d) 46.75 e) 0.05 f) 0.0074

3. Express each of the following quantities as a number between 1 and 10 multiplied or divided by a power of 10:

a) $\frac{5 \cdot 10^3 \cdot 11 \cdot 10^4}{3 \cdot 10^2}$ b) $\frac{6 \cdot 10^2}{3 \cdot 10^4 \cdot 5 \cdot 10}$ c) $\frac{9 \cdot 10^4 \cdot 12 \cdot 10^5}{3 \cdot 10^7 \cdot 5 \cdot 10^2}$

d) $\frac{4 \cdot 10^3 \cdot 3 \cdot 10^7}{5 \cdot 10^4 \cdot 11 \cdot 10^6}$ e) $\frac{10^7}{3 \cdot 10^5 \cdot 12 \cdot 10^3}$ f) $\frac{10^{12} \cdot 3 \cdot 10^8}{5 \cdot 10^{19}}$

4. The frequency at which a frequency-modulation (FM) station broadcasts is 91 million cycles per second. Write the frequency as a number between 1 and 10 multiplied by a power of 10.
5. The mass of the earth is $13.1 \cdot 10^{24}$ lb. A gram of mass is 0.002205 lb. Find the mass of the earth in grams.
6. The mass of the sun is $4.40 \cdot 10^{30}$ lb. Use the data of Exercise 5 to compute the mass of the sun in grams.

In the following exercises you may use the fact that when M is the mass of the earth, $GM = 32 \cdot (4000)^2(5280)^2$.

7. Calculate the acceleration which gravity imparts to an object
   a) 2000 mi above the surface of the earth,
   b) 10,000 mi above the surface of the earth.
8. Suppose a man weighs 200 lb at the surface of the earth. Calculate his weight when he is
   a) 2000 mi above the surface of the earth,
   b) 10,000 mi above the surface of the earth.
9. What does a man who weighs 200 lb on the earth weigh on the moon if the weight there is due only to the attraction of the moon. The acceleration which the moon imparts to objects near its surface is 5.3 ft/sec^2.

Topics for Further Investigation

1. The astronomical work of Copernicus. The books by Armitage, Dreyer, Koyre, Kuhn, Wolf, and any number of others listed in the Recommended Reading would be fine source material.
2. The astronomical work of Kepler. The books by Caspar, Dreyer, Koyre, Kuhn and Wolf listed in the Recommended Reading would be fine source material.
3. Show how the history of the heliocentric theory exemplifies the influence of mathematics on western European culture. The books by Kline in the Recommended Reading will provide material.

Recommended Reading

ARMITAGE, ANGUS: *Sun, Stand Thou Still,* Henry Schuman, New York, 1947. Also in paperback under the title *The World of Copernicus.*

ARMITAGE, ANGUS: *Copernicus,* W. W. Norton and Co., New York, 1938.

BAUMGARDT, CAROLA: *Johannes Kepler, Life and Letters,* Victor Gollancz Ltd., London, 1952.

BELL, E. T.: *Men of Mathematics,* Chaps. 6, 9, 10, and 11, Simon and Schuster, New York, 1937.

BONNER, FRANCIS T. and MELBA PHILLIPS: *Principles of Physical Science,* Chaps. 1 and 4, Addison-Wesley Publishing Co., Inc., Reading, Mass., 1957.

Burtt, E. A.: *The Metaphysical Foundations of Modern Physical Science*, rev. ed., Chap. 2 and pp. 202–262, Routledge and Kegan Paul Ltd., London, 1932.

Butterfield, Herbert: *The Origins of Modern Science*, Chaps. 2 and 8, The Macmillan Co., New York, 1951.

Caspar, Max: *Johannes Kepler*, Abelard-Schuman, New York, 1960.

Cohen, I. Bernard: *The Birth of a New Physics*, Chap. 7, Doubleday and Co., Anchor Books, New York, 1960.

Dampier-Whetham, Wm. C. D.: *A History of Science and Its Relations with Philosophy and Religion*, pp. 160–195, Cambridge University Press, London, 1929.

De Santillana, Giorgio: *The Crime of Galileo*, University of Chicago Press, Chicago, 1955.

Drake, Stillman: *Discoveries and Opinions of Galileo*, Doubleday & Co., Anchor Books, New York, 1957.

Dreyer, J. L. E.: *A History of Astronomy From Thales to Kepler*, 2nd ed., Dover Publications, Inc., New York, 1953.

Dreyer, J. L. E.: *Tycho Brahe, A Picture of Scientific Life and Work in the Sixteenth Century*, Dover Publications, Inc., New York, 1963.

Gade, John A.: *The Life and Times of Tycho Brahe*, Princeton University Press, Princeton, 1947.

Galilei, Galileo: *Dialogue on the Great World Systems*, The University of Chicago Press, Chicago, 1953. Other editions of this work, originally published in 1632, also exist.

Hall, A. R.: *The Scientific Revolution*, Chap. 9, Longmans, Green and Co., Inc., New York, 1954.

Holton, Gerald and Duane H. D. Roller: *Foundations of Modern Physical Science*, Chaps. 4, 5, 8 through 12, Addison-Wesley Publishing Co., Inc., Reading, Mass., 1958.

Jeans, Sir James: *The Growth of Physical Science*, 2nd ed., Chap. 6, Cambridge University Press, London, 1951.

Jones, Sir Harold Spencer: "John Couch Adams and the Discovery of Neptune," in James R. Newman: *The World of Mathematics*, Vol. II, pp. 820–839, Simon and Schuster, Inc., New York, 1956.

Kline, Morris: *Mathematics: A Cultural Approach*, Chapter 12, Addison-Wesley Publishing Co., Inc., Reading, Mass., 1962.

Kline, Morris: *Mathematics in Western Culture*, Chap. 9, Oxford University Press, N.Y., 1953. Also in paperback.

Koyre, Alexandre: *From the Closed World to the Infinite Universe*, Chaps. 1 through 4, The Johns Hopkins Press, Baltimore, 1957.

Kuhn, Thomas S.: *The Copernican Revolution*, Harvard University Press, Cambridge, 1957.

Mason, S. F.: *A History of the Sciences*, Chaps. 17 and 25, Routledge and Kegan Paul Ltd., London, 1953.

More, Louis T.: *Isaac Newton*, Dover Publications, Inc., New York, 1962.

NEWMAN, JAMES R.: *The World of Mathematics,* Vol. I, pp. 254–285, Simon and Schuster, Inc., New York, 1956.

SMITH, PRESERVED: *A History of Modern Culture,* Vol I, Chap. 2 and Vol. II, Chap. 2, Holt, Rinehart and Winston, Inc., New York, 1934.

SULLIVAN, JOHN WM. N.: *Isaac Newton,* The Macmillan Co., New York, 1938.

TAYLOR, LLOYD WM.: *Physics, The Pioneer Science,* Chaps. 9, 10, and 13, Dover Publications, Inc., New York, 1959.

WIGHTMAN, WM. P. D.: *The Growth of Scientific Ideas,* Chaps. 8, 10, and 11, Yale University Press, New Haven, 1951.

WOLF, ABRAHAM: *A History of Science, Technology and Philosophy in the Sixteenth and Seventeenth Centuries,* 2nd ed., Chaps. 2, 3, 6, and 7, George Allen and Unwin Ltd., London, 1950. Also in paperback.

CHAPTER 16 *

THE DIFFERENTIAL CALCULUS

No nature except an extraordinary one could ever easily formulate a theory.

PLATO

16–1 INTRODUCTION

The mathematical ideas explored in the preceding chapters, arithmetic, algebra, Euclidean geometry, trigonometry, coordinate geometry, and the various types of functions, comprise a considerable amount of mathematics. Of course, the development of each of these ideas is far more extensive than we have indicated or than school courses usually cover. But the seventeenth century, which inspired and initiated the modern scientific movement, provided the problems and suggestions for new branches of mathematics which dwarf in extent, depth, and power the mathematics we have examined thus far. The most significant mathematical creation of that century and the one which proved to be most fruitful for the modern development of mathematics and science is the calculus. Like Euclidean geometry, it is a landmark of human thought.

16–2 THE PROBLEMS LEADING TO THE CALCULUS

The mathematicians of the seventeenth century who were gradually developing the ideas and processes which now comprise the calculus were beset by several problems. We have seen that the seventeenth century was primarily concerned with the study of motion, the motion of objects on or near the earth and the motion of heavenly bodies. In this study the problem of determining the speed and acceleration of moving bodies is, of course, quite important. Now speed, we usually say, is the rate at which distance changes with time, but if an object moves with varying speed, then to determine its speed, one must compute the rate of change of distance with time at any instant, or its instantaneous speed. The same remarks apply to acceleration. We shall see that the determination of such instantaneous rates presents a new kind of difficulty. It is true that we did determine and work with speed and acceleration of falling bodies, but we treated simple motions and so circumvented the essential difficulty. The problem is no longer simple when, for example, one seeks the speed and acceleration of a planet moving on an elliptical path.

The converse problem is equally important. Suppose one knows the acceleration of a moving body at each instant of time. How does one find the speed and distance traveled at any instant? When the acceleration is constant, one can multiply the acceleration by the time of travel and obtain the speed acquired, but this procedure does not yield correct results when the acceleration is variable.

Another problem of motion is that of determining the direction in which an object is moving at any instant of its flight. Depending upon its direction, a projectile may make a direct hit on a target or merely strike a glancing blow. Also, the direction in which a projectile is fired determines the horizontal and vertical components of its velocity (see Chapter 14). Hence it is desirable to know the direction in which an object is moving. Generally this direction varies from one instant to another, and therein lies the difficulty.

The third major problem was that of finding the maximum and minimum values of a function. When a bullet is fired straight up, one may wish to know how high it will go. For simple motions near the surface of the earth, we were able to find the maximum height. But the methods used will not suffice to compute, for example, the maximum or minimum distance of a planet from the sun or from another planet. Nor do they suffice to discuss the motion of a rocket which travels sufficiently far up so that one must take into account the variation in the acceleration due to gravity.

The fourth major problem confronting the seventeenth century was that of determining lengths, areas, and volumes. Let us consider, for example, the volume of the earth. The true shape of the earth is that of an oblate spheroid, a sphere flattened somewhat at the top and bottom.* How does one find the volume of such a figure? Or let us consider the motion of a planet along its elliptical path. How does one find the length of the path over which the planet travels in a given period of time? This information is important if one wishes to predict the position of the planet at some future instant of time. One could also ask, What is the total distance traveled by a planet in one complete revolution; in other words, What is the length of a given ellipse?

All these questions and many others that we shall encounter in the present and later chapters bedeviled the mathematicians of the seventeenth century, and hundreds of capable men worked on them. When Newton and Leibniz made their contributions to the calculus, it became clear that all of the above problems and others too could be solved by means of one basic concept, the instantaneous rate of change of one variable with respect to another. Hence we shall begin with this concept.

---

* Recent observations made from satellites indicate that this description is not quite accurate.

16-3 THE CONCEPT OF INSTANTANEOUS RATE OF CHANGE

There are three closely related ideas: change, average rate of change, and instantaneous rate of change. These three ideas should be carefully distinguished. The concept of change itself is by now a familiar one. When a ball is thrown up into the air, its height above the ground changes. The pursuit of physical problems involving functions soon obliges one to consider not just the mere fact of change but the rate of change of one variable with respect to another. In the case of a ball thrown up into the air, one might wish to know what initial speed will enable the ball to reach a height, say of 100 feet, or what speed the ball has on returning to the ground; that is, information about the speed, which is the rate of change of height with respect to time, is desirable. The statement that the earth travels around the sun in one year is a fact about rate of change rather than about mere change. Our great concern in this age for faster transportation and communication is a concern with rate of change. Circulation of the blood in one's body means quantity of blood per unit time passing through a specific artery or a collection of arteries, and here, too, it is rate of change which counts. The rate of physiological activity, that is the metabolic rate, measured in terms of the rate of consumption of oxygen per second, is a rate of change. To sum up: the rate of change of one variable with respect to another is a physically useful quantity in many situations.

The rates of change which are of interest to laymen and even to many specialists are average rates. Thus, if a motorist travels 500 miles in 10 hours, the average speed, i.e., the distance traveled divided by the time of travel, is 50 miles per hour. This average speed is what usually matters, and in most instances it is quite irrelevant that the driver may occasionally have stopped for food and thus had no speed at all during those periods of the trip. Most people like to increase their wealth and are satisfied if the rate of growth, that is the growth in wealth per month or per year, is appreciable. The increase of a country's population is usually measured per year because this average rate tells the story which is of importance for most purposes.

However, the average rate of change is not the significant quantity in many practical and scientific phenomena. If a person traveling in an automobile strikes a tree, it is not his average rate of speed for the time he has traveled from the starting point to the tree that matters. It is his speed at the instant of collision which determines whether or not he will survive the accident. Here we have an instantaneous speed or instantaneous rate of change of distance with respect to time.

There are two mathematical and physical facts involved in this event which require some elaboration. First of all there is the matter of time. As the person travels, time elapses. Mathematically this time is represented by a variable, t, say, and the values of t increase continually as the trip goes on. If time is measured from the instant at which the man starts out and if he has been traveling for 20 minutes, say, then t varies from 0 to 20. We also speak of the 20 minutes as an interval of time or an amount of time. We have, of course, referred to and used this mathematical representation of time right along. It is important now, however, to recognize that the collision of automobile and tree does not last an interval of time but occurs in what is called an instant. Many other events take place at an instant or are instantaneous. A lightning flash is instantaneous or at least happens so fast that we describe it as happening at an instant. The clock strikes at an instant. A bullet strikes a target at an instant.

The mathematical representation of an instant is simple. Mathematically, we say that when $t = 20$ or some other value, we are dealing with an instant of time; that is, an instant is merely one value of t, whereas an interval is some range of t-values, as, for example, from $t = 0$ to $t = 20$. Just as we have used the notion of an interval of time in past work, so have we used the notion of an instant. For example, we have spoken of the height of a ball at the end of the third second of flight, that is, when $t = 3$.

The second fact which must be clearly understood about the phenomenon of the automobile striking the tree is that the automobile has a speed at the instant of collision. This physical fact is apparent enough, and yet when we pursue the notion, we find that it presents difficulties. There is no difficulty in defining and calculating average speed, which is simply the distance traveled during some *interval* of time divided by that amount of time. But suppose we were to try to carry over this concept to instantaneous speed. The distance the automobile travels in one instant is zero, and the time that elapses during one instant also is zero. Hence the distance divided by the time is 0/0, and this expression is meaningless (Chapter 4). Thus, although instantaneous speed is a physical reality, there seems to be a difficulty in stating precisely what it means, and unless we can do so, we shall not be able to work with it mathematically.

16-4 THE CONCEPT OF INSTANTANEOUS SPEED

The problem of defining and calculating instantaneous rates such as speed and acceleration attracted almost all the mathematicians of the seventeenth century. Descartes, Fermat, Newton's teacher Isaac Barrow, Newton's friend John Wallis, Huygens, and hosts of other scholars worked on this and related problems. The men who finally grasped, formulated, and applied the general ideas of the calculus, which their predecessors had only partially understood, were Newton and Gottfried Wilhelm Leibniz, about whom we shall learn more later. The fact that every major mathematician of the century took up the problem of instantaneous rates of change is in itself of interest. It illustrates how even the best minds become absorbed in the problems of their times. Genius makes its contributions to the advancement of civilization, but the substance of its thoughts is determined by its age.

To explain the concept and method of finding instantaneous speeds and accelerations, we shall begin with the problem of determining the instantaneous speed of a falling body. Let us take the simplest case, that of a body which is dropped near the surface of the earth. Our method presupposes that we know the formula relating distance and time. We know from our work in Chapter 13 that this formula is $d = 16t^2$, where d is the distance fallen and t, the time elapsed. Let us seek the speed at the end of the fourth second of fall, that is, the speed at the instant $t = 4$. We have already pointed out (Section 16–3) that we cannot obtain this speed in the same manner in which we calculate the average speed over some interval of time since it is meaningless to divide the zero distance covered at $t = 4$ by the zero time elapsed. A practical solution of the difficulty might be to calculate the average speed *during the fourth second.* Though this solution will not yield the desired result, let us see what it yields. At the beginning of the fourth second, that is, when $t = 3$, the distance covered by the falling body is obtained by substituting 3 for t in the formula $d = 16t^2$. This distance is then $16 \cdot 3^2$, or 144. The distance covered by the end of the fourth second, that is, when $t = 4$, is $16 \cdot 4^2$, or 256. Hence the ratio of distance covered during the fourth second to the time elapsed is

$$\frac{256 - 144}{1}, \quad \text{or} \quad \frac{112}{1}.$$

The average speed during the fourth second is then 112 ft/sec.

As we have already stated, the average speed during the fourth second is not the speed at $t = 4$ itself, for during the fourth second the speed of the body keeps changing. Hence the quantity 112 can be no more than an approximation to the instantaneous speed. We may, however, improve the approximation by calculating the average speed in the interval of time from 3.9 to 4 seconds, for during this interval the average speed can, on physical grounds, be expected to approximate more closely the speed actually possessed by the body at $t = 4$. We therefore repeat the procedure of the preceding paragraph, this time using the values 3.9 and 4 for t. Thus for $t = 3.9$,

$$d = 16(3.9)^2 = 16(15.21) = 243.36;$$

and for $t = 4$,

$$d = 16 \cdot 4^2 = 256.$$

The average speed during the interval $t = 3.9$ to $t = 4$ seconds is then

$$\frac{256 - 243.36}{0.1} = \frac{12.64}{0.1} = 126.4 \text{ ft/sec.}$$

We note that the average speed during this one-tenth of a second is quite different from the value 112 for the fourth second.

Of course, the average speed during the interval $t = 3.9$ to $t = 4$ is not yet the speed at $t = 4$ because even during one-tenth of a second the speed of the falling body changes and the average is not the value finally attained at $t = 4$. We can obtain a still better approximation to the speed at $t = 4$ if we calculate the average speed during the one-hundredth of a second from $t = 3.99$ to $t = 4$, because the speed during this short interval of time near $t = 4$ ought to be almost equal to that at $t = 4$. Hence we shall apply our previous procedure once more. For $t = 3.99$,

$$d = 16(3.99)^2 = 16(15.9201) = 254.7216;$$

and for $t = 4$,

$$d = 16 \cdot 4^2 = 256.$$

Thus the average speed during the interval $t = 3.99$ to $t = 4$ is

$$\frac{256 - 254.7216}{0.01} = \frac{1.2784}{0.01} = 127.84 \text{ ft/sec.}$$

We could continue the above argument and process. The speed during the interval $t = 3.99$ to $t = 4$ is not the exact speed at $t = 4$ because the speed of the falling body changes even in one-hundredth of a second. We could therefore calculate the average speed in the interval $t = 3.999$ to $t = 4$ and expect that the average would be even closer to the speed at $t = 4$ than the preceding averages. The result incidentally would be 127.989 ft/sec. Of course, no matter how small the interval over which the average speed is calculated, the result is *not* the speed at the instant $t = 4$. How far then should the process be continued? The answer to this question is the core of the new idea supplied by the seventeenth-century mathematicians. The new thought is that one should compute average speeds over smaller and smaller intervals of time and note whether these average speeds get closer and closer to one *fixed* number. If so, *this number is taken to be the instantaneous speed at* $t = 4$. Let us pursue this idea.

In our case the average speeds over the intervals of time 1, 0.1, 0.01, and 0.001 proved to be 112, 126.4, 127.84, and 127.989. These numbers seem to be approaching, or getting closer to, the fixed number 128. Hence we take 128 to be the speed of the falling body at $t = 4$. This number is called the *limit* of the set of average speeds. We should note that the instantaneous speed is *not* defined as the quotient of distance and time. Rather it is the limit approached by average speeds as the intervals over which these average speeds are computed approach zero.

Two objections to what we have done may occur. The first is, What right do we have to take the number approached by the average speeds to be the speed at $t = 4$? The answer is that mathematicians have adopted a definition which makes good physical sense. They argue that the smaller the interval of time bordering $t=4$ over which the average speeds are computed, the closer must the behavior of the falling body be to that at $t=4$. Hence the number approached by average speeds over the smaller and smaller intervals of time bordering $t=4$ should be the speed at $t=4$. Since mathematics seeks to represent physical phenomena, it quite naturally adopts definitions that seem to be in accord with physical facts. It can then expect that the results obtained by mathematical reasoning and calculations will fit the physical world.

The second possible objection to our definition of instantaneous speed is a more practical one. Apparently, one must calculate average speeds over many intervals of time and attempt to discern what number these average speeds seem to be approaching. But there appears to be no guarantee that the fixed number chosen is the correct one. Thus, if in our above calculations one had obtained only the average speeds 112, 126.4, and 127.84, he might decide that these speeds are approaching the number 127.85, and his result would then be in error by 0.15 ft/sec. The answer to this objection is that we can generalize the entire process of obtaining the instantaneous speed, so that it can be carried out more quickly and with certainty. We shall now illustrate how the new method operates.

16-5 THE METHOD OF INCREMENTS

Let us again calculate the instantaneous speed of a dropped body at the end of the fourth second of fall, that is, at the instant $t=4$. The formula which relates distance fallen and time of travel is, of course,

$$d = 16t^2. \tag{1}$$

Again, as in our earlier work, we can calculate at once the distance fallen by the end of the fourth second. This distance, which we shall denote by d_4, is $16 \cdot 4^2$, or

$$d_4 = 256. \tag{2}$$

The generality of our new process consists in calculating the average speed, not over a specific interval of time such as 0.1 of a second, but over an arbitrary interval of time. That is, we introduce a quantity h which is to represent any interval of time beginning at $t=4$ and extending before or after $t=4$. The quantity h is called an increment in t because it is some additional interval of time before or beyond $t=4$. If h is positive, then it represents an interval after $t=4$; if it is negative, then it denotes an interval before $t=4$.

We shall first calculate the average speed in the interval 4 to $4+h$ seconds. To do this, we must find the distance traveled in this interval of time. We therefore substitute $4+h$ for t in (1) and obtain the distance fallen by the body in $4 + h$ seconds. This distance will be denoted by $d_4 + k$. Here d_4 is the distance the body falls in four seconds, and k is the additional distance fallen, or the increment in distance, in the interval of h seconds. Thus

$$d_4 + k = 16(4 + h)^2.$$

Multiplying $4 + h$ by itself gives

$$d_4 + k = 16(16 + 8h + h^2).$$

Application of the distributive axiom of algebra yields

$$d_4 + k = 256 + 128h + 16h^2. \tag{3}$$

To obtain k, the distance traveled in the interval of h seconds, we have but to subtract equation (2) from equation (3). The result is

$$k = 128h + 16h^2. \tag{4}$$

The average speed in the interval of h seconds is the distance traveled in that time divided by the time, that is, k/h. Let us therefore divide both sides of equation (4) by h. Then

$$\frac{k}{h} = \frac{128h + 16h^2}{h}. \tag{5}$$

When h is *not* zero, it is correct to divide the numerator and denominator on the right-hand side of (5) by h. The result is

$$\frac{k}{h} = 128 + 16h. \tag{6}$$

Hence (6) is also a correct expression for the average speed in the interval h.

To obtain the instantaneous speed at $t = 4$, we must determine the number approached by the average speeds as the interval h of time over which these speeds are computed becomes smaller and smaller. From (6) we can now readily obtain what we seek. If h decreases, $16h$ must also decrease, and when h is very close to zero, $16h$ is also close to zero. In view of (6), then, the fixed number which the average speed approaches is 128. This number is the speed at $t = 4$.

The process we have just examined, called the method of increments, is basic in the calculus. It is more subtle than appears at first sight. One should not expect to note and appreciate the finer points on first contact, any more than one gets to know another person well on the basis of one meeting. As a step in the right direction, however, we shall make one or two observations. First we wish to emphasize the fact that we sought the number or limit approached by average speeds as the intervals of time during which the average speeds were computed became smaller and smaller and close to zero. The correct expression for the average speed in any time interval h is given by (5). Since h is not zero, we may divide numerator and denominator in (5) by h. The resulting expression for the average speed, namely (6), happens to be especially simple, and from (6) we can easily determine what the limit of the average speeds is; that is, we observe that as h approaches zero, so does $16h$, and thus the number approached by the average speeds is readily seen to be 128.

In the present case of the rather elementary function $d = 16t^2$, we may let h be zero in (6) and find that the result is also 128. This agreement between the value given by the right side of (6) when h is zero and the number approached by k/h as h gets closer to zero will show up in a number of fairly simple functions. However, let us not lose sight of the fact that what we seek is a limit of k/h as h approaches zero rather than the value of the expression for k/h when h is zero. If the two values happen to be the same in some cases, as in (6), we are lucky, but let us not press this luck too hard.* The reader who wishes to tempt fate may substitute zero for h in simplified expressions such as (6).

The main point that emerges from this section is the possibility of finding instantaneous speed by a general process, the method of increments. No tedious arithmetical calculations are necessary, nor is there any doubt about what the limit approached by the average speeds is.

To appreciate what the limit process achieves we might consider an analogy. Suppose that a marksman seeks to hit a particular spot on a target. Even if he is a good shot, he is not likely to hit the given spot squarely, but will hit all around it and indeed come close. A bystander observing the location of the hits will readily determine the exact spot at which the marksman is aiming, by noting the concentration of the hits. This process of inferring the precise location at which the marksman is aiming is analogous to determining the instantaneous speed from a knowledge of the average speeds. We note what number the average speeds are approaching by examining (5) or the simplified form (6), and this limit is taken to be the instantaneous speed.

EXERCISES

1. Distinguish between the change in distance which results when an object moves for some interval of time and the rate of change of distance compared to time in that interval.
2. Distinguish between average speed and instantaneous speed.

---

* We could pursue the point further and learn just when the limit approached by k/h must agree with the value of the expression for k/h when h is zero. But to do so would involve a long digression into theory which, at the moment, is of secondary importance.

3. What mathematical concept is used to define instantaneous speed?
4. If the distance d, in feet, which a body falls in t seconds is given by the formula $d = 16t^2$, calculate the average speed of the body during the first five seconds of fall and during the fifth second of fall.
5. If the distance d, in feet, which a body falls in t seconds is $d = 16t^2$, calculate the instantaneous speed of the body at the end of the fifth second of fall, that is, when $t = 5$.
6. If the formula which relates the height above the ground and the time of travel of a ball thrown up into the air is $d = 128t - 16t^2$, calculate the speed at the instant $t = 3$.

16-6 THE METHOD OF INCREMENTS APPLIED TO GENERAL FUNCTIONS

We have calculated the instantaneous speed at the end of the fourth second for an object which falls according to the law $d = 16t^2$. Obviously the process used would have limited value if it were applicable to just the fourth second and to the formula $d = 16t^2$. Let us investigate the possibility of generalizing the procedure and see whether it might apply to any instant of time and perhaps to other formulas. We begin by considering the formula

$$y = ax^2, \tag{7}$$

where a is some constant and y and x are any variables related by (7). (After all, the fact that d represented distance and t time in the formula $d = 16t^2$ played no role in the purely mathematical process of calculating the rate of change of d with respect to t at $t = 4$.) By using the letters y and x and the constant a we emphasize the fact that we are considering a strictly mathematical relationship, and we shall calculate the rate of change of y with respect to x at a given value of x. Such rates, incidentally, are also called instantaneous rates, even though x does not always represent time. The word "instantaneous" has been carried over because the original and many current applications of the calculus contain time as the independent variable.

Let x_1 denote the value of x at which we are to compute the instantaneous rate of change of y compared to x. Thus x_1 is analogous to the value 4 of t used in the preceding section. To compute the desired rate of change, we shall repeat the process employed there. We first compute the value of y when x has the value x_1. This value of y, which we shall call y_1, is obtained by substituting x_1 for x in (7). Then

$$y_1 = ax_1^2. \tag{8}$$

We now consider an increase or increment h in the value of x, so that the new value of x is $x_1 + h$. To compute the new value of y, which we shall denote by $y_1 + k$, we must substitute the new value of x in (7). Then

$$y_1 + k = a(x_1 + h)^2.$$

Since

$$(x_1 + h)^2 = x_1^2 + 2x_1h + h^2,$$

it follows that

$$y_1 + k = ax_1^2 + 2ax_1h + ah^2. \tag{9}$$

Our next step is to determine the change k in y which results from the change h in x, by subtracting equation (8) from (9). Thus

$$k = 2ax_1h + ah^2. \tag{10}$$

To arrive at the *average* rate of change of y in the interval h, we must find k/h. Accordingly, we divide both sides of (10) by h and obtain

$$\frac{k}{h} = \frac{2ax_1h + ah^2}{h}. \tag{11}$$

Equation (11), which gives the average rate of change of y with respect to x in the interval h, is the generalization of equation (5).

To secure the instantaneous rate of change of y compared to x at the value x_1 of x, we must now determine the limit of the right side of (11) as h approaches zero. We are again fortunate in that we may divide the numerator and denominator of (11) by h and obtain

$$\frac{k}{h} = 2ax_1 + ah. \tag{12}$$

As h becomes smaller and smaller, the quantity ah, which is merely a constant times h, also becomes smaller, and the quantity k/h approaches the value $2ax_1$. This last quantity is the limit approached by the average rates of change, k/h, and so is the rate of change of y with respect to x at the value x_1 of x. Just to check our result, we note that when $a = 16$ and $x_1 = 4$, the quantity $2ax_1$ is 128, and this is the limit we obtained in the special case treated earlier.

Since y and x are variables which have no physical meaning, we cannot speak of the limit $2ax_1$ as an instantaneous speed. Instead we must describe it as the instantaneous rate of change of y compared to x at the value x_1 of x. To avoid stating this lengthy phrase, the quantity is called the *derivative* of y with respect to x at the value x_1. We shall denote it by $\dot{y}$, the notation used by Newton. (Leibniz devised the notation dy/dx. However, this notation, though suggestive of what takes place, can be misleading, for the instantaneous rate of change of y with respect to x is not a quotient but rather the limit approached by the quotient k/h.) Thus we have established that at the value x_1 of x

$$\dot{y} = 2ax_1. \tag{13}$$

Actually, we have arrived at a more general result. The quantity x_1 was any value of x. Hence we may as well emphasize this fact by dropping the subscript and writing

$$\dot{y} = 2ax. \tag{14}$$

Equation (14) states that when $y = ax^2$, the instantaneous rate of change of y compared to x at any value of x is $2ax$, or the *derivative* of y with respect to x is $2ax$. Since (14) holds at any value of x, it is a function; that is, the derivative of y with respect to x is itself a function of x. The process of deriving (14) from (7) is called *differentiation.*

The result (14) holds regardless of the physical meaning of y and x. Hence in any situation in which the formula $y = ax^2$ applies, we may conclude at once that the instantaneous rate of change of y compared to x is $2ax$. The generality of this result is immensely valuable, since a general mathematical result can always be applied to many different physical situations. To illustrate this point for the derivative (14), let us first reconsider our old friend $d = 16t^2$. In this case, d plays the role of y; t plays the role of x; and 16 is the value of a. Hence

$$\dot{d} = 2 \cdot 16t = 32t. \tag{15}$$

Now the instantaneous rate of change of distance compared to time is the instantaneous speed, and since speed occurs so often in applications, it is usually denoted by a special symbol, v; that is, $\dot{d} = v$. Hence (15) says that

$$v = 32t. \tag{16}$$

Knowing the formula which relates distance and time of a dropped object, we have derived the formula for the instantaneous speed. Thus from one formula we may derive another significant formula by applying the process of determining the instantaneous rate of change, i.e., by differentiation.

Now let us apply (14) to the formula for the area of a circle, namely, $A = \pi r^2$. Here A plays the part of y; r plays the part of x; and the constant π is the value of a. Formula (14) then tells us that

$$\dot{A} = 2\pi r. \tag{17}$$

The result, (17), has a very simple geometrical meaning (Fig. 16–1). It says that the instantaneous rate of change of the area of a circle with respect to the radius at any given value of the radius is the circumference. More loosely stated, the rate at which the area increases when r increases is the size of the circumference. This result is very reasonable. When the radius r is increased by an amount h, the area A of the circle increases by an amount k. Roughly speaking, we may think of k as a sum of circumferences and of h as the number of such circumferences. The ratio k/h is then an average circumference in the region k. As h approaches zero, this average circumference approaches the circumference for the radius r. This latter circumference is the instantaneous rate with which the area increases at the given value of r.

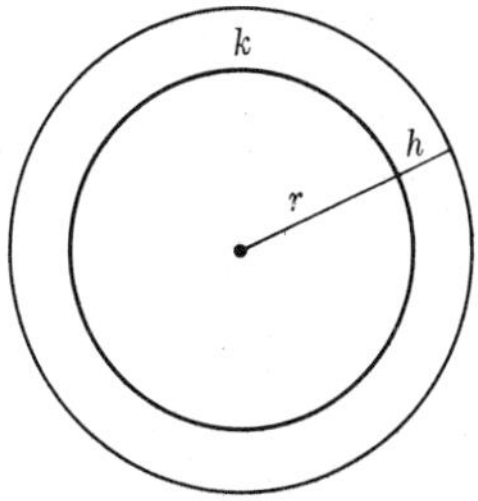

Fig. 16–1

Of course, the process of finding the instantaneous rate of change can be applied to all functions and not just to the simple function $y = ax^2$. For example, if y represents the pressure of the atmosphere and x represents the height above the surface of the earth, then $\dot{y}$ represents the rate of change of pressure with respect to height at a given height. If y represents the price level of a commodity and x represents time, then $\dot{y}$ represents the rate of change of price with respect to time at a given instant. Various other examples will be presented in the course of our subsequent work.

To make effective use of the calculus, one must learn how to determine the instantaneous rate of change for many types of formulas, because the variety of functions occurring in applications is very great. Since our purpose is primarily to gain some idea of what the calculus has to offer, we shall limit ourselves to just the simplest ones. Thus if

$$y = bx, \tag{18}$$

where b is any constant, then by using the method of increments, we would find that the instantaneous rate of change of y with respect to x is

$$\dot{y} = b. \tag{19}$$

This result applies, for example, to a body which falls with the velocity

$$v = 32t. \tag{20}$$

Formula (20) is just a special case of (18) where y becomes v, x is replaced by t, and b is 32. Hence (19) tells us that

$$\dot{v} = 32. \tag{21}$$

Since $\dot{v}$ is the instantaneous rate of change of velocity with respect to time, it is the instantaneous acceleration. Hence (21) tells us that a body which falls with a velocity $v = 32t$ has an acceleration at each instant of 32, that is, $a = 32$.

Were we to go through the process of determining the instantaneous rate of change of y compared to x when

$$y = ax^3, \tag{22}$$

where a is any constant, we would find that

$$\dot{y} = 3ax^2. \tag{23}$$

Occasionally we shall also treat a formula consisting of a sum of two terms instead of a single term. Thus suppose that the functional relationship between the variables y and x is given by the formula

$$y = ax^2 + bx, \tag{24}$$

where a and b are constants. The method of increments can, of course, still be applied to find the instantaneous rate of change of y with respect to x. Actually the work amounts to treating simultaneously a formula such as (7) and a formula such as (18). The result can be anticipated. In view of the rate of change (14) which applies to $y = ax^2$ and the rate of change (19) which applies to $y = bx$, we should expect that

$$\dot{y} = 2ax + b. \tag{25}$$

This is the correct result.

EXERCISES

1. By going through the full process of finding the instantaneous rate of change, that is, the method of increments, prove that
   a) if $y = bx$, then $\dot{y} = b$;
   b) if $y = ax^3$, then $\dot{y} = 3ax^2$;
   c) if $y = c$, where c is a constant, then $\dot{y} = 0$.
2. Apply the method of increments to find the instantaneous rate of change of $y = x^2 + 5$ and compare the result with the instantaneous rate of change of $y = x^2$. Does this example suggest a general conclusion?
3. Find the derivative, or the instantaneous rate of change of the dependent variable compared to the independent variable, for the following functions. [You may use formulas (14), (19), (23), and (25).]

   a) $y = 2x^2$ b) $d = 2t^2$ c) $y = (\frac{1}{2})x^2$ d) $y = 4x^3$

   e) $y = -2x^2$ f) $d = -16t^2$ g) $h = -16t^2 + 128t$ h) $h = 128t - 16t^2$
4. If an object is thrown downward with the initial velocity of 100 ft/sec, then the distance it falls in t seconds is given by the formula $d = 100t + 16t^2$. Calculate the speed of the object at the end of the fourth second of fall. [*Suggestion:* Apply formula (25).]

5. In geometrical terms the instantaneous rate of change of the area of a circle compared to the radius is the circumference.
   a) What is your guess as to the geometrical interpretation of the instantaneous rate of change of the volume of a sphere compared to the radius?
   b) Now determine $\dot{V}$ mathematically by applying formula (23) of the text to the formula for the volume of a sphere, $V = (\frac{4}{3})\pi r^3$, and check your answer to part (a).
6. a) When $y = ax^2$, then $\dot{y} = 2ax$; when $y = ax^3$, then $\dot{y} = 3ax^2$. Now suppose $y = ax^4$. What would you expect $\dot{y}$ to be?
   b) Verify your conjecture in part (a) by applying the method of increments to $y = ax^4$.
7. Find the rate of change of the area of a square with respect to a side at a given value of the side. Is the result intuitively reasonable?
8. The area of a rectangle is given by the formula $A = lw$, where l and w are the length and width, respectively. Suppose that l is kept fixed. What is the rate of change of area with respect to width? Interpret the result geometrically.

16-7 THE GEOMETRICAL MEANING OF THE DERIVATIVE

The instantaneous rate of change of y with respect to x can be interpreted geometrically. This interpretation not only clarifies the meaning of such a rate, but at the same time points the way to new uses of the concept. Let us consider the function

$$y = x^2, \tag{26}$$

and let us interpret geometrically the instantaneous rate of change of y with respect to x at $x = 2$. To find this rate of change by the method of increments we first calculate y at $x = 2$. This value of y, denoted by y_2, is

$$y_2 = 2^2 = 4.$$

The values 2 for x and 4 for y are, of course, the coordinates (2, 4) of a point, denoted by P in Fig. 16–2, on the curve which represents $y = x^2$. The second step in the method of increments is to increase the independent variable by an amount h so that its value becomes $2 + h$. The dependent variable then changes by an amount k so that its new value is $4 + k$. Now the quantities $2 + h$ and $4 + k$ can be interpreted as the coordinates of another point on the curve which represents $y = x^2$, because when x is $2 + h$, y becomes $4 + k$. The new point is shown as the point Q in Fig. 16–2. Next we calculate the average rate k/h. As the figure shows, k is the difference in the y-values of P and Q, whereas h is the difference in the x-values of P and Q. The ratio k/h is the slope of the line PQ, which, as in plane geometry, is called a secant. Thus far, then, we see that for any value of h and the corresponding value k, the ratio k/h is the slope of the secant through two points of the curve representing $y = x^2$.

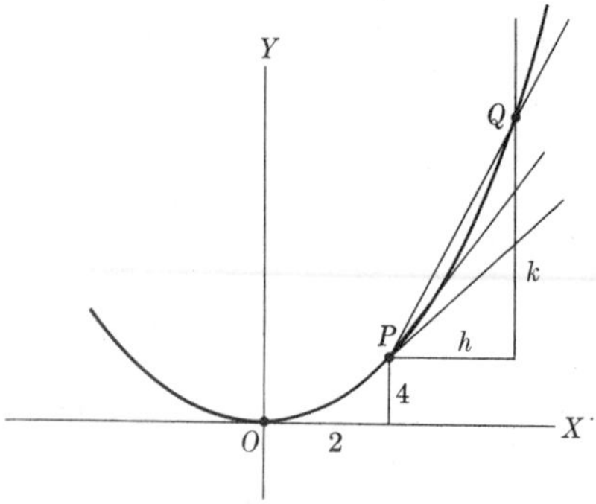

Fig. 16–2.
The secant PQ approaches the tangent at P as Q approaches P along the curve.

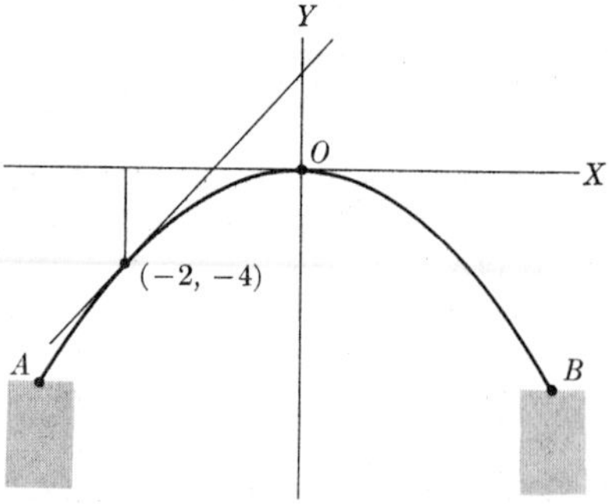

Fig. 16–3.
The slope of the roadway of a bridge at $x = -2$.

Finally we consider the limit approached by the ratio k/h as h gets closer and closer to zero. As h decreases, the point Q on the curve of Fig. 16–2 moves closer to the point P. The secant through P and Q changes position, always, of course, going through the fixed point P and the point Q, wherever the latter happens to be. As h approaches zero, the point Q approaches the point P, and the secant PQ comes closer and closer to the line which just touches the curve at P; that is, PQ approaches the tangent at P. Since k/h is the slope of PQ, the limit approached by k/h must be the slope of the line approached by PQ. In other words, *the instantaneous rate of change of y with respect to x at $x=2$ is the slope of the tangent to the curve at P, the point whose coordinates are* (2, 4). Of course, the value of 2 for x has been arbitrarily chosen to present a typical, yet concrete example. We could have been more general and have carried through the entire discussion for the value a, say, of x; that is, the rate of change of y with respect to x at any given value of x is the slope of the tangent to the corresponding curve at the point having that given value of x as abscissa.

We see therefore that the derivative of a function has a precise geometrical counterpart: the slope. Since slope is the rise (or fall) of a line per unit of horizontal distance (Chapter 12), the geometrical meaning is a rather simple one. Thus, since the value of the derivative of $y=x^2$ at $x=2$ is 4, the slope of the tangent at $x=2$ is 4; Fig. 16–2 does not show this because the scale on the Y-axis is not the same as that on the X-axis.

From the standpoint of application, the fact that the derivative is the slope of the tangent is very significant. The slope of a curve at a point on that curve is, very reasonably, defined to be the slope of the tangent at that point. Knowing the slope of the tangent thus means knowing the slope of the curve.

Just to get some idea of how useful this information is, let us consider for a moment the roadway of a bridge which is pictured as the arc AOB in Fig. 16–3. For the purpose of our illustration we can assume that this arc is part of the parabola $y = -x^2$. Now the slope of the curve at $x = -2$ is given by the derivative. Since the derivative of $y = -x^2$ at an arbitrary value of x is $-2x$, the derivative at $x = -2$ is $+4$. This then is the slope of the roadway at $x = -2$; that is, the roadway is rising at the rate of 4 feet for every foot of horizontal distance. This rate of climb is totally impractical, since no automobile or truck has the power to climb at such a rate. Our example thus makes the general point that the derivative enables us to calculate the slope of a curved roadway and to determine whether the slope is or is not too steep for the vehicles that are to use that route.

As another illustration suppose a projectile shot up and out from the point O (Fig. 16–4) is to strike the wall BC at the point B. Knowing the equation of the path of the projectile (Chapter 14), we can calculate the slope at the point B. This slope amounts to the direction that the projectile possesses at the point B, because the projectile is headed in the direction of the tangent.* One might want the direction of the projectile at B to be perpendicular to the wall because such an impact would damage the wall more effectively than a hit in a glancing direction. If necessary one could adjust the angle of fire and initial velocity to achieve the desired direction at B.

A third example illustrating the usefulness of knowing the slope is furnished by the phenomena of reflection and refraction of light. Let us consider the

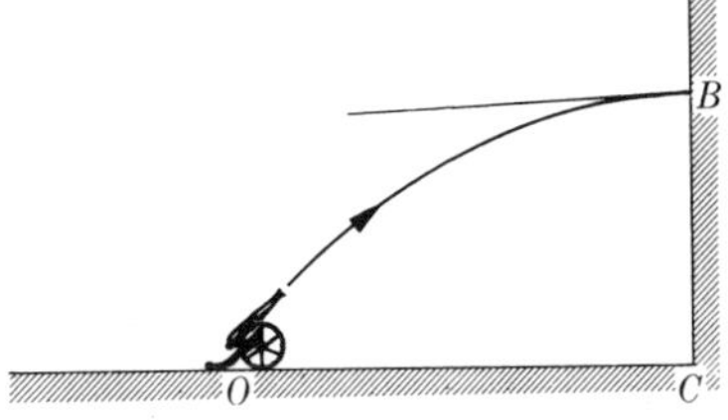

Fig. 16–4.
The slope of a projectile's path when it strikes a wall at B is the slope of the tangent at B.

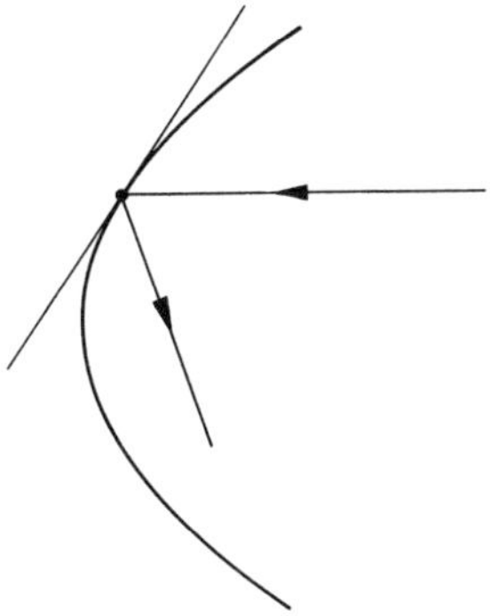

Fig. 16–5.
The angles which light rays make with a curve are determined by the slope of the curve.

* Sometimes the word direction is taken to mean the angle which the tangent to the curve makes with the horizontal. However the slope is an equally good indication of direction.

case of reflection. Suppose one wishes to design a mirror such that all rays of light coming from some source are reflected to one point. We know from Chapter 6 that when a light ray strikes a mirror, the angle of reflection equals the angle of incidence. Suppose that we consider a plane section of the mirror which contains the incident and the reflected rays (Fig. 16–5). This plane section is a curve. The angle which the incident ray makes with the mirror is, in fact, the angle between the incident ray and the tangent. To discuss this angle as well as the corresponding angle of reflection, we must know the direction, and hence the slope, of the tangent.

EXERCISES

1. Suppose that an uphill path can be represented by the equation $y = (\frac{1}{100})x^2$.
   (a) What is the slope of the hill at $x = 3$?
   (b) Is the slope steeper or more gradual at $x = 3$ or at $x = 5$?
   (c) Determine the slope at $x = 0$, and interpret the result geometrically.
2. Suppose that the path of a projectile is represented by the equation $y = 4x - x^2$.
   (a) What direction does the projectile have when $x = 1$?
   (b) At what value of x is the direction of the projectile horizontal?

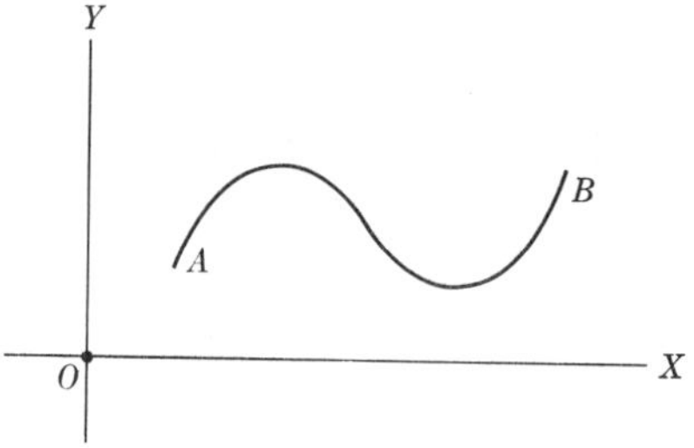

Fig. 16–6

3. The variation of y with x of a certain function is illustrated in Fig. 16–6. Describe how the derivative of y with respect to x varies as x increases from A to B.
4. Can you explain geometrically why the functions $y = x^2$ and $y = x^2 + 5$ should have the same derivative at, say, $x = 2$?

16–8 THE MAXIMUM AND MINIMUM VALUES OF FUNCTIONS

We have had occasion to apply our elementary algebra and geometry to problems in which the objective was to maximize or minimize some important physical quantity. For example, in Chapter 6 we found the dimensions of the rectangle having maximum area among all rectangles with the same perimeter. In Chapter 13 we found the maximum height attained by an object thrown or shot up into the air. The methods used to solve these problems were rather limited; they worked for the problems in question, but could hardly be applied to other types. One of the advantages of the calculus is that the concept of instantaneous rate of change of a function proves to be the key to a general method of finding the maximum or minimum values of variable quantities.

Let us reconsider the problem of determining the maximum height reached by a ball which is thrown up into the air. If the ball leaves the hand with a speed or velocity* of 128 ft/sec, then, according to Chapter 13, the formula which relates d, the height of the ball, and t, the time that the ball has been in motion, is

$$d = 128t - 16t^2. \tag{27}$$

During our earlier discussion of the motion represented by formula (27), we had to resort to an independent physical argument, to prove that the velocity of the ball at any instant is given by

$$v = 128 - 32t, \tag{28}$$

whereas now the purely mathematical process of differentiation immediately yields (28) as the formula for the instantaneous speed of the ball.

To determine the maximum height reached by the ball, we argued in Chapter 13 that the velocity of the ball at the highest point must be zero or the ball would continue to rise. Hence, to find the instant t_1 at which $v=0$, we set v equal to zero, that is, we set

$$128 - 32t_1 = 0, \tag{29}$$

and, by solving this equation for t_1, we found that $t_1=4$. We then substituted this value of t in (27) to arrive at the maximum value of d.

We can now see that, translated into the language of the calculus, our above procedure of determining the maximum value of the variable d given by formula (27) consisted in setting the instantaneous rate of change, $\dot{d}$, equal to zero and finding the value (or values) of the independent variable, t in the present case, at which the rate of change is zero. This example suggests a general procedure. If y is a function of x, and if we wish to find the maximum value of y, we set the instantaneous rate of change of y with respect to x equal to zero; find the value of x for which this rate of change, or derivative, is zero; and then substitute this value of x in the formula for y. The resulting value of y is the maximum value of y.

Of course, we really do not know that this general procedure is justified. For a ball thrown up into the air, we used the *physical* argument that the velocity must be zero at the highest point. This argument might be suitable

* We remind the reader that the words speed and velocity are often used interchangeably. Strictly the word velocity includes the sign.

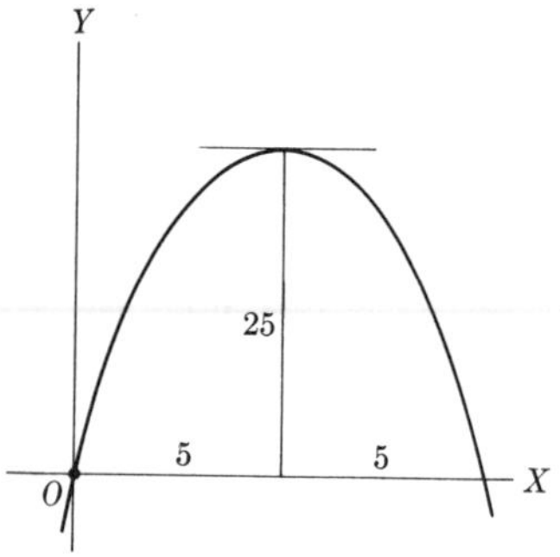

Fig. 16-7

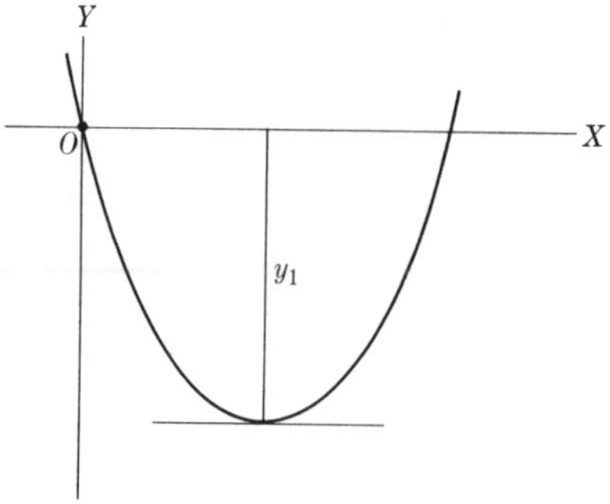

Fig. 16-8

for the motions of balls, but it certainly is not applicable to formulas which represent quite different phenomena. However, we shall now introduce a geometrical argument which proves that the procedure is indeed justified.

Let us use a specific function to illustrate the idea. The argument we shall give could be phrased in general terms. Suppose that we wish to find the maximum value of the function

$$y = 10x - x^2, \tag{30}$$

represented by the curve in Fig. 16–7. We observe that at the point on the curve where y has a maximum value, the tangent is horizontal; that is, the slope of the tangent is zero. Now the slope of the curve at any value of x is the value of the derivative, or instantaneous rate of change of y with respect to x at that value of x. Hence to determine the value x_1 of x at which the slope of the curve is zero, we find the derivative of y in (30), that is, we find $\dot{y}$, and set this derivative equal to zero. Thus for formula (30), we have

$$10 - 2x_1 = 0.$$

We see at once that $x_1 = 5$. To find the maximum y-value of (30), we substitute the value 5 for x and find that y_1, the maximum y-value, is 25.

This geometrical argument proving that the derivative of a function is zero at the function's maximum value also applies to its minimum value. The minimum value of the function $y = x^2 - 10x$ is the length y_1 of Fig. 16–8. The slope of the curve at the point where y has this minimum value is zero. Hence, as before, at this point the derivative $\dot{y}$ must be zero, and we may use the process already described in connection with maxima of functions to determine minima also.

The question arises, If the same process yields a maximum and a minimum, how do we know in a particular problem whether we are obtaining one or the other? For physical problems the answer is given by the sense of the problem. But there are also purely mathematical criteria which enable us to determine whether we have found the maximum or the minimum value of a function.

EXERCISES

1. Calculate the instantaneous speed at $t=4$ of a body whose height d above the ground at time t is given by the formula $d=128t-16t^2$. Interpret the result physically and geometrically.
2. To illustrate the power of the calculus, Fermat showed how it can be used to prove that of all rectangles with the same perimeter the square has maximum area. Carry out this task. [*Suggestion:* Let p be the perimeter common to all the rectangles. If x and y are the dimensions of any one of the rectangles, then $2x+2y=p$ or $y=(p/2)-x$. The area A of any rectangle is given by $A=xy$. Express A as a function of x only and apply the calculus.] Which method do you prefer, that of Euclidean geometry or the calculus?
3. A farmer wishes to enclose a rectangular piece of land which borders on a river, so that no fence will be required along the bank. He has 100 ft of fencing at his disposal. What dimensions should he choose to obtain maximum area? [*Suggestion:* If y is the side which parallels the river, then the amount of fencing required is $y+2x$. This must equal 100. The area A of the rectangle is $A=xy$. Replace y by its value from $y+2x=100$ and find the maximum value of A.] Do you prefer this method or the method of Euclidean geometry?
4. A farmer wishes to use 100 ft of fencing to enclose a rectangular area and to divide the area into two rectangles by running a fence down the middle (Fig. 16–9). What dimensions should he choose to enclose the maximum total area?

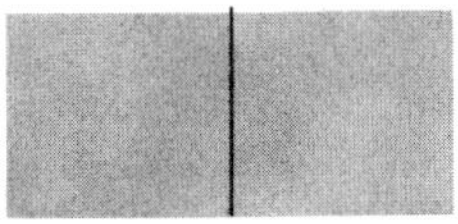

Fig. 16–9

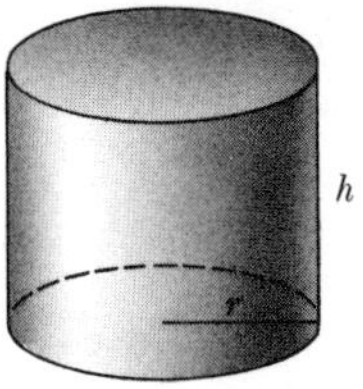

Fig. 16–10

5. A manufacturer wishes to construct cylindrical tin cans (Fig. 16–10) such that each can is made of a fixed amount of tin, say 100 sq in, and has maximum volume. What should the radius r of the base and the height h of the cylinder be? [*Suggestion:* The amount of tin used equals the surface area of the can, that is the sum of the area of the side, $2\pi rh$, and the area of the top and bottom, $2\pi r^2$. Hence

$$2\pi rh + 2\pi r^2 = 100. \qquad (1)$$

The volume V of the can is

$$V = \pi r^2 h. \qquad (2)$$

Solving (1) for h, we have

$$h = \frac{50 - \pi r^2}{\pi r}. \tag{3}$$

If we substitute this value of h in (2), we obtain

$$V = \pi r^2 \left(\frac{50 - \pi r^2}{\pi r} \right) = r(50 - \pi r^2) = 50r - \pi r^3.$$

Now apply the calculus.]

REVIEW EXERCISES

1. If the distance d in feet which a body falls in t sec is given by the formula $d = 16t^2$, calculate
   a) the average speed during the first 6 sec of fall,
   b) the average speed during the sixth second of fall,
   c) the instantaneous speed at the end of the sixth second of fall.
2. Suppose you were asked to calculate the average speed of a body which falls according to the law $d = 16t^2$ for the interval of time $t = 5.9$ to $t = 6$. How could you obtain an approximate answer quickly?
3. If $y = 10x^2$ what is the instantaneous rate of change of y with respect to x

   a) at $x = 2$, b) at $x = 3$, c) at $x = a$?

4. The instantaneous acceleration of a moving object is by definition the instantaneous rate of change of the speed with respect to the time. Suppose an object falls a distance d given by the formula $d = 16t^2$, where d is in feet and t is in seconds. What is the instantaneous acceleration of the object at any time t?
5. Find the instantaneous rate of change of y with respect to x for the functions below:

   a) $y = x^2 + 10x$ at $x = 2$, b) $y = x^2 - 10x$ at $x = 2$,
   c) $y = -x^2 + 10x$ at $x = 2$, d) $y = -x^2 - 10x$ at $x = 2$.

6. Find the slope of the curve $y = x^2 + 2x$ at $x = 2$, and illustrate the slope graphically.
7. Find the value of x at which the slope of the curve corresponding to $y = x^2 + 2x$ is 0.
8. Find the maximum or minimum values of the following functions and use the graphs of the functions to decide whether the calculated value is a maximum or a minimum.

   a) $y = x^2 + 10x$ b) $y = x^2 - 10x$ c) $y = -x^2 + 10x$
   d) $y = -x^2 + 6x$ e) $y = -x^2 + 6x + 2$

9. Find the value of x at which the derivative of $y = x^3$ is 0 (see Exercise 1 of Section 16–6). Does the function have a maximum or a minimum at that value of x? Is the following statement correct? At a value of x for which the derivative of a function is 0, the function has a maximum or a minimum.
10. A projectile is shot up into the air so that its height after t sec is given by the formula $h = 144t - 16t^2$. What is the maximum height attained by the projectile and how long does it take to attain that height?
11. An open box with a square base is to be made from 400 sq. in of lumber. What dimensions should be chosen so that the box has maximum volume?
12. A box with a square base and a cover is to be made from 400 sq in of lumber. What dimensions should be chosen so that the box has maximum volume?

Recommended Reading

See the list at the end of Chapter 17.

THE INTEGRAL CALCULUS

More laws are vain where less will serve.

ROBERT HOOKE

17-1 DIFFERENTIAL AND INTEGRAL CALCULUS COMPARED

The material examined in the preceding chapter belongs to the differential calculus. The basic process in this subject is to start with the formula relating two variables and to find the instantaneous rate of change of one variable with respect to the other. Suppose, however, that one began with the rate of change of one variable with respect to another and wished to find the formula which relates the two variables. For example, if we should happen to know that $\dot{y} = 2x$, could we find the relation between y and x? One might expect that the answer is affirmative because it would seem that among the various functions whose derivatives we have obtained, there should surely be one whose derivative is $2x$, and this function is the answer to our question. Except for a minor difficulty which we shall consider later, this expectation is correct. In this connection one might also ask whether there is any point in determining functions whose derivatives are given. The answer decidedly is yes. As we shall see, in numerous physical problems the most readily available information is an instantaneous rate of change, whereas the information sought can be best obtained from the function which relates the variables in question. Hence the process of finding the function from its derivative is immensely valuable—indeed, even more valuable than the basic process of finding derivatives from given formulas.

The major idea characterizing the integral calculus is the inverse to that underlying the differential calculus: namely, instead of finding the derivative of a function from the function, one proceeds to find the function from the derivative. Of course, all really significant ideas prove to have extensions and applications far beyond what is immediately apparent, and we shall find this to be true of the integral calculus also.

17-2 FINDING THE FORMULA FROM THE GIVEN RATE OF CHANGE

The key concern, then, of the integral calculus is to determine the formula which relates two variables from the given instantaneous rate of change of one variable with respect to another. Before we can see how useful this idea is, we must examine and learn a few facts about the mathematical process itself.

Suppose we happen to know that the instantaneous rate of change of some variable y with respect to another variable x is $2x$, that is $\dot{y}=2x$. What formula relates y and x? The mathematician's method of answering this question is to survey all the rates of change of functions obtained in the past and to locate the function whose rate of change he has previously found to be $2x$. In our case, his eye will soon light on the function

$$y=x^2.$$

Hence this function is the answer to the problem of finding the relation between y and x such that $\dot{y}=2x$. The function $y=x^2$ is called the *indefinite integral*, or antiderivative, or often just the integral of the derivative $\dot{y}=2x$, and the process of obtaining it is called *integration* or antidifferentiation.

However, the formula $y=x^2$ is not the only integral of $\dot{y}=2x$. We had occasion to observe in the preceding chapter that the presence of a constant term in a formula has no effect on the instantaneous rate of change. For example, $y=x^2$ and $y=x^2+5$ both lead to $\dot{y}=2x$. Hence $y=x^2+5$ is as much an integral of $\dot{y}=2x$ as $y=x^2$ is. In fact, $y=x^2+C$, where C is any constant, is an integral of $\dot{y}=2x$. If C is chosen to be zero, we obtain $y=x^2$, and if C is chosen to be 5, we obtain $y=x^2+5$. It may seem unfortunate that there should be more than one answer, but we shall see in a moment that the reverse is the case.

The general problem of finding the formula relating y and x when we are given $\dot{y}$ as a function of x is handled by the method illustrated in our example of $\dot{y}=2x$; that is, we must examine the formulas whose rates of change we have previously determined and try to locate among these derivatives the rate of change we are concerned with. Since this rate of change has been previously derived from some formula relating y and x, that formula is the answer to our problem; in addition, we can add any constant to the formula and still have the correct answer. The process of searching among all formulas whose rates of change have previously been found may seem to be haphazard. But in practice mathematicians tabulate these formulas according to distinctive properties, so that a little experience with the tables usually enables one to find the desired formula. Since we are limiting the variety of formulas and their derivatives to a few cases, we shall not bother to become acquainted with a table. Instead we shall seek to recall the formulas and their derivatives which were calculated in the preceding chapter.

EXERCISES

For the following problems, find the formula which relates the variables whose instantaneous rate of change is given:

a) $\dot{y} = 3x^2$ b) $\dot{y} = 5$ c) $\dot{y} = x$ d) $\dot{y} = 3x$

e) $\dot{d} = 2t$ f) $\dot{d} = 32t$ g) $\dot{v} = 32$ h) $\dot{d} = 2t + 10$

i) $\dot{d} = -32t + 128$ j) $\dot{v} = -32$ k) $\dot{v} = 32t$

17-3 APPLICATIONS TO PROBLEMS OF MOTION

We shall now present some examples of the usefulness of integration in physical problems. Galileo had found that all objects falling to earth from points near the surface of the earth possess the same acceleration, namely 32 ft/sec². This acceleration is constant; that is, it is the same at each instant of the fall. Now the acceleration at any one instant is the instantaneous rate of change of velocity with respect to time. Hence, instead of writing $a = 32$, we can equally well write

$$\dot{v} = 32. \tag{1}$$

The physically important question is, What formula relates v and t? By reviewing the formulas for which we obtained rates of change [see formula (20) of Chapter 16] we find that $v = 32t + C$, where C is any constant.

In a particular physical problem, the quantity C can be chosen to fit the situation. Thus suppose that the object is merely dropped to earth; that is, at the instant it begins to fall its velocity is zero. If time is measured from the instant the object begins to fall, then the velocity v at $t = 0$ is zero. Hence, to make the formula

$$v = 32t + C \tag{2}$$

fit the physical fact that v must be zero when $t = 0$, we must have

$$0 = 32 \cdot 0 + C,$$

or $C = 0$. Hence

$$v = 32t \tag{3}$$

is the answer to this particular problem in which the object is dropped, and time is measured from the instant it begins to fall.

Physical problems often require knowledge of the distance which an object falls in time t. Since the instantaneous velocity is the rate of change of distance with respect to time, then if d denotes the distance the object falls, $\dot{d} = v$. In view of (3), which applies when an object is dropped, we may state that

$$\dot{d} = 32t. \tag{4}$$

We now wish to find the formula which relates d and t. Again we appeal to our experience with formulas and their derivatives [see formula (15) of Chapter 16] and note that the formula $d = 16t^2$ has the derivative given by (4). However, the formula:

$$d = 16t^2 + C, \tag{5}$$

where C is any constant, also has the derivative (4). Since we have no reason to ignore the constant, we must accept (5) as the formula for the distance fallen in time t. However, if we agree to measure the distance fallen from the point where the object happens to be at the instant it starts to fall, and if time of fall is also measured from this instant, then it follows that $d = 0$ when $t = 0$. Substituting these values in (5) yields

$$0 = 16 \cdot 0 + C,$$

and we see that C must be zero if formula (5) is to represent our situation. Hence

$$d = 16t^2 \tag{6}$$

gives the distance the dropped object falls in time t if time is measured from the instant the object begins to fall and if distance is measured from the point where the object is at $t = 0$.

We have been able to reverse or invert the process of finding the rate of change of a function and thus proceed from a knowledge of acceleration to velocity as given by (3), and from velocity to distance fallen as given by (6). Before we comment further, let us consider some other situations.

Suppose that an object is *thrown* downward and leaves the hand with a velocity of 100 ft/sec. The acceleration is still given by (1), and so the speed is still given by (2). However, if time is measured from the instant the object leaves the hand, then at the instant $t = 0$, $v = 100$. To make the formula

$$v = 32t + C$$

fit this new situation, we must have $v = 100$ at $t = 0$, or

$$100 = 32 \cdot 0 + C,$$

or $C = 100$. Hence

$$v = 32t + 100 \tag{7}$$

is the final formula for the velocity of an object thrown downward with an initial speed of 100 ft/sec.

Now let us seek the distance covered in time t. We know that the instantaneous velocity is the instantaneous rate of change of distance with respect to time. Thus, if we denote distance by d, we may write $\dot{d} = v$. In view of (7), we have

$$\dot{d} = 32t + 100. \tag{8}$$

We must now ask, What formula relates d and t? By reviewing the derivatives and the functions from which they were obtained, we find that the term $32t$ in (8) must come from the term $16t^2$ and the term 100 must come from $100t$. The formula for d therefore is presumably $d = 16t^2 + 100t$. However, we must recall that the formula

$$d = 16t^2 + 100t + C, \tag{9}$$

where C is any constant, also has the derivative (8). Hence, so far (9) is the general formula for distance fallen. If we agree to measure distance from the point at which the object happens to be when it begins to fall, and if time is measured from the instant the object begins to fall, then $d = 0$ when $t = 0$. Substituting these values in (9), we have

$$0 = 16 \cdot 0 + 100 \cdot 0 + C,$$

whence $C = 0$, and

$$d = 16t^2 + 100t \tag{10}$$

is the final formula for our situation.

We see from the examples already presented that the occurrence of the constant C in the integration is not a disadvantage but rather an advantage. It permits us to adjust the formulas for velocity and distance to the specific situation we wish to describe, although the basic fact in all instances is $\dot{v} = 32$.

The applications of integration made thus far have involved proceeding from the constant acceleration of 32 ft/sec^2 to the formula for distance. But this we were also able to do in Chapter 13 without depending upon the calculus. It might seem that, thus far at least, the process of integration has not added at all to the power of mathematics. However, there are two points to be taken into consideration. The derivation of formula (10), for example, from the basic physical fact, $a = 32$, is much more readily done by integration than by the argument given in Chapter 13. But the second and more important point is that the method displayed here for the derivation of the formula for velocity from that for acceleration and of the formula for distance from that for velocity applies to all formulas, whereas the argument given in Chapter 13 is limited to constant acceleration. Thus, if an object should move with variable acceleration, as is the case when an object falls to the earth from a great height, then the method of Chapter 13 no longer applies, whereas integration does. We shall treat such problems later.

Since the motions of objects *thrown up* into the air are very important, let us note that our present method applies to them also except for minor modifications. We shall again restrict ourselves to objects which do not rise very far from the surface of the earth, so that we can continue to use the physical fact that the acceleration is constant and equal to 32 ft/sec². When we studied the motion of a freely falling object, we decided, for convenience, to consider the acceleration to be positive. As a consequence, the velocity at any instant of time and the distance fallen turned out to be positive. However, an object thrown up into the air will, of course, rise and then fall. Hence, if we regard the velocity in the upward direction to be positive, then we must take the acceleration to be negative because it causes speed in the downward direction. We start then with the basic fact that

$$\dot{v} = -32. \tag{11}$$

By integration we obtain

$$v = -32t + C. \tag{12}$$

To fit the value of C to our situation, we shall use the physical fact that at $t = 0$, that is, at the instant at which the object is thrown upward, the hand or possibly a gun imparts to the object a velocity of, say 100 ft/sec. Thus at $t = 0$, $v = 100$. If we substitute these values in (12), we have

$$100 = -32 \cdot 0 + C.$$

Hence $C = 100$, and the final formula for velocity is

$$v = -32t + 100. \tag{13}$$

Since, as we know, instantaneous velocity is the instantaneous rate of change of distance with respect to time, we can now apply integration to find the distance traveled. Let us use d to represent the height above the ground reached by the object in time t. Then by integrating (13) we obtain [cf. (9)]

$$d = -16t^2 + 100t + C. \tag{14}$$

We now wish to adjust the value of C to fit the physical situation. At the instant $t = 0$, the object is about to be thrown up, and at this instant, $d = 0$. If we substitute 0 for d and 0 for t in (14), we obtain

$$0 = -16 \cdot 0 + 100 \cdot 0 + C,$$

and so $C = 0$. Thus the final formula for height above ground is

$$d = -16t^2 + 100t. \tag{15}$$

Having obtained various formulas for velocity and distance such as (13) and (15) or (7) and (10), we can now proceed to solve problems of the type considered in Chapter 13. We shall not repeat this work here, but instead propose to show in the following sections how integration, or the inverse of differentiation, produces useful formulas from basic physical facts.

EXERCISES

In all of the following problems the motions involved take place near the surface of the earth. Hence you may assume that the acceleration is constant.

1. Suppose that an object is thrown up into the air with an initial velocity of 150 ft/sec. Derive the formulas for the velocity and height above the ground.
2. Given that an object is dropped and falls to earth. Suppose that distance is measured from a point 50 ft above the point at which the object is dropped, but that time is measured from the instant the object begins to fall. What formula relates distance fallen and time of fall?
3. Suppose that an object is dropped from a point 75 ft above the ground. Derive the formulas for the velocity and height above the *ground*.
4. Suppose that an object is thrown up into the air from the roof of a building 50 ft high. The initial velocity is 100 ft/sec. Derive the formulas for the velocity and for the height above the *ground*.
5. Suppose an object is thrown downward from the roof of a building 50 ft high and that the initial velocity is 100 ft/sec. Derive the formulas for the speed and height above the *ground*.

17–4 AREAS OBTAINED BY INTEGRATION

The derivation of formulas useful in the study of motion was one of the seventeenth-century problems which motivated the creation of the calculus. Another basic class of problems was concerned with finding the lengths of curves, the areas bounded by curves, and the volumes bounded by surfaces. In Section 16–2 of the preceding chapter, we mentioned a few problems which called for the determination of lengths, areas, and volumes. The expansion of science and technology has brought about literally thousands of new uses of curves and surfaces for which the very same quantities are required. The distance a ship travels along the spherical surface of the earth is the length of a curve. Cables and roadways of bridges are curves, and in planning the construction one must know the lengths of these cables and roadways. The weights of various objects employed in scientific and engineering projects are easily obtained once the volumes are known. For example, if a steel beam of some particular shape is to be used in the framework of a building, then, since the material is the same throughout the beam, the weight is merely the volume multiplied by the weight per cubic foot of the metal. Hence volume is the essential quantity to be determined.

But problems concerning lengths of curves, areas, and volumes had already been solved in Euclidean geometry. Why should they then have presented special difficulties to the scientists of the seventeenth century? The answer is that Euclidean geometry is adequate only to treat figures bounded by straight line segments and by circles. This limitation is inherent in the subject. Examination of the axioms of Euclidean geometry shows that they state properties of lines and circles. Naturally the theorems which can be deduced readily must also be limited to such figures. Although the Greeks managed to compute a few areas and volumes of figures bounded by other geometrical shapes, they were able to do so only with great difficulty and by introducing special methods limited to the figures in question. The variety and number of problems which arose in the seventeenth century demanded more general and more easily applicable methods.

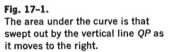

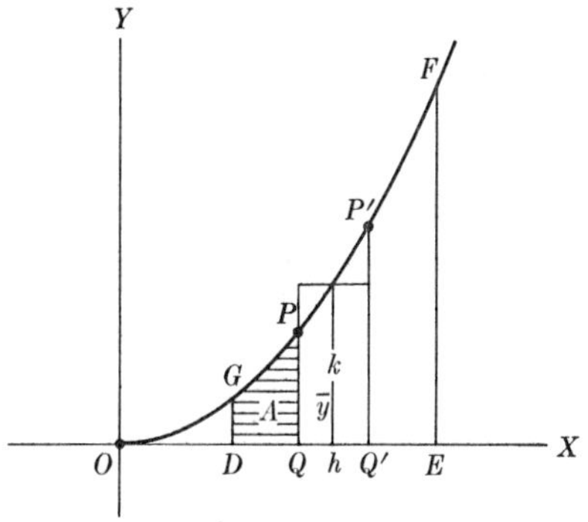

Fig. 17–1.
The area under the curve is that swept out by the vertical line *QP* as it moves to the right.

Though it is by no means evident, the calculus proves to be the very mathematical tool which enables us to calculate the lengths of curves, the areas bounded by curves, and the volumes bounded by surfaces. We shall illustrate this fact by treating the problem of area. Let us try to determine the area *DEFG* of Fig. 17–1. This area is bounded by the vertical line segments *DG* and *EF*, by the segment *DE* and by the arc *FG* of the curve whose equation is, say $y = x^2$. We may think of this area as being swept out by a vertical line segment *PQ* which starts at the position *DG* and moves to the right. Naturally the length of *PQ* varies as it moves. Let us suppose that *PQ* has reached the position shown in the figure. The area swept out by this moving segment depends, of course, upon the position it has reached. This position can be specified by the x-value of the point *Q*. Hence the variable area, which we shall denote by A, is a function of x, the abscissa of the point *Q*. We now propose to find the formula which relates A and x. Our procedure is as follows: We begin by determining the rate of change of A with respect to x at any given x and integrate this derivative to arrive at the desired formula.

Our first task therefore is to find the rate of change of A with respect to x.* To do so, let us suppose that PQ has moved a little farther to the position $P'Q'$. The abscissa of Q' is, of course, somewhat larger than that of Q. Let us denote the abscissa of Q' by $x + h$, so that the increase in the abscissa from Q to Q' is h. Obviously, the variable area A also increases when PQ moves to $P'Q'$. Let us use k to denote this increase which geometrically is the area $QQ'P'P$. It is immediately evident that the increase is equal to the area of a rectangle whose base is h and whose height is an ordinate, $\bar{y}$, which is larger than PQ and smaller than $P'Q'$. (We do not know how large $\bar{y}$ is, but we shall see in a moment that this does not matter.) We have, then,

$$k = \bar{y}h. \tag{16}$$

Let us divide both sides of (16) by h. Then

$$\frac{k}{h} = \bar{y}. \tag{17}$$

Now k/h is the average rate of change of area with respect to the abscissa in the interval h. By the very definition of an instantaneous rate, the rate of change of area with respect to abscissa at the x-value of Q should be the limit of the average rate of change as h approaches zero. But as h approaches zero, $\bar{y}$ approaches the y-value of P, or the length PQ. Thus

$$\dot{A} = y. \tag{18}$$

Since the y in (18) is the ordinate of the point P and P lies on the curve OGF, it follows that $y = x^2$. Then

$$\dot{A} = x^2. \tag{19}$$

We now have the rate of change, with respect to x, of the variable area A. To find A itself, we must ask ourselves what formula has the derivative x^2. A review of previously obtained derivatives tells us that the derivative of x^3 is $3x^2$, and that therefore $A = x^3/3$. We know, however, that the integral may contain a constant term and that this will not affect the derivative, which will remain unchanged. Hence the full answer is:

$$A = \frac{x^3}{3} + C. \tag{20}$$

* At this stage, the reader might reconsider the example in the preceding chapter dealing with the rate of change of the area of a circle with respect to the radius and try to guess the answer to our present problem of rate of change.

To determine the value of C, we make use of the fact that when PQ is at DG, the area is zero because DG was the starting position of PQ. Suppose that the x-value of D is 3. Then, by substituting 0 for A and 3 for x in (20) we have

$$0 = \frac{3^3}{3} + C,$$

or $C = -9$. Thus

$$A = \frac{x^3}{3} - 9, \tag{21}$$

and this formula gives the area between DG and the variable position of the moving line segment PQ. If we wish to determine the area from DG to EF, we may assume that PQ has reached the position EF. Let us suppose that the x-value of E is 6. If we now substitute 6 for x in (21), we obtain the area $DEFG$. Hence

$$\text{area } DEFG = \frac{6^3}{3} - 9 = 72 - 9 = 63. \tag{22}$$

Thus we have found the area bounded by a curve through the process of integration. We have, of course, used the equation of the curve, which, thanks to Descartes and Fermat, should be known to us.

In working problems, one can eliminate some writing by neglecting to introduce the constant C in (20) and using just the formula $A = x^3/3$. We then substitute 6, which is the abscissa of the point E, in this formula; next, we substitute 3, which is the abscissa of the point D; and finally we subtract the second result from the first. These steps lead to the result given by (22).

EXERCISES

1. Find the area bounded by the curve $y = x^2$, the X-axis, and the ordinates at $x = 2$ and $x = 6$.
2. Find the area bounded by the curve $y = x^2$, the X-axis, and the ordinates at $x = 4$ and $x = 6$.
3. Find the area bounded by the straight line $y = x$, the X-axis, and the ordinates at $x = 4$ and $x = 6$.
4. Find the area bounded by the curve $y = x^2 + 9$, the X-axis, and the ordinates at $x = 3$ and $x = 6$.

17-5 THE CALCULATION OF WORK

An important quantity for scientific and engineering purposes is the work done in various physical operations. When a person raises an object for some distance he does work. The definition of the word "work" as used in the physical sciences is the product of the force applied and the distance through which the force acts. This quantity is important, for example, in the operation of machinery. One must know how much work a machine is capable of doing, to decide whether it is suitable for a particular task. A train pulling a load over some distance and an airplane carrying freight or passengers do work, and again the capacity of these carriers and the fuel requirements must be known for proper design.

We shall now consider how the calculus can be used to calculate work. Let us suppose that we wish to compute the work required to raise a 500-pound load to a height of 100 miles. This problem arises, for example, in determining the quantity of fuel required to raise a rocket to some desired height. Now the force that will accomplish this goal must be great enough to offset the force of gravity, which pulls the object down. The force of gravity, as we know, is given by

$$F = \frac{GMm}{r^2}, \tag{23}$$

where G is the gravitational constant, M is the mass of the earth, m is the mass of the object, and r is the variable distance between the position of the object and the center of the earth. In our problem, we shall regard G, M, and m as constant, so that the only variables in (23) are F and r. (However, in actual rocket problems, the fuel itself is part of the load which must be raised, and, since the fuel is gradually burned up as the rocket rises, the mass m also is a variable.) Since the object is to travel a distance of 100 miles up, the force which must be applied varies over the distance. Hence it is not possible to calculate the work by merely multiplying the force by the distance.

Let W be the work required to raise the object from the surface to some distance r from the center of the earth. Of course, W is a function of r and is unknown. Let us suppose that the object is raised an additional distance h, that is, from r to $r + h$ (Fig. 17–2). The corresponding extra work, k, again depends upon the force which must be applied [given by (23)] and the distance, h, over which it operates. However, the force varies even as r increases to $r + h$. Let $\bar{r}$ be some value of r between r and $r + h$ such that the corresponding force $GMm/\bar{r}^2$ is the average force required during the interval r to $r + h$. This last-mentioned force is entirely analogous to the quantity $\bar{y}$ introduced in the treatment of area as an average ordinate in the interval h on the X-axis. We shall see in a moment that the precise value of $\bar{r}$ plays no role.

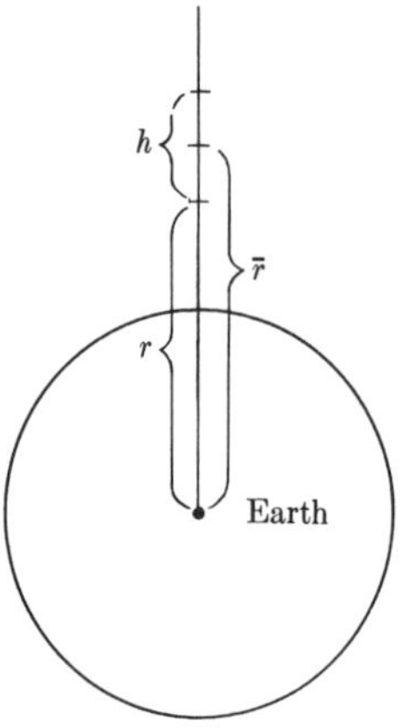

Fig. 17–2

Then the work done in raising the mass m from r to $r + h$ is

$$k = \frac{GMm}{\bar{r}^2} h. \tag{24}$$

Equation (24) yields the additional work required to raise the object the distance h. We shall determine next the average rate of change of work with respect to distance. This quantity, k/h, is obtained from equation (24) by dividing both sides by h. Thus

$$\frac{k}{h} = \frac{GMm}{\bar{r}^2}.$$

We compute next the instantaneous rate of change of work with respect to distance. This rate is obtained by letting h approach zero and finding the limit approached by k/h. However as h approaches zero, the quantity $\bar{r}$ must approach r because $\bar{r}$ is always intermediate between r and $r + h$. Then the instantaneous rate of change of work with respect to distance is given by

$$\dot{W} = \frac{GMm}{r^2}. \tag{25}$$

Since we now know $\dot{W}$, we can find W by integration; that is, we examine the various functions we have differentiated, to find one which yields (25) as its derivative. It so happens that in our work we did not encounter the rate of change given by (25), but we can take for granted now and check later that the function corresponding to this derivative is

$$W = -\frac{GMm}{r} + C. \tag{26}$$

To determine the constant of integration, we recall that when $r = 4000$ miles or $4000 \cdot 5280$ feet, the object is on the surface of the earth, and hence, $W = 0$. Since we do not wish to manipulate large numbers at the moment, we denote the radius of the earth by R. Then when $r = R$, $W = 0$. We substitute these values in (26) and obtain

$$0 = -\frac{GMm}{R} + C$$

or

$$C = \frac{GMm}{R}.$$

Hence

$$W = -\frac{GMm}{r} + \frac{GMm}{R}. \tag{27}$$

We now have the function which expresses the work done in raising an object of mass m from the surface of the earth to a height r units from the center. By substituting the numerical data at our disposal we can calculate the work done in the example proposed at the outset. We know G and the mass M of the earth. The quantity R is $4000 \cdot 5280$ feet, and the value of r is 4100 miles or $4100 \cdot 5280$ feet. The mass m, in our example, is the mass of an object which weighs 500 pounds or $500 \cdot 32$ poundals at the surface of the earth. Hence the mass m is 500 pounds.

We can simplify the arithmetic somewhat since we know that the acceleration due to the earth's gravitational attraction is [formula (5) of Chapter 15]

$$a = \frac{GM}{r^2},$$

and that when $r = R$, then

$$a = 32 \text{ ft/sec}^2.$$

Thus

$$32 = \frac{GM}{R^2},$$

whence

$$GM = 32R^2.$$

If we substitute this result in (27), we obtain

$$W = -\frac{32R^2m}{r} + \frac{32R^2m}{R}$$

or

$$W = -\frac{32R^2m}{r} + 32Rm. \tag{28}$$

By applying the distributive axiom we may write

$$W = 32Rm\left(1 - \frac{R}{r}\right). \tag{29}$$

Formula (29) is the useful form for the calculation of the work, except for one detail. The unit of work is usually taken to be foot-pounds. However if the mass m is given in pounds, then the formulas we have employed here and elsewhere express the force in poundals. Hence formula (29) yields the work done in foot-poundals. To obtain an answer in foot-pounds we just ignore the factor 32 in (29).

EXERCISES

1. Calculate the work done in raising the 500-lb weight to a height of 100 mi.
2. Suppose that one neglected the variation in gravitational force with height and assumed that the 500-lb weight remains constant over the 100 mi that it is raised. What is the work required to raise it?
3. A cable weighing 2 lb/ft is suspended in a well 100 ft deep; a tool weighing 300 lb is attached at the cable's lower end. Find the work done to raise the tool to the surface. [*Suggestion:* Let W be the work done to raise the tool x ft. Since now only $100 - x$ ft of cable remain, the work k done in raising the tool h feet more is $k = [300 + 2(100 - \bar{x})]h$, where $\bar{x}$ is some value of x between x and $x + h$. Now find $\dot{W}$, and then W. Determine the constant and compute the work done.]
4. Use the method of increments to show that the function $W = c/r$, where c is any constant, has the derivative $\dot{W} = -c/r^2$.

17-6 THE CALCULATION OF ESCAPE VELOCITY

We can use the theory of the preceding section to answer a question which is of special interest today, namely, what velocity one must give to a rocket to ensure that it just reaches a specified height. The condition intended here is that the rocket will have zero velocity when it reaches this height, for if it still possessed some velocity, it would continue to rise.

As the rocket travels upward, it loses velocity because the acceleration of gravity, which is directed downward, continually decreases the velocity. However, if the initial velocity V is properly chosen then the rocket will have zero velocity at the required height. We wish to determine V.

In the preceding section we calculated the work done in raising an object to a height of d feet above the surface of the earth. However the result did not involve the initial velocity V. We shall therefore obtain another expression for this work. It is physically clear that the work done against gravity in the process of sending an object up with some initial velocity V to reach a height of d feet should equal the work done by gravity acting on the object when it falls d feet. Hence let us calculate the latter. We have an object which begins its fall with zero velocity and falls d feet. Since it gains velocity on the way down in just the reverse order in which it loses velocity on the way up, it strikes the ground with a velocity of V ft/sec.

If an object falls from a great height above the surface of the earth, it does not fall with constant acceleration. However, there is some average acceleration which would produce the same final velocity in any time of fall. Let us denote this average acceleration by a. In Section 13–5 we found the formula for the speed of and the distance fallen by an object which is dropped and falls to earth with an acceleration of 32 ft/sec^2. If the acceleration were a

instead of 32, the formulas would be (Exercise 14 of Section 13–5)

$$v = at \qquad \text{and} \qquad d = \tfrac{1}{2}at^2.$$

Now, if we take the value of t from the first formula and substitute it in the second one, we obtain the relationship between d and v. Thus since $t = v/a$,

$$d = \frac{1}{2}a\left(\frac{v}{a}\right)^2 = \frac{1}{2}a\frac{v^2}{a^2} = \frac{1}{2}\frac{v^2}{a}.$$

Then

$$V^2 = 2ad, \tag{30}$$

wherein we have written V merely to denote that it is the final velocity after the object has fallen d feet.

Since the object falls with a constant acceleration a, the force which gravity applies, by Newton's second law of motion, is ma, where m is the mass of the rocket. Then the work done by gravity is

$$W = mad.$$

However by (30) we see that $ad = V^2/2$. If we substitute this value in the formula for W we have

$$W = \frac{mV^2}{2}. \tag{31}$$

Formula (31), then, is an expression for the work that gravity does in causing an object to fall a distance d and to acquire, at the end of the fall, the velocity V. As we have already noted, this is the work we must do to *raise* the object from the surface of the earth where it has velocity V to the point where it has velocity zero.

Formula (29), namely,

$$W = 32Rm\left(1 - \frac{R}{r}\right),$$

also gives the work required to raise an object from the surface of the earth to the distance r from the center. Let $r = R + d$ so that the object is raised to the height d above the surface. Then

$$W = 32Rm\left(1 - \frac{R}{R+d}\right). \tag{32}$$

We now have two expressions, formulas (31) and (32), for the work. We equate them and obtain

$$\frac{mV^2}{2} = 32Rm\left(1 - \frac{R}{R + d}\right).$$

Dividing both sides by m and multiplying by 2 yield

$$V^2 = 64R\left(1 - \frac{R}{R + d}\right). \tag{33}$$

This then is the expression for the initial velocity required to send an object up so that it just reaches the height of d feet above the surface. In applications, the quantities occurring in formula (33) must be expressed in feet.

Formula (33) has a very interesting consequence. If we wished to send an object farther and farther up so that d becomes indefinitely large, then, since R is fixed, the quantity $R/R + d$ approaches zero. The result is

$$V^2 = 64R$$

or

$$V = \sqrt{64R} = 8\sqrt{R}. \tag{34}$$

This velocity is often described as the velocity required to reach infinity and is called the escape velocity. Of course, infinity is not a geographical location, and what is really meant is that the object will keep going out indefinitely and never return. If the initial velocity is less than the escape velocity, then the object will attain zero velocity at some finite, though possibly large, distance from the earth and fall back to earth.

We can readily calculate the escape velocity. Since

$$R = 4000 \cdot 5280 = 21{,}120{,}000 \text{ ft},$$

we have

$$\begin{aligned} V &= 8\sqrt{21{,}120{,}000} = 8\sqrt{2112} \cdot \sqrt{10{,}000} \\ &= 8 \cdot 4600 = 36{,}800 \text{ ft/sec} \\ &= 7 \text{ mi/sec, approximately.} \end{aligned}$$

This is the velocity necessary to escape from our trouble-infested earth.

EXERCISES

1. Calculate the velocity required to send an object 240,000 mi up (this is the distance to the moon) and arrive there with zero velocity. [*Suggestion:* Use (33).]
2. Does formula (34) give the escape velocity from the moon, that is, does it give the velocity required to send an object from the moon to infinity? If not, how should it be modified?

17-7 THE INTEGRAL AS THE LIMIT OF A SUM

In our discussion of integration as a means of finding areas bounded by curves, we mentioned that from Greek times up to the seventeenth century the efforts of mathematicians to determine such areas had not been very successful. The reason, already noted, was that these men tried to use Euclidean geometry, and this geometry is limited in power. To prove theorems about the areas of figures bounded by curves, they had to overcome great difficulties by a special method known as the method of exhaustion. They approximated the area in question by figures bounded by straight lines—for such figures the area could readily be found—and then considered what happened as the approximation was improved more and more. Although, for the moment, it may seem that we are taking a step backward, we shall nevertheless reapproach the problem of area by adopting the Greek view. We shall find that our re-examination will have fruitful results for a new and wide class of problems.

Let us consider the problem [treated in Section 17–4] of finding the area *DEFG* (Fig. 17–3) which is bounded by the arc *FG* of the curve whose equation is $y=x^2$, by *DE*, and by the vertical line segments *DG* and *EF*. We subdivide the interval *DE* into three equal parts, each of length h, and denote the points of subdivision by D_1, D_2, and D_3, where D_3 is the point *E*. Let y_1, y_2, and y_3 be the ordinates at the points of subdivision. Now y_1h, y_2h, and y_3h are the areas of three rectangles shown in Fig. 17–3, and the sum

$$y_1h + y_2h + y_3h \tag{35}$$

is the sum of the three rectangular areas and thus an approximation to the area *DEFG*.

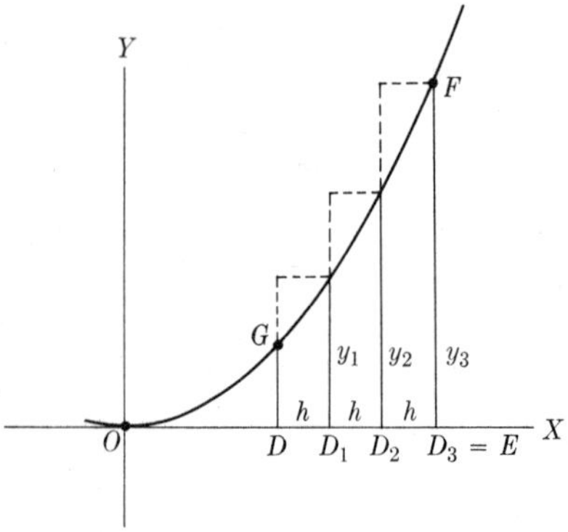

Fig. 17–3.
The area under a curve approximated by a sum of rectangular areas.

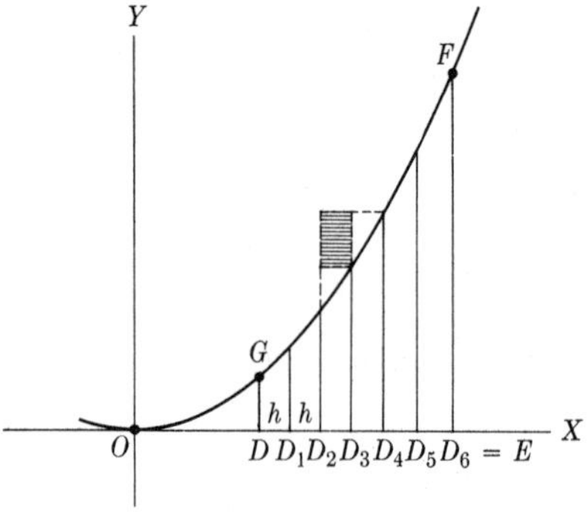

Fig. 17–4.
Decreasing the widths of the rectangles improves the approximation provided by the sum of the rectangular areas.

We can obtain a better approximation to the area *DEFG* by using smaller rectangles and more of them. To illustrate this point, suppose that we subdivide the interval *DE* into six parts. Figure 17–4 shows what happens to the middle rectangle of Fig. 17–3. This rectangle is replaced by two, and because we use the y-value of each point of subdivision as the height of a rectangle, the shaded area in Fig. 17–4 is no longer a part of the sum of the areas of the six rectangles which now approximates the area *DEFG*. Therefore the sum

$$y_1h + y_2h + y_3h + y_4h + y_5h + y_6h \tag{36}$$

is a better approximation to the area *DEFG* than the sum (35).

We can make a more general statement concerning this process of approximation. Suppose that we divide the interval *DE* into n parts. There would then be n rectangles, each of width h. The ordinates at the points of subdivision are $y_1, y_2, \ldots, y_n$, where the dots indicate that all intervening y-values at points of subdivision are included. The sum of the areas of the n rectangles is then

$$y_1h + y_2h + \cdots + y_nh, \tag{37}$$

and the dots again indicate that all intervening rectangles are included. In view of what we said above about the effect of subdividing *DE* into smaller intervals, the approximation to the area *DEFG* given by the sum (37) improves as n increases. Of course, as n gets larger, h gets smaller because $h = DE/n$.

We see so far how figures formed by line segments—rectangles in the present case—can be used to provide better and better approximations to an area bounded by a curve. Thus far we have utilized the Greek idea. We now depart from it somewhat and introduce the concept of a limit. Specifically the area *DEFG* is the limit approached by the sum of the rectangles as the number of rectangles becomes larger and larger, or, one says, as the number of rectangles becomes infinite. Thus the number of rectangles might be successively 3, 6, 12, 24, 48, . . . , where the dots indicate that we continue to double the number indefinitely. Of course, h, the width of each rectangle, approaches zero. In symbols, we write

$$\text{area } DEFG = \underset{h \to 0}{\text{limit}} \, (y_1h + y_2h + \cdots + y_nh); \tag{38}$$

the symbol "$\underset{h \to 0}{\text{limit}}$" means that what we wish to obtain is the number approached by the sum in parentheses as h approaches zero.

What have we accomplished? We seem to have made a simple thing difficult. The innocent area *DEFG* has been approximated by a sum of rectangles and, as the number of rectangles increases (while each rectangle becomes thinner), the sum becomes a better and better approximation of the area *DEFG*. However, we know from Section 17–4 that the area *DEFG* may be obtained in the following way: If the equation of the curve *FG* is $y=x^2$, we find the formula whose derivative is x^2 or, in other words, we find the integral of x^2. This happens to be $x^3/3$. We substitute the abscissa of the point *E* and obtain a number. We next substitute the abscissa of the point *D* and obtain a number. Finally, we subtract the latter result from the former. We see, then, that limits of the kind expressed in (38) can be determined by integration and the subsequent numerical work just described.

Now insofar as obtaining areas is concerned, we do not seem to have accomplished very much. Actually, were we to pursue the subject of area a little further than we shall in this book, we would find the new point of view significant in this very connection. However, we shall go on to other applications in which the fact that a limit of the form (38) can be obtained by integration will be the key to the solution.

It is helpful to shorten the writing of an expression such as (38). The notation used in calculus books is

$$\text{area } DEFG = \int_a^b y\,dx. \tag{39}$$

This notation must not be taken too literally. The symbol $\int$ is an abbreviated *S* and is intended to denote that we are dealing with the limit of a sum. For areas this sum is a sum of rectangles. The y in (39) indicates that the heights of these rectangles are ordinates of some curve, and the dx indicates that the base of each rectangle is a small interval along the *X*-axis. The number a is the abscissa of the left-hand end point of the interval *DE*, and the number b is the abscissa of the right-hand end point. The entire expression on the right side of (39) is called the *definite integral* of the function represented by y. The words "definite integral" denote that we are interested in the integral regarded as the limit of a sum.

Whereas Newton had concentrated on finding the derivatives of given functions and on the inverse process, the recognition that limits of sums, such as that expressed by (38), can be obtained by reversing differentiation is due primarily to Gottfried Wilhelm Leibniz (1646–1716). Leibniz's career contrasts sharply with Newton's. Newton, as we know, had undertaken the study of mathematics and physics early in life and had pursued these two fields almost exclusively, although he did make minor contributions to chemistry and theology. His career as a professor gave him the opportunity to concentrate. Leibniz started by studying law at the University of Leipzig, the city in which he was born and lived as a youth. He secured a bachelor's degree at Leipzig and in 1666 a doctor's degree at the University of Altdorf. His first position was that of ambassador for the Elector of Mainz, and until 1672 his interest in mathematics was secondary. In 1672, during a trip to Paris on behalf of his employer, he met Huygens, who acquainted Leibniz with current scientific problems and activities. Leibniz's interests were deeply stirred and thereafter he devoted much time to mathematics. In 1676 he was appointed librarian and councillor to the Elector of Hannover and, although this position also entailed many administrative duties, he nevertheless had more leisure for academic pursuits. In 1700 he went to Berlin to work for the Elector of Brandenburg and, while there, founded the Berlin Academy of Sciences. What is amazing about the man is the vast quantity of first-rate contributions to many fields. Although his profession was jurisprudence, his work in mathematics and philosophy ranks among the best the world has produced. He also did major work in mechanics, nautical science, optics, hydrostatics, logic, philology, and geology, and was a pioneer in historical research. Throughout his life he tried to reconcile the Protestant and Catholic faiths. We may recall also his previously noted activities—his efforts to organize a society devoted to the dissemination of the new scientific knowledge and to turn the German language into a suitable vehicle for the new ideas. No subject pursued by intellectuals of his age was neglected; only Leibniz himself went unrecognized and neglected by his contemporaries.

From our present point of view, Leibniz's emphasis on area as a limit of a sum may seem to be no blessing. But the full import of what he taught, namely that such limits can be evaluated by reversing differentiation, is of vast significance because limits of sums arise naturally in physical problems. Let us consider an example. In Newton's and Leibniz's time and for one hundred years thereafter, one of the major problems was to calculate the gravitational force exerted by one mass on another. If these masses are so compact that they can be regarded as concentrated at points, then the distance between them is the definite distance between these points, and the force of attraction is given by the usual formula. If, however, one mass is the earth and the other is some small object at, say a distance of a few hundred miles from the earth, then, although the latter may in many cases be considered as concentrated at a point, the earth itself cannot be so regarded. The difficulty is that the mass of the earth is distributed over an enormous volume and hence cannot be said to be separated from the mass m by a definite distance.

We can, however, regard the volume of the earth (Fig. 17–5) as broken up into small cubes numbered from 1 to n.* Since each cube is small, a good approximation of the distance of a cube from the mass m is the distance from the center of the cube to the mass m. Thus if this distance is r_1 for the first

* Strictly speaking, there will be pieces left over since a sphere is not a sum of cubes. However, we shall see that these pieces become negligible.

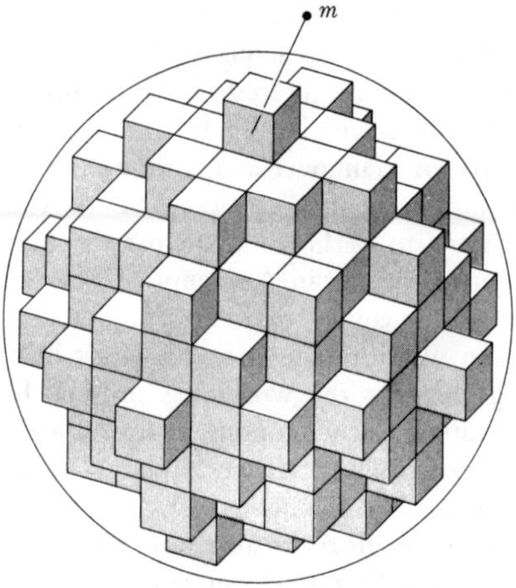

Fig. 17–5.
The volume of sphere approximated by a sum of cubical volumes.

cube, then the gravitational attraction exerted by the cube on the mass m is given by

$$F_1 = \frac{Gmh}{r_1^2},$$

where h now stands for the mass of the cube. The same applies to each cube up to the nth cube. Then the total gravitational attraction F of the n cubes is

$$\frac{Gm}{r_1^2}h + \frac{Gm}{r_2^2}h + \cdots + \frac{Gm}{r_n^2}h. \tag{40}$$

Formula (40) is entirely analogous to (37). The quantity y_1 in (37) has now become Gm/r_1^2; y_2 has become Gm/r_2^2; and so forth. But (40) is not the exact expression for the attraction exerted by the earth because it assumes that each cube acts as though its mass were concentrated at its center. However, if we make each cube smaller, which means that h will be smaller, and increase the number of cubes, n, so that they continue to fill as much of the sphere as possible, the sum of attractions exerted by the n cubes will be a better approximation to the force of attraction exerted by the entire sphere. The reason is that the smaller the cube, the more appropriate it is to assume that its mass can be regarded as concentrated at its center. The exact value of F is

$$F = \operatorname*{limit}_{h \to 0} \left(\frac{Gm}{r_1^2}h + \frac{Gm}{r_2^2}h + \cdots + \frac{Gm}{r_n^2}h \right). \tag{41}$$

Formula (41) is exactly like (38). It is now clear that to determine the total gravitational attraction exerted by a distributed mass, the earth in our discussion, we must calculate the limit of a sum. We know that such limits can be calculated by reversing differentiation. The particular limit in (41) cannot be computed with the mathematics at our disposal. The reason is that h in (41) is three-dimensional (because it is mass per cubic foot times volume), whereas the h in (38) is a segment. Hence (41) is a little more complicated than (38). But our example makes the point that limits of sums arise in physical problems, and that we can evaluate them by reversing differentiation.

As a matter of history, Newton solved the very problem we have been considering and proved that the earth attracts a small mass *as if* the entire mass of the earth were concentrated at its center. In other words, although the earth's mass is distributed over a large region, it happens to be true that a spherical mass attracting a small mass can be treated as if its mass were concentrated at its center. The solution of this problem enabled Newton to make further advances in the theory of gravitation. In our discussion of the law of gravitation, we also considered the quantity r to be measured from the center of the earth; that is, we made implicit use of Newton's result.

17-8 SOME RELEVANT HISTORY OF THE LIMIT CONCEPT

We could continue to study extensions and further applications of the calculus, but there are other features of this mathematical development which take precedence in view of the time we can devote to the subject. In the first place, it is most important to note that the calculus rests on a new concept, the concept of the limit of a function. We employed this concept in two essential ways. In the differential calculus we introduced the instantaneous rate of change of a function. This rate is the limit of the average rate of change of speed, that is of k/h, as h approaches zero. In the integral calculus we used the limit concept to speak of the quantity approached by the sum

$$y_1h + y_2h + \cdots + y_nh$$

as h approaches zero. This limit can represent area, the gravitational force exerted by an extended body, or other quantities, depending upon the physical or geometrical interpretation of the function relating y and x, and of h. Thus it is this new concept which distinguishes the calculus from the branches of mathematics previously studied.

The limit was defined as the number approached by some function of h as h approaches zero. This description is admittedly vague. In particular, the word "approach" is suspect. If for smaller and smaller values of h the ratio k/h should have the values $\frac{1}{4}$, $\frac{3}{8}$, $\frac{7}{16}$, $\frac{15}{32}$, . . . , are these values approaching 1? They are, in the sense of getting closer to 1, but it is also clear that they are always less than $\frac{1}{2}$, and so the limit might very well be $\frac{1}{2}$. In other words, how closely must the values of k/h approach a particular number before we can decide that that number is the limit of k/h? We shall not attempt to give a precise formulation of the limit concept. However, it may be a comfort to know that a precise definition can be given today.

The history of the efforts of mathematicians to grasp this concept properly is instructive as to how mathematics develops. The trouble started early in the seventeenth century. We have already mentioned that many mathematicians of that century made contributions to the calculus, even before Newton and Leibniz began to work on the subject. These forerunners realized that they were unable to give satisfactory expositions of their ideas and, in fact, hardly comprehended the significance of what they were creating. Despite the long tradition of rigorous proof in mathematics, the early workers in the calculus did not hesitate to advance their crude and imprecise ideas and defended themselves in ways that seem strange for mathematicians. Rigor, said Bonaventura Cavalieri, a pupil of Galileo and professor at the University of Bologna, is the concern of philosophy and not of geometry. Pascal argued that the heart intervenes to assure us of the correctness of mathematical steps. Proper finesse rather than logic is what is needed to do the correct thing, just as, he added, the appreciation of religious grace is above reason.

Although Newton and Leibniz made the most significant advances in the formulation of the ideas and methods of the calculus, neither contributed much to the rigorous establishment of the subject. They both realized that they had not presented clearly and precisely the basic ideas of instantaneous rate of change and the definite integral. Yet they were sure that their ideas were sound because they made sense physically and intuitively, and because the methods gave results which agreed with observations and experiments. Both gave many versions in the attempt to hit upon the precise concepts, but neither was successful. In some writings Leibniz went so far as to say that the methods of the calculus were only approximate, but since its errors were smaller than observational or measurable errors, the subject was useful.

The work of Newton and Leibniz was criticized even by their contemporaries. Newton did not reply to the criticisms, but Leibniz did. In addition to defending the methods by an appeal to the agreement of the results with experience, he attacked the critics as overprecise—a strange stand for a mathematician. He also said that we should not lose the fruits of an invention by excessive scruples. Of course, such replies did not provide the missing clarity and rigor. Some writers on the calculus proceeded as though there were no difficulties. Their attitude seemed to be that what was incomprehensible needed no further explanation.

The successors of Newton and Leibniz attempted to supply better foundations for the calculus. However their efforts were blocked in two ways. First of all, the formulations presented by Newton and Leibniz were different, but both yielded correct results; hence the rigorous construction of the calculus had to reconcile the two formulations. Secondly, the whole situation became complicated by an argument between Newton and Leibniz on the question of whether Leibniz had stolen ideas from Newton. Newton's friends, and English mathematicians in general, sided with him, while continental mathematicians defended Leibniz. The quarrel between the two groups became so bitter that they stopped corresponding with each other for about one hundred years. English mathematicians continued to talk about fluents and fluxions, Newton's terms for functions and their derivatives, whereas continental scientists talked about infinitesimals, Leibniz's name for h and k or, in his notation, dx and dy.

Two of the greatest mathematicians of the eighteenth century, Leonhard Euler and Joseph Louis Lagrange, worked on the problem of clarifying the calculus, but without success. Both arrived at the conclusion that as it stood, the calculus was unsound, but that somehow errors were offsetting one another so that the results were correct. A more drastic opinion was offered by the mathematician Michel Rolle (1652–1719). He taught that the calculus was a collection of ingenious fallacies. Voltaire called the calculus "the art of numbering and measuring exactly a Thing whose existence cannot be conceived." Near the end of the eighteenth century the distinguished mathematician Jean le Rond d'Alembert (1717–1783) felt obliged to advise his students that they should persist in their study of the calculus; faith would eventually come to them. All eighteenth-century attempts to supply rigorous foundations for the calculus failed. In the first quarter of the nineteenth century, Augustin-Louis Cauchy (1789–1857), the leading French mathematician, gave the first satisfactory definitions of the derivative and definite integral. Gradually other concepts of the calculus, which we have not considered, were clarified. The differences in notation were also eliminated.

This history of the development of the calculus is significant because it illustrates the way in which mathematics progresses. Ideas are first grasped intuitively and extensively explored before they become fully clarified and precisely formulated even in the minds of the best mathematicians. Gradually the ideas are refined and given the polish and rigor which one encounters in textbook presentations. In the instance of the calculus, mathematicians recognized the crudeness of their ideas and some even doubted the soundness of the concepts. Yet they not only applied them to physical problems, but used the calculus to evolve new branches of mathematics, differential equations, differential geometry, the calculus of variations, and others. They had the confidence to proceed so far along uncertain ground because their methods yielded correct physical results. Indeed, it is fortunate that mathematics and physics were so intimately related in the seventeenth and eighteenth centuries—so much so that they were hardly distinguishable—for the physical strength supported the weak logic of mathematics. Of course, mathematicians were selling their birthright, the surety of results obtained by strict deductive reasoning from sound foundations, for the sake of scientific progress, but it is understandable that the mathematicians succumbed to the lure.

It may be clear from this account of the difficulties that mathematicians experienced with the concept of limit that the two chapters devoted to the calculus do not provide a complete description of the concept and all its ramifications. The reader may justifiably feel some vagueness and uneasiness about what has been presented. Further study of the calculus would eliminate these objections. We must also point out that in illustrating the ideas of the calculus we have confined ourselves to the simplest functions. The subject is a vast one and far more powerful than we can indicate with the technique employed here. Indeed an enormous number of new branches of mathematics rest on the calculus. All of these, together with the calculus, constitute a division of mathematics, called analysis, which is considerably more extensive than algebra or geometry. But some glimpse of this most significant mathematical creation of modern times may in itself afford a rich enough reward.

EXERCISES

1. What essentially new idea does the calculus treat?
2. Mathematicians are logical thinkers; they reason directly and flawlessly to the desired conclusions. Discuss this assertion in the light of the history of the calculus.

17-9 THE AGE OF REASON

The men of the Renaissance had turned to new sources of knowledge—nature, reason, and mathematics—but were rather vague as to the specific methods by which they were to reconstruct the various branches of thought. Fortunately, a series of developments and creations not only gave strong impetus to the urge to rebuild knowledge, but actually supplied the method by which positive truths were to be attained. Galileo formulated this method clearly and applied it to obtain the laws of motion of objects moving on and near the surface of the earth. Newton conclusively demonstrated the efficacy of Galileo's method and supplied the doctrine which was to be the keystone in any new system of thought: All bodies in the universe were subject to one set of physical axioms, and their behavior could be deduced from these axioms. Nature was mathematically designed, and natural phenomena adhered strictly to universal mathematical laws which described the behavior of a speck of dust and of the most distant stars.

The intellectual leaders of the seventeenth and eighteenth centuries believed that they now had the tenets which vouchsafed a totally new outlook on the universe and which justified a reconstruction of all of mankind's systems of thought, institutions, and way of life. The right foundation in the form of a mathematical-mechanical explanation was available. New ideas, always far more effective in determining the course of cultures than wars or political events, began to work their influence, and the growing use of books permitted the leaders to reach large groups of people.

The intellectuals were confident that reason, based on the new truths made manifest by the mathematical and scientific work of the seventeenth century and cleansed of the metaphysical and theological presuppositions of the medieval period, could rebuild philosophy, religion, literature, art, political thought, and economic life. They saw clearly that mathematics and science were offering not just a few theorems and isolated results, but a new approach to truths and a new interpretation of the universe. Thus thinking men were impelled to a sweeping reorganization of all knowledge and institutions.

The eighteenth century has been called the Age of Reason. It was not the first period of history in which reason played a dominant role, but it was the first in which the intellectual elite emboldened by some successes in physical science dared to apply reason to the reconstruction of an entire civilization. The spirit and outlook of the leaders are indicated by their reference to their own age as the Enlightenment.

We cannot examine in the text the new doctrines which the Age of Reason fashioned in the social sciences, humanities, and the arts. The reader who would like to pursue this subject will find several survey chapters in the books by the author listed in the Recommended Reading. Further references are given in those chapters.

REVIEW EXERCISES

1. Suppose an object is thrown up from the ground with an initial velocity of 200 ft/sec. Using 32 ft/sec^2 as the downward acceleration of gravity,
   a) find the formula for its height above the ground in t sec;
   b) calculate the maximum height to which the object will rise;
   c) find the velocity with which the object will hit the ground.
2. Suppose that an object is thrown upward from the roof of a building which is 100 ft high and is given an initial velocity of 200 ft/sec.
   a) Find the formula for its height above the *roof* in t sec.
   b) Find the formula for its height above the *ground* in t sec.
3. The acceleration which the moon's gravity imparts to objects near the surface of the moon is 5.3 ft/sec^2. Suppose an object is thrown up from the surface with an initial velocity of 200 ft/sec. Find
   a) the formula for the height of the object after t sec,
   b) the maximum height the object reaches.
   c) Comparing the answers to Exercises 1(b) and 3(b) should show that the object reaches a greater maximum height on the moon. Can you explain in physical terms why this must be so?
4. a) Find the area bounded by the straight line $y = 3x$, the x-axis, and the ordinate at $x = 4$ by using the calculus method.

b) The figure described in part (a) is a triangle. Find its area by using the appropriate theorem of Euclidean geometry. Does your result agree with the answer to part (a)?

5. a) Find the area bounded by the straight line $y = 2x + 7$, the x-axis, and the ordinates at $x = 4$ and $x = 6$.

   b) The figure described in part (a) is a trapezoid. According to a theorem of Euclidean geometry the area of a trapezoid is one-half the altitude times the sum of the bases. Calculate the area by this formula and see whether it checks with the answer to part (a).

6. Calculate the work required to raise an object whose mass is 200 lb from the surface of the earth to a height of 100 mi. The variation in the earth's gravitational force must be taken into account.

7. Suppose you wish to shoot a rocket up from the surface of the earth to a height of precisely 100 mi. What initial velocity must you give the rocket?

Topics for Further Investigation

1. The predecessors of Newton and Leibniz in the creation of the calculus.
2. The work of Newton on the calculus.
3. The controversy between Newton and Leibniz on priority in founding the calculus.

Recommended Reading

BALL, W. W. ROUSE: *A Short Account of the History of Mathematics*, Chaps. 16 and 17, Dover Publications, Inc., New York, 1960.

BELL, ERIC T.: *Men of Mathematics*, Chap. 7, Simon and Schuster, Inc., New York, 1937.

COLERUS, EGMONT: *From Simple Numbers to the Calculus*, Chaps. 24 through 34, Wm. Heinemann Ltd., London, 1954.

EVES, HOWARD: *An Introduction to the History of Mathematics*, 2nd ed., Chap. 11, Holt, Rinehart and Winston, N.Y., 1964.

KASNER, EDWARD and JAMES R. NEWMAN: *Mathematics and the Imagination*, Chap. 9, Simon and Schuster, Inc., New York, 1940.

KLINE, MORRIS: *Mathematics in Western Culture*, Chaps. 16, 17 and 18, Oxford University Press, N.Y., 1953. Also in paperback.

KLINE, MORRIS: *Mathematics: A Cultural Approach*, Chaps. 20 through 22, Addison-Wesley Publishing Co., Reading, Mass., 1962.

SAWYER, W. W.: *Mathematician's Delight*, Chaps. 10 through 12, Penguin Books Ltd., Harmondsworth, 1943.

SAWYER, W. W.: *What Is Calculus About?* Random House, New York, 1961.

SCOTT, J. F.: *A History of Mathematics*, Chaps. 10 and 11, Taylor and Francis, Ltd., London, 1958.

SINGH, JAGIT: *Great Ideas of Modern Mathematics: Their Nature and Use*, Chap. 3, Dover Publications, Inc., New York, 1959.

SMITH, DAVID EUGENE: *History of Mathematics*, Vol. II, Chap. 10, Dover Publications, Inc., New York, 1958.

WIENER, PHILIP P. and AARON NOLAND: *Roots of Scientific Thought*, pp. 412–442, Basic Books, Inc., New York, 1957.

WIGHTMAN, WM. P. D.: *The Growth of Scientific Ideas*, Chap. 9, Yale University Press, New Haven, 1953.

CHAPTER 18

TRIGONOMETRIC FUNCTIONS AND OSCILLATORY MOTION

All the effects of nature are only mathematical results of a small number of immutable laws.

P. S. LAPLACE

18-1 INTRODUCTION

In the seventeenth century one of the most pressing problems of the times was time itself. The increasing scientific activity, particularly in an age which had decided to measure and to seek quantitative laws, created the need for convenient, accurate methods of measuring time. Moreover, as we have already had occasion to mention in other connections, the seventeenth and eighteenth centuries were concerned with the very practical problem of improving the method by which ships determined their longitude at sea. Here a good clock is the simplest answer. Suppose that the longitude of a given place on land is known and that a ship has on board a clock set to agree with the time prevailing at that given locality. Since the earth turns through 360° of longitude in one day, it turns through 15° in each hour. Hence for each 15° that a ship is west, say, of the fixed locale, midday occurs one hour later compared to the time at the fixed position on land. If a ship's officer notes (by means of the sun's position) when midday occurs at his position at sea and finds, for example, that his clock reads 3 o'clock whereas it should, of course, read 12 o'clock, he knows that the longitude of his position is 45° west of the given reference locality on land. We can see then why scientists decided to search for a reliable and accurate clock.

The thought which suggested itself almost at once was to look for some physical phenomenon which repeated itself regularly. The day contains 24 hours; hence when the number of repetitions per day is known, the duration of each repetition is readily calculated. Where then could one find a repetitive or periodic physical phenomenon? Two prospects attracted the attention of seventeenth-century scientists. The first of these was the motion of a mass, called a bob, attached to a spring and oscillating up and down, and the second was the motion of a pendulum, that is, a bob attached to a string and swinging to and fro. Now first reactions to the possibility of using the motion of a bob on a spring or a pendulum as a measure of time are apt to be negative. The bob on a spring, for example, does go through each cycle, that is, each complete up and down motion, in the same time so far as the eye can judge, but the motion soon dies down. The same is true for the pendulum. But, if air resistance could be minimized or perhaps compensated for, then these motions might become truly periodic and should therefore merit investigation. The scientist or mathematician who expects to see at once the solution of a problem he sets out to study will never accomplish much. The best he can hope for at the outset is an idea or a clue to pursue.

In this chapter we shall examine first the physical problem of the motion of a bob on a spring, a prime example of oscillatory motion. To study such motions mathematicians created a new class of functions, the trigonometric functions. We shall then discuss these functions and see how they are used to derive some knowledge about the physical problem which motivated their introduction. Surprisingly, trigonometric functions proved to be admirably suited for the study of sound, electricity, radio, and a host of other oscillatory phenomena. Of these latter developments we shall learn more in the next chapter.

18-2 THE MOTION OF A BOB ON A SPRING

The problem of investigating the motion of a bob on a spring was undertaken by one of the greatest experimentalists in the history of physics, the Englishman Robert Hooke (1635–1703). Hooke was professor of mathematics and mechanics at Gresham College. His claim to fame also rests upon his success as an inventor. To his credit are a telescope moved by a clock mechanism and devices for measuring the moisture in the atmosphere, the force of the wind, and the amount of rainfall. He improved the microscope, the barometer, the air pump, and the telescope. One of his findings, namely, that white light passed through thin sheets of mica breaks into many colors, parallels Newton's work on light. He also discovered the cell structure of plants. Hooke was very much interested in designing a useful clock and thought that springs would furnish the essential device. While working on the action of springs, he discovered a basic law, still known as Hooke's law, which we shall discuss later.

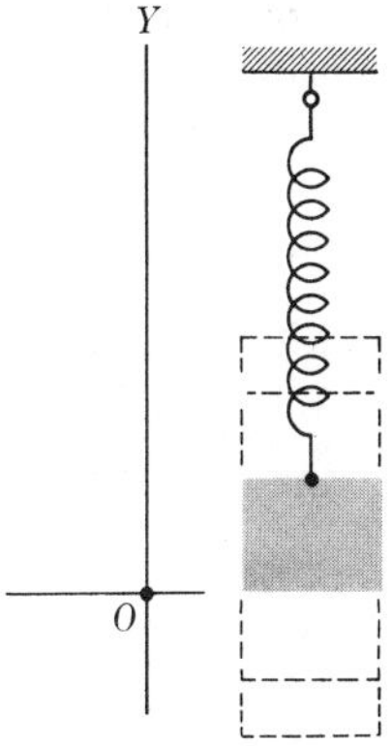

Fig. 18-1. A bob on a spring.

Let us follow Hooke in studying the motion of a bob on a spring. The upper end of the spring is attached to a fixed support, and a bob is attached to the lower end. Because gravity pulls the bob downward, the spring will be extended until the tension in the spring offsets the force of gravity. The bob then comes to rest in some position which is called the *rest* or *equilibrium position* (Fig. 18–1). If one now pulls the bob downward some definite distance below the rest position and then releases it, the bob moves up to the rest position, continues past that point to some highest position, and then moves downward. When it reaches the point to which it had been pulled down, it starts upward and repeats its former motion. Following Galileo's plan of idealizing the physical situation, let us suppose that air resistance is negligible. (Strictly speaking, energy is also lost in the expansion and contraction of the spring, but this loss is negligible.) Then the bob will continue to move up and down endlessly.

To begin to get some mathematical description of this motion let us introduce a Y-axis alongside the bob (Fig. 18–1) and suppose that $y=0$ corresponds to the rest position of the bob. When the bob is above or below the rest position, the bob is said to be displaced and the distance that it is above or below the rest position is called its *displacement.* To distinguish displacements above from those below the rest position, we shall call the former positive and the latter negative. Each displacement may then be described by a value of y. Thus $y = -\frac{1}{2}$ means that the bob is $\frac{1}{2}$ unit below the rest position.

To study the motion of the bob mathematically, it would be most helpful if we could find the formula which relates the displacement of the bob and the time it is in motion. Let us therefore seek such a formula.

18-3 THE SINUSOIDAL FUNCTIONS

No one of the formulas that we have considered thus far would be useful to represent the motion of the bob, for the peculiarity of the present phenomenon is that after each up and down motion, or oscillation, has been completed, the displacements go through their former sequence of values. Hence we apparently must seek a new type of formula which expresses the periodic character of the motion of the bob. We do not seem to have any clue, but a little imagination may supply one.

Suppose a point P moves around a circle of unit radius at a constant speed. Let us denote some of its positions by P_1, P_2, . . . (Fig. 18–2). We can, if we wish to, introduce a point Q on the vertical line through the center O such

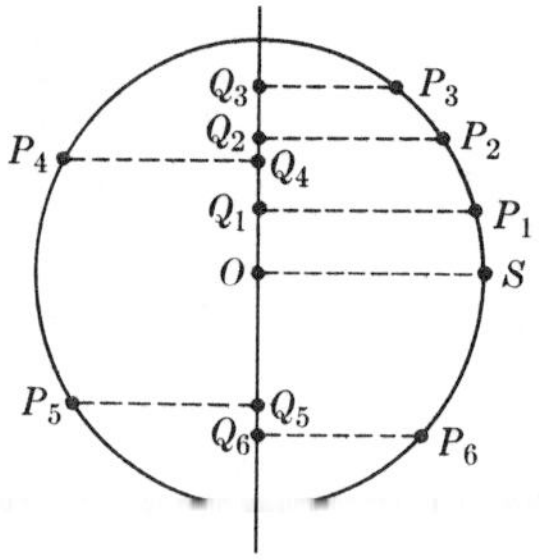

Fig. 18–2.
Successive positions of a point P which moves around a circle at a constant velocity, and the corresponding positions of Q.

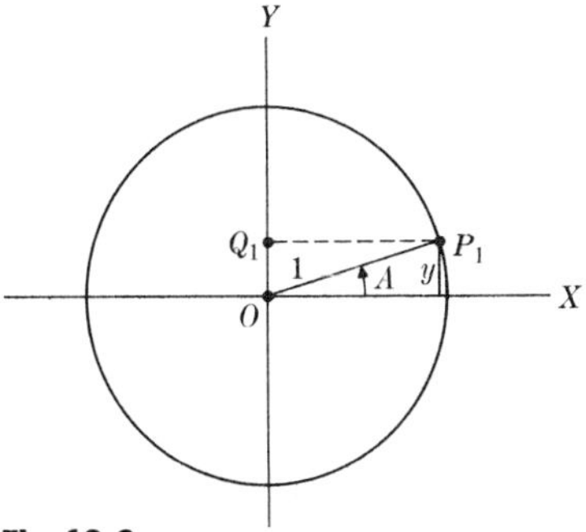

Fig. 18-3

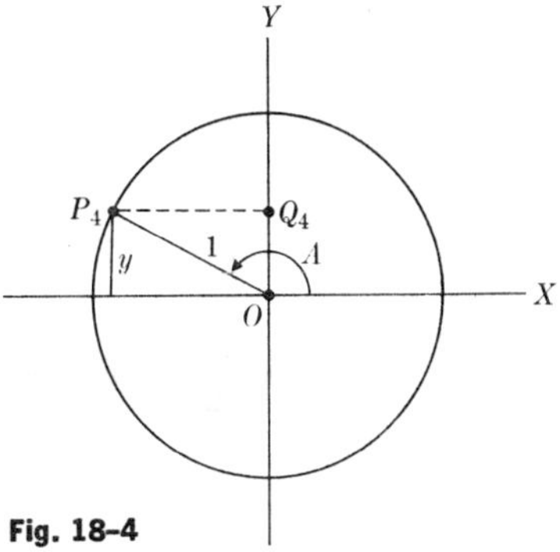

Fig. 18-4

that Q always has the same height that P has above or below the horizontal through O. The point Q is called the *projection* of P on the vertical line. Thus to the position P_1 of P there corresponds Q_1; to P_2 there corresponds Q_2; and so on. Why should we introduce the point Q? Well, let us imagine P moving around the circle through many revolutions starting from the position S at the right. What does its "shadow" Q do? It moves up from O to a highest position, moves down again to O, moves past O to a lowest position on the vertical line, moves up again to O, and then repeats this up and down motion. The motion of Q certainly seems to have the essential characteristics of the motion of the bob on the spring. Hence perhaps by pursuing further the motion of Q we may obtain the function we are seeking.

Let us introduce coordinate axes as shown in Fig. 18–3. If P starts from the X-axis and reaches, say the position P_1, then we may describe the position of P by the angle A shown in the figure. The height of Q above the X-axis *is the same* as the y-value of P. Now

$$\sin A = \frac{y}{1}.$$

Hence

$$y = \sin A. \tag{1}$$

Thus if the position of P is described by the angle A, then the position of the corresponding point Q on the vertical line is given by (1).

But now suppose P has moved to the position P_4 shown in Fig. 18–4. The angle A which describes the position of P_4 is the obtuse angle shown in the figure. This angle is no longer an acute angle of a right triangle, and we therefore have no right to speak of $\sin A$. However, let us extend the meaning of sine so that, by definition, $\sin A$ is the y-value of P_4. Since the height of Q_4 above O equals the y-value of P_4, we may continue to write $y = \sin A$ to describe the position of Q. That is, the distance of Q from O on the vertical line will be given by $y = \sin A$.

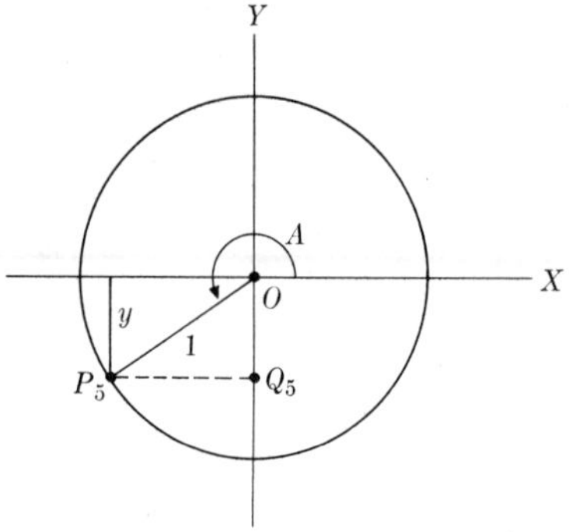

Fig. 18–5

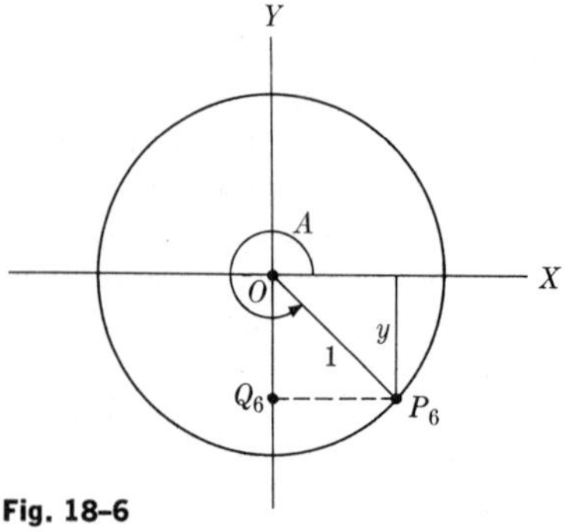

Fig. 18–6

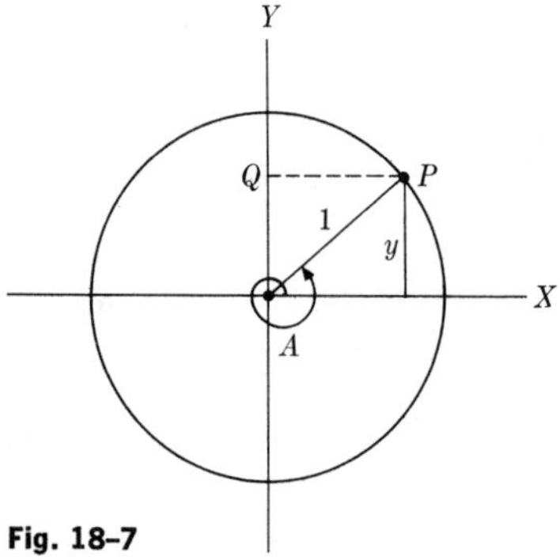

Fig. 18–7

Suppose next that P occupies the position P_5 shown in Fig. 18–5. The angle A which describes how far around the circle P has moved is now the angle shown. Let us agree again that by sin A we shall mean the y-value of P_5, which is also the distance below the X-axis of the point Q_5. Then we again may write $y = \sin A$ to describe the position of Q. Note that y is now a negative quantity.

If P reaches the position P_6 shown in Fig. 18–6, then its position is represented by the angle A shown, and if we again agree to mean by sin A the y-value of P_6, we shall be able to say here too that $y = \sin A$ describes the position of Q. In this instance also, y is a negative quantity.

As P returns to the X-axis and starts to repeat its revolution, the angle A which describes the position of P will now be 360° plus some additional angle (Fig. 18–7). It is only by including 360° for each revolution of P that we can keep track of the number of revolutions. However, let us note that the y-values of P will recur in precisely the same order in which they appeared on the first revolution. Despite the fact that on the second revolution the values of A are larger than 360°, we shall continue to mean by sin A the y-value of P. Thus sin 390° will be the same as sin 30°. As P goes through

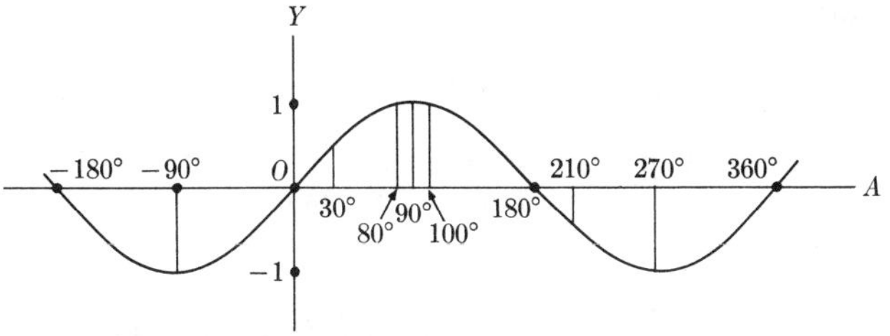

Fig. 18–8. The graph of $y = \sin A$.

its second revolution, Q repeats the motions of the first revolution. Hence it will still be true that $y = \sin A$ describes the position of Q on the vertical line. With each revolution of P, the y-values repeat, although the angle A increases by 360°. Since the motion of Q also repeats, its position on the vertical line will continue to be represented by $y = \sin A$.

If P were to revolve in the clockwise direction, then we would make only one change, i.e., call the values of A negative. The y-value of P, wherever it is, is still, by definition, $\sin A$ and this y-value would represent the position of Q.

Let us survey what we have done. To describe mathematically the position of the point Q, we have introduced a new function: $y = \sin A$. When A is an acute angle, as in Fig. 18–3, then $\sin A$ has the old meaning; that is, it is the ratio of the side opposite angle A to the hypotenuse of the right triangle in which A lies. (In the present case the hypotenuse is 1.) But when A is larger than 90°, then the equation $y = \sin A$ is a definition of what we mean by $\sin A$. Since there is a definite y-value for each value of A, positive or negative, we do indeed have a function.

To appreciate the nature of this function let us graph it. Figure 18–8 shows the graph. The values of A are plotted along the horizontal axis, and the corresponding y-values are plotted in the usual way.

Do we know the precise numerical value of y for each value of A? We do. For values of A which are between 0° and 90°, the y-values are the ordinary sine values which we find in our trigonometric table. In the interval from 90° to 180°, the values of $\sin A$ repeat, but in *reverse* order, the values which $\sin A$ has when A varies from 0° to 90°. This statement implies that $\sin 100° = \sin 80°$, $\sin 110° = \sin 70°$, and so forth. Stated in more general terms:

$$\sin A = \sin (180° - A). \tag{2}$$

In the interval from 180° to 360°, $\sin A$ has the same numerical values as when A varies from 0° to 180°. However, now $\sin A$ is negative. Thus sin

$210° = -\sin 30°$; $\sin 220° = -\sin 40°$; and in general:

$$\sin A = -\sin (A - 180°). \tag{3}$$

Since for each 360°-interval beyond the interval 0° to 360° sin A repeats the values that it has in the interval from 0° to 360°, $\sin 390° = \sin 30°$; $\sin 400° = \sin 40°$; and so on. In symbolic form,

$$\sin A = \sin (A - 360°). \tag{4}$$

The values of sin A for negative values of A are also shown in Fig. 18–8. If we look at the figure, we see that for any negative A-value, sin A is the negative of the sine of the corresponding positive A-value. That is, $\sin(-30°) = -\sin(30°)$; $\sin(-50°) = -\sin(50°)$, and in general:

$$\sin A = -\sin(-A). \tag{5}$$

Thus we have arrived at a definition of the function $y = \sin A$ for all values of A. Since we know quantitatively what sin A is for values of A between 0° and 90°, formulas (2) through (5) enable us to calculate sin A for all other values of A. The function we have just introduced is called a *periodic function* because the y-values repeat themselves in every 360°-interval of A-values. The interval of 360° is called the period of $y = \sin A$, and the entire set of y-values in one period is called the *cycle* of y-values.

EXERCISES

1. Using formulas (2) through (5) or Fig. 18–8, express the following sine values as sines of angles between 0° and 90°.

a) sin 120°	b) sin 150°	c) sin 210°	d) sin 260°
e) sin 270°	f) sin 300°	g) sin 350°	h) sin 370°
i) sin −50°	j) sin 750°		

2. What is the largest value of sin A? What is the smallest value of sin A?
3. At what value of A between 0° and 360° does the function $y = \sin A$ reach a maximum?
4. Why is $y = \sin A$ called a periodic function?
5. What purpose does the function $y = \sin A$ serve with respect to the location of Q, the projection of P?
6. What is the relationship between the function $y = \sin A$ and the trigonometric ratio sin A studied in Chapter 7?
7. Describe how sin A varies as A varies from 0° to 360°; from 360° to 720°.
8. For how many values of A between 0° and 360° is $\sin A = 0.5$?
9. Distinguish between the period and the cycle of $y = \sin A$.

Thus far we have described the size of angles in degrees. There is, however, no need to stick to this unit. Let us return to the motion of the point P in Figs. 18–3 through 18–7. The size of angle A can be specified by describing the *arc length* traversed by P from its starting point on the X-axis. This arc length is as much a measure of the size of angle A as the rather arbitrary agreement that a complete revolution of one side of A should be 360°.

Suppose that we agree to use the arc length traversed by P as a measure of A. How do we express in this new unit an angle of 90°, for example? When A is 90°, P has traversed one-quarter of the entire circumference. But the entire circumference of a circle of unit radius is 2π. Then the size of A in the new unit is $\pi/2$, that is about 1.57. We call this new unit *radians*. Thus an angle of 90° is also one of $\pi/2$ or 1.57 radians.

The advantage of radians over degrees is simply that it is a more convenient unit. Since an angle of 90° is of the same size as an angle of 1.57 radians, we now have to deal only with 1.57 instead of 90 units. The point involved here is no different from measuring a mile in yards instead of inches. If yards are just as good on other grounds, then it is far more convenient to speak of 1760 yards than 63,360 inches.

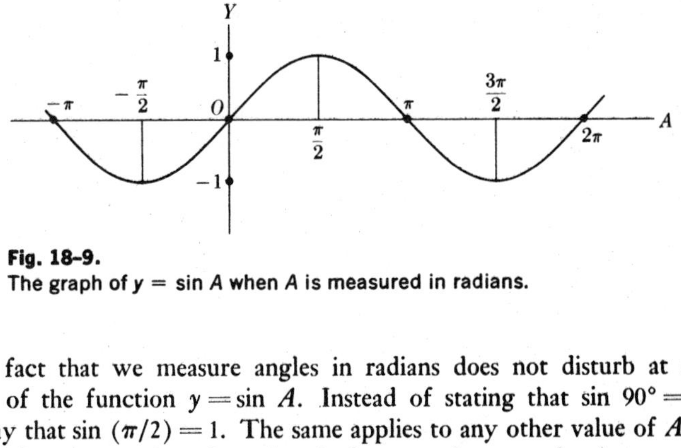

Fig. 18–9.
The graph of $y = \sin A$ when A is measured in radians.

The fact that we measure angles in radians does not disturb at all the meaning of the function $y = \sin A$. Instead of stating that $\sin 90° = 1$, we simply say that $\sin(\pi/2) = 1$. The same applies to any other value of A in the sinusoidal function we have introduced. Suppose, for example, that we wished to find the value of $y = \sin A$ when $A = \pi/6$. Because our table is set up in degrees, we note first that an angle of $\pi/6$ is of the same size as 30°, for $\pi/2$ radians is the same as 90°. Now from our tables we see that $\sin 30° = 0.5$, and so $\sin(\pi/6) = 0.5$.

Since we shall be using radians a good deal, we may as well become familiar with the function $y = \sin A$ when A is expressed in radians. Figure 18–9 shows the same function as Fig. 18–8 except that the units of A are now radians.

EXERCISES

1. Express the sizes of the following angles in radians: 90°, 30°, 180°, 270°, 360°, 420°.
2. The sizes of the following angles are in radians. Express the same angles in degrees.

$$\pi/2 \qquad 2\pi/3 \qquad 5\pi/2 \qquad 3\pi \qquad -\pi/2 \qquad 1$$

3. Find the value of

a) $\sin \pi$ b) $\sin (\pi/2)$ c) $\sin (\pi/3)$
d) $\sin (3\pi/2)$ e) $\sin 3\pi$ f) $\sin (5\pi/2)$.

4. Describe how $\sin A$ varies as A varies from 0 to 2π, as A varies from 2π to 4π.

The function $y = \sin A$ has a maximum value of $+1$ and a minimum value of -1. The maximum y-value, incidentally, is called the *amplitude* of the function. Such a function, even if it were suitable in all other respects, could not represent the motion of a bob whose maximum displacement is 2 or 3, say. This difficulty is easily obviated. Now that we have $y = \sin A$, we can readily manufacture hundreds of new functions whose amplitudes are whatever we choose to make them. Consider, for example, $y = 2 \sin A$. How does this function behave compared to $y = \sin A$? The answer is immediate. For any value of A, $y = 2 \sin A$ is twice as much as $y = \sin A$. Thus when $A = \pi/4$ or 45°, $\sin A = 0.71$, and $2 \sin A$ is 1.42. Figure 18–10 shows how $y = 2 \sin A$ looks compared to $y = \sin A$. If we want a sine function with amplitudes 3, $\frac{1}{2}$, or any other number, we can write one down immediately. As is evident from the nature of the function $y = 2 \sin A$, the function

$$y = D \sin A$$

has amplitude D.

Before we can use functions such as $y = \sin A$ or $y = 3 \sin A$ to represent the motion of a bob on a spring, we must clear one more hurdle. The func-

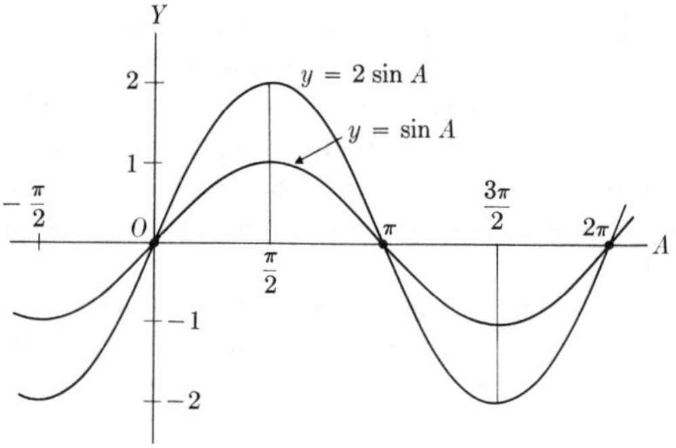

Fig. 18–10. Comparison of $y = \sin A$ and $y = 2 \sin A$.

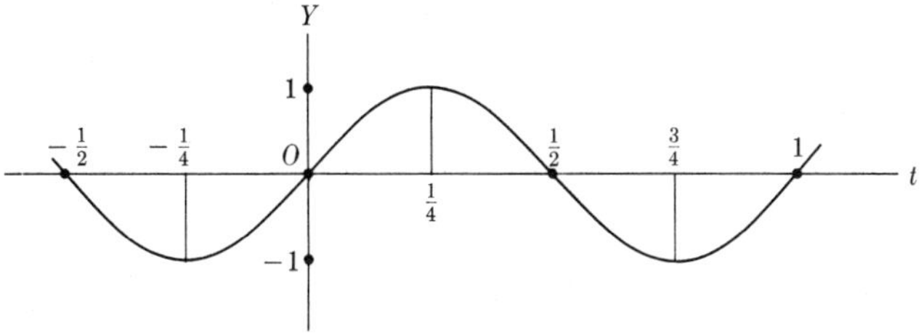

Fig. 18–11. The graph of $y = \sin 2\pi t$.

tion we seek should represent a relationship between displacement and time. The y-values of our functions do indeed represent the displacement of a point Q which moves up and down on a line, but our independent variable is an angle. Suppose, however, that the point P revolves around the circle f times in one second. Then, since for each revolution the angle A increases by 2π radians, the size of the angle which describes the amount of revolution of P in one second is $2\pi f$. If the point P revolves for t seconds and makes f revolutions per second, it will make ft revolutions in t seconds. The angle generated during these ft revolutions will be $2\pi ft$. Hence, the value of A in t seconds will be $2\pi ft$. Thus the function $y = \sin A$ becomes

$$y = \sin 2\pi ft. \tag{6}$$

This function requires some study. Suppose the point P makes one revolution per second. Then $f = 1$. The function (6) then is $y = \sin 2\pi t$. As t increases from 0 to 1, the quantity $2\pi t$ will increase from 0 to 2π. We must now ask, How will $\sin 2\pi t$ vary as $2\pi t$ varies from 0 to 2π? Since the *angle* which $2\pi t$ describes now varies from 0 to 2π, the function will go through the entire cycle of sine values. However, if we now label our horizontal axis with time values, we obtain the graph shown in Fig. 18–11.

Next let us consider a slightly more difficult case. Suppose the point P makes 2 revolutions per second so that $f = 2$. As t increases from 0 to $\frac{1}{2}$, $2\pi \cdot 2t$ will increase from 0 to 2π and $\sin 2\pi \cdot 2t$ will go through the entire cycle of sine values. As t increases from $\frac{1}{2}$ to 1, $2\pi \cdot 2t$ increases from 2π to 4π. Then $\sin 2\pi \cdot 2t$ takes on the values corresponding to angles from 2π to 4π. But in this range the sine function takes on the same values as it does in the range from 0 to 2π. Hence in the entire interval 0 to 1 for t, the graph will be as shown in Fig. 18–12. The conclusion, which emerges clearly from the graph, is that

$$y = \sin 2\pi \cdot 2t$$

goes through 2 complete cycles in one second or, as one says, it has a *frequency* of 2 cycles per second.

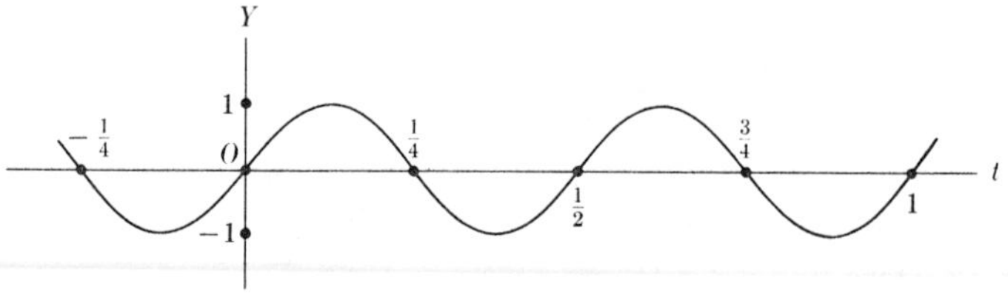

Fig. 18–12. The graph of $y = \sin 2\pi \cdot 2t$.

We can now anticipate what happens for any f. The function

$$y = \sin 2\pi ft$$

will go through f cycles in one second, or it has a frequency of f cycles per second.

To increase the amplitude of any of these functions, we have but to introduce the factor D. Thus the function

$$y = D \sin 2\pi ft \tag{7}$$

will have a frequency of f cycles per second and an amplitude of D. Let us note that while f is the number of revolutions per second of P, it is also the number of oscillations per second of Q.

The y-values of formula (7) oscillate above and below the zero value as t varies. We can make the number of oscillations per second what we please by merely inserting the proper value of f and we can do the same with respect to the amplitude by inserting the proper value of D. Of course, we do not know the proper values of f and D which fit the motion of a bob, but we shall see in the next section that it is not difficult to determine them.

Let us summarize what we have accomplished. We sought to represent the motion of a point which oscillates back and forth on a straight line. We were able to do so by introducing a point Q which is the projection of a point P moving around a circle at a constant velocity. Because the y-value of P equals the displacement of the oscillating point Q and because the y-value of P is expressible as a sinusoidal function, we can represent the motion of the oscillating point Q by such a function. That the approach to the oscillating point Q through the circle should be successful may be surprising, but, as Aristotle pointed out, "There is nothing strange in the circle being the origin of any and every marvel."

EXERCISES

1. Find the value of $2 \sin A$ when A is $30°$, $90°$, $\pi/2$, $\pi/3$.
2. What is the maximum value of $3 \sin A$? the minimum value?
3. What is the amplitude of $y = 4 \sin A$?

4. Find the value of

a) $\sin 2t$ when $t = \pi/4, \pi/2, 3\pi/4, \pi$;

b) $\sin 3t$ when $t = \pi/6, \pi/3, \pi/2, 2\pi/3$.

5. What is the shape of the graph of $y = \sin 2\pi \cdot 2t$ as t varies from 1 to 2?
6. Graph the function $y = \sin 2\pi \cdot 3t$ as t varies from 0 to 1.
7. Graph the function $y = 2 \sin 2\pi \cdot 2t$ as t varies from 0 to 1.
8. What is the frequency (in one second) of $y = \sin 2\pi \cdot 10t$?
9. Find the value of

a) $y = \sin 2\pi \cdot 2t$ when $t = \frac{1}{8}, \frac{1}{4}, \frac{1}{3}$;

b) $y = \sin 2\pi \cdot 4t$ when $t = \frac{1}{6}, \frac{1}{2}, 1$;

c) $y = 2 \sin 2\pi \cdot 3t$ when $t = \frac{1}{6}, \frac{1}{4}, \frac{1}{12}$.

18-4 ACCELERATION IN SINUSOIDAL MOTION

What we have seen in the preceding article is that if a point P moves around a circle of unit radius at a constant speed and makes f revolutions per second, then the projection Q of P onto the vertical diameter moves up and down this diameter, and the displacement y of Q from its central position at O can be represented by formula (6), namely,

$$y = \sin 2\pi ft. \tag{8}$$

Moreover, if we wished to represent the same kind of oscillatory motion but with an amplitude D instead of 1, we have merely to modify (8) to read

$$y = D \sin 2\pi ft. \tag{9}$$

Our goal, however, is to represent the motion of the bob on the spring. Before we can do this we must learn one more fact about the motion of Q, namely, the acceleration of Q. The motion of Q, as we approached it, was determined by the motion of P which travels around the unit circle at a constant speed, say v. If an object moves along a circular path, then we know from our work in Chapter 15 that it must be subject to a centripetal acceleration, and by formula (24) of Chapter 15 this centripetal acceleration is v^2/r, where r is the radius of the circle. In our case, since P moves on a circle of unit radius, $r = 1$, and so the centripetal acceleration of P is v^2. This acceleration is directed toward the center of the circle.

We are, however, interested not in the motion of P, but in the motion of Q, which moves in the same way as the vertical motion of P. Hence we should seek the vertical acceleration of P. We learned in Chapter 14 that even though an object moves along a curve, we can study its motion by considering the horizontal motion and the vertical motion separately. By Galileo's principle these two motions are independent of each other. What we should like then is the vertical acceleration of P. When we considered the motion of a shell shot from a cannon inclined at an angle A to the ground, we found the horizontal and vertical velocities by dropping perpendiculars from the end point of the line segment representing velocity onto the horizontal and vertical axes (Section 14–4).

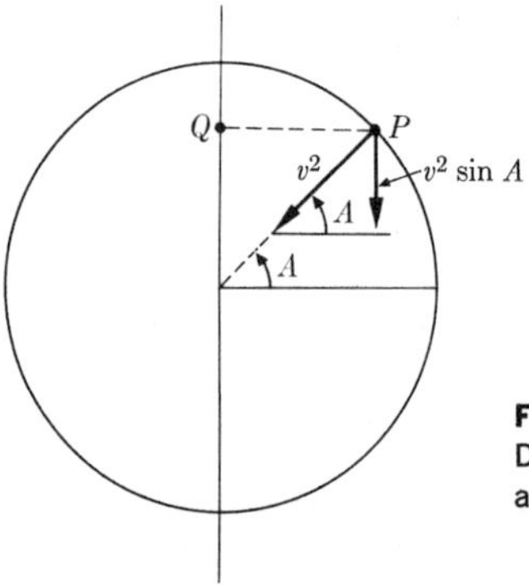

Fig. 18–13. Determination of the vertical acceleration of P.

Now acceleration, like velocity, is a directed quantity or a vector. Moreover, the acceleration determines the velocity. Hence we should compute the vertical acceleration of the point P in the same manner as we computed the vertical component of the velocity of the shell. Figure 18–13 shows the centripetal acceleration v^2 of P as a line segment directed toward the center of the circle. If we drop a perpendicular from the end point of this line segment onto the vertical line through P, we obtain the vertical component of the acceleration. Angle A in the figure determines the position of P. We see, then, that the vertical component of the acceleration, which we shall denote by a, is

$$a = v^2 \sin A.$$

We know, however, from (1) that

$$\sin A = y,$$

where y is the ordinate of P. Hence

$$a = v^2 y.$$

However, since the acceleration is directed downward when y is positive, we must write

$$a = -v^2 y. \tag{10}$$

If now the moving point P makes f revolutions per second, then P covers f

circumferences per second; that is, $v = 2\pi f$ and $v^2 = 4\pi^2 f^2$. We substitute this result in (10) and obtain

$$a = -4\pi^2 f^2 y \tag{11}$$

as the acceleration of the vertical motion of P or of its shadow Q on the Y-axis.

What we have shown then is that the motion of Q, which is described by the formula

$$y = \sin 2\pi ft, \tag{12}$$

is subject to an acceleration of

$$a = -4\pi^2 f^2 y. \tag{13}$$

18-5 THE MATHEMATICAL ANALYSIS OF THE MOTION OF THE BOB

We may now undertake to represent mathematically the motion of the bob on the spring. We know that this motion is periodic and has a definite frequency and amplitude. However, we do not know that the motion is really sinusoidal; that is, as t varies, do the displacements of the bob follow precisely the variation of y in a function of the form

$$y = D \sin 2\pi ft? \tag{14}$$

If, for example, the motion of the bob should be faster on the upper half of its path than on the lower half, it could still have the same period for each complete oscillation and perform a fixed number of oscillations per second. Yet the motion would not be of the form (14). We need a little more insight into the motion of the bob than we now have.

This insight into the action of bobs on springs was supplied by Robert Hooke. The principle he discovered, still known as Hooke's law, is very simple. We all know that if we stretch or compress a spring, the spring seeks to restore itself to its normal length; that is, when stretched or compressed the spring exerts a force. Hooke's law says that the force is a constant times the amount of compression or extension. In symbols, if L is the increase or decrease in length of the spring and F is the force exerted by the spring, then $F = kL$, where k is a constant for a given spring. The quantity k is called the spring or stiffness constant and it represents the stiffness of the spring. If k is large, the spring exerts considerable force even for small L.

We shall now see what we can deduce from Hooke's law. Suppose that a bob of mass m is attached to a spring. Then we know that gravity pulls the bob downward some distance d where the bob comes to rest (Fig. 18–14). The rest position is reached when the force of gravity acting on the bob, or the weight of the bob, just offsets the upward force exerted by the spring. Now the force of gravity is $32m$, and, according to Hooke's law, the upward force exerted by a spring which is pulled downward a distance d is kd. Since

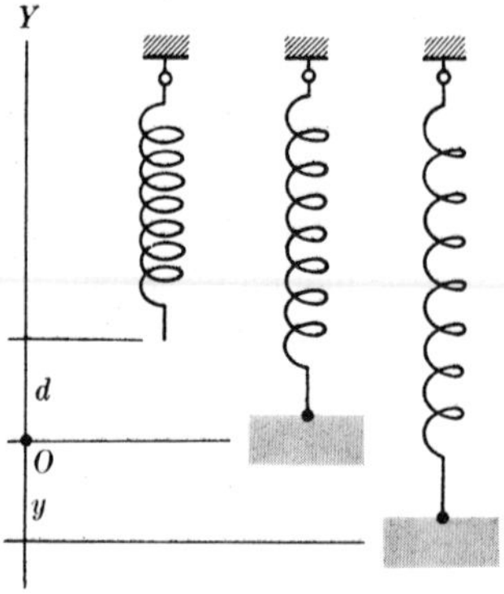

Fig. 18-14.
A bob on a spring in the rest position (center) and pulled down a distance *y* (right).

at the rest position these two forces just offset each other, we have

$$32m = kd. \tag{15}$$

Now suppose the spring is pulled downward an additional distance y. If we use the convention agreed upon in Section 18–2 that displacements above the rest position are to be positive and below the rest position negative, then the total extension of the spring is now $d - y$ because y itself is negative. The force that the spring exerts in an upward direction is, by Hooke's law,

$$k(d - y) \quad \text{or} \quad kd - ky. \tag{16}$$

However, the weight of the bob, or $32m$, exerts a constant downward force. Hence the net upward force is $kd - ky - 32m$. In view of equation (15) the *net* upward force is $-ky$. We now apply Newton's second law of motion, which says that when a force is applied to a mass, the force equals the mass times its acceleration. Thus we have

$$ma = -ky \tag{17}$$

or, by dividing both sides of this equation by m,

$$a = -\frac{k}{m}y. \tag{18}$$

Formula (18) is the basic law governing the motion of the bob on the spring. For, suppose the bob is pulled down some distance and then released. The spring exerts a force which pulls the bob back toward its rest position. The acceleration created by this force is precisely that given by (18). The acceleration now determines the velocity of the bob and the velocity determines the distance covered by the bob in any specified interval of time. The argument we are presenting here is in principle the same as the one we used in Chapter 13, where we discussed the motion of a body which is raised some distance from the surface of the earth and then dropped. In this situation, gravity immediately exerts an acceleration of 32 ft/sec², and thereafter the velocity and distance fallen by the object are determined. Of course, in the present case the acceleration is a more complicated expression, and the subsequent motion is not simply in one direction, but the argument is of the same nature.

We should now compare formula (13), namely,

$$a = -4\pi^2 f^2 y \tag{19}$$

and formula (18),

$$a = -\frac{k}{m} y. \tag{20}$$

In both cases the acceleration is a constant times the displacement. The constant is $4\pi^2 f^2$ in the former case and k/m in the latter. When the acceleration is given by (19), the motion itself [see (12) and (13)] is represented by

$$y = \sin 2\pi f t. \tag{21}$$

Since the acceleration (20) is precisely of the same form as (19), except for the label of the constant, and since the acceleration determines the motion, the motion of the bob must also be representable by a formula of the form (21).

However we do not know what f is in the case of the bob. But k/m in the case of the bob plays the role of $4\pi^2 f^2$ in the case of the motion of the point Q. That is

$$4\pi^2 f^2 = \frac{k}{m},$$

so that

$$f^2 = \frac{1}{4\pi^2} \cdot \frac{k}{m}$$

and

$$f = \sqrt{\frac{1}{4\pi^2} \cdot \frac{k}{m}} = \sqrt{\frac{1}{4\pi^2}} \sqrt{\frac{k}{m}}$$

or

$$f = \frac{1}{2\pi} \sqrt{\frac{k}{m}}. \tag{22}$$

In other words, if we let f in (21) be the value given by (22), we can write the formula for the bob's motion in terms of the quantities k and m. Thus if we substitute this value of f in (21), we obtain

$$y = \sin 2\pi \left(\frac{1}{2\pi}\sqrt{\frac{k}{m}}\right) t$$

or

$$y = \sin \sqrt{\frac{k}{m}}\, t. \tag{23}$$

We made one misleading statement in the preceding discussion. We said that the acceleration of the bob determines the motion of the bob, and so the formula of the bob's motion must be of the form (21). The acceleration does determine the essential characteristics of the motion, but the initial velocity and initial displacement do have some effect. This point may become clearer if we compare the present case with the vertical motion of objects. All bodies rising or falling near the surface of the earth are subject to an acceleration of 32 ft/sec^2, and this fact determines the essential nature of the motion. But if a body is thrown up into the air, the final formula depends also on the initial velocity given to the object and on the position from which it is thrown up. In the case of the bob, if it is pulled down to a distance D below the rest position and then released, this initial position must enter into the formula for the motion. To complete our determination of the formula we are obliged, with the limited mathematics at our disposal, to call upon observation, which tells us that in each oscillation the bob will rise to a height of D above the rest position and then descend a distance D below it. That is, the amplitude of the motion, D, is determined by the initial displacement. Thus the final formula for the motion of the bob is

$$y = D \sin \sqrt{\frac{k}{m}}\, t. \tag{24}$$

We can now draw several conclusions about the motion of the bob. Formula (22) gives the frequency of the bob's motion in terms of k and m. We see that the spring constant k, which represents the stiffness of the spring, and the mass m of the bob determine the frequency of the motion. If we wished to have the bob make, for example, two complete oscillations per second, we could pick values of k and m so that f in (22) should be 2. The period of the bob's motion, that is, the time required to make one oscillation, is

$$T = \frac{1}{f} = \frac{2\pi}{\sqrt{k/m}} = \frac{2\pi}{\sqrt{k/m}}\frac{\sqrt{m/k}}{\sqrt{m/k}} = \frac{2\pi\sqrt{m/k}}{\sqrt{(k/m)\cdot(m/k)}}$$

or

$$T = 2\pi\sqrt{\frac{m}{k}}. \tag{25}$$

As Hooke observed, this formula for the period is immensely significant. The period is independent of the amplitude of the motion; that is, whether one pulls the bob down a great distance or a short distance and then releases it, the time for the bob to go through each complete oscillation will be the same.

This fact is immensely useful. At the very outset of our treatment of the bob's motion we pointed out that the resistance of the air and internal energy losses in the spring will cause the motion to die down. At the time we decided to ignore this fact and to suppose that there was no loss of energy. But there is. Suppose, however, that we were to give the bob a little upward push every time it reached its lowest position, i.e., add energy to the motion and keep the bob moving. Such an action might alter the amplitude, but would *not* affect the period, and each successive oscillation of the bob would therefore continue to take the same amount of time. Hence the oscillations of the bob on the spring can be used to measure time or to regulate the motion of some hands on a dial which would show time elapsed.

Of course, the motion of a bob on a spring is not quite the practical device for a clock. The device actually used can be found in every modern pocket or wrist watch. There a spring coiled in a spiral and carrying a weight called the balance wheel expands and contracts regularly. Each second the wheel is given a little "kick" which restores the energy the spring loses on each oscillation. (The energy comes from a mainspring which is wound up by hand usually once a day.) The spiral spring regulator was invented and patented by Christian Huygens in 1675. The first chronometer which was sufficiently accurate to be used by ships to determine longitude was invented by John Harrison, who in 1772 won a prize of £20,000 offered by the British government for such a device.

EXERCISES

1. If a mass of 2 lb pulls a spring down 6 in., what is the spring constant? [*Suggestion:* Use (15)]
2. Suppose that one attaches a mass of 2 lb to a spring whose stiffness constant is 50. Calculate the number of oscillations per second which the mass would make if set into vibration.
3. What is the period of a mass vibrating at a rate of 100 oscillations per second?
4. Suppose a mass is set to vibrating on a spring at the rate of 50 oscillations per second. If the mass has a maximum displacement of 3 in., what formula describes the motion?
5. Suppose a mass of 3 lb is attached to a spring whose stiffness constant is 75. The mass is pulled down 3 in. below the rest position and then released. Write a formula relating displacement and time.
6. Suppose that you are given a spring with a stiffness constant of 50. Calculate the mass that you would have to place on the spring to produce a period of oscillation of one second.

7. Suppose that a mass oscillates on a spring so that the relation between displacement and time is

   a) $y = 4 \sin 2\pi \cdot 5t$ b) $y = 4 \sin 10t$.

   Describe the motion of the mass for (a) and (b).

8. Suppose that you wished to decrease the number of oscillations per second which a mass makes on a given spring. How would you alter the mass?

9. Suppose that a tunnel is dug through the earth and a man of mass m steps into the tunnel. Inside the earth the force of gravity on a mass m at a distance r from the center is $F = GmMr/R^3$, where M is the mass and R is the radius of the earth. This force is directed toward the center. To distinguish distances above and below the center, let r be positive above and negative below the center. Then the acceleration acting on the mass m is $a = -GMr/R^3$. Discuss the subsequent motion of the man (Fig. 18–15).

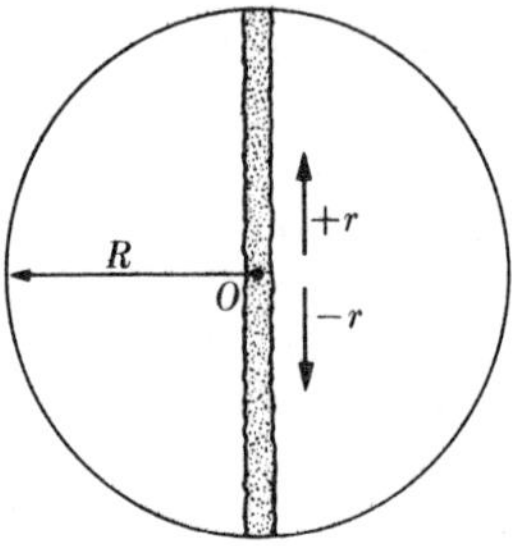

Fig. 18–15

18-6 SUMMARY

The mathematical objective of this chapter was to introduce a new type of mathematical function, the sinusoidal function. There is not just one sinusoidal function, for all functions of the form $y = D \sin 2\pi ft$, no matter what D and f may be, are sinusoidal. The sinusoidal functions are also called trigonometric functions because they are obtained by extending the concept of the sine of an angle, a concept which was first created and studied in trigonometry. Other trigonometric functions can be derived from an extension of the concepts of cosine and tangent of an angle and other trigonometric ratios which we did not study. All trigonometric functions are highly useful in scientific work.

The creation of trigonometric functions was motivated by the study of vibratory or oscillatory motion. We have used the motion of a bob on a spring to illustrate such a motion, and we have shown how the mathematical description of this motion can be used to deduce information about it. We have yet to see some of the major uses of sinusoidal functions.

Topics for Further Investigation

1. The mathematics of pendulum motion.
2. The trigonometric function $y = \cos A$.

Recommended Reading

BROWN, LLOYD A.: "The Longitude," in James R. Newman: *The World of Mathematics*, Vol. II, pp. 780–819, Simon and Schuster, Inc., New York, 1956.

KLINE, MORRIS: *Mathematics and the Physical World*, Chap. 18, T. Y. Crowell Co., New York, 1959. Also in paperback, Doubleday and Co., New York, 1963.

RIPLEY, JULIEN A., JR.: *The Elements and Structure of the Physical Sciences*, Chap. 15, John Wiley and Sons, Inc., New York, 1964.

TAYLOR, LLOYD WM.: *Physics, The Pioneer Science*, Chap. 15, Dover Publications, Inc., New York, 1959.

WHITEHEAD, ALFRED N.: *An Introduction to Mathematics*, Chaps. 12 and 13, Holt, Rinehart and Winston, Inc., New York, 1939.

THE TRIGONOMETRIC ANALYSIS OF MUSICAL SOUNDS

Motion appears in many aspects—but there are two obvious kinds, one which appears in astronomy and another which is the echo of that. As the eyes are made for astronomy so are the ears made for the motion which produces harmony: and thus we have two sister sciences, as the Pythagoreans teach, and we assent.

PLATO

19-1 INTRODUCTION

In this chapter we intend to show how trigonometric functions have given man his first real insight into the nature of musical sounds, and how this knowledge is utilized in the design of such devices as the telephone, the phonograph, the radio, and sound films.

The mathematical study of musical sounds did not start with the application of trigonometric functions. Indeed, it goes back to the very first emergence of any real mathematics and science, namely the beginning of the classical Greek period. For example, the Pythagoreans discovered that the lengths of two equally taut plucked strings whose sounds harmonize are related by simple numerical ratios such as 2 to 1, 4 to 3, and 3 to 2. The lower note in each case originates with the longer string. They also designed musical scales whose notes, as measured quantitatively by the lengths of the vibrating strings, possessed precise numerical values. From Pythagorean times onward, mathematicians and scientists were convinced that musical sounds had important mathematical properties, and music, along with arithmetic, geometry, and astronomy, became part of the quadrivium. These four subjects were studied together right through the medieval period. Although Greek, Arab, and medieval mathematicians continued to investigate musical sounds and wrote books on music, their work was essentially limited to the construction of new systems of scales for instrumental and vocal music.

It was the mathematicians and scientists of the seventeenth century who initiated other investigations and made the next series of important discoveries. Familiar names, such as Galileo, his French pupil and colleague Father Marin Mersenne (1588–1648), Hooke, Halley, Huygens, and Newton, obtained significant new results. Whereas the Pythagoreans had studied strings of different length but equal tension, Mersenne studied the effect of changing tension and mass of a string and found that an increase in mass and a decrease in tension produce lower notes in a string of given length. This discovery was very important for stringed instruments such as the violin and the piano; to secure the range of pitch which these instruments possess by variations in length only would require exceedingly long strings. Galileo and Hooke demonstrated experimentally that each musical sound is characterized by a definite number of air vibrations per second, a statement which will mean more to us in a few moments. The determination of the velocity of sound (about 1100 feet per second in air) was another achievement. It is of interest that the clocks which some of these men designed and constructed were essential to the progress made in the study of sound because, as we can see from the results cited, the ability to measure small intervals of time was an indispensable condition for any work in this field.

The best mathematicians of the eighteenth century, Leonhard Euler, Daniel Bernoulli (1700–1782), Jean le Rond d'Alembert (1717–1783), and Joseph Louis Lagrange, studied vibrating strings, such as the violin string, and vigorously disputed whether trigonometric functions were adequate to represent the vibrations. The mathematical analysis of sound waves soon followed and proved to be the chief tool in the theoretical mastery of musical sounds. We can readily see why mathematics was invaluable in these investigations, for observation of the air, even of air in the process of propagating sound, reveals nothing.

Before we undertake to study just what the nineteenth-century mathematicians and scientists learned, we must make some distinctions. The first is a matter of terminology. We shall be interested in the analysis of musical sounds as opposed to noise. However, in the present context, the term "musical sound" is used in a technical sense and includes not only those sounds commonly understood to be music, but also the sounds of ordinary speech. As a matter of fact, the physicist's meaning might be more appropriately represented by the term *intelligible sound*. Just what is meant by either phrase will be clear in a few moments.

The second distinction one must make is between sound as a motion of air and sound as a sensation which human beings experience. The former is a physical phenomenon which takes place in space and whose physical and mathematical properties are fixed. On the other hand, the sensations which human beings may receive because moving air strikes their ears and stimulates certain nerves depend upon their auditory mechanism and may vary from one person to another. There are, for example, physical sounds which humans cannot hear at all. Though we shall have something to say about the perception of sound, our first and main concern will be to understand the physical phenomenon.

19-2 THE NATURE OF SIMPLE SOUNDS

The variety of sounds given off by musical instruments, the human voice, phonographs, radios, and whirring machinery, for example, is so great that one cannot hope to study all of them in one swoop. Hence it would seem wise to start one's investigation with simple sounds. But which sounds are simple? If we rely upon our ears to decide this question, then the sounds given off by tuning forks seem to be simple. The ear may, indeed, be deceived here, but let us follow up this suggestion.

If either prong of a tuning fork is struck, both prongs will move inward and then outward very rapidly and will repeat this motion for a long time. Let us consider one prong, say the right one shown in Fig. 19–1. Before the prong is struck, it occupies what we might call the rest position. After being struck, the tip is displaced some distance to the right. It then moves to the left, to a position somewhat to the left of the rest position, and then moves to the right. The sequence then repeats itself many times. The displacement of the tip varies with time, and the first question one might raise is, What is the relationship between displacement and time? There are two considerations which suggest that the formula is sinusoidal: First of all, the prong resembles a spring-and-bob arrangement. The spring is the prong itself, though the motion is a sidewise oscillation rather than an expansion and contraction. The mass which corresponds to the bob is the mass of the prong itself, though admittedly this mass is not concentrated in one place as it is in the case of a bob on a spring. The second consideration is that as the tip of the prong moves farther and farther out from the rest position, the force which the prong exerts to return to the rest position may be expected to increase with the displacement. The simplest assumption one might make in this case is that the force increases directly with the displacement. From formula (17) of the preceding chapter we can see that this is indeed the mathematical law which underlies and determines the sinusoidal motion of the bob. Hence it seems reasonable to expect that the relation between displacement of the tip of either prong and time is sinusoidal. The amplitude of this relation is the maximum displacement of the tip, and the frequency is the frequency per second with which the prong oscillates.

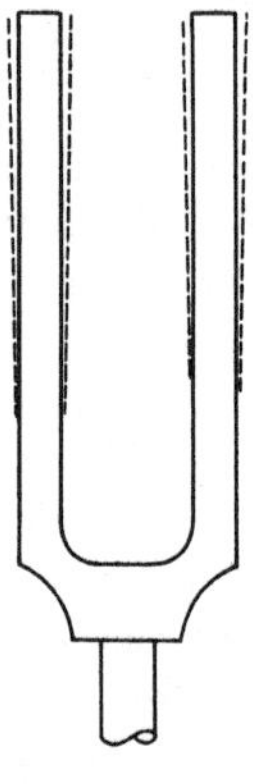

Fig. 19–1. A vibrating tuning fork.

Of course, we are not so much interested in the motion of the tuning fork as we are in the sound it creates. Hence what matters next is, How does the air respond to the vibration of the tuning fork? The fundamental fact about the behavior of air which is of importance in this connection is that air pressure seeks to become uniform everywhere. This means that if the air

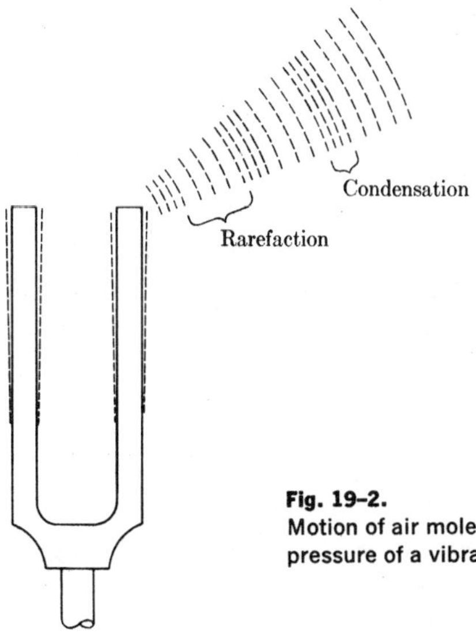

Fig. 19–2.
Motion of air molecules under pressure of a vibrating tuning fork.

pressure for any reason should become high in one place, the air will spread out from that place into neighboring regions where the pressure is lower and so try to equalize the pressure in the entire region under consideration. With this physical fact in mind, let us see what happens when the right prong of the tuning fork moves, say to the right. The prong pushes the molecules of air near it to the right and thus crowds them into a place occupied by other molecules. The pressure becomes high in this place, and since the molecules of air cannot move to the left because the prong is there, they will move off farther to the right (and in other directions) in order to equalize the pressure. But this motion means that the crowding now occurs a little farther away from the tuning fork and again, to equalize the pressure, the molecules move farther to the right. The process continues, and the crowding, or *condensation*, as it is usually called, moves off to the right.

The prong, having moved as far to the right as it can, will now move back not only to its rest position but farther to the left. This motion leaves an empty region—the place that the prong had occupied—and so the molecules of air on the right rush into this empty space. Molecules still farther to the right also move to the left because the pressure has become less to their left. Thus a state of low pressure, or *rarefaction*, as it is called, moves to the *right* as molecules move to the left to equalize the pressure in their neighborhood. With each successive vibration of the prong, a condensation and rarefaction move off to the *right* (Fig. 19–2). The successive condensations and rarefactions also move out in other directions, but it is sufficient for our purposes to follow what happens in one direction.

The action of the air is somewhat complicated because it consists of billions of molecules, and they do not all behave in exactly the same way. However, there is an average effect. It is convenient to speak of a series of typical molecules to the right of the prong which represent the average behavior of the entire collection. If we consider the action of any one typical molecule, say one near the prong, then what it does is to move to the right when the prong moves to the right. When the prong moves to the left, the typical molecule will also move to the left because the air pressure has been lowered. Like the prong, it will move past its rest position, and continue to the left. Then, as the prong moves to the right, the molecule will be pushed to the right, will pass its rest position, go farther to the right, and, from this time on, it will continue to oscillate.

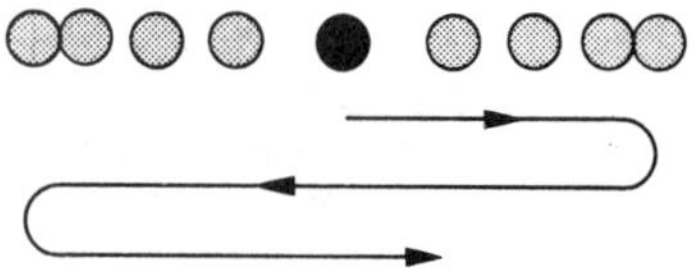

Fig. 19–3.
Motion of a typical molecule.

Typical molecules farther to the right will behave like the typical molecule near the prong; however, their reaction will be slightly delayed since condensations and rarefactions reach them a little later. Figure 19–3 illustrates the motion of a typical molecule reacting to a series of prong oscillations.

Two important facts emerge from the above discussion. The first is that the average, or typical, molecule follows, in effect, the motion of the prong. Any one molecule acts as if it were attached to the prong by a spring. When the prong moves to the right, it contracts the spring. The latter seeks to restore its length and so pushes the molecule to the right. While the molecule moves to the right, the prong moves to the left, and hence the spring is extended. It now seeks to contract and so pulls the molecule to the left. The molecule moves to the left, and the spring contracts. But now the prong is ready to move to the right, and consequently the motion of prong and molecule repeats itself. The action of air pressure is indeed like the action of the spring. In fact, Hooke used the phrase, "the spring of the air," to describe the effect of air pressure.

The second fact is that the sound wave which moves from the prong to some person's ear, say, consists of the series of condensations and rarefactions induced by the prong's motion. Each molecule merely oscillates about its rest position, but in doing so it produces the increase and reduction of pressure which cause the neighboring molecules to oscillate.

The nature of the sound wave may perhaps be made clearer by comparing it with a water wave. If the end of a stick is quickly moved back and forth in still water, a series of waves will spread out from the end of the stick.

However, the individual water molecules do not move out. Each oscillates about its original position, but the increase and decrease in pressure which the stick creates cause the molecules farther away to duplicate the motion of the molecules near the stick.

Since the motion of any typical molecule whether near or far from the prong is the same, let us study the motion of any one of these molecules. Specifically, let us seek the relationship between the displacement from rest position and the time that the molecule is in motion. What formula relates displacement and time? We have already produced two crude physical arguments suggesting that for the prong, displacement and time are related by a sinusoidal formula. Since the motion of any typical air molecule duplicates the action of the prong, the formula which relates displacement and time for the typical air molecule should also be sinusoidal. Actually these physical arguments do not really prove that the formula is sinusoidal. However, this fact can be established either by a rather complex mathematical analysis of air motion or, experimentally, by converting the air pressure to electric current (by means of a microphone, for example) and by then displaying the current on a cathode-ray tube (television tube).

We shall take for granted that for a typical air molecule, the formula relating displacement and time is sinusoidal. Hence, if y is the displacement and t is the time, then, in view of what we have learned in the preceding chapter, the formula is

$$y = D \sin 2\pi f t, \tag{1}$$

where D is the amplitude, or maximum displacement, and f is the number of oscillations, or the frequency of cycles per second. We wish to emphasize that the formula applies to the sounds produced by tuning forks or to what we have reason to believe are simple sounds.

To use formula (1), we must know D and f. The value of f is the frequency with which the tuning fork oscillates. A frequency commonly used to standardize the pitch of sounds is 440 per second. This then is a typical value of f. The value of D, the amplitude of the motion of a typical air molecule, is *not* the amplitude of the prong's motion, but depends upon the medium in which sound spreads out or is propagated. It depends, so to speak, on the "springiness" of the medium. In air, 0.001 inch can be considered to be a reasonable value for D. Hence a typical formula for a simple sound is

$$y = 0.001 \sin 2\pi \cdot 440t. \tag{2}$$

Thus, a typical air molecule oscillating in accordance with formula (2) shuttles back and forth about its mean, or rest, position 440 times per second or, as we say, it goes through 440 complete cycles in one second. The farthest distance from the mean position that it reaches, that is, the amplitude of its motion, is 0.001 inch.

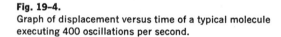

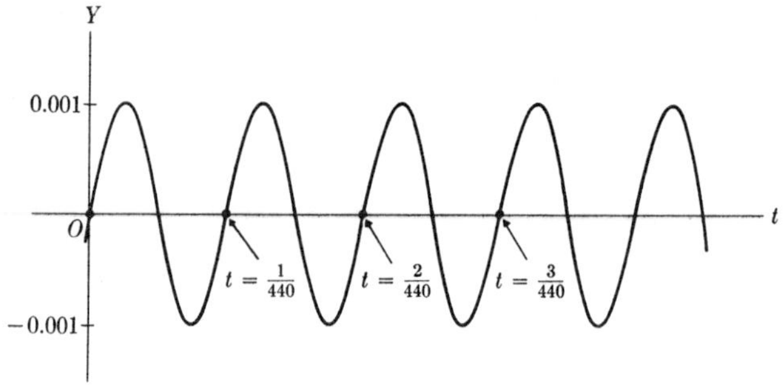

Fig. 19–4.
Graph of displacement versus time of a typical molecule executing 400 oscillations per second.

Figure 19–4 illustrates the relationship between displacement and time for a simple sound such as formula (2) represents. Of course the typical molecule shuttles backward and forward, but on the graph its displacements are plotted as ordinates and the time elapsed is shown by the corresponding abscissas.

Although formula (2) represents only simple sounds—we have yet to discuss the formulas describing more complicated sounds—it enables us to understand what we meant earlier by the phrase "intelligible sounds." We see that a simple sound has a regularity or periodicity. The motion of the air molecules repeats itself a number of times a second. When the ear receives many cycles of this motion, it can identify the sound. If, on the other hand, the motion of the air molecule is not regular but varies irregularly with time, the ear still hears sound, but sound that does not convey any meaning, i.e., noise.

EXERCISES

1. What is the basic mathematical formula which represents simple sounds? State the physical meaning of the various letters in the formula.
2. State the formula which describes the relationship between displacement and time for a simple sound whose frequency is 300/sec and whose amplitude is 0.0005 in.
3. If $y = 0.002 \sin 2\pi \cdot 540t$ is the mathematical description of a sound, what are the frequency and amplitude of this sound?
4. If a sound has a frequency of 400 cycles/sec, how many cycles would the ear receive in 1/20 sec?

19–3 THE METHOD OF ADDITION OF ORDINATES

We have now a good mathematical representation of simple sounds. But interesting musical sounds, whether vocal or instrumental are, as a rule, not simple, and the really significant contribution of mathematics to the understanding of musical sounds lies in the analysis of more complex sounds. To comprehend this contribution, we must first examine a relevant mathematical idea. Instead of considering simple sinusoidal functions such as (2), let us take the function

$$y = \sin 2\pi t + \sin 4\pi t. \tag{3}$$

What sort of relationship between y and t does formula (3) represent?

A good way to investigate this question is to draw a graph of the above function. Since we wish to obtain merely some general idea of how y varies with t, we shall seek only a sketch rather than a very accurate graph. We could proceed by selecting values of t, calculating the corresponding values of y, and then plotting the points whose coordinates have thus been determined. However, there is a quicker method which is also more perspicuous. Let us consider the two functions:

$$y_1 = \sin 2\pi t \tag{4}$$

and

$$y_2 = \sin 4\pi t. \tag{5}$$

We have used the notation y_1 and y_2 to distinguish the dependent variables in (4) and (5) from the y in formula (3). Formulas (4) and (5) are easily graphed. Formula (4) is the ordinary sine function which goes through the regular cycle of sine values in each unit of t. Formula (5) has a frequency of 2 in each unit of t; that is, the y-values go through the complete cycle of sine values twice in each unit of t. Let us sketch both functions on the same set of axes (Fig. 19–5).

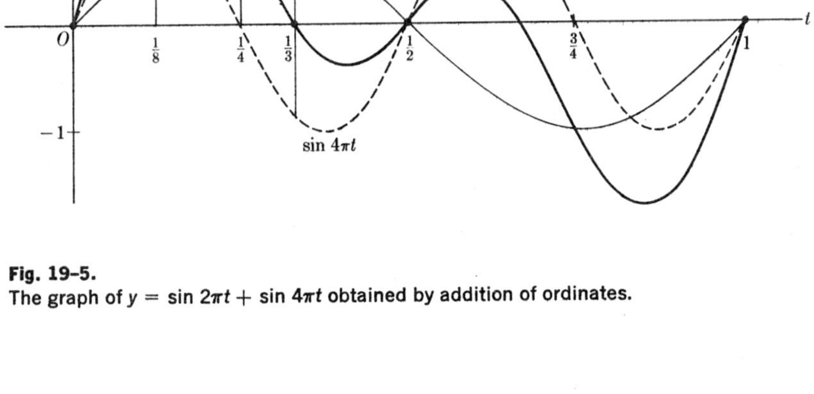

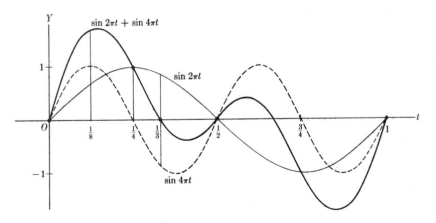

Fig. 19–5.
The graph of $y = \sin 2\pi t + \sin 4\pi t$ obtained by addition of ordinates.

Now the y of formula (3) is clearly the sum of y_1 and y_2. Hence adding the values of y_1 and y_2 at various values of t will yield y. Since we are interested only in a sketch, let us perform the addition by using Fig. 19–5 to obtain the values of y_1 and y_2. Thus for $t=0$, the graphs show that y_1 and y_2 are both zero. Hence y, the sum of y_1 and y_2, also is zero. At $t=\frac{1}{8}$, we see from the graph that y_1 is about 0.7 and y_2 is 1. Hence $y=1.7$ when $t=\frac{1}{8}$. At $t=\frac{1}{4}$, we find that $y_1=1$ and $y_2=0$. Hence $y=1$ when $t=\frac{1}{4}$. At $t=\frac{1}{3}$, y_1 is about 0.85 and y_2 is about -0.85. In adding the last two values for y_1 and y_2, we must take into account that one is positive, the other negative, and their sum zero. Hence at $t=\frac{1}{3}$, $y=0$. By selecting a few more values of t and estimating the corresponding y_1- and y_2-values, we can obtain more y-values. Finally, we join the various points which belong to the graph of formula (3) by a smooth curve. The result is the heavy-lined curve shown in Fig. 19–5. The method just described for graphing y as a function of t provides a rough sketch. If one wishes to obtain a more accurate graph, he can calculate the value of y for each value of t.

How far need we carry this process of determining y-values corresponding to various t-values? We note that the function $y_1=\sin 2\pi t$ repeats itself when t becomes larger than 1. The function $y_2=\sin 4\pi t$ goes through two full cycles in the interval from $t=0$ to $t=1$ and begins its third cycle of sine values as soon as t increases beyond 1. Thus at $t=1$ both functions begin to repeat the values which they had taken on at $t=0$ and, in the interval from $t=1$ to $t=2$, both functions will repeat the behavior exhibited in the interval from $t=0$ to $t=1$. Since y_1 and y_2 repeat their former behavior, it follows that y, which is the sum of y_1 and y_2, will also repeat its former behavior. In other words, in the interval from $t=1$ to $t=2$, y will behave precisely as it did in the interval from t=0 to t=1. And in each succeeding unit interval of t-values, the function will repeat the behavior exhibited in the interval from $t=0$ to $t=1$. If we therefore determine the behavior of y in the interval from 0 to 1, we know how it behaves for all larger values of t.

There are several major facts to be learned from this example. First of all, since the function (3) repeats its behavior in every unit of t-values, it is periodic. Moreover, because the term $\sin 4\pi t$ goes through two cycles of sine values in exactly the t-interval in which $\sin 2\pi t$ goes through one cycle, the entire function repeats itself with the frequency with which $y=\sin 2\pi t$ repeats itself. Hence the frequency of formula (3) is one cycle per unit of t. Thirdly, the shape of the graph of formula (3) shows that the formula is *not* sinusoidal even though it is periodic. In other words, the sum of two sine functions can yield a function whose shape is quite different from that of a sine function, but the sum can nevertheless repeat itself.

We might expect that functions built up of three or more sine functions could have quite strange shapes and yet be periodic if the summands *all* began to repeat at some value of t, say $t=1$, the values they had taken on at $t=0$.

EXERCISES

1. By following the method described in the text sketch the graph of

   a) $y = \sin 2\pi t + \sin 6\pi t$, b) $y = \sin 2\pi t + \frac{1}{2}\sin 4\pi t$, c) $y = \sin 2\pi t + \sin 3\pi t$.

2. What is the frequency, in one unit of t, of the function

   a) $y = \sin 2\pi t + \sin 8\pi t$,

   b) $y = 2 \sin 2\pi t + \sin 4\pi t$,

   c) $y = \sin 2\pi t + \sin 4\pi t + \sin 6\pi t$,

   d) $y = \sin 2\pi \cdot 100t + \sin 2\pi \cdot 200t + \sin 2\pi \cdot 300t$?

19-4 THE ANALYSIS OF COMPLEX SOUNDS

We have already mentioned that the sounds given off by almost all musical instruments and by the human voice are not simple sounds; that is, they are not representable by functions of the form (1). Yet these sounds are intelligible, which means that they must be periodic or that the pattern of displacement versus time must repeat itself. The shapes of the curves which represent such sounds are, however, quite varied. In fact, to each sound there corresponds a characteristic shape. For example, Fig. 19-6 shows the shape corresponding to the sound of a piano note C. To obtain this graph, the sound is converted to electric current and the vibration of the current is made visible by means of a cathode-ray tube. In view of the variety of musical sounds it may seem that we have reached an impasse in our attempt to analyse all such sounds mathematically. But by a stroke of good luck mathematics provided the very theorem which gives us remarkable insight into all complex sounds. The stroke of good luck was the mathematician Joseph Fourier (1768-1830).

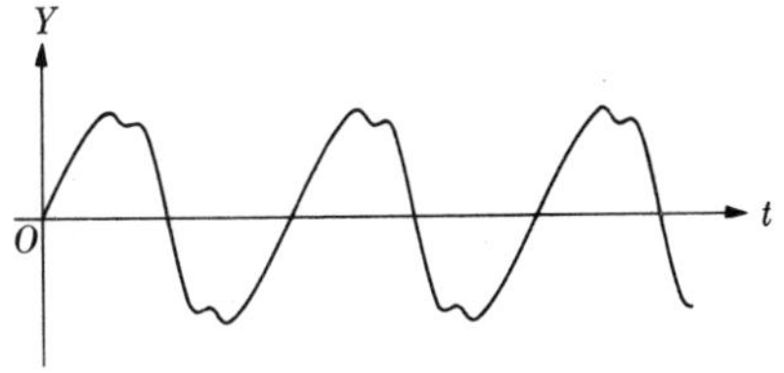

Fig. 19-6. Displacement versus time of a typical molecule for the note C on a piano.

Fourier was the son of a French tailor. While attending a military school he became intrigued with mathematics. Since he realized that his low birth would not permit him to become an army officer, he let himself be persuaded by members of the Church to study for the priesthood. However, he abandoned the priesthood to accept a professorship of mathematics at the military school that he had attended. Later he became a professor at the École Normale and at the École Polytechnique, universities founded by Napoleon.

Fourier's main interest was mathematical physics, and his most important work in that domain concerned the conduction of heat; for example, he studied how heat travels along metals. His chief contribution, a book entitled *The Analytical Theory of Heat* (1822), is one of the great classics of mathematics. In the development of the theory of heat Fourier established a mathematical theorem whose value extends far beyond the physical application for which it was intended. Our interest in the theorem lies in what it does to analyze complex musical sounds.

Fourier's celebrated theorem says that any periodic function is a sum of simple sine functions of the form $D \sin 2\pi ft$. Moreover, the frequencies of these component functions are all integral multiples of one frequency. To illustrate the significance of this theorem, let us suppose that y is a periodic function of t. Then the formula which relates y and t must be of the form

$$y = \sin 2\pi \cdot 100t + 0.5 \sin 2\pi \cdot 200t + 0.3 \sin 2\pi \cdot 300t + \cdots \qquad (6)$$

The numbers in this formula depend, of course, on the choice of the initial periodic function, but let us suppose that they are correct and see what they stand for. The numbers 1, 0.5, 0.3 are the amplitudes of the respective sinusoidal components of the entire periodic function. The lowest frequency per second, that of the first term, is 100. The second term has frequency 200, or twice the lowest frequency. The third term has frequency 300, or three times the lowest frequency, and so on. The dots at the end of formula (6) imply that we might need additional terms like the ones shown, to represent any given periodic function. In accordance with the theorem, all frequencies occurring in such additional terms must be multiples of 100.

Before we consider the significance of Fourier's theorem for the study of musical sounds, we should satisfy ourselves that formulas such as (6) do represent periodic functions. In this connection, two results of our work in Section 19–3 should be helpful. We learned there that the sum of two sine functions can produce a rather peculiarly shaped but nevertheless periodic graph. Moreover, because the second term in formula (3) had twice the frequency of the first one, the frequency of the *entire* function was the lower of the two frequencies. The situation in (6) is very much the same. It is a sum of sine terms, and the graph of this sum may indeed have a peculiar or irregular shape. But the shape will repeat itself because during the time that the first term goes through one cycle, namely the interval $t = 0$ to $t = \frac{1}{100}$, the second term will go through two cycles, and the third term through three, so that the entire function will repeat itself as soon as the first term does. Since the frequency of the first term is 100 in one unit of t, the entire function has the frequency of the first term.

And now what does Fourier's theorem have to do with the analysis of musical sounds? The application of this theorem to music was made by a German, Georg S. Ohm, a teacher of mathematics and physics, who lived in the first half of the nineteenth century. As pointed out earlier in this section, every musical sound is a periodic function; that is, the relation between displacement and time of a typical air molecule oscillating under the pressure exerted originally by the source of the sound is a periodic function of t. But Fourier's theorem says that every such function is a sum of simple sine functions of the type illustrated in (6). Each simple sine function corresponds to a simple sound such as is given off by a tuning fork. Hence one arrives at the important conclusion that *every* musical sound is a sum of simple sounds. Moreover, the frequencies per second of these simple sounds are all multiples of one lowest frequency. To put the matter differently, every musical sound can be duplicated by a combination of tuning forks, each vibrating with the proper frequency and amplitude.

The musical sound whose graph is shown in Fig. 19–6, for example, is a sum of five simple sounds. The frequencies of these sounds and their respective amplitudes are tabulated below. The amplitudes are expressed in terms of the first one entered, which is chosen to be 1.

Frequency	512	1024	1536	2048	2560
Amplitude	1	0.2	0.25	0.1	0.1

We should note that the frequencies are all multiples of the lowest one, which is 512. The formula representing this sound is then

$$y = \sin 2\pi \cdot 512t + 0.2 \sin 2\pi \cdot 1024t + 0.25 \sin 2\pi \cdot 1536t + 0.1 \sin 2\pi \cdot 2048t + 0.1 \sin 2\pi \cdot 2560t.$$

The assertion that every musical sound is no more than a combination of simple sounds is so surprising that, although it is backed by unassailable mathematics, one wishes to see it confirmed by experimental evidence. Such evidence is available. First of all, a trained ear can recognize the simple sounds present in a complex sound. Secondly, if one releases the dampers on the strings of a piano and then strikes a note, a number of other strings will also begin to vibrate, namely those whose basic frequencies are the same as the component frequencies present in the note struck. The physical explanation is that the note struck gives off several frequencies—the frequencies of its component simple sounds. Each of these frequencies sets off air vibrations which in turn force into vibration all other strings whose basic frequencies are the same as those of the simple sounds.

Perhaps the best experimental evidence is furnished by some specially designed instruments. The distinguished nineteenth-century physician, physicist, and mathematician Hermann von Helmholtz (1821–1894) gave two kinds of demonstrations. In the first one he designed special pipes, called resonators, each of which selected and rendered audible only that frequency which was suited to the dimensions of the pipe. A resonator in the neighborhood of a complex sound will pick up and render audible any component of the sound whose frequency excites the resonator. By using resonators of different sizes Helmholtz was able to show that the frequencies present in the complex sound were just those called for by Fourier's theorem. Then Helmholtz demonstrated the reverse. He set up electrically driven tuning forks of the proper frequency and amplitude such that the combination of simple sounds duplicated a given complex sound. A modern version of this latter device is the electronic music synthesizer.

There is no question, then, that any musical sound is no more than a sum of simple or sinusoidal sounds. The simple sound of lowest frequency is called the fundamental, or first partial, or first harmonic. The simple sound whose frequency is twice that of the lowest one is called the second partial or second harmonic; and so on. The frequency of the entire complex sound is the frequency of the first harmonic for the reason already given in our discussion of Fourier's theorem. The amplitudes of the individual sine terms are the amplitudes or strengths of the harmonics present.

EXERCISES

1. State Fourier's theorem.
2. Suppose that a complex sound is representable by the function

$$y = 0.001 \sin 2\pi \cdot 240t + 0.003 \sin 2\pi \cdot 480t + 0.01 \sin 2\pi \cdot 720t.$$

What is the frequency of the complex sound? What is the amplitude of the third harmonic?

3. Write the formula for a musical sound whose frequency is 500/sec and whose first, second, and third harmonics have amplitudes of 0.01, 0.002, and 0.005, respectively.
4. If the relationship between displacement and time for the fundamental of a musical sound is $y = 3 \sin 2\pi \cdot 720t$, what is the frequency of the third harmonic?
5. Explain why the frequency of a complex musical sound is always that of the first harmonic.

19-5 SUBJECTIVE PROPERTIES OF MUSICAL SOUNDS

Musical sounds as received by the ear seem to possess three essential properties; that is, the ear recognizes what are commonly called the pitch, the loudness, and the quality of a sound. One of the major values of the mathematical analysis of musical sounds is that it clarifies and makes precise just what we mean by these properties. We shall consider them in turn.

In our subjective judgment, sounds vary from low or deep tones to high or piercing ones. Verbal descriptions of the pitch of sounds are, of course, qualitative and vague. If one experiments with tuning forks of different pitch, he readily discovers that high pitch means high frequency of fork vibration and therefore high frequency of oscillation of the air molecules. Correspondingly, low pitch means that the fork and the air molecules vibrate with low frequencies. Prior to the availability of the analysis examined in the preceding section, the notion of pitch was not clear for complex sounds. But we now know that all musical sounds have a definite frequency, namely the frequency of the fundamental. Thus, although complex sounds contain other frequencies, that is, the frequencies of the higher harmonics, it is the over-all frequency of the composite sound which determines whether it appears high- or low-pitched to the ear. For example, as one strikes the notes on a piano going from left to right, the fundamental frequency steadily rises.

The loudness of a musical sound is determined by the amplitude of the corresponding molecular motion, but the relationship between loudness and amplitude is not quite so simple as that between pitch and frequency. Let us note, first of all, that amplitude means the maximum displacement of the typical air molecule or the largest y-value of the corresponding graph. Physicists call the square of this amplitude the intensity of the sound. Thus intensity is still a physical or objective property of a musical sound. Among sounds of a given frequency, the more intense sound will seem louder to the ear. However, this is no longer true if the frequencies of the sounds differ. The average ear is most sensitive to a frequency of about 3500 per second and less so to frequencies above and below this value. Hence a very intense sound at a frequency of 1000 per second may sound softer to the ear than a less intense one at a frequency of 3500 per second. As a matter of fact the average human ear does not hear at all sounds above about 16,000 vibrations per second, no matter how intense they are. Loudness depends not only on the intensity and the frequency of the sound but also on the shape of the graph within any one period. Two sounds may possess the same frequency and the same amplitude, but may have differently shaped graphs. Such sounds will, in general, not sound equally loud to the ear.

The most interesting and from an aesthetic standpoint the most important aspect of musical sounds is their quality. It is this property which determines whether or not a sound is pleasing. The quality of a sound depends upon which harmonics are present in the sound and the amplitudes of these harmonics. Thus a sound emitted by a piano and a sound of the same frequency emitted by a violin create different effects on the ear because they differ in the harmonics present and in the amplitudes of these harmonics. Since the harmonics and their amplitudes determine the shape of the graph, it follows that, mathematically, the quality of a sound is the shape of the graph within any one period.

Sounds or tones vary greatly with respect to harmonics and their amplitudes. Some sounds, for example, the sounds of tuning forks, some notes on the flute, and sounds produced by wide-stopped organ pipes, possess only a few harmonics or, in effect, merely the first. On the other hand, most instruments give off sounds containing many harmonics, but some of these may have small or almost zero amplitude. For example, the sounds of organ pipes are, in general, weak in the higher harmonics. The sounds of a violin possess a great number of harmonics and are usually strong in the first six harmonics. The relative amplitudes of the harmonics present in violin sounds are about the same for all notes; however, there are enough differences for the ear to distinguish, say the A- from the D-string, even though both are sounded at the same frequency. The uniformness of quality may explain why the sounds of a violin are so pleasing. The sounds of a piano also contain many harmonics, but the relative amplitudes of the harmonics in any one sound depend upon the velocity with which the hammer strikes the string.

The vowel sounds of the human voice are rich in harmonics. For example, the sound of "oo" as in tool, expressed at a fundamental frequency of 125 vibrations per second, has as many as 30 detectable harmonics. The relative amplitudes of the first six are 0.4, 0.7, 1, 0.2, 0.2, and 0.2, respectively. The higher harmonics, though present, have lower amplitudes. However, not only do the number of harmonics present and their relative amplitudes vary considerably from one vocal sound to another, but even the same sound issued at two different pitches will have different harmonics and amplitudes.

The physical reason for the differences in quality among the many types of musical instruments is, of course, the nature of the device itself. The piano and violin both use vibrating strings, but piano strings are struck whereas violin strings are bowed. The clarinet, oboe, and bassoon are operated by forcing air against vibrating reeds. Air is forced past the edge of an opening in the organ pipe also, but here the edge or lip is rigid. In addition, each instrument possesses a resonance device which emphasizes certain harmonics. The sounding board of the piano, the hollow box of the violin, and the pipes of an organ are resonance devices.

Although two people may not quite agree about their reactions to sounds, it is on the whole true that the qualities of sounds which we describe by such words as soft, piercing, rich, dull, braying, hollow, bright, and the like, are due to the harmonics and their relative amplitudes. Sounds which contain only the first harmonic are soft but dull. For brightness and acuteness the higher harmonics are essential. Sounds which possess the first six harmonics are grand and sonorous. If harmonics beyond the sixth or seventh are present and have appreciable amplitudes, the tones are piercing and rough. In general, the amplitudes of harmonics decrease as the frequencies increase. However, if the amplitudes of higher harmonics are too large compared with that of the fundamental, the tone is described as poor rather than rich.

EXERCISES

1. Suppose that two sounds are represented respectively by $y = 0.06 \sin 2\pi \cdot 200t$ and $y = 0.03 \sin 2\pi \cdot 250t$. Which one is louder? Which one is higher pitched?
2. Explain in mathematical terms the meaning of a simple sound.
3. What is the mathematical criterion of a musical sound as opposed to noise?
4. Which mathematical properties of the formula for a complex musical sound represent the pitch of the sound and the quality of the sound?
5. Discuss the assertion that music is basically just mathematics.

Topics for Further Investigation

1. The construction of musical scales. Include, in particular, the work of J. S. Bach on the equal-tempered scale.
2. The human voice as a source of musical sounds.
3. The functioning of the human ear.

Recommended Reading

BENADE, ARTHUR H.: *Horns, Strings and Harmony*, Doubleday and Co., New York, 1960.

FLETCHER, HARVEY: *Speech and Hearing*, D. Van Nostrand Co., Princeton, 1929.

HELMHOLTZ, HERMANN VON: *On the Sensations of Tone*, Dover Publications, Inc., New York, 1954.

JEANS, SIR JAMES H.: *Science and Music*, Cambridge University Press, London, 1937.

MILLER, DAYTON C.: *The Science of Musical Sounds*, 2nd ed., The Macmillan Co., New York, 1926.

OLSON, HARRY F.: *Musical Engineering*, McGraw-Hill Book Co., Inc., New York, 1952.

REDFIELD, JOHN: *Music, A Science and an Art*, A. A. Knopf, Inc., New York, 1926.

SEARS, FRANCIS W. and MARK W. ZEMANSKY: *University Physics*, 3rd ed., Chaps. 21 to 23, Addison-Wesley Publishing Co., Inc., Reading, Mass., 1964.

TAYLOR, LLOYD WM.: *Physics, The Pioneer Science*, Chaps. 24 to 28, Dover Publications, Inc., New York, 1959.

VON BERGEIJK, WILLEM A., JOHN R. PIERCE and EDWARD E. DAVID, JR.: *Waves and the Ear*, Doubleday and Co., New York, 1960.

WOOD, ALEXANDER: *The Physics of Music*, 6th ed., Dover Publications, Inc., New York, 1961.

CHAPTER 20

NON-EUCLIDEAN GEOMETRIES AND THEIR SIGNIFICANCE

One must do no violence to nature, nor model it in conformity to any blindly formed chimaera; . . .

JOHN BOLYAI

20-1 INTRODUCTION

The most significant revolutions in this world are not political. Political revolutions hardly change the daily life of man or, if they do, exert a short-term effect which may even be reversed by subsequent revolutions. The significant upheavals are caused by new ideas; these far more effectively, powerfully, and lastingly alter the lives of civilized human beings. For example, the beliefs in the importance of the human being and in his right to life, liberty, and the pursuit of happiness have permanently and radically changed the lives and aspirations of hundreds of millions of human beings. Indeed, many political revolutions were inspired by the desire to realize these ideas. The two concepts which have most profoundly revolutionized our intellectual development since the nineteenth century are evolution and non-Euclidean geometry. The theory of evolution is generally well recognized as a prime influence, but non-Euclidean geometry, despite its more fundamental and more far-reaching effects, seems to escape attention.

In this chapter we shall examine the nature of non-Euclidean geometry, its value for science, its implications for the nature of mathematics, and, finally, its influence on our culture.

20-2 THE HISTORICAL BACKGROUND

Euclidean geometry, as well as developments such as arithmetic, algebra, and calculus, rests upon axioms. The Greeks, who formulated the axioms of Euclidean geometry, believed that human minds immediately recognized some truths about the geometrical properties of physical objects and of space. Thus, it seemed indubitable that two points determined one and only one line, and that equal line segments added to another pair of equal line segments gave equal sums. For two thousand years the entire intellectual world accepted the Greek doctrine that the axioms of Euclidean geometry and of mathematics in general, were truths about the physical world, truths so clear and so evident that no one in his right mind could question them. Of course, since the axioms of geometry were truths, and since the theorems were logically necessary consequences of the axioms, the entire body of Euclidean geometry constituted a collection of indubitable truths about idealized objects and phenomena of the physical world.

One slight blemish seemed to mar the collection of axioms. Euclidean geometry deals with parallel lines. By definition, two lines in the same plane are parallel if they do not meet, that is, if they do not contain any point in common. This last statement expresses what we mean by parallel lines and so is not objectionable. In itself it does not assert that there are any parallel lines. But Euclidean geometry contains an axiom which implies the *existence* of parallel lines, and our present discussion centers about the nature of this axiom. As stated by Euclid, the axiom asserts that if two lines m and l (Fig. 20–1), meet a third line, n, so as to make the sum of angles 1 and 2 less than 180°, then the lines m and l meet on that side of the line n on which the angles 1 and 2 lie. Euclid then proves, for example, that if the sum of angles 1 and 2 is 180°, then m and l are parallel. This axiom is a bit involved, and there is some reason to believe that even Euclid himself was not too happy about it. Neither he nor later mathematicians up to the nineteenth century really doubted the truth of this axiom; that is, they had no doubt that it was a correct idealization of the behavior of actual or physical lines. What bothered Euclid and his successors was that the axiom was not quite so self-evident as the axiom, say, that any two right angles are equal.

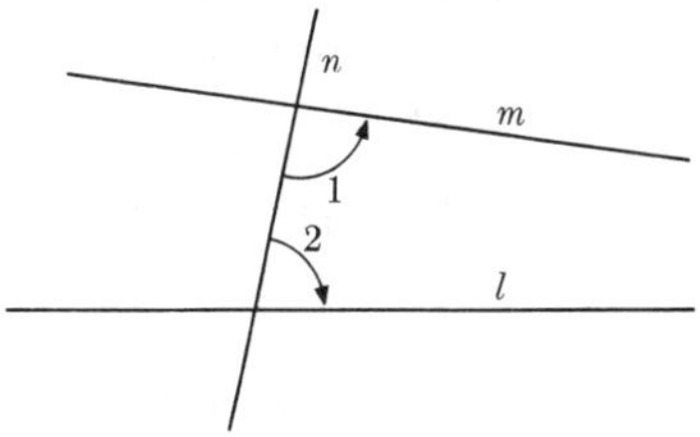

Fig. 20–1.
Euclid's parallel axiom.

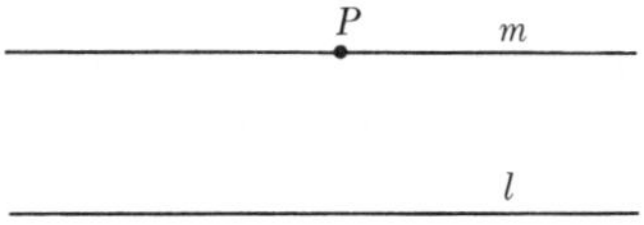

Fig. 20–2.
Playfair's parallel axiom.

From Greek times on mathematicians sought to replace this axiom with an equivalent one. An equivalent axiom is one which, together with the other nine axioms of Euclid, will make it possible to derive the same body of theorems that Euclid deduced. Many equivalent axioms were proposed. One of these, known as Playfair's axiom [it was adopted by the mathematician John Playfair (1748–1819)], is the one we usually learn in high school. Playfair's axiom states that, given a line l (Fig. 20–2) and a point P not on that line, there is one and only one line m in the plane of P and l which passes through P and does not meet l. From Playfair's parallel axiom and the other nine axioms of Euclid one can deduce all the theorems of Euclidean geometry.

Playfair's axiom appears intuitively convincing. That is, it does seem as though straight lines in physical space possess the property asserted. However, mathematicians were not satisfied with Playfair's axiom or other proposed equivalents of Euclid's parallel axiom. They recognized that every proposed substitute involved directly or indirectly an assertion about what happened far off in space. Thus, Playfair's axiom asserts that the line m through P will not meet l even in the very distant space to which these lines extend. As a matter of fact, Euclid's axiom is superior in this respect because it does not assert that lines will not meet, but states conditions under which they will meet at some finite distance.

What is objectionable about axioms which assert what happens far out in space? The answer is that they transcend experience. The axioms of Euclidean geometry are supposed to be immediately convincing statements about the properties of space. But how can one be sure of what happens millions of miles away? How can one be sure that it is possible to extend straight lines indefinitely far out into physical space, without the lines ever being forced to meet? Thus the efforts made to find a simpler statement than Euclid's did succeed insofar as simplicity of statement was concerned, but bred doubts about the truth of any assertion regarding the existence of parallel lines.

By the eighteenth century some mathematicians decided to try a new tack. Euclid's set contained ten axioms. Perhaps nine sufficed; that is, perhaps one *could* prove an assertion about parallel lines by deduction from the other nine axioms. If this should be possible, then there would be no further problem, because the assertion about parallel lines would be a necessary consequence of the nine entirely acceptable axioms. However, all these efforts failed.

One of these efforts, though, deserves special attention. The Jesuit priest Girolamo Saccheri (1667–1733) decided to apply the indirect method of proof. Euclid's parallel axiom asserts, in effect, the existence of one and only one line through P and parallel to l. To establish the truth of this statement by contradiction, two alternatives are available: no line parallel to l through P or more than one. Saccheri's plan was to assume in turn that each alternative to Euclid's parallel axiom was true and, with this alternative and the other nine axioms of Euclid, show that deductions would lead to a contradiction. He could then proclaim that the only questionable assumption, namely the alternative to Euclid's parallel axiom, must be false. After doing this with each of the two alternatives, he could then assert that the only remaining possibility, Euclid's axiom, must be true. As a matter of history, the alternative asserting that there were no parallels to l through P produced a contradiction. However, from the second alternative (there is more than one parallel to l through P) Saccheri derived a number of strange theorems but no contradiction.

The strangeness of the theorems he obtained was enough to convince Saccheri that this second alternative could not be true, and hence that Euclid's parallel axiom must be true because it was the only possible alternative. And so in 1733 he published a book entitled *Euclid Vindicated From All Defects.* Of course, strangeness of theorems and logical contradiction are quite different matters, and Saccheri was not justified in substituting one for the other. But he was tied to his time and, in drawing the conclusion that Euclid's assertion on parallel lines was a necessary consequence of the other nine axioms, he showed merely that if a man sets out to establish something of which he is already convinced, he will satisfy at least himself that he has proved its truth regardless of the facts.

The first to draw the conclusion which Saccheri should have drawn was Karl Friedrich Gauss (1777–1855). Gauss was one of the greatest mathematicians of all times. His father was a bricklayer who expected his son to adopt the same trade. However, Gauss showed his precocity in elementary school, and his teachers saw to it that he received a good education. At the University of Göttingen his teachers were hard put to keep up with him. At 22 Gauss submitted his doctoral thesis to the University of Helmstedt. In this thesis he proved what is often called the fundamental theorem of algebra, namely, that every algebraic equation of any degree has at least one root. At the age of 30 he was appointed professor of astronomy at the University of Göttingen. He discouraged students from attending his lectures.

Gauss's scientific interests, like those of Archimedes and Newton, were unbelievably broad. He was, for example, a great inventor. He designed many instruments for use in geodesy and was one of the inventors of the electric telegraph. He devised methods of making maps, was an excellent astronomical observer, and set up systems of insurance. At the request of the Elector of Hannover he made a survey of the principality. In addition to devising instruments for the measurement of the earth's magnetic field, he himself studied the variation in the strength of this field over the earth. The unit now used to measure the strength of magnetic fields is called the gauss.

Inspired by his scientific and practical interests, Gauss contributed to a number of major branches of mathematics. In fact, he would not think of separating mathematics from its application to science. Though his greatest achievements were in mathematics and he is therefore most often described as a mathematician, it would be more appropriate to call him a student of nature. His motto read:

Thou, nature, art my goddess; to thy laws
My services are bound. . .

A famous phrase, Mathematics, Queen of the Sciences, is due to him.

One of Gauss's greatest creations and certainly the most momentous from the standpoint of its implications is his non-Euclidean geometry. Gauss started his thinking on this subject as a boy and, like others before him, he began by trying to replace Euclid's parallel axiom by a more acceptable one, that is, one which did not involve any assertion about what must happen far off in space. But he did not succeed. He also appreciated at the age of 15, as he told his friend Schumacher, that one could not hope to prove an assertion about parallel lines on the basis of the other nine axioms. That is, he realized that these nine axioms did not in themselves dictate the form of the parallel axiom. Gauss was too brilliant a man to overlook the implication of this fact. If there was some freedom in the choice of a parallel axiom, then one might choose an axiom different from Euclid's and build a new kind of geometry. Gauss did just this. He pursued the logical implications of a system of axioms which included the assumption that *more than one* parallel to a given line passed through a given point, and thus created non-Euclidean geometry. (Gauss himself finally adopted the term non-Euclidean after having called his system anti-Euclidean geometry, and later astral geometry.)

Though Gauss realized that such a geometry could conceivably apply to physical space and hence was indeed significant, he did not publish his results. Gauss was far ahead of his times in concluding that Euclidean geometry was not necessarily the correct description of physical space and that some non-Euclidean geometry might prove as accurate. Hence he feared that he would be laughed at. In a letter written in 1829 to his friend, the mathematician Friedrich Wilhelm Bessel, Gauss confessed that he feared the clamor of the Boeotians, a figurative reference to the dull-witted, for the Boeotians had been one of the more simple-minded Greek tribes. Gauss's work on non-Euclidean geometry was found among his papers after his death in 1855.

The men who usually receive credit for creating non-Euclidean geometry because they published their results are Nicholas I. Lobachevsky (1793–1856) and John Bolyai (1802–1860). Lobachevsky, born to a poor family and displaying brightness even as a youngster, attended the University of Kazan in Russia. One of his teachers was the German mathematician J. M. C. Bartels, a friend of Gauss, and it is very likely that the problem of the parallel axiom was called to Lobachevsky's attention by Bartels. Lobachevsky became a professor at Kazan at the age of 23 and continued to work on the problem. By 1823 he realized that there can be other geometries and that Euclidean geometry need not be the correct description of physical space. He said later that two thousand years of fruitless attempts to put the parallel axiom on an unquestionable basis had led him to suspect that it could not be done. From 1829 on Lobachevsky published books and papers in which he expounded the theorems that hold in his non-Euclidean geometry. Despite valuable services to mathematics, his university, and the Russian government, he was dismissed in 1846, but continued to work in his field until his death.

Bolyai, a Hungarian, was an Austrian army officer. He learned mathematics from his father, Wolfgang Bolyai, who set his son thinking about the parallel axiom. In 1823 John arrived at the same conclusion that had been reached by Gauss and Lobachevsky, namely, that the Euclidean parallel axiom cannot be proved and that, in fact, it was but one alternative. He proceeded to develop a non-Euclidean geometry and published his work as an appendix to his father's book on mathematics which appeared in 1833. Knowing Gauss's interest in the subject of non-Euclidean geometry, Wolfgang sent a copy of his son's work to Gauss. The latter replied,

If I commenced by saying that I am unable to praise this work, you would certainly be surprised for a moment. But I cannot say otherwise. To praise it would be to praise myself. Indeed, the whole contents of the work, the path taken by your son, the results to which he is led, coincide almost entirely with my meditations, which have occupied my mind partly for the last thirty or thirty-five years.

Though Wolfgang was pleased to learn that his son's thinking paralleled that of the great Gauss, John was not, for, like others who had found that Gauss had beaten them to a discovery, he felt cheated of the glory. The irony of the situation was compounded further. Although Lobachevsky and Bolyai are now credited with the discovery of non-Euclidean geometry because they were the first to publish on the subject, the mathematical world of the 1830's and 1840's ignored their publications until Gauss's notes on non-Euclidean geomtry were found among his papers after his death. The name of Gauss attached to the idea made the mathematical world grant the proper importance to the subject.

On the surface, the history of non-Euclidean geometry seems remarkable because after two thousand years of futile work, three men suddenly saw the parallel axiom and Euclidean geometry in the proper light. However, while such coincidences of history are by no means uncommon, the present one is perhaps less remarkable than is ordinarily believed. Gauss did not hesitate to tell his mathematical friends of his radical views on the parallel axiom and of his doubts about the necessary truth of Euclidean geometry. Both Bartels and Wolfgang Bolyai were in this circle of friends, and they may have communicated Gauss's views to their respective charges. The mere derivation of theorems from new parallel axioms, though a considerable technical achievement, goes back to Saccheri, and numerous minor and major mathematicians had done similar work in the intervening one hundred-year period. But the correct evaluation of the significance of these logical developments was a new step, and this seems to have been made first by Gauss. However, Wolfgang Bolyai, perhaps to defend the originality of his son's thoughts, said of the work by the three men,

. . . because it seems to be true that many things have, as it were, an epoch in which they are discovered in several places simultaneously, just as the violets appear on all sides in the springtime.

EXERCISES

1. What is the definition of parallel lines?
2. What objection did Euclid's successors have to Euclid's axiom on parallel lines?
3. What objections were there to axioms such as Playfair's which replaced Euclid's parallel axiom by simpler assumptions?
4. Describe Saccheri's plan to establish the truth of Euclid's parallel axiom.
5. Did Saccheri arrive at the concept of a non-Euclidean geometry?
6. What advances did Gauss, Lobachevsky, and Bolyai make over Saccheri with respect to the nature of geometry?

20-3 THE MATHEMATICAL CONTENT OF GAUSS'S NON-EUCLIDEAN GEOMETRY

To appreciate the significance of what Gauss, Lobachevsky, and Bolyai created, we must look into the specific mathematical facts of their work. For brevity we shall, at present, refer to their geometry as Gauss's geometry. The main idea conceived by the three men asserted that one was logically free to adopt a parallel axiom which differs fundamentally from Euclid's, and that one could construct a new geometry which would be as valid as Euclid's and which might even be a good description of physical space.

What was the new parallel axiom? Euclid's parallel axiom, at least in the equivalent form which Playfair had given, stated that given a line l and a point P, there is one and only one line in the plane of l and P which passes through P and does not meet l (Fig. 20–2). Gauss, Lobachevsky, and Bolyai assumed that there are two lines, m and n, through P (Fig. 20–3) which are parallel to l, that is, these two lines through P do not meet l, and that any line through P falling within the angle MPN does meet l. Of course, any line such as q which passes through P and lies within the angle NPR cannot meet l because to do so, q would have to cross m or n and therefore meet m or n in a second point. But since two lines can intersect in at most one point, it follows that all lines through P which lie within the angle NPR will not meet l. Hence the assumption that there are at least two parallels to l implies that there is an infinite number of parallels.

The term *parallel lines* was reserved by Gauss, Lobachevsky, and Bolyai for m and n, whereas lines such as q were called nonintersecting lines. For

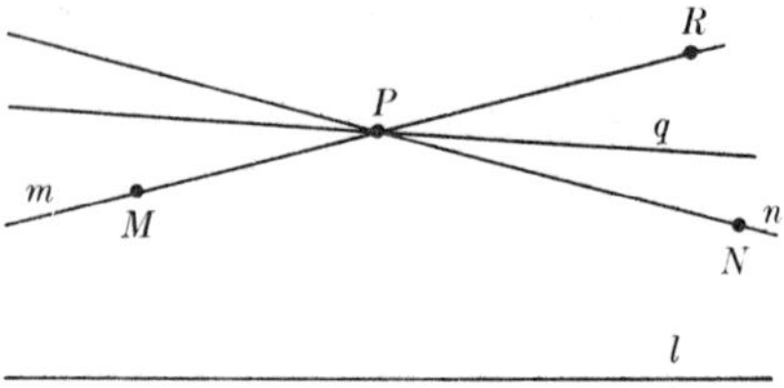

Fig. 20–3. Gauss's parallel axiom.

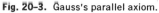

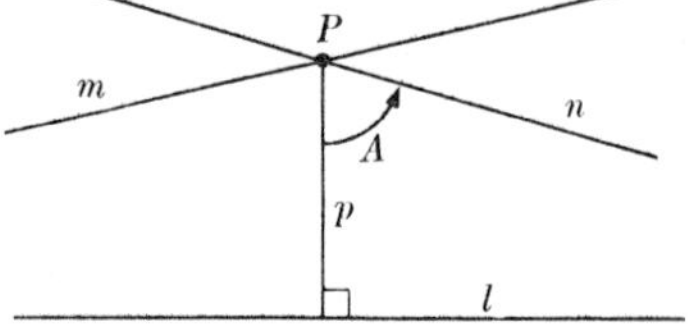

Fig. 20–4

this reason, their geometry is often described as a geometry with two parallel lines, although, in the Euclidean sense of the term *parallel lines* (two lines in a plane which have no point in common), it contains an infinite number of parallels. We shall use the terminology chosen by Gauss, Bolyai, and Lobachevsky.

Before proceeding to draw conclusions from this new parallel axiom and the other nine Euclidean axioms, let us dispel in advance any doubts about the common sense of the material we are about to present. We are so accustomed to Euclidean geometry (those of us who did not take our high-school geometry very seriously may be in a better position) that the idea of adopting a new parallel axiom and of proving theorems seems ridiculous. Euclid's geometry is truth to us, and to defy the truth and pursue the consequences of this folly seem a waste of time. There are two counterarguments. We can admit that from a practical standpoint the effort required to develop such a new system is pointless. Nevertheless, we should be able to comprehend that it is *logically* possible to investigate the consequences of a new set of axioms. What we are about to do is analogous to changing one of the articles in the Constitution of the United States while retaining the others, just to see what changes might follow in our laws. Thus we might decide to replace an elected president by a monarch and yet continue to elect congressmen and retain the institution of a Supreme Court. The second counterargument is based on the fact that Gauss, Bolyai, and Lobachevsky doubted the truth of Euclidean geometry and were ready to consider an alternative geometry as a description of physical space. To what extent their doubts were justified and whether the new geometry could serve in physical applications will be clearer when we know what theorems actually resulted from the new set of axioms.

We shall describe, but not prove, the theorems of the new geometry. The methods of proofs are precisely the same as in Euclidean geometry; however, one must be careful to use only the new axioms to support any assertions.

Many theorems of the new geometry are precisely the same as those of Euclidean geometry. Indeed those theorems of Euclidean geometry which are proved with the aid of only the first nine axioms of Euclid, that is, those theorems which do not depend upon the Euclidean parallel axiom, must also be theorems of the new geometry because these nine axioms are retained in the new system. Thus the theorems which assert when two triangles are congruent, the theorem asserting that the base angles of an isosceles triangle are equal, the theorem stating that an exterior angle of a triangle is greater than either remote interior angle, and the theorem postulating that from a point off a line, there is one and only one perpendicular to that line, are all valid in the new geometry.

But now let us examine some theorems of the new geometry which are different from those of Euclidean geometry. Suppose that the point P and the line l (Fig. 20–4) are given. In the new geometry, the parallel axiom states that there are two lines, m and n, through P which are parallel to l. Let us now drop a perpendicular p from P to l. Then in this new geometry the angle A between p and n is no longer a right angle, although n and l are parallel lines. Angle A is in fact an acute angle. Moreover, the size of angle A depends upon the length of p. The shorter p is, the larger does angle A become, and as p approaches zero, angle A approaches 90° in size. Of course, in Euclidean geometry angle A is a right angle and this holds no matter how large p is. A key theorem of the new geometry asserts that the sum of the angles of a triangle is always less than 180°, whereas it is exactly 180° in Euclidean geometry. Further, in the new geometry the sum of the angles of a triangle varies with the area of the triangle. The smaller the area, the closer is the angle sum to 180°. A very surprising theorem of the new geometry states that if the three angles of one triangle equal, respectively, the three angles of another, the triangles are congruent. Of course, in Euclidean geometry two such triangles would be similar, and one could be very much larger than the other. There are many more interesting consequences of the new axioms, but we have seen enough to obtain some indication of the nature of the new geometry.

EXERCISES

1. Describe the parallel axiom adopted by Gauss, Bolyai, and Lobachevsky.
2. Distinguish between the Euclidean use of the term parallel lines and the usage in Gauss's geometry.
3. State three theorems which are common to Euclidean geometry and Gauss's non-Euclidean geometry.
4. State three theorems of Gauss's geometry which do not hold in Euclidean geometry.
5. Suppose that one triangle has an angle sum of 170° and another, of 175°. Which has the larger area in Gauss's geometry?
6. Given two triangles, one with angles of 30°, 40°, and 100° and another with angles of 35°, 45°, and 90°. What does Gauss's geometry say about their area?
7. Suppose that the angles of one triangle are equal, respectively, to the angles of another. What may you conclude about these triangles on the basis of Gauss's geometry?

20-4 RIEMANN'S NON-EUCLIDEAN GEOMETRY

We saw in the preceding section that it is possible to investigate a new set of geometrical axioms and to deduce logical consequences. Perhaps we are not satisfied as yet that such an investigation is of value, but we shall reserve our final judgment until we have looked into one more mathematical creation in the field of non-Euclidean geometry.

Investigations of the parallel axiom had led mathematicians to question its truth because it made assertions about what must happen in physical space far beyond man's experience. Having become aware of this shortcoming, mathematicians examined the remaining axioms and soon found another which suffered from the same failing. Euclid's second axiom asserts that a straight line extends indefinitely far in either direction. This axiom attracted the attention of the nineteenth-century mathematical giant, Georg Friedrich Bernhard Riemann (1826–66). Riemann, the son of a Lutheran pastor, was born sickly and precocious. He became a student of Gauss at the University of Göttingen, and was later appointed professor at the same institution. Like Gauss, and most mathematicians for that matter, he was keenly interested in science and in the applicability of mathematics to the physical world.

Riemann observed that experience suggests not the infinite extent of the straight line but rather its endlessness. He accordingly distinguished between endlessness, or unboundedness, and infinite length. The simplest example of that distinction is furnished by the circle. One can traverse the circle endlessly, yet its length is finite. Hence Riemann proposed to replace the Euclidean axiom that a straight line extends indefinitely far by the axiom that it is unbounded.

Riemann also observed that experience does not vouch for the existence of any parallel lines. Within the limits of experience, we could equally well assume that any two lines meet. Hence Riemann proposed this axiom as an alternative to Euclid's parallel axiom. It may be recalled that Saccheri had considered the same possibility, but had found that it led to contradictions. However, Saccheri had combined this axiom with the other nine Euclidean axioms, whereas Riemann proposed an additional change, namely the unboundedness of the straight line, and had therefore reason to believe that he, unlike Saccheri, would not encounter any contradictions.

Since Riemann did retain some of Euclid's axioms, he arrived at some theorems identical to those of Euclidean geometry. Thus the theorem that two triangles are congruent when two sides and the included angle of one are equal to the corresponding parts of the other is a theorem of Riemann's geometry, as are other familiar congruence theorems.

The striking theorems of Riemann's geometry are, of course, those which differ markedly from Euclid's results. One Riemannian theorem asserts that every straight line has the same finite length. Another asserts that all perpendiculars to a line meet in one point. In Riemann's geometry the sum of the angles of a triangle is always *greater* than 180°. Moreover the sum varies with the area of the triangle and *decreases* to 180° as the area approaches zero. Two similar triangles are necessarily congruent (this is also the case in Gauss's geometry). These theorems are, of course, just a representative selection from a vast number.

EXERCISES

1. Why did Riemann question the Euclidean axiom that a straight line extends indefinitely far in either direction?
2. What axiom on parallel lines did Riemann adopt?
3. Some theorems of Gauss's and Riemann's non-Euclidean geometries are identical with Euclidean theorems. Why does this occur?
4. Which of the two non-Euclidean geometries would you expect to have more theorems in common with Euclidean geometry and why?
5. Compare the Euclidean, Gaussian, and Riemannian theorems about the sum of the angles of a triangle.
6. State three theorems peculiar to Riemann's geometry.
7. What is non-Euclidean geometry?
8. What happens to the Euclidean distinction between congruent and similar triangles in non-Euclidean geometry?

20-5 THE APPLICABILITY OF NON-EUCLIDEAN GEOMETRY

The very fact that there can be geometries other than Euclid's, that one can formulate axioms fundamentally different from Euclid's and prove theorems, was in itself a remarkable discovery. The concept of geometry was considerably broadened and this suggested that mathematics might be something more than the study of the implications of the self-evident truths about number and geometrical figures. However, the very existence of these new geometries caused mathematicians to take up a deeper and more disturbing question, one which had already been raised by Gauss. Could any one of these new geometries be applied? Could the axioms and theorems fit physical space and perhaps even prove more accurate than Euclidean geometry? Why should one continue to believe that physical space was necessarily Euclidean?

At first blush the idea that either of these strange geometries could possibly supersede Euclidean geometry seems absurd. That Euclidean geometry is *the* geometry of physical space, that it is the truth about space is so ingrained in people's minds that any contrary thoughts are rejected. The mathematician Georg Cantor spoke of a law of conservation of ignorance. A false conclusion once arrived at is not easily dislodged. And the less it is understood, the more tenaciously is it held. In fact, for a long time non-Euclidean geometry was regarded as a logical curiosity. Its existence could not be denied, but mathematicians maintained that the real geometry, the geometry of the physical world, was Euclidean. They refused to take seriously the thought that any other geometry could be applied. However, they ultimately realized that their insistence on Euclidean geometry was merely a habit of thought and not at all a necessary belief. Those few who failed to see this were shocked into the realization when the theory of relativity actually made use of non-Euclidean geometry.

It is important to see how and why a non-Euclidean geometry can fit physical space. Let us recall, first of all, why Gauss, Bolyai, and Lobachevsky doubted the truth of the Euclidean parallel axiom. They realized that this axiom and all of the simpler forms which might serve as substitutes contained assertions about what happens in space far beyond the range of man's experience. Hence experience does not support such axioms. As a matter of fact, if the lines m and n of Fig. 20–3 and all lines, such as q, falling within the angle NPR should have almost the direction of l, then they would certainly not meet l within a short distance from the point P. Hence Gauss's axiom is as much in accord with experience as Euclid's. Our intuition is seemingly violated by such a thought, but this intuition may be conditioned by our familiarity with Euclidean geometry. In other words, insofar as experience is concerned, one can adopt Gauss's or Riemann's alternative axioms.

In view of our inability to decide *a priori* which of the several alternative axioms fits physical space, we might consider another approach to the problem. The theorems of any geometry are logical consequences of the axioms. Perhaps it would be easier to discriminate among these geometries by seeing how well their respective theorems fit physical space. This thought had already occurred to Gauss. He noted that in his geometry the sum of the angles of a triangle must be less than 180°, whereas in Euclidean geometry it is exactly 180°. Hence he had three observers stand on three mountain peaks and directed each one to measure the angle between his lines of sight to the other two observers. The sum of the angles of the triangle formed by the three peaks turned out to be 170°59′58″; that is, it was within 2″ of 180°.

This result might be interpreted as a victory for Euclidean geometry, but the situation is not so simple. In Gauss's geometry the sum of the angles of a triangle increases as the area of the triangle decreases, and the sum approaches 180° as the area approaches zero. In Riemann's geometry, the sum of the angles of a triangle is always larger than 180°, but the sum again appproaches 180° as the area of the triangle decreases. Hence for *small* triangles, all three geometries call for an angle sum close to 180°. Had Gauss's measurement of the angles formed by the mountain peaks been exact, he might have been able to assert that his result was less than, equal to, or more than 180°, and so his test would have been decisive. However, every measurement contains an error because the eye and hand are not precise. We may safely suppose that the error in Gauss's measurement was more than 2″. Thus, in view of his result, the true angle sum could have been anything from a little more than 180° to a little less. Hence the result was consistent with any one of the three geometries.

For a large triangle, the angle sum should be significantly less than 180° in Gauss's geometry and significantly more in Riemann's. Hence, by using a large triangle, say one formed by three celestial bodies at some instant, one might be able to obtain a result which would fit only one of the three geometries, provided the error of measurement were kept small. Thus, a result of 175°, with an error of measurement of less than 5°, would certainly establish the physical correctness of Gauss's geometry. However, in the nineteenth century, no such result was obtained. Gauss, Lobachevsky, and Bolyai realized that, at least in their day and with the instruments at their disposal, measurement would not decide the question.

Since a test based on the sum of the angles of a triangle does not succeed in establishing which of the geometries fits physical space, one might cast about for another theorem which would serve the purpose. He might then hit upon the theorem which holds in both non-Euclidean geometries, namely the theorem stating that two similar triangles must be congruent. Here a decisive test seems likely, for it appears quite obvious that one can construct a small and a large triangle and make them similar. One only has to ensure that the angles of the small triangle equal the angles of the large one, and the similarity of the two triangles necessarily follows. Only Euclidean geometry fits this physical situation. The argument, however, has a flaw since one cannot be sure that the corresponding angles of the two triangles are really equal. After all, measurement is approximate, and at least one angle of one triangle might differ from the corresponding angle of the other triangle. If this happened, the triangles would not be similar.

The substance of the above arguments is that there is no simple test which points to one geometry rather than to another. However, if an application involved a truly large triangle, as astronomical investigations do, one might be able to determine which geometry fits better. As a matter of fact, this possibility has been realized. In the theory of relativity, Einstein employed a non-Euclidean geometry (though one more complicated than those we have studied), and the agreement between his predictions and observations was better than the result obtained by means of Euclidean geometry.

EXERCISES

1. Does measurement help us to show that objects in physical space are better described by the theorems of one rather than another geometry?
2. Could we use Gauss's or Riemann's non-Euclidean geometry for engineering and architecture?
3. Why did Gauss's measurement of the sum of the angles of a triangle fail to show that space is or is not Euclidean?

20-6 THE APPLICABILITY OF NON-EUCLIDEAN GEOMETRY UNDER A NEW INTERPRETATION OF LINE

Thus far we have considered the applicability of non-Euclidean geometry to our physical space, with the understanding that the physical meaning of a straight line is the stretched string or the ruler's edge. Even when we visualize a straight line so long that we could never construct it, for example, a line from the earth to the moon, we picture a long taut string or an imaginary long straight stick. However, we must now realize that the mathematical straight line is not limited to this physical or geometrical figure. Let us consider an example.

Suppose that we begin with a flat sheet of paper which can be imagined to extend indefinitely far in all directions. Now let us suppose that the paper is bent so that it curves upward to the right and to the left (Fig. 20–5) and, of course, continues to extend indefinitely far out in all directions. In mathematics, incidentally, this shape is known as a cylindrical surface. (One can form a limited cylindrical surface by merely bending upward the sides of a sheet of paper.) As a consequence of the change from a plane to a cylindrical surface, many of the straight lines of the plane become curves on the surface. Thus AC in Fig. 20–5 is an arc of a curve which, if the surface were flattened, would become a segment of a straight line. Curves which derive from straight lines we shall continue to call lines. Angles in the plane become, under the bending, angles on the surface. A triangle in the plane corresponds to a triangle formed by arcs of curves on the surface. A circle in the plane becomes a new curve on the surface which we shall continue to call a circle.

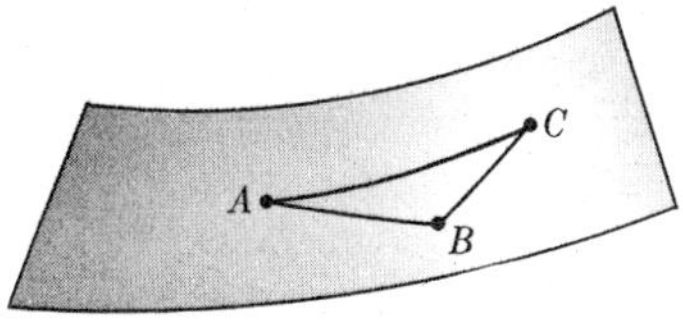

Fig. 20–5.
Euclidean geometry on a cylindrical surface.

Now let us note that the bending of the plane into the cylindrical surface does not change lengths or angles, because the plane neither stretches nor contracts as one forms the surface. Hence a distance between two points measured *along* the surface is the same as the distance between the two points when on the plane.

Since lengths and angles do not change, it follows that if the figures in the plane satisfy the axioms of Euclidean geometry, then so do the figures on the surface, provided the meaning of lines, angles, and circles on the surface is as specified above. And since theorems are logical consequences of axioms, it follows that the Euclidean theorems, too, retain their validity for figures on the surface. Thus the sum of the angles of triangle ABC in Fig. 20–5 is 180°. These last few assertions appear at first to be incredible. Our triangles are no longer formed by straight line segments, and yet the Euclidean theorems continue to hold. How is it possible to apply the word *line* of the Euclidean axioms and theorems to the curves that are no longer straight lines?

The answer to this question involves a major point about geometry and indeed about mathematics in general. In our review of Euclidean geometry in Chapter 6, we briefly mentioned that Euclid's definitions of point, line, plane, and other concepts were not quite satisfactory. The full story is that his definitions of these concepts are meaningless. He defined curve as length without breadth, but he did not say what he understood by length and breadth. He then defined the straight line as a curve which lies evenly between the ends, but he did not indicate what "lies evenly" meant. Euclid was relying on our intuitive understanding of these terms, but intuitive understanding cannot be part of a logical treatment. Mathematics does not rest logically on physical meanings. What then is the alternative? It is not possible to begin the process of developing any new branch of mathematics by defining the initial notions, for definition requires that one provide descriptions in terms of other concepts whose meaning is already established. But obviously the initial concepts cannot be defined in terms of prior ones since there are no prior ones. The point, then, is that the initial concepts cannot be defined. They must remain undefined.

This assertion raises another question. If point, line, plane, and other basic concepts are undefined, how shall we know what we mean by them and how shall we know how to treat them? The answer is that the axioms of Euclidean geometry tell us what properties these concepts possess. These properties and only these may be used in establishing the proofs. The concepts are therefore limited only to the extent that they must satisfy the axioms.

Now, as far as we can determine, physical straight lines, such as rulers' edges and stretched strings, do satisfy the axioms of Euclidean geometry. In fact, historically, the axioms were formulated by observing the properties of rulers' edges and stretched strings. What is surprising, however, is that one may encounter other "lines" which also satisfy the axioms. If they do, they are legitimate interpretations or realizations of the axioms and hence of the theorems. This is the case for the lines of our cylindrical surface. We know that they satisfy the Euclidean axioms because in deforming the plane into the cylindrical surface, we did not disturb any of the properties asserted by the axioms. Consequently, Euclidean geometry as a whole applies to the points, lines, triangles, polygons, circles, and other figures on the cylindrical surface.

The significance, then, of the entire discussion about the cylindrical surface is that it is possible to associate a totally new picture with Euclidean geometry. In other words, we see that Euclidean geometry applies to more physical situations than we had previously suspected. But if one can put Euclidean geometry to new uses by adopting a new view of the straight line, perhaps one can do the same for the non-Euclidean geometries, and hence these, too, may be more valuable and more meaningful than we have hitherto suspected. We shall now show that it is possible to give a very simple and practical interpretation of Riemann's non-Euclidean geometry.

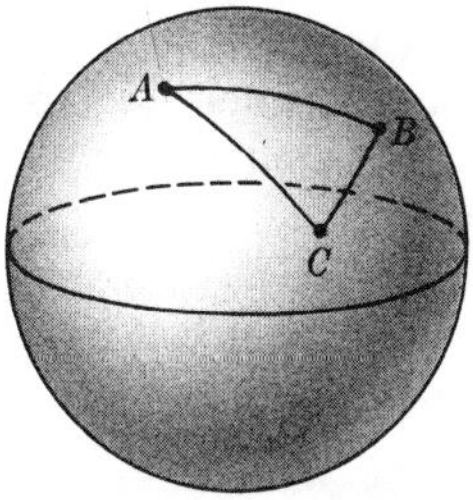

Fig. 20–6.
A spherical triangle formed by arcs of great circles.

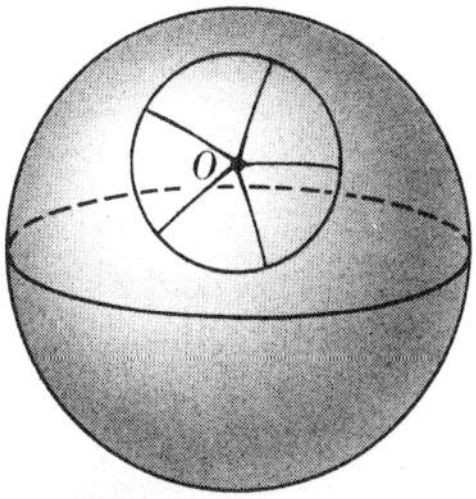

Fig. 20–7.
A circle on a sphere.

Let us consider the surface of the earth and let us suppose that we were assigned the task of developing a geometry that would fit the surface of the earth. We would, of course, choose as points the idealization of the usual physical points. We would most likely cast about next for the figure which would play the role of the straight line. Now straight lines in the usual sense do not exist on the surface of the earth because the surface is curved. Hence our line cannot be the rulers' edge or a stretched string. The most useful curve to choose would be the curve which connects two points by the shortest path. The shortest path between any two points on a sphere is the shorter arc of the great circle through these points.* Hence it seems reasonable to choose great circles on the sphere as our lines. A triangle on the sphere would be the figure formed by three arcs of three great circles (Fig. 20–6). A circle on the sphere would be the set of all points at a given distance from a fixed point (Fig. 20–7). (This distance between the fixed point and any point on the circle would, of course, be measured along the great circle through the two points.) We could, in fact, describe a variety of geometrical figures in terms of points and lines on the sphere.

The next step in the construction of our geometry would be to determine the axioms which our points, lines, and geometrical figures satisfy. Now our lines are great circles. These lines are not infinite in extent. However, each is unbounded; that is, there is no beginning or end. Hence our axiom should assert merely that a line is unbounded. We note next that two points which are not on opposite ends of a diameter determine a unique great circle. However two points which are on opposite ends of a diameter do not determine a unique great circle. Consequently we cannot adopt the axiom that any two points determine a unique line. Let us look next into the subject of an axiom on parallel lines. Any two great circles on a sphere do meet; in fact, they meet at two points which are diametrically opposite. Hence the parallel axiom should read that any two lines meet; that is, there are no parallel lines. We shall not

* The concept of great circle was explained in Chapter 7.

go further in selecting axioms for our system of geometry, for the conclusion toward which we are heading is now obvious. The axioms we would have to adopt to make our geometry fit the surface of the sphere would be exactly those which Riemann adopted for his non-Euclidean geometry.

Since the axioms of Riemann's geometry hold for the sphere, the theorems must also hold. Hence the geometry on the surface of a sphere is an application of Riemann's non-Euclidean geometry. Merely to satisfy ourselves, let us see whether a few theorems of Riemann's geometry really do apply. Let us consider the theorem that all perpendiculars to a line meet in a point. Figure 20–8 illustrates this theorem, and we see at once that it does apply. Another theorem asserts that the sum of the angles of a triangle is always greater than 180°. To verify that this theorem holds for every triangle on a sphere, we would have to call upon a theorem proved in spherical geometry which asserts that the sum of the angles of any spherical triangle is always between 180° and 540°. However, we can see at once in particular cases that Riemann's theorem does apply. Let us consider triangle *ABC* of Fig. 20–8. Angles *A* and *B* are each 90°. Hence, whatever the size of angle *C*, the sum of the three angles is larger than 180°. We could, of course, verify that all theorems of Riemann's geometry hold for the sphere.

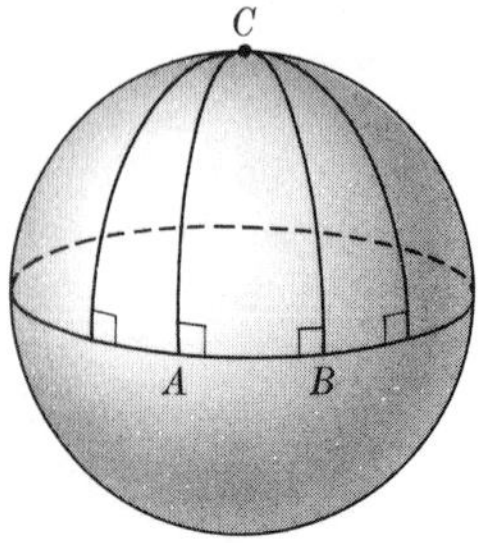

Fig. 20–8.
The sum of the angles of a spherical triangle is greater than 180°.

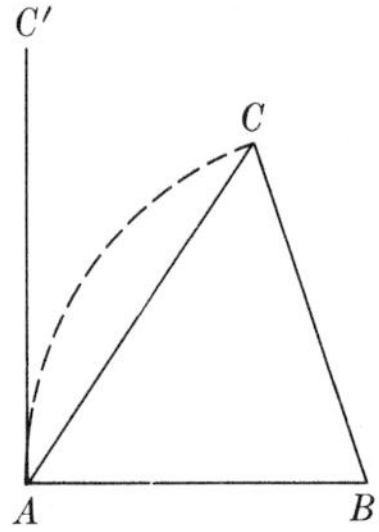

Fig. 20–9

The major point, then, which emerges from our discussion of the geometry on a sphere is that we have found a new use for Riemann's non-Euclidean geometry. It applies directly to the surface of a sphere, provided that straight line means great circle on the sphere.

Consideration of this last fact does lead to a question: Since Riemann's geometry applies so naturally to the surface of a sphere, why did not mathematicians hit upon non-Euclidean geometry in Greek times? Why did the realization that there can be non-Euclidean geometries take so long to strike home? The answer is that the Greeks, no doubt influenced by the Egyptians and Babylonians, had chosen the stretched string or ruler's edge as the physical straight line and the Euclidean axioms as the basis for their geometry. Such choices were very natural for people whose experiences were limited to a small part of the earth's surface. When they came to consider the sphere among the various surfaces, they had to approach it through concepts and axioms already adopted and hence described its properties in Euclidean terms. Thus great circles were treated as curves. As we have already noted in another connection, the Greeks and all mathematicians up to 1800 were so sure that Euclidean geometry was the true geometry of physical space that the idea of approaching the sphere directly and building a special geometry for it would have seemed nonsensical. They already possessed the true geometry and could not break what we can now see clearly was just a habit of thought.

It is possible to exhibit a surface to which the axioms and theorems of Gauss's non-Euclidean geometry apply, provided that the curves chosen as the lines on that surface are, as in the case of the sphere, those which connect two points by the shortest path. However, this surface is not widely used, and so we shall not devote any time to it. It will be more profitable to turn to the reconsideration of our familiar physical space.

We have grown accustomed to the idea that the reasonable and convenient physical interpretation of the straight line is the ruler's edge or a stretched string. Even with this interpretation we found that experience in limited regions did not enable us to exclude the non-Euclidean geometries as possible descriptions of physical space. We should now note that the ruler's edge and stretched string are, indeed, not the major physical interpretation of the mathematical straight line, and that we commonly and necessarily use another. Let us consider how surveyors determine distances. They begin by adopting a convenient base line AB (Fig. 20–9) whose length is measured by actually applying a tape measure. To determine the distance AC, a surveyor measures angle A by sighting point C in his telescope stationed at A and then swings the telescope around until he sights point B. On his theodolite he has a scale which tells him how much he has rotated his telescope, and hence he knows angle A. In a similar manner, he measures angle B. By means of trigonometry he can now calculate AC and BC. The surveyor proceeds on the assumption that the light rays which travel from C to A and from B to A follow the straight-line (stretched string) paths between those pairs of points, and, since the axioms of Euclidean geometry fit stretched strings, he applies Euclidean geometry or trigonometry to calculate AC and BC. However, the surveyor may be mistaken. The light ray from C to A may have followed the broken-line path shown in Fig. 20–9, and the surveyor at A would have to point his telescope tangentially to the light ray in order to receive the light. Hence the telescope would really be pointed to C' although the surveyor sees the point C in his telescope. Consequently, the angle he actually measures is $C'AB$

and not CAB. Thus the use of Euclidean geometry may have led to erroneous results for AC and BC.

What then should he have done? He should have applied a geometry whose axioms fit the behavior of light rays since these are the straight lines he really used. But do not light rays follow truly straight paths? We certainly know of situations in which they do not. We have had occasion to note (Chapter 1) that when light from the sun passes through the earth's atmosphere, it is bent by the refractive effect of the atmosphere. Hence there is some question of what path light rays do follow. Observation and measurement show that over short distances along or near the earth's surface light rays follow straight paths closely enough, but over larger distances this is certainly not true.

The above example may serve to suggest a major point. In astronomical measurements we necessarily depend entirely on light rays to measure angles. Since these rays travel over long distances, the paths they follow may not be truly straight, and we cannot check the true paths by laying down a tape measure or ruler. Hence we cannot be sure whether we should use Euclidean geometry.

We could try to solve this difficulty in one of two ways. We could attempt to determine by physical investigations what paths light rays follow, and, even though we may find that these are curved, treat them as if they were curves in Euclidean geometry. This is what physicists do when they study the behavior of light near the surface of the earth. The law of refraction tells us how light behaves, and all the reasoning in formulating and in applying this law is based on Euclidean geometry. Alternatively we can regard the paths of light rays as lines and construct a geometry which fits the convention that the light rays are to be the lines of that geometry. How to construct such a geometry is not obvious. One would have to determine some facts about the behavior of light rays, which would become the axioms of that geometry, and then deduce theorems. The resulting geometry might very well be non-Euclidean and might even differ from those developed by Gauss and Riemann.

All discussion of the *possibility* of applying non-Euclidean geometry is in a sense outdated. One of the basic current theories of science, the theory of relativity, does presuppose that our space is non-Euclidean and on this basis obtains better agreement between theory and experiment than the older theory of Newtonian mechanics based on Euclidean geometry was able to do. For reasons too lengthy to be discussed in this brief survey of the applicability of non-Euclidean geometry, the theory of relativity requires that the position as well as the time of events be treated together so that an object or event is described not only by the coordinates x, y, and z which denote position, but also by the value of the time t at which the object or event occurs at position x, y, z. In other words, the relevant geometry is four-dimensional. This fact in itself does not imply that we must resort to a non-Euclidean geometry.

However, in attempting to explain the phenomenon of gravitational attraction, it is necessary to regard the four-dimensional space-time as a nonhomogeneous geometry. This means that the nature of space-time varies from region to region, just as a two-dimensional mountainous surface changes its geometrical character from region to region. In the geometry of general relativity, it is also the presence of physical masses which determines the character of the geometry in any region, these masses being the earth, the moon, the sun, and other heavenly bodies. Moreover, the geometry is constructed in such a way that the "straight lines" are the shortest "paths" of that space-time. Light rays do take the shortest "paths" in the space-time system. Hence, while the straight line is not defined in terms of light rays in the non-Euclidean geometry of relativity, it is nevertheless significant that light rays do take such *paths*.

EXERCISES

1. Can you explain how it is that Euclidean geometry applies to a cylindrical surface even though the lines of that geometry do not have the shape of stretched strings?
2. Why must there be undefined concepts in the initial stages of any branch of mathematics?
3. Under what conditions does Riemann's non-Euclidean geometry prove to be the correct geometry on the surface of a sphere?
4. Should we distinguish between the geometries created by mathematicians and the geometry of physical space?
5. Does science really use the stretched string as its physical model of the mathematical straight line?
6. Imagine people living in a mountainous region who wish to construct a geometry for that region. They agree to consider the shortest path between two points as the line of that geometry. What kind of geometry might they arrive at?
7. Imagine people living in a mountainous region who wish to construct a geometry for that region. Since they travel by foot, using the shortest path to get from one place to another might require more time than some indirect path, because the shortest path may involve difficult mountain climbing. These people therefore agree to take as the line joining two points the path requiring least travel time. What kind of geometry might these people arrive at?

20-7 NON-EUCLIDEAN GEOMETRY AND THE NATURE OF MATHEMATICS

The existence of non-Euclidean geometries which can fit physical space, to say nothing of the actual use of one of these non-Euclidean geometries in the theory of relativity, has had profound implications for mathematics itself, for science, and for some segments of our culture. In this section we shall discuss the implications for mathematics.

The most important effect of this creation has been the realization that mathematics does not offer truths. The Greeks adopted the axioms of

Euclidean geometry because they believed that they were self-evident truths about our physical space. The axioms appealed to their minds as necessary truths which anyone must grant, even without experience. Since theorems are obtained by deductive reasoning and so are inescapable consequences of the axioms, the Greeks believed that the theorems, too, were truths. The observed agreement between the theorems deduced from these axioms and experience reinforced their certainty that the axioms were truths. The belief that mathematics offers truths was firmly held by every thinking being until the creation of non-Euclidean geometry. But if several geometries which contradict one another all fit physical space, then it becomes very obvious, indeed, that all of these cannot be the truth, and, worse yet, one can no longer be sure that any of these is the truth.

We see more clearly now that one must distinguish between mathematical space and physical space. Mathematicians and scientists believe that a physical world exists outside of and independently of human beings, and they seek to understand it by adopting axioms which seem to fit this physical space and then deducing theorems from these axioms. We now recognize that we have no reason to identify the mathematical construction with physical space. Indeed, several different mathematical theories may fit equally well. Mathematical theories of space are like any scientific theories; that is, the mathematical system used is the one which at the time best fits experience. If, as experience widens, it becomes clear that another geometry will fit experience better, then the older theory of space is discarded, and a new one adopted. This is precisely what happened when the theory of relativity sought to account for phenomena to an accuracy which would surpass that of the older scientific work.

The remarks we have made thus far about truth in mathematics and the provisional character of mathematical theories have been based on the developments in geometry. But the reader may have an objection. He may admit that geometry no longer offers truths, but continue to be convinced that our arithmetic, algebra, and other developments based on our number system do constitute truths. We shall devote Chapter 21 to this point and perhaps then see more clearly that with respect to truth the domain of number does not differ from geometry.

If the various branches of mathematics have only a more or less useful correspondence with physical experience, does then mathematics differ in any way from science? Mathematics had always been distinguished from science because mathematical axioms were regarded as truths, whereas the axioms of science were clearly recognized to be generalizations from limited experience or experiments. Scientists recognized that their researches produced only theories which did not provide a veridical description of what occurred in nature, but which might have to be altered to fit new facts. Is this not precisely what we now perceive to apply also to mathematics?

Insofar as the study of the physical world is concerned, mathematics has the same character as any of the sciences. It offers nothing but theories. And, as in science, new mathematical theories may replace older ones when experience or experiment shows that a new theory provides closer correspondence than an older one. As Einstein put it,

> *So far as the theories of mathematics are about reality, they are not certain; so far as they are certain, they are not about reality.*

Yet there are basic distinctions between mathematics and science. Mathematics confines its work to numbers, geometrical figures, and relationships which obtain among numbers and among geometrical figures. All other concepts and relationships of mathematics are derived from numbers and geometry. Science deals with mass, velocity, force, energy, molecular structure, chemical processes, the structure of plants, animals, and humans, and hundreds of other concepts. That is, the subject matter of mathematics differs from that of science.

A second difference is that mathematics will always insist on deductive proof, whereas the sciences, even though they aim to be deductive, will continue to utilize any experimental or observational fact as a basis for conclusions; that is, the sciences do not insist on a thoroughly deductive structure based on a fixed number of axioms stated at the outset. There are many hypotheses of mathematics, such as, for example, that every even number is the sum of two primes, for which the inductive evidence is most conclusive. No scientist would hesitate to use an assertion so well supported by evidence. But the mathematician continues to search for a deductive proof. This difference between mathematics and science is perhaps one of degree or method of operation. Having chosen his axioms, the mathematician proceeds to derive as many conclusions as possible from them. The scientist will not hesitate to introduce new axioms if such a step seems warranted by inductive evidence.

A third difference between mathematics and science was, peculiarly, accentuated by the creation of non-Euclidean geometry. Mathematics, like science, had been devoted primarily to the exploration of nature. Yet mathematicians had always felt free to develop the implications of the axioms of number and Euclidean geometry even though there was no immediate application for any of the results pursued. Number and Euclidean geometry were regarded as so important in the study of nature that almost any information about them was welcomed. However, non-Euclidean geometry, which, at the outset and for a long time thereafter, seemed to concern axioms which could not possibly apply to the physical world, had finally proved useful in the study of the physical world. Thus history teaches us that mathematicians should feel free to investigate axioms which have no immediate or obvious bearing on the physical world. Consequently, mathematics has been given a new dimension of freedom, the freedom to explore what the mind wishes to explore, and has been released from bondage to the axioms of number and Euclidean geometry. One could say that the creation of non-Euclidean geometry had the effect of divorcing mathematics from science. By the end of the nineteenth century, Georg Cantor, one of the great mathematical minds of modern times and the creator of a strange and revolutionary theory, the theory of transfinite numbers, was able to say, "The essence of mathematics is its freedom." The enormous expansion in mathematical activity in the last century is partly the consequence of the new freedom.*

EXERCISES

1. Why does the existence of non-Euclidean geometry show that mathematics does not offer truths?
2. What distinction should one make between a mathematical space and physical space?
3. Is it proper to regard mathematics as one of the sciences?

20-8 THE IMPLICATIONS OF NON-EUCLIDEAN GEOMETRY FOR OTHER BRANCHES OF OUR CULTURE

In view of the role which mathematics plays in science and the implications of scientific knowledge for all of our beliefs, revolutionary changes in man's understanding of the nature of mathematics could not but mean revolutionary changes in his understanding of science, doctrines of philosophy, religious and ethical beliefs, and, in fact, all intellectual disciplines.

Let us consider first the effect on scientific thought. Although scientists had more or less recognized that their theories in various branches of science were not the final word, in the back of their minds they continued to believe that true accounts of the various phenomena of nature were possible and that they were working toward these goals. Indeed, in the fields of astronomy and mechanics, the eighteenth-century thinkers proclaimed with certainty that they had found the true laws of nature. The influence of Newtonian mechanics on almost all thought was profound, precisely because the intellectual leaders of the eighteenth century were convinced that the mathematical account of nature's behavior was correct. The creation of non-Euclidean geometry affected scientific thought in two ways. First of all, the major facts of mathematics, i.e., the axioms and theorems about triangles, squares, circles, and other common figures are used repeatedly in scientific work and had been for centuries accepted as truths—indeed, as the most accessible truths. Since these facts could no longer be regarded as truths, all conclusions of science which depended upon strictly mathematical theorems also ceased to be truths. Or, to broaden our statement, since scientific structures were and are in large part just

* However, see Section 24–4.

series of mathematical chains of reasoning, the appearance of non-Euclidean geometries raised doubts about the very framework of these structures.

Secondly, the debacle in mathematics led scientists to question whether man could ever hope to find a true scientific theory. The Greek and Newtonian views put man in the role of one who merely uncovers the design already incorporated in nature. However, scientists have been obliged to recast their goals. They now believe that the mathematical laws they seek are merely approximate descriptions and, however accurate, no more than man's way of understanding and viewing nature.

Even on the level of engineering a serious question emerged. Since bridges, buildings, dams, and other works were based on Euclidean geometry, was there not some danger that these structures would collapse? Actually there is no guarantee that they will not. But this thought did not alarm the scientists and engineers of the nineteenth century, who, despite the existence of non-Euclidean geometry, did not believe that the geometry of physical space could be other than Euclidean. The other geometries they dismissed as logical curiosities. The behavior of these scientists and even mathematicians illustrates what has been called the law of inertia in the world of ideas. Just as a body at rest or in motion exhibits inertia or unwillingness to change its velocity, so do human beings balk at changing their ideas. However, the advent of the theory of relativity drove home the point that Euclidean geometry is not necessarily the best geometry for applications. Why then do engineers continue to use Euclidean geometry for ordinary projects? They do so because on the basis of experience Euclidean geometry has been known to be reliable. This is their only assurance. For engineering involving motion with high velocities such as modern accelerators of electrons or neutrons develop, the theory of relativity is used.

In the realm of philosophy, all doctrines built on science were necessarily affected. The most majestic development of the seventeenth and eighteenth centuries, Newtonian mechanics, fostered and supported the view that the world is designed and determined in accordance with mathematical laws. The discovery of more laws in fields such as electricity and light during the early nineteenth century reinforced the belief in a highly mechanistic and deterministic universe. But once non-Euclidean geometry destroyed the belief in mathematical truth and revealed that science offered merely theories about how nature might behave, the strongest reason for belief in determinism was shattered.

Perhaps even more devastating to philosophy was the realization that man can no longer be sure of his ability to acquire truths. Through philosophy man has sought knowledge of ultimate realities, knowledge which would enable him to live wisely, and knowledge which would answer irrepressible questions about the meaning and purpose of his existence on this earth. All people, prior to non-Euclidean geometry, had shared the fundamental belief that man can obtain certainties. The solid basis for this belief had been that man had already obtained some truths—witness, mathematics. No system of thought has ever been so widely and completely accepted as Euclidean geometry. To preceding generations it was the "rock of ages" in the realm of truth. Tradition buttressed self-evidence, and experience bolstered "common sense." Men such as Plato and Descartes were convinced that mathematical truths were innate in human beings. Kant based his entire philosophy on the existence of mathematical truths. But now philosophy is haunted by the specter that the search for truths may be a search for phantoms.

The implication of non-Euclidean geometry, namely, that man may not be able to acquire truths, affects all thought. Past ages have sought absolute standards in law, ethics, government, economics, and other fields. They believed that by reasoning one could determine the perfect state, the perfect economic system, the ideals of human behavior, and the like. The standards sought were not just the most effective ones, but the unique, the correct ones. This belief in absolutes was based on the conviction that there were truths in the respective spheres. But in depriving mathematics of its claim to truth, the non-Euclidean geometries destroyed the shining knight of truth and shattered man's hope of ever attaining any truths. When the anchor of truth was lost, all bodies of knowledge were cast adrift. Apparently the intellectual process does not lead to certainties. In Henri Bergson's words, "One can always reason with reason."

Our own century is the first to feel the impact of non-Euclidean geometry because the theory of relativity brought it into prominence. It is very likely that the abandonment of absolutes has seeped into the minds of all intellectuals. We no longer search for the ideal political system or ideal code of ethics but rather for the most workable. It is almost commonplace to hear people say that one cannot expect perfection. This attitude contrasts sharply with those of the eighteenth century and the Victorian age.

Perhaps the greatest import of non-Euclidean geometry is the insight it offers into the workings of the human mind. No episode of history is more instructive. The view that mathematics is a body of truths, which obtained prior to non-Euclidean geometry, was accepted at face value by every thinking being for 2000 years, in fact, practically throughout the entire existence of Western culture. This view, of course, proved to be wrong. We see therefore, on the one hand, how powerless the mind is to recognize the assumptions it makes. It would be more appropriate to say of man that he is surest of what he believes, than to claim that he believes what is sure. Apparently we should constantly re-examine our firmest convictions, for these are most likely to be suspect. They mark our limitations rather than our positive accomplishments. On the other hand, non-Euclidean geometry also shows the heights to which the human mind can rise. In pursuing the concept of a new geometry, it defied intuition, common sense, experience, and the most firmly entrenched philosophical doctrines just to see what reasoning would produce.

EXERCISES

1. What would you regard as the most serious implication of the creation of non-Euclidean geometry?
2. How does the existence of non-Euclidean geometry affect the goals of scientists?
3. Develop the analogy between different systems of geometry and different bodies of law.
4. Does the existence of non-Euclidean geometry augment the power of science to provide rational comprehension of natural phenomena?

Topics for Further Investigation

1. The history of attempts to find a simpler parallel postulate.
2. The work of Girolamo Saccheri.
3. The life of Carl Friedrich Gauss.
4. The use of non-Euclidean geometry in the theory of relativity.

Recommended Reading

BELL, ERIC T.: *Men of Mathematics*, Chaps. 14 and 16, Simon and Schuster, Inc., New York, 1937.

BONOLA, ROBERTO: *Non-Euclidean Geometry, A Critical and Historical Study of Its Development*, Dover Publications, Inc., New York, 1955.

CARSLAW, H. S.: *Non-Euclidean Plane Geometry and Trigonometry*, Chelsea Publishing Co., New York, 1959.

DUNNINGTON, G. W.: *Carl Friedrich Gauss: Titan of Science*, Exposition Press, New York, 1955.

DURELL, CLEMENT V.: *Readable Relativity*, G. Bell and Sons Ltd., London, 1931.

FRANK, PHILIPP: *Philosophy of Science*, Chap. 3, Prentice-Hall, Inc., Englewood Cliffs, N.J., 1957.

GAMOW, GEORGE: *One Two Three . . . Infinity*, Chaps. 4 and 5, The New American Library, Mentor Books, New York, 1947.

KLINE, MORRIS: *Mathematics in Western Culture*, Chap. 27, Oxford University Press, New York, 1953.

POINCARÉ, HENRI: *Science and Hypothesis*, Chaps. 3 to 5, Dover Publications, Inc., New York, 1952.

RUSSELL, BERTRAND: *The ABC of Relativity*, Harper and Bros., New York, 1926.

SOMMERVILLE, D. M. Y.: *The Elements of Non-Euclidean Geometry*, Dover Publications, Inc., New York, 1958.

WOLFE, HAROLD E.: *Introduction to Non-Euclidean Geometry*, The Dryden Press, New York, 1945.

YOUNG, JACOB W. A.: *Monographs on Topics of Modern Mathematics*, Chap. 3, Dover Publications, Inc., New York, 1955.

CHAPTER 21

ARITHMETICS AND THEIR ALGEBRAS

And wisely tell what hour o' the day
The clock doth strike, by Algebra

SAMUEL BUTLER

21-1 INTRODUCTION

We have seen that mathematics contains several geometries and that the very existence of these geometries has profound implications for the nature of mathematics and for the relationship of geometry to our physical world. It is therefore only natural to ask whether there are also many algebras, and whether the existence of these algebras has comparable implications for mathematics and its relation to the physical world. This question is vital not only because algebra plays a most important role in physical applications but because, after non-Euclidean geometry had taught mathematicians that geometry does not offer truths, many turned to the ordinary number system and the developments built upon it and maintained that this part of mathematics still offered truths. The same thought is often expressed today by people who, wishing to give an example of an unquestionable truth, quote $2 + 2 = 4$.

Examination of the relationship between our ordinary number system and the physical situations to which it is applied will show that it does not offer truths. We shall then see that other algebras do exist and are useful, just as non-Euclidean geometries are useful.

21-2 THE APPLICABILITY OF THE REAL NUMBER SYSTEM

Mathematicians are, of course, free to introduce the symbols 1, 2, 3, 4, . . . , where 2 means $1 + 1$, 3 means $2 + 1$, 4 means $3 + 1$, and so on. Moreover, as we pointed out in Chapter 4, experience suggests that for any three numbers a, b, and c, $(a + b) + c = a + (b + c)$, and so this associative property is adopted as an axiom. We may now prove readily that $2 + 2 = 4$ because, first of all, by the very meaning of 2, we have

$$2 + 2 = 2 + (1 + 1).$$

From the associative axiom it follows that

$$2 + (1 + 1) = (2 + 1) + 1.$$

According to the definition of 3,

$$(2 + 1) + 1 = 3 + 1,$$

and, according to the definition of 4,

$$3 + 1 = 4.$$

And now, by applying the axiom that things equal to the same thing are equal to one another, we can assert that $2 + 2 = 4$.

Thus by a purely logical process which employs definitions and axioms we have proved that $2 + 2 = 4$. But the question we seek to answer is not whether the mathematician can set up definitions and axioms and deduce conclusions. We grant that arithmetic is a valid deductive system. We wish to know whether this system necessarily expresses truths about the physical world.

A possible denial might be entered at once on the ground that the only justification for the associative axiom was limited experience with simple numbers, whereas the axiom asserts something about all whole numbers. There is force to this argument because a generalization on the basis of limited experience may be erroneous. However, there are many thinkers who would assert that whether or not the associative axiom was suggested by experience, it is clearly a truth. Of course, the burden of proof then rests on those who proclaim truths. We shall not insist on this point, since there are weaker links between arithmetic and the physical world.

If a farmer has two herds consisting of 10 and 25 heads of cows, respectively, he knows by adding 10 and 25 that the total number of cows is 35. That is, he need not count the cows. Suppose, however, he brings the two herds of cows to market where they are selling for $100 apiece. Will a herd of 10 cows which might bring in $1000 and a herd of 25 cows which might bring in $2500 together bring in $3500? Every businessman knows that when supply exceeds demand, the price may drop, and hence 35 cows may bring in only $3000. In some idealized world the value of the cows may continue to be $3500, but in actual situations this need not be true.

Let us consider next whether some of the slightly deeper results of arithmetic apply to the physical world. Certainly the statement that $2 \cdot \frac{1}{2} = 1$ is arithmetically correct. But do two half-sheets of paper make one whole sheet and do two half-shoes make one whole shoe? Clearly two physical halves never make one whole unless they can be joined in such a way that the halves merge into one whole. Two half-dollars, in general, equal one whole dollar in purchasing power, but in areas where silver is preferred to bills two half-dollars are worth more. To know whether the arithmetic is applicable, we must examine the physical situation.

Let us consider the addition of velocities. If a river flows at the rate of 3 miles per hour and a man capable of rowing at 5 miles per hour in still water rows downstream, his velocity relative to some fixed point in the river is the sum of 3 and 5, that is, 8 miles per hour. But if Mr. *A* walks along a road at the rate of 3 miles per hour and Mr. *B* walks along at the rate of 5 miles per hour, then *B*'s velocity relative to some fixed point is not 8 miles per hour. Of course not, we would say. But why do we add the 3 and 5 in one case and not in the other? It is the physical situation which tells us when to add and when not to do so.

In all of the above examples we must examine the particular physical situation to determine whether the mathematical result fits. But if we must resort to experience to decide when the results of our arithmetic apply, then it is not the truth of arithmetic on which we rely.

Let us test further the applicability of arithmetic. Suppose that we measure two boards and find them to be 3 and 4 feet long, respectively. If we place these boards end to end, will the result be 7 board feet? Probably not. All measurement is approximate, and our statement that the individual boards are 3 and 4 feet long merely means that we are unable to detect any difference between the actual lengths of the boards and the 3- and 4-foot marks on our measuring device. But the first may be 3.01 feet, and the second 4.01 feet. Together they are then 7.02 feet, and we may be able to detect a difference of 0.02 feet. One may object here and say that the trouble is due to the limitations of our senses. This is indeed true; however, can we continue to claim that $3 + 4 = 7$ applies to the physical world, insofar at least as situations involving measurement are concerned?

We learn in chemistry that when one mixes hydrogen and oxygen, he obtains water. More precisely, if one takes 2 volumes, say 2 cubic centimeters, of hydrogen and 1 volume of oxygen, he obtains 2 volumes of water vapor. Likewise 1 volume of nitrogen and 3 volumes of hydrogen yield 2 volumes of ammonia. We happen to know the physical explanation of these surprising arithmetic relationships. By Avogadro's hypothesis, equal volumes of any gas, under the same conditions of temperature and pressure, contain the same number of *particles*. If, then, a given volume of oxygen contains 10 molecules, the same volume of hydrogen will also contain 10 molecules. Then there are 20 molecules in 2 volumes of hydrogen. Now it happens that the molecules of oxygen and hydrogen are diatomic; that is, each contains two atoms. Each of these 20 diatomic hydrogen *molecules* combines with one *atom* of oxygen to form 20 molecules of water or 2 volumes of water.* The chemistry is interest-

* This phenomenon is clearly explained in Francis T. Bonner and Melba Phillips: *Principles of Physical Science*, Addison-Wesley Publishing Co., Inc., 1957, p. 149.

ing, but the main point we wish to make is that ordinary arithmetic fails to describe correctly the result of combining gases by volume.

Suppose, next, that one raindrop is added to another raindrop. Do we now have two raindrops? If one cloud is joined to another cloud do we now have two clouds? One may protest that in these examples the merged objects have lost their identity, and that the addition process of arithmetic does not contemplate such loss. And precisely for this reason, arithmetic in the normal sense no longer applies.

All of the above examples lead to two general conclusions. One is that there are many physical situations where ordinary arithmetic does not apply; that is, ordinary arithmetic is unable to express proper quantitative truths about these situations. The second conclusion is that even though there are a few situations to which ordinary arithmetic does apply, such as, for example, adding herds of cattle, we must depend upon experience with those very situations to know this fact. If herds of cattle behaved like volumes of gases or like raindrops, then the arithmetic would not apply, and it is only through experience that we learn how they do behave. Hence we have no guarantee that arithmetic *per se* represents truths about the physical world.

EXERCISES

1. If we place the length of one 10-ft ladder on top of the length of another, do we obtain a 20-ft ladder? What is the point of the question?
2. Since measurement is approximate, can we say that pouring two 10-lb packages of flour into one bag will produce one 20-lb package of flour?
3. If we balance two objects in the pans of a scale and then add 5 lb to each pan, will the scale still balance? What axiom of arithmetic is applicable to this situation?
4. If an object is thrown downward with a velocity of 100 ft/sec and acquires a velocity due to gravity of $32t$ ft/sec, is its total downward velocity $(100 + 32t)$ ft/sec? Justify your answer.
5. If we superimpose a sinusoidal sound wave of frequency 100 cycles per second on one which has a frequency of 50 cycles per second, do we obtain a sound wave of frequency 150 cycles per second?
6. If one mixes two equal volumes of water, one having a temperature of 40°F and the other of 50°F, what is the temperature of the mixture?

21-3 BASEBALL ARITHMETIC

Mathematics does not have at its disposal special arithmetics to treat all of the situations in which ordinary arithmetic fails. For example, there is no arithmetic which tells us how volumes of gases combine. Each distinct combination must be analyzed on the basis of physical knowledge about the molecules involved. But there are situations which warrant the introduction of special arithmetical concepts and operations. If an arithmetic, that is, the concepts and operations, accurately describes physical events and permits prediction of future behavior, just as ordinary addition predicts the result of combining two herds of cattle, then it is worth creating.

Our first example of an arithmetic different from the ordinary arithmetic does not serve any deep scientific purposes, but it does fill a need which many millions of Americans experience. These people seem to be very much excited by the batting averages of baseball players. Let us look into this matter. Suppose a player goes to bat 3 times in one game and 4 times in another. How many times in all did he go to bat? There is no difficulty here. He went to bat a total of 7 times. Suppose he hit the ball successfully, that is, got to first base or farther, 2 times in the first game and 3 times in the second. How many hits did he achieve in both games. Again there is no difficulty. The total number of hits is 2 + 3 or 5.

What the audience and the player himself are usually most interested in is the batting average, that is, the ratio of the number of hits to the number of times at bat. In the first game this ratio was 2/3; in the second game the ratio was 3/4. And now suppose the player or a baseball fan wishes to use these two ratios to compute the batting average for both games. One would think that all one had to do would be to add the two fractions. That is,

$$\tfrac{2}{3} + \tfrac{3}{4} = \tfrac{17}{12}.$$

Of course this result is absurd. The player could not get 17 hits in 12 times at bat. Evidently the ordinary method of adding fractions does not give the batting average for both games by adding the batting averages for the separate games.

How can we obtain the correct batting average for the two games from those of the separate games? The answer is to use a new method of adding fractions. We know that the average for both games is 5/7 and the separate batting averages are 2/3 and 3/4. We see that if we add the numerators and add the denominators of the separate fractions and then form the new fraction, we get the correct answer. That is,

$$\tfrac{2}{3} + \tfrac{3}{4} = \tfrac{5}{7},$$

provided this plus sign means adding the numerators and adding the denominators.

We now have the makings of a new arithmetic. The addition of integers will be the usual one. However, the addition of fractions is to obey the definition

$$\frac{a}{b} + \frac{c}{d} = \frac{a+c}{b+d}$$

We can introduce subtraction of one integer from another and subtraction of one fraction from another. In the latter case, the definition is

$$\frac{a}{b} - \frac{c}{d} = \frac{a - c}{b - d}.$$

Though we have no significant physical interpretation for negative integers and negative fractions, at least in terms of the baseball situation, there would be no objection mathematically to introducing them. Likewise, though the physically significant fractions so far as baseball is concerned would be less than 1 or 1, we could introduce any quotient a/b even when a is greater than b. One could even introduce the usual multiplication of fractions and all of the ideas associated with this operation.

The laws of this new arithmetic would, in many instances, be the familiar ones. As to the basic operation of addition, whether of integers or fractions, let us note that the commutative law (see Section 4–5) holds, that is,

$$\frac{a}{b} + \frac{c}{d} = \frac{c}{d} + \frac{a}{b};$$

the associative law holds, that is

$$\frac{a}{b} + \left(\frac{c}{d} + \frac{e}{f}\right) = \left(\frac{a}{b} + \frac{c}{d}\right) + \frac{e}{f};$$

and, if we were to introduce the usual multiplication of fractions, then the distributive law holds, that is

$$\frac{a}{b}\left(\frac{c}{d} + \frac{e}{f}\right) = \frac{ac}{bd} + \frac{ae}{bf}.$$

Nevertheless, the arithmetic will have some peculiarities. Let us note a few of these. Ordinarily $\frac{4}{6} = \frac{2}{3}$. However in adding two fractions, say,

$$\tfrac{2}{3} + \tfrac{3}{5},$$

it would not do to replace $\frac{2}{3}$ by $\frac{4}{6}$, for the answer in one case is $\frac{5}{8}$ and in the other $\frac{7}{11}$, and these two answers are not equal. Further, in the normal arithmetic, fractions such as $\frac{5}{1}$ and $\frac{7}{1}$ behave exactly as do the integers 5 and 7. However, if we add $\frac{5}{1}$ and $\frac{7}{1}$ as fractions in the new arithmetic, we do not obtain $\frac{12}{1}$ but $\frac{12}{2}$.

We shall not pursue further this arithmetic of baseball because it does not have any broad mathematical significance. What we should learn from it is that given a physical situation we can invent an arithmetic to fit it. The arithmetic is man-made. Moreover, when we apply this arithmetic to any physical situation we have no guarantee in advance, any more than we do in the case of ordinary arithmetic, that the arithmetic will work, that is, will predict what happens. Insofar as the arithmetic has been designed to fit a particular situation, the computation of batting averages, it will work, but this is not surprising or due to any inherent truth in arithmetic.

EXERCISES

1. Is a batting average of $\frac{2}{3}$ the same as $\frac{4}{6}$ or $\frac{8}{12}$ for the purpose of computing the batting average for several games?
2. Show that under the definition of addition of fractions discussed in the text and the usual product definition, the distributive law holds.
3. What fraction added to $\frac{1}{2}$ gives 0 in this baseball arithmetic?
4. Suppose a salesman selling vacuum cleaners on a house-to-house basis makes 3 sales in 10 calls on one day, 4 sales in 11 calls the next day, and so on. Could he compute his "batting average" by adding fractions in the manner described in the above text?

21-4 MODULAR ARITHMETICS AND THEIR ALGEBRAS

An entirely different arithmetic from the normal one or the arithmetic of baseball is suggested by our system of recording the time of the day. Six hours after 10 o'clock is not 16 o'clock but 4 o'clock; that is, in this system,

$$10 + 6 = 4.$$

Similarly, 6 hours before 3 o'clock is 9 o'clock; that is,

$$3 - 6 = 9.$$

The idea which this system of telling time suggests is that if two numbers differ by 12 or a multiple of 12, then they are equal. Thus $26 = 2$ because $26 - 2 = 2 \cdot 12$, and $9 = -3$ because $9 - (-3) = 12$. Clearly the equality sign here does not mean the same as in ordinary arithmetic; hence we use the symbol $\equiv$ and write our new equations, which are now called *congruences*, as follows:

$$26 \equiv 2, \text{ modulo } 12; \qquad 9 \equiv -3, \text{ modulo } 12.$$

The phrase "modulo 12" after each equation repeats, in shorthand form, the condition stated above, namely, that the equation holds if and only if we neglect multiples of 12.

In this arithmetic, which is usually limited to whole numbers, any number larger than 12 is congruent to some number less than 12 because we can always subtract from the larger number some multiple of 12 to obtain a number less than 12. Thus, if we start with 35, we can subtract $2 \cdot 12$ and obtain 11. Then

$$35 \equiv 11, \text{ modulo } 12.$$

The number 12 itself is congruent to 0 because $12 - 0 = 1 \cdot 12$. Hence

$$12 \equiv 0, \text{ modulo } 12.$$

Similarly, any negative number is congruent to some positive number less than 12. For example, $-25 \equiv 11$ because $11 - (-25) = 3 \cdot 12$, or if one prefers, $-25 - 11 = -3 \cdot 12$. Thus in the arithmetic modulo 12 we need deal only with positive integers from 0 to 11. We also regard any positive integer less than 12 to be congruent to itself because the difference of the two integers is zero times 12. Thus

$$7 \equiv 7, \text{ modulo } 12.$$

Let us see what the results of simple addition and multiplication are in this modular arithmetic. For example,

$$9 + 6 \equiv 3, \text{ modulo } 12.$$

Also

$$9 + 3 \equiv 0, \text{ modulo } 12,$$

and

$$9 \times 4 \equiv 0, \text{ modulo } 12.$$

Here, then, is an arithmetic in which the sum and the product of two positive numbers can be zero, although the summands and the factors are not zero.

We see that addition and multiplication are possible in this arithmetic modulo 12. It is mathematically significant to ask, Do the inverse operations exist? That is, can one subtract any number from another or divide one number by another in this arithmetic? In discussing the answers to these questions let us keep in mind that the only numbers we need to consider are from 0 to 11.

Let us consider subtraction. We should like to know, given a and b, whether there is some quantity x such that

$$a - b \equiv x, \text{ modulo } 12$$

or, by the meaning of subtraction as the inverse of addition,

$$a \equiv b + x, \text{ modulo } 12.$$

If we take x to be $12 - b + a$, then we shall have

$$a \equiv b + 12 - b + a \quad \text{or} \quad a \equiv 12 + a \quad \text{or} \quad a \equiv a.$$

Just to see what this means, suppose $a = 3$ and $b = 7$. Then $x = 12 - 7 + 3$ or 8. Hence

$$3 - 7 \equiv 8, \text{ modulo } 12$$

which is certainly correct because $-4 \equiv 8$ modulo 12. Thus we can subtract any number from any other in the system of numbers 0 to 11, modulo 12.

And now let us consider division. Here we ask, given a and b with b not zero, is there a number in our collection such that

$$\frac{a}{b} \equiv x, \text{ modulo } 12$$

or, by the meaning of division as the inverse of multiplication,

$$a \equiv bx, \text{ modulo } 12?$$

If we were to try to prove that there always is such an x, we would not succeed. Indeed we can show by an example, that such an x does not always exist. Consider

$$1 \equiv 3x, \text{ modulo } 12.$$

The possible values of x are 0 to 11, and trial of all these values shows that not one will satisfy the equation. Thus we cannot divide 1 by 3.

This example does not prove that division is never possible. For example, it is possible to divide 3 by 9 because

$$3 \equiv 9 \cdot 3, \text{ modulo } 12,$$

so that 3 divided by 9 is 3, modulo 12. Let us also note that

$$3 \equiv 9 \cdot 7, \text{ modulo } 12,$$

so that in this case of 3 divided by 9 there are at least two answers: 7 and 3.

Modular arithmetics are numerous. Another one is suggested by the fact that there are 7 days in the week. If one wanted to know what the day of the week will be 26 days from a Wednesday, say, he could of course determine it by pursuing the sequence, Thursday, Friday, etc., until he had counted 26 days. However, he could recognize that every 7 days brings him back to Wednesday and that all he need to do is count 5 days from Wednesday. In other words, so far as days of the week are concerned, any multiple of 7 can be neglected. This fact suggests an arithmetic modulo 7.

Thus we have

$$5 + 4 \equiv 2, \text{ modulo } 7 \qquad \text{and} \qquad 5 \cdot 4 \equiv 6, \text{ modulo } 7.$$

In the system of arithmetic modulo 12, it was necessary to consider only the numbers from 0 to 11. Analogously, in the system modulo 7 it is necessary to consider only the numbers from 0 to 6. The properties of this latter system are very similar to those of the former but they differ in one essential respect. It was not possible to divide any specific number by any other given number in the modulo 12 system and where division was possible the answer was in some cases not unique. It is possible to divide any number from 0 to 6 by any other from 1 to 6 (division by 0 is excluded) in the modulo 7 system and the answer is unique. Thus 2 divided by 5 is 6 because

$$2 \equiv 5 \cdot 6, \text{ modulo } 7.$$

The fact that division is always possible is easy enough to check by actual trial and so we shall not prove the general fact. We note that the operation of division behaves more normally in the modulo 7 system than in the modulo 12 system. The difference is due to the fact that 7 is a prime number, whereas 12 is not.

When a mathematician investigates an arithmetic and its algebra he attempts to ascertain what basic properties the members of the system obey. We have already done some of this. For example, we have investigated subtraction and division for the modular systems. The mathematician would also investigate whether the basic operations of addition and multiplication are commutative and associative and whether multiplication is distributive with respect to addition. We shall leave some of these matters to the exercises and consider others.

We know that in ordinary arithmetic we can add one equality to another and the sums are equal. Is this true for congruences? Thus if

$$a \equiv b, \text{ modulo } m \tag{1}$$

and

$$c \equiv d, \text{ modulo } m, \tag{2}$$

is it true that

$$a + c \equiv b + d, \text{ modulo } m?$$

Let us see if we can prove this.

The statement that $a \equiv b$, modulo m means that

$$a - b = pm,$$

where p is some whole number. Likewise the statement that $c \equiv d$, modulo m means that

$$c - d = qm,$$

where q is some whole number. If we add these two ordinary equations we find that

$$a - b + c - d = pm + qm$$

or

$$a + c - (b + d) = (p + q)m.$$

This last statement says that the difference between $a + c$ and $b + d$ is a multiple of m. But this can be written as

$$a + c \equiv b + d, \text{ modulo } m. \tag{3}$$

Thus we can add congruences with respect to the same modulus.

It is likewise possible to prove that if

$$a \equiv b, \text{ modulo } m$$

and

$$c \equiv d, \text{ modulo } m,$$

then

$$ac \equiv bd, \text{ modulo } m. \tag{4}$$

We shall leave this proof to an exercise.

We can use these two simple theorems on addition and multiplication of congruences to make an application to ordinary arithmetic. We know that

$$10 \equiv 1, \text{ modulo } 9. \tag{5}$$

Multiplying this congruence by itself, we obtain

$$100 \equiv 1, \text{ modulo } 9. \tag{6}$$

Multiplication of the two above congruences yields

$$1000 \equiv 1, \text{ modulo } 9. \tag{7}$$

Obviously we could continue to higher and higher powers of ten.

Now let us consider any number, say 457. This number actually is $4 \cdot 100 + 5 \cdot 10 + 7$. We may certainly state that

$$7 \equiv 7, \text{ modulo } 9. \tag{8}$$

Since

$$5 \equiv 5, \text{ modulo } 9, \tag{9}$$

we can multiply equations (9) and (5) and obtain

$$5 \cdot 10 \equiv 5, \text{ modulo } 9. \tag{10}$$

Similarly, from $4 \equiv 4$, modulo 9, and from (6), we obtain by multiplication of congruences:

$$4 \cdot 100 \equiv 4, \text{ modulo } 9. \tag{11}$$

Statement (3) says that we can add congruences, modulo the same number; hence, let us add the congruences (8), (10), and (11). The result is

$$4 \cdot 100 + 5 \cdot 10 + 7 \equiv 4 + 5 + 7, \text{ modulo } 9,$$

or

$$457 \equiv 4 + 5 + 7, \text{ modulo } 9.$$

What this result says is that a number and the sum of its digits are congruent. Hence *a number minus the sum of its digits must be a multiple of* 9. To test this statement for the number 457 itself, we note that $457 - (4 + 5 + 7)$, or 441, is 49 times 9.

This result is, of course, of interest to those who like to play with numbers, but it also provides a useful method of checking the ordinary arithmetic operations of addition, subtraction, and multiplication. For example, let us consider the product of 457 and 892. We know that

$$457 \equiv 4 + 5 + 7, \text{ modulo } 9, \qquad \text{and} \qquad 892 \equiv 8 + 9 + 2, \text{ modulo } 9.$$

Moreover, $4 + 5 + 7$, or 16, is congruent to 7, modulo 9, and $8 + 9 + 2$, or 19, is congruent to 1, modulo 9. Hence

$$457 \equiv 7, \text{ modulo } 9,$$

and

$$892 \equiv 1, \text{ modulo } 9.$$

In view of (4), we may multiply these congruences and state that

$$457 \cdot 892 \equiv 7 \cdot 1, \text{ modulo } 9.$$

Then the product of 457 and 892 is congruent to 7, modulo 9. That is, the actual product and the product of the sums of the digits in the two factors are congruent. But the actual product is also congruent to the sum of its digits. Hence the sum of the digits in the product is congruent to the product of the sums of the digits in the factors. We have therefore a *check* on the correctness of the multiplication, which is known as the rule for casting out nines.* But

* If two numbers are congruent modulo 9, then they must have the same remainders upon division by 9, for the difference must be an exact multiple of 9. Hence the rule for casting out nines is sometimes stated as follows: the product of two numbers and the product of the sums of the digits must have the same remainder upon division by 9.

we did not prove and so cannot conclude that if the congruence does hold, then the multiplication is correct.

We shall not pursue further the subject of modular arithmetics. These are studied extensively in the branch of mathematics known as the *theory of numbers*, wherein they are often classified under congruences. Our objective in studying them is primarily to see that there are alternative arithmetics and their algebras. We have also been concerned to see that there is no one necessary arithmetic whose applicability to the physical world is guaranteed. Let us note that in arithmetic modulo 4, $2 + 2 = 0$. Hence $2 + 2$ does not always make 4.

EXERCISES

1. Make up the addition table for the arithmetic, modulo 7.
2. Make up the multiplication table for the arithmetic, modulo 7.
3. By trials with actual numbers decide whether it is possible to subtract any whole number from another and obtain an answer in the arithmetic, modulo 7.
4. Are there any answers to the problem of dividing 4 by 2 in the arithmetic, modulo 6? of dividing 3 by 2?
5. Answer the same questions as in Exercise 4, but applied to the arithmetic, modulo 7. Does any significant conclusion suggest itself from a comparison of the answers in this exercise with those of Exercise 4?
6. Solve the equation $x + 5 \equiv 2$ in the arithmetic, modulo 12.
7. Prove that if $a \equiv b$, modulo m, and $c \equiv d$, modulo m, then $a - c \equiv b - d$, modulo m.
8. Prove that if $a \equiv b$, modulo m, and $c \equiv d$, modulo m, then $ac \equiv bd$, modulo m. [*Suggestion*: Since $a \equiv b$, modulo m, $a = b + pm$.]
9. We know that $16 \equiv 4$, modulo 6, and $4 \equiv 4$, modulo 6. By dividing the first congruence by the second one, we obtain $4 \equiv 1$, modulo 6. What conclusion do you draw?
10. What is the analogue for congruences of the usual axiom that things equal to the same thing are equal to each other?
11. Check the addition of 578 and 642 by the rule of casting out nines.
12. Check the multiplication of 578 and 642 by the rule of casting out nines.
13. a) Does the commutative law of addition hold for congruences modulo m, that is, is $a + b \equiv b + a$, modulo m?
    b) Does the associative law of addition hold for congruences modulo m?
    c) Does the distributive law of multiplication with respect to addition hold for congruences modulo m?
14. What is the answer to each of the following division problems in the arithmetic modulo 7?

    a) $\frac{1}{3}$ b) $\frac{2}{3}$ c) $\frac{3}{4}$ d) $\frac{5}{6}$ e) $\frac{6}{5}$ f) $\frac{5}{4}$

15. Obtain all possible solutions of the following equations in the arithmetic system modulo 6.

a) $2x \equiv 2$ b) $3x \equiv 0$ c) $3x \equiv 3$

21-5 THE ALGEBRA OF SETS

We shall now examine still another algebra. Suppose that a man has inherited two different libraries of books. It would be very natural for him to merge them into one library in which, contrary to the case of inheriting dollar bills, he would not want any duplicates. He is, then, combining the two libraries and hence, mathematically speaking, he is adding one to the other. However, since he will reject any duplicates, he will not add in the usual arithmetic sense. For example, if there are 100 books in one library and 200 in the other, the combined library may contain fewer than 300 books. Indeed, if there are 50 titles in one of the collections that duplicate titles in the other, the unified library will contain only 250 books. Thus, to represent the operation of combining two libraries, we need an addition which permits $100 + 200$ to equal 250.

Mathematicians have devised an arithmetic and algebra whose addition processes represent precisely what happens in the combining of the two libraries. The system is called the algebra of sets. The arithmetic and algebra are both so simple that we may as well discuss the algebra at once.

Let A and B be any two sets of objects. Thus A and B might be the two libraries discussed above. To indicate the addition of B to A in the sense in which the libraries were to be joined, that is, an object common to A and B is to be taken only once, we write $A \cup B$ and this combination of A and B is called the union of A and B. The operation we have just introduced implies: a book is in $A \cup B$ if it is in A, or in B, or in both, but if it is in both, it is counted only once.

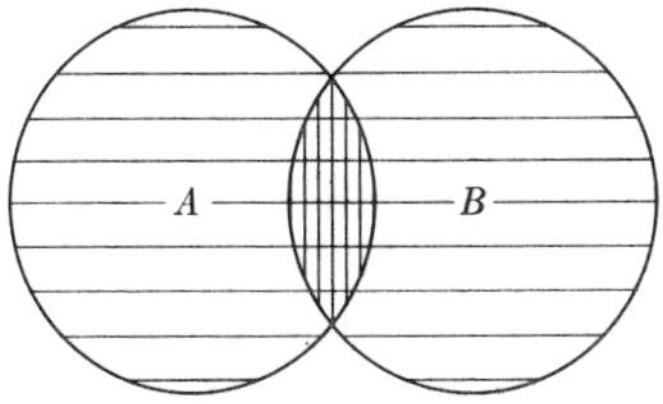

Fig. 21-1.
The union of two sets.

Before proceeding, let us illustrate what the new operation of addition amounts to. Let us suppose that the books in library A are represented as points, and that the entire library consists of all points inside some curve. The set of books in library B can also be represented as the set of points inside some other curve (Fig. 21–1). Since the two libraries contain duplicate titles, these two regions will overlap. The union $A \cup B$ in the sense defined above will then be represented by the collection of points inside both curves, i.e., by the entire shaded area in Fig. 21–1. Of course, the points common to the two regions, the crosshatched area in Fig. 21–1, count only once in the union, but this just means that the union is represented by the points in the entire shaded area.

We thus have a new concept, the union of sets. Just as in the case of ordinary addition, for any A and B,

$$A \cup B = B \cup A$$

because the same collection is formed whether we add collection B to collection A or A to B, that is, the commutative property (Chapter 4) holds for this new operation of union.

Further, the associative property holds for the union of sets. Thus if we have three collections, A, B, and C, we may form $A \cup B$ and then add C to $A \cup B$ to obtain $(A \cup B) \cup C$, or we may begin with $B \cup C$ and add this sum to A to obtain $A \cup (B \cup C)$. Both procedures yield the same total collection. In other words, for this new concept of addition, we have

$$(A \cup B) \cup C = A \cup (B \cup C),$$

whether or not the sets overlap (Fig. 21–2).

It is because these familiar properties of commutativity and associativity hold for the new concept of union that some people use the term addition and the usual plus sign, although the operation of combining the two libraries while rejecting duplicate titles is not the same as the usual combination of collections of cows or dollars. We should, however, note that there are essential differences between the new concept of union and the usual addition. Suppose, for example, that all titles in library B are also in A. Then our new concept of union requires that (Fig. 21–3)

$$A \cup B = A.$$

Moreover, if we add collection A to itself, we obtain A. That is

$$A \cup A = A.$$

The algebra of sets also has a concept somewhat analogous to multiplication. Suppose that the man combining the two libraries were interested in determining how many titles were common to the two libraries. For example, he might wish to know how many books can be sold. The set of titles common to the two libraries A and B is called the intersection of A and B and is denoted by $A \cap B$ or simply, as in ordinary algebra, by AB. If the books in the sets A and B are again pictured as the points inside two curves (Fig. 21–4), then the intersection is represented by the area common to the sets of points A and B, the shaded area in the figure.

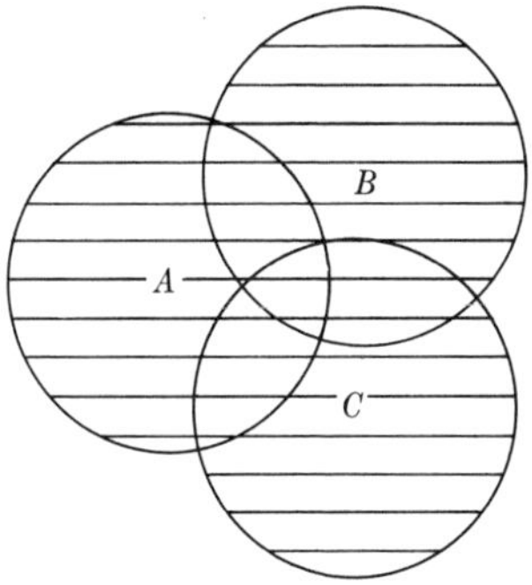

Fig. 21–2

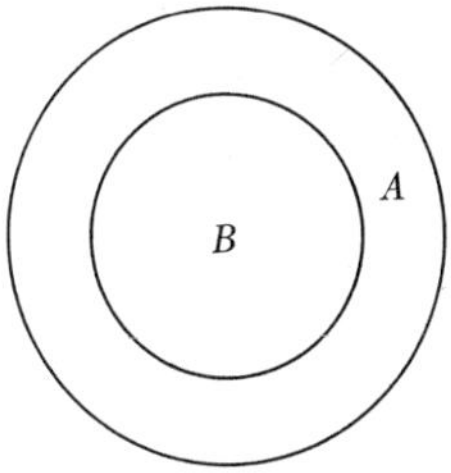

Fig. 21–3. $A \cup B = A$.

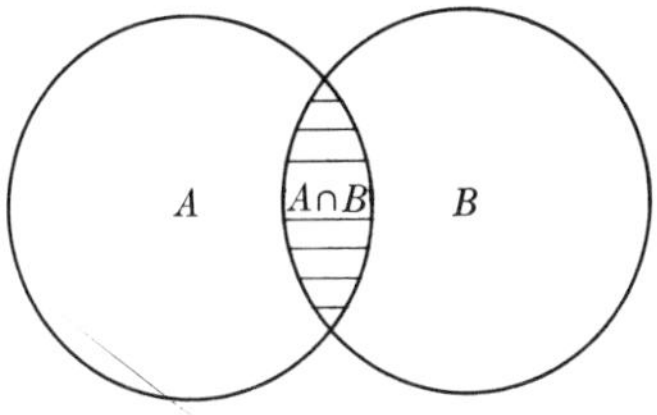

Fig. 21–4. $A \cap B$.

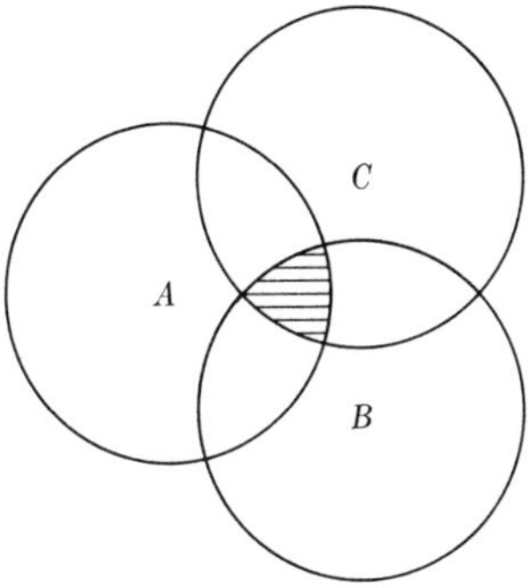

Fig. 21–5

This concept of intersection differs from the ordinary concept of multiplication in that intersection in set algebra usually produces a much smaller set than is contained in either of the factors A or B. Yet the fundamental properties of the intersection of sets are the same as for ordinary multiplication. For example, it is certainly true for the intersection of sets that

$$A \cap B = B \cap A$$

because the same final set, for example, the books common to both libraries, is obtained whether we consider the objects common to A and B or to B and A. Similarly, it is true that

$$(A \cap B) \cap C = A \cap (B \cap C)$$

for, if we select the objects common to A and B (Fig. 21–5) and then those common to $A \cap B$ and C, we surely obtain the same set as if we selected the objects common to B and C and then the titles common to A and $B \cap C$. In either case, we obtain the set of objects in all three sets, A, B, and C. Thus the commutative and associative properties hold for this new concept of intersection.

There are, however, essential conceptual differences between the intersection of sets of objects and the product for ordinary numbers. If in our library example the set B consists of titles which are all in A, then (Fig. 21–3)

$$A \cap B = B,$$

because the books common to A and B are those in B. Also, the intersection of A and A is A; that is,

$$A \cap A = A.$$

Finally, suppose that A and B have no objects in common. What is the set of objects common to A and B? Physically, there is none. Mathematically we introduce the symbol 0 to represent an empty set of objects* and write

$$A \cap B = 0.$$

The symbol 0 possesses many of the usual properties of the number zero. Thus, by the above definitions of union and intersection for sets, it is true that

$$A \cup 0 = A$$

and

$$A \cap 0 = 0.$$

There are other interesting concepts and operations in the algebra of sets. For example, one may be concerned with all books that exist in the world. Then this entire collection of books is called the universe of discourse, and the entire collection is denoted by 1. It follows, unlike ordinary algebra, that

$$A \cup 1 = 1.$$

On the other hand, as in ordinary algebra, we have

$$A \cap 1 = A.$$

We need not explore the entire theory of the algebra of sets to appreciate that this algebra is quite distinct from the algebra of ordinary numbers. However, we shall note that while there is a concept of subtraction, it is not the inverse of addition. Normally $7 - 4$ is a number which when added to 4 gives 7, and we say that subtraction is the inverse of addition because the result of the subtraction when added to the subtrahend or the quantity subtracted gives the original quantity. Given any two sets A and B, the difference $A - B$ is, by definition, the set of objects in A which are not in B. Thus if A and B are the

* Some books use the symbol ϕ.

two sets of points enclosed by the respective circles in Fig. 21–6, $A - B$ consists of the shaded portion of A. If we join B to $A - B$, the union is not A but $A \cup B$. Thus the concept of subtraction is not the usual one. As for division of one set by another it is not possible to make a definition which will yield a quotient for any two sets A and B.

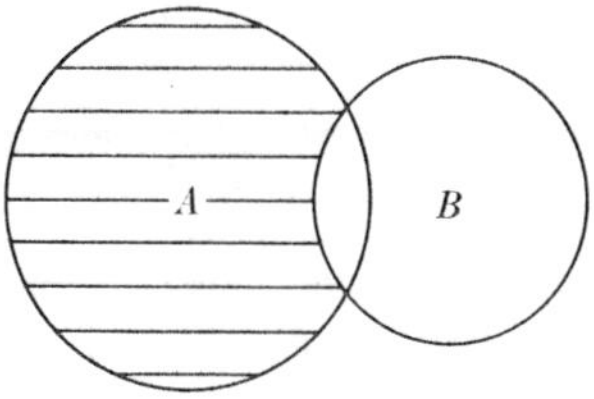

Fig. 21–6

Historically, one of the motivations for the study of the algebra of sets was provided by the study of logic. Mathematicians, notably Descartes and Leibniz, were so much impressed by the usefulness of ordinary algebra that they conceived the idea of inventing an algebra for reasoning in all fields of thought. The concepts of ethics, politics, economics, and philosophy would be the analogues of numbers, and the relationships among these concepts would be the analogues of the operations of arithmetic. They referred to this plan as a universal algebra. The work of Descartes and Leibniz was not successful because they undertook too much. (One can hardly imagine learned Republican and Democratic algebraists sitting down to calculate the solution of a vexing political problem by means of some system of algebra.) It is not likely that the ideas of disciplines such as philosophy and economics can be compactly represented by symbols and the reasoning performed by suitable algebraic operations. However, about 1850, George Boole, one of the founders of mathematical logic, showed that the reasoning processes themselves, which are studied in logic, can be formalized and carried out by an algebra of logic which is identical with the algebra of sets.

Boole's first idea was that in ordinary reasoning we deal with classes or sets of objects. The statement that all students are wise deals with the class of students and the class of wise people. Moreover, the statement itself says that the class of students is included in the class of wise people. If we let A be the class of all students and B the class of all wise people, then the statement that A is included in B can be expressed by the fact that the intersection of A and B is A. Thus the symbolic equivalent is

$$A \cap B = A. \tag{12}$$

The statement that no wise people ignore mathematics can also be expressed symbolically. We let C denote the class of people who ignore mathematics. Since the statement says that there is no person common to the class B of wise people and the class C of those who ignore mathematics, then, in symbols, the statement says

$$B \cap C = 0. \tag{13}$$

These two premises should lead to a conclusion about students and people who ignore mathematics. Hence let us derive an equation involving A and C. Since the left and right sides of (12) and (13) are identical sets, we have

$$(A \cap B) \cap (B \cap C) = A \cap 0. \tag{14}$$

The associative property of intersection tells us that we may arbitrarily group any two factors, just as in the product $(3 \cdot 4)\ (5 \cdot 6)$ we may group the 4 and 5 and write $3(4 \cdot 5)6$. Then

$$(A \cap B) \cap (B \cap C) = A \cap (B \cap B) \cap C.$$

However, $B \cap B = B$. Hence

$$(A \cap B) \cap (B \cap C) = A \cap B \cap C.$$

But by (12) $A \cap B = A$. Therefore

$$(A \cap B) \cap (B \cap C) = A \cap C. \tag{15}$$

From (14) and (15) it now follows that

$$A \cap C = 0. \tag{16}$$

Translated into words, this conclusion states that the class of students and the class of people who ignore mathematics have no members in common, or no student ignores mathematics. We have thus arrived at a conclusion by purely algebraic means.

This example illustrates how Boole used the symbols and algebraic operations of the algebra of sets to perform ordinary reasoning. With his algebra of logic Boole hoped not only to facilitate reasoning but to impart precision to the logical methods of reasoning. His ideas were taken up by others and became the basis of the subject now known as *symbolic logic*.

The algebra of sets and symbolic logic will not be pursued further. Neither is central in mathematics. The algebra of sets is used in a few advanced branches of mathematics, although even there it is a subsidiary notion. Symbolic logic is another very specialized field, somewhat apart from the main body of mathematics. It is used mainly by logicians who are seeking to clarify problems of logic and the relationship of mathematics to logic. The algebra of sets has been presented here only to help illustrate the variety of algebras in mathematics.

EXERCISES

1. If A and B are sets and $A \cup B = B$, what may you infer about the objects in A and B?
2. If A and B are sets and $A \cap B = A$, what may you infer about the objects in A and B?
3. The operation of union of sets has a different meaning from the addition of ordinary numbers. Why might one use the word addition and the symbol "+" for sets?
4. Evaluate in the algebra of sets: $A \cup (A \cap A)$.
5. Suppose A and B contain no objects in common. Evaluate: $A \cap (B \cup A)$.
6. Given the premises that all professors are intelligent and that no students are intelligent, translate these premises into the algebra of sets and deduce a conclusion about the relationship of students to professors.
7. Show that the distributive law $A \cap (B \cup C) = A \cap B \cup A \cap C$ applies to set multiplication and addition.
8. Show that for any two sets A and B,

   a) $A \cup (A \cap B) = A$,
   b) $A \cap (A \cup B) = A$,
   c) $(A - B) \cup B = A \cup B$.

9. Show that if A and B are any two sets, it is impossible to find a set X such that $A/B = X$ or $A = B \cap X$.

21-6 MATHEMATICS AND MODELS

Our study of non-Euclidean geometries and of exotic algebras may have prepared us to see that mathematics is a somewhat different activity from what man had presumed on the basis of the study of ordinary arithmetic and Euclidean geometry alone. The development of the latter two subjects had given rise to the belief that mathematics takes over certain truths about the physical world, adopts these as axioms, and then proceeds to study the physical world by deducing the implications of the axioms. Men did not question that the axioms were truths about the world but tried instead to account for their possession of truths by theories of knowledge or by crediting God with implanting these truths in human minds. However, as mathematicians undertook to study and encompass new classes of physical phenomena or to represent more accurately a previously studied phenomenon such as physical space, they were forced to recognize the need for new concepts and new sets of axioms.

The mathematician really creates models of reality. The concepts, axioms, and theorems of an algebra or a geometry are a model with which to think about some aspect of the physical world. Each model has a limited applicability. Moreover, one must distinguish between the mathematical model and the physical world or between mathematical theories and physical reality.

Topics for Further Investigation

1. Modular arithmetics.
2. The properties of the algebra of sets.
3. The algebra of logic.
4. The nature of symbolic logic.

Recommended Reading

BELL, ERIC T.: *Men of Mathematics,* Chap. 23, Simon and Schuster, Inc., New York, 1937.

BOOLE, GEORGE: *An Investigation of the Laws of Thought,* Chaps. 1 to 7, Dover Publications, Inc., New York, 1951.

COURANT, R. and H. ROBBINS: *What is Mathematics?*, pp. 31–40, 108–116, Oxford University Press, New York, 1941.

LANGER, SUSANNE K.: *An Introduction to Symbolic Logic,* 2nd ed., Dover Publications, Inc., New York, 1953.

NEWMAN, JAMES R.: *The World of Mathematics,* Vol. III, pp. 1852–1900 (selections on symbolic logic), Simon and Schuster, Inc., New York, 1956.

SAWYER, W. W.: *Prelude to Mathematics,* Chaps. 7, 8, 13, and 14, Penguin Books Ltd., Harmondsworth, England, 1955.

THE STATISTICAL APPROACH TO THE SOCIAL AND BIOLOGICAL SCIENCES

People who don't count won't count.

ANATOLE FRANCE

22-1 INTRODUCTION

The success of the deductive approach employed by mathematics and the physical sciences depends upon the acquisition of correct and significant basic principles. In mathematics proper these principles are the axioms of number and geometry. In the physical sciences they are, for example, the laws of motion and gravitation. Though the social scientists sought such principles, they did not succeed in finding them.

The social scientists' inability to find fundamental principles is undoubtedly due to the immense complexity of the phenomena that they wish to study. Human nature is a more complicated structure than a mass sliding down an inclined plane or a bob vibrating on a spring. A phenomenon such as national prosperity is even more complicated; not only are millions of human wills and rapacities involved, but so are natural resources, relationships with other nations, the disruptions of war, and a dozen other major factors. The difficulties which harass the social scientists are also encountered by the biologists. Although the physical sciences have provided some insight into the functioning of the eye, the ear, the heart, and into muscular action, and although chemistry is making rapid advances in the study of complex molecular structures, the operation of the human body and the brain remains, on the whole, a great mystery.

If one attempts to simplify these problems by making assumptions about some of the factors involved or by neglecting what appear to be minor factors, just as Galileo, for example, neglected air resistance, one is likely to make the problem so artificial that its solution no longer has any bearing on real situations.

Very fortunately the social and biological sciences have acquired a totally new mathematical method of obtaining information about their respective phenomena—the method of statistics. By resorting to numerical data and by applying techniques which distill the essential content of those data, these sciences have made striking progress in the past one hundred years. However, the use of statistical methods has also given rise to the problem of determining the reliability of the results, and this aspect of statistics is treated by means of the mathematical theory of probability. In the present chapter we shall survey some of the concepts of statistics and in the next one we shall study the concept and applications of probability.

22-2 A BRIEF HISTORICAL REVIEW

The realization that statistics could serve as a method of attack on major social problems came first to a prosperous seventeenth-century English haberdasher, John Graunt (1620–74). Purely out of curiosity Graunt studied the death records in English cities and noticed that the percentages of deaths due to accidents, suicides, and various diseases were about the same in the localities studied and scarcely varied from year to year. Thus occurrences which superficially seemed to be a matter of chance possessed surprising regularity. Graunt also was the first to discover the excess of male over female births. On this statistic he based an argument: since men are subject to occupational hazards and war service, the number of men available for marriage approximately equals the number of women, and so monogamy has natural sanction. He also noticed the high mortality rate of children and the higher death rate in urban as compared to rural areas. In 1662 Graunt published his *Natural and Political Observations . . . upon the Bills of Mortality*, a book which might be said to have launched the trend toward scientific method in the social sciences and which certainly founded the science of statistics.

Graunt's work was followed and supported by his friend Sir William Petty (1623–85), professor of anatomy at Oxford, professor of music at Gresham College, and later an army physician. Petty wrote on medicine, mathematics, politics, and economics. His *Political Arithmetic*, written in 1676 and published in 1690, did not contain any more striking facts than Graunt's work, but is of particular significance because it calls specific attention to the method of statistics. The social sciences, he insisted, must become quantitative. He says,

> *The method I use is not yet very usual; for, instead of using only comparative and superlative words, and intellectual arguments, I have taken the course . . . to express myself in terms of number, weight, and measure; to use only arguments of sense, and to consider only such causes as have visible foundations in nature.*

To the infant science of statistics he gave the name of "Political Arithmetic," defining it as "the art of reasoning by figures upon things relating to the government." In fact, he regarded all of political economy as just a branch of statistics.

The work of Graunt and Petty was followed by studies of population and income and by extensive studies of mortality rates among which those made by the astronomer Edmond Halley are famous. Life insurance companies formed at the end of the seventeenth century and in the eighteenth century explored further data on mortality. However, though the subject of statistics did come to be known in the eighteenth century as data for statesmen, no mathematical methods for extracting significant implications from the data were developed.

Undoubtedly it was the aggravated social ills brought on by the Industrial Revolution in Europe which prompted a number of men to wade farther into important statistics, such as birth and death records, national and individual incomes, mortality, unemployment, and incidence of diseases, and to seek solutions for major problems through statistical methods. The man who revived Graunt's and Petty's basic thought that statistical methods might produce significant laws for the social sciences was a Belgian, L. A. J. Quetelet (1796–1874). Inspired by the successes of the physical sciences and conscious of the failure of the deductive approach to the social sciences, Quetelet undertook to construct and apply statistical methods suitable for social and sociological investigations. Quetelet was professor of astronomy and geodesy at the École Militaire and in 1820 became director of the Royal Belgian Observatory, which he founded. In 1835 he published his *Essay on Social Physics.* In 1848 he presented to the Royal Belgian Academy a memoir, *On Moral Statistics,* which contained conclusions on the science of government. Ironically, the publication of this memoir coincided with the outbreak of the revolution of 1848 in Paris. Prince Albert of Belgium remarked that the law governing the causes which led to revolutions had unfortunately come a little late.

In the latter half of the nineteenth century a number of well-known scientists, attracted by the already evident power of statistical methods, entered the field. We must content ourselves with mentioning Francis Galton (1822–1911) and Karl Pearson (1857–1936). It so happened that statistical techniques were already proving to be highly important in astronomy and in the theory of gases, and so the physical and social scientists accelerated the creation and application of statistical methods.

Before we examine the mathematics of extracting information from data, we should be clear as to how the method of statistics differs from the deductive approach. To put the matter crudely, the statistical approach to a problem is first of all a confession of ignorance. When crucial experiments, observation, or intuition fails to give us fundamental principles which can be used as premises for a significant chain of reasoning, we turn to *data* and seek to cull whatever information we can from what has happened. If we lack the knowledge which permits us to *deduce* what a new medical treatment should achieve, we apply the treatment, note results, and then attempt to draw some conclusions. Even if we come to the conclusion that the treatment is remarkably successful and should be widely applied, we still do not know what physical or chemical factors are operative. Perhaps the most important difference between the deductive approach and statistical methods is that the latter tell us what happens to large groups and do not provide definite predictions about any one given case, whereas the former predicts precisely what must happen in individual instances.

22-3 AVERAGES

The task of the science of statistics is to summarize, digest, and extract information from large quantities of data. Our illustrations and discussions of various statistical techniques will be based on somewhat artificial and limited classes of data. Real problems usually involve large collections of data whose handling then becomes so encumbered by arithmetic that one loses sight of the essential mathematical idea.

About the simplest mathematical device for the distillation of knowledge from data is the average. A housewife who buys a 5-pound bag of potatoes once a week at prices which vary throughout a year can add the sums spent and divide by 52. She then has an average, known as the arithmetic *mean*, which represents fairly well what potatoes cost during that year.

Such an average can be meaningful in some situations and quite misleading in others. Suppose that we wish to study the wages of workers in an industry, that we have selected 1000 people as our representative sample, and that we have listed the wages of each worker. Suppose further that the mean wage turns out to be $1200. This figure gives some information about the earnings of the people, but not too much. For example, 990 people of the 1000 could be earning $1100 each; 10 people could be earning $11,000 each; and the mean of these earnings would be $1200. Hence our mean figure tells us nothing about the inequalities in the distribution of these wages.

Another type of average commonly used is called the *mode*. In a study of wages, for example, it is that wage which is earned by most people. Suppose that the distribution of wages is such that 25 people happen to receive the wage of $1150, whereas any other salary, larger or smaller, is received by fewer than 25 people. In this case, the mode is $1150. What does this figure tell us about the wage distribution if we do not know what the remaining 975 people earn? Are these others receiving wages near $100,000 a year or near $100 a year? Obviously, the mode may not be the average to represent such a situation.

A third type of average is called the *median*. Let us examine its meaning with the help of the following table:

Salary	*Number of People*
1,000	1
1,100	3
1,200	4
1,300	2
10,000	2
20,000	2
50,000	1

The table lists a total of 15 salaries. The median salary is the salary of the middle man, so to speak, or the eighth person; that is, there are as many who earn less than this middle person as there are who earn more. Since the eighth person occurs among the four who earn \$1200, the median salary is \$1200. To obtain the median of a set of data, one must arrange the data (salaries in the above example) in order of increasing magnitude and then find the datum which occurs in the middle. Of course, one must take into account how many times each datum occurs. Determining the median is a clumsy procedure. But more objectionable is the median's failure to provide any information about the level of the salaries above and below the median.

Although not one of the three averages we have discussed, or others we could discuss, is particularly informative, the mean is nevertheless the best one, for it, at least, takes into account the actual salaries earned by all the people involved. As we shall see later, the mean also proves to be the most useful concept in other statistical techniques.

EXERCISES

1. State the meanings of mean, mode, and median.
2. Calculate the mean and mode for the salaries listed in the above table.
3. A class of 20 students was graded as follows:

Number of Students	6	1	2	3	3	3	2
Grade	10	8	7	5	4	3	2

   What are the mean, median, and mode of these data? Which average best represents the data?

4. The following weekly wages were paid to the employees of a company:

Number of Employees	4	18	10	9	13
Wages	50	40	35	30	10

   Answer the same question as in Exercise 3.

5. Criticize the assertion, "Obviously there must be as many people with above-average intelligence as there are with below-average intelligence."
6. Is it safe for an adult to step into a pool whose mean depth is 4 ft?

22-4 DISPERSION

We have already pointed out that no one of the averages provides detailed information about a set of data. Insofar as the mode and median are concerned, we can certainly change the salaries above and below either of these averages as much as we want to without affecting them. The mean has similar shortcomings. For example, the numbers 3, 5, and 7, each taken once, have a mean of 5, but so do the numbers 0, 5, and 10. Thus one may again change the data, though not at will, and still obtain the same mean.

To obtain further information about a set of data, statisticians seek to measure how closely the data are grouped about the mean; i.e., they seek to determine the *dispersion* of the data about the mean. Thus the data 0, 5, 10 are more widely dispersed about the mean of 5 than are the data 3, 5, 7. Various measures of dispersion might be introduced, but the one which has proved to be the most useful and also the best from the point of view of mathematical manipulation is the *standard deviation.*

Let us note first what is meant by a deviation. Suppose that the grades of six students in a class are:

Number of Students	1	1	1	1	1	1
Grade	3	4	5	6	8	10

We calculate first the mean grade; that is, we multiply each grade by the number of students earning that grade and divide by the number of students. In the present example, the mean grade is 6. The deviation of any grade from the mean is merely the difference between that grade and the mean. Thus the deviations for the above set of grades are

$$3,\ 2,\ 1,\ 0,\ 2,\ 4.$$

To obtain the *standard deviation,* one squares each deviation, calculates the mean of these squares, and then takes the square root. The squares of the deviations are

$$9,\ 4,\ 1,\ 0,\ 4,\ 16.$$

To compute the mean of these squares, we must add them *taking each with the frequency with which it occurs,* and divide by the total number of data. Thus

$$\frac{9 + 4 + 1 + 0 + 4 + 16}{6} = 5.66.$$

The standard deviation, denoted by σ (sigma), is the square root of this mean. Then

$$\sigma = \sqrt{5.66} = 2.4.$$

The standard deviation of 2.4 is merely a convenient and yet somewhat arbitrary measure of how close the various grades are to the mean. Had the grades of the six students been

Number of Students	2	2	2
Grade	2	6	10

,

the mean would again be 6, but the standard deviation would turn out as follows. The deviations are

$$4, \quad 0, \quad 4.$$

The squares of these deviations are

$$16, \quad 0, \quad 16.$$

The mean of these squares is

$$\frac{2 \cdot 16 + 2 \cdot 0 + 2 \cdot 16}{6} = 10.66,$$

and the standard deviation is

$$\sigma = \sqrt{10.66} = 3.3.$$

In other words, the fact that more grades in this latter distribution are farther from the mean of 6 is reflected in the change of the standard deviation from 2.4 to 3.3.

EXERCISES

1. The grades of eight students on a quiz were 1, 2, 4, 5, 8, 9, 9, 10. Calculate the standard deviation of this distribution of grades.
2. Calculate the standard deviation of the grades listed in Exercise 3 of Section 22–3. You may take the mean to be 6.
3. Calculate the standard deviation of the wages listed in Exercise 4 of Section 22–3.
4. Suppose that the mean height of men in a certain city is 5 ft 7 in. and the standard deviation is 2 in., while in another city the mean is the same but the standard deviation is 3 in. What fact is revealed by the difference in standard deviation?
5. Suppose that a student made a grade of 75 in an examination for which the mean grade was 65 and the standard deviation of the grades was 5, and another student made the same grade in an examination for which the mean was also 65, but the standard deviation was 15. Which student did better?
6. What is the significance of the standard deviation?

22–5 THE GRAPH AND THE NORMAL CURVE

A better knowledge about some collections of data than the mean and standard deviation afford can be obtained by means of a graph. Consider, for example, the wages paid in an industry. In this case, the graph might show the wages

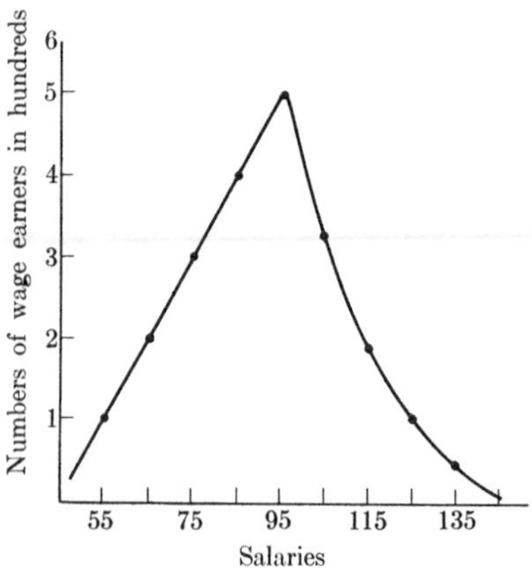

Fig. 22–1.
Frequency distribution of numbers of employees earning different salaries.

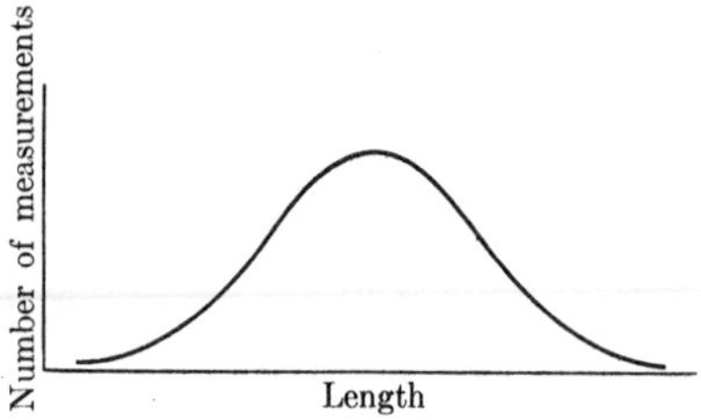

Fig. 22–2.
The normal frequency curve.

paid as the abscissas and the numbers of people earning those wages as the ordinates. Although there may be thousands of different salaries, it is not necessary to plot that many points. One might group the salaries in $10-intervals, and enter into one interval all those earning from $50 to $60, in the next interval those earning from $60 to $70, and so on. One might then regard all those grouped in the first interval as earning $55, those in the second as earning $65, and so forth. Thus one point on the graph will have an abscissa of 55 and and an ordinate equal to the number of people whose salary is somewhere between $50 and $60 or, to be precise, from $50 to $59.99. When a smooth curve is drawn through the points plotted, its shape clearly exhibits the gradual variation in the number of people earning salaries from $50 to the maximum salary paid (Fig. 22–1).

Of course, the process of lumping together all those with earnings from $50 to $60 and using a representative figure of $55 has introduced some element of error into the data and graph. If this error should matter for the purposes of the study, it might be necessary to use a smaller interval of, say $5 or $2, instead of $10. The smooth curve may also be misleading. The graph in Fig. 22–1, for example, seems to show that to *every* possible salary in the range from $50 to the maximum there correspond some wage earners. Actually the graph represents only the number of people earning salaries of $55, $65, $75, and so on. However, the shape of the graph is reasonably accurate with respect to the *relative* frequencies. For example, the smooth curve shows more people earning $60 than, say $55; that is, it shows a distribution which very likely reflects the actual situation, especially if the number of wage earners is large.

The value of the graph is apparent. The mean and standard deviation of the salaries involved would not reveal the rather sharp drop in number from medium- to high-salaried employees and the small number of people earning very high salaries. Graphs, then, do provide a useful picture. A person who reads the newspapers and magazines can hardly fail to observe how commonly they are employed. Not all of these graphs are smooth curves. Bar graphs and pie charts are also frequently used.

The variables whose relationship is represented by a graph may be time and stock prices, production and consumption of coal, or hundreds of other similarly related data. The relationship which plays a central role in statistical work is called a *frequency distribution*. Thus the graph plotting wages versus the number of people earning the various wages illustrates a frequency distribution. The heights of people and the numbers of people possessing these various heights, for example, or intelligence quotients and the numbers of people possessing the various quotients, or grades on an examination and the numbers of students earning those grades are all frequency distributions.

Among frequency distributions one type is of particular importance. Consider the rather simple problem of measuring a length. A scientist interested in the exact length of a piece of wire, say, measures it not once but, if need be, fifty times. Partly because no measurement is exact and partly because environmental conditions such as temperature affect the length, these fifty measurements will differ from each other, sometimes perceptibly and sometimes imperceptibly. A graph plotting the results of all fifty measurements against the number of times that each measurement occurs will look like the curve in Fig. 22–2. In fact, the more measurements that are made, the more nearly will their frequency distribution follow this curve.

The graph in Fig. 22–2 has been well known to physical scientists since about 1800 because it is obtained from almost all measurements of physical quantities. The measurements cluster about one central value, just as the shots of a rifleman at a target will, if he is a marksman, cluster about the bull's-eye. The similarity of the two situations, the measurement of a length and shots at a target, suggests that the exact length should be the central value. The other lengths apparently represent random or accidental variations from the true length just as the shots near, but not on, the bull's-eye represent accidental errors in marksmanship. Because the shape of the graph in Fig. 22–2 occurs repeatedly in connection with errors of measurement, it has come to be known as the *error curve* or the *normal frequency curve*. Its very existence affirms the seemingly paradoxical but nonetheless true conclusion that accidental errors in measurements do not follow any chance pattern, but always follow the error curve. Humans may not even err at will.

The normal frequency curve is not just one curve but a class of curves possessing common mathematical properties, just as the parabola is not a single

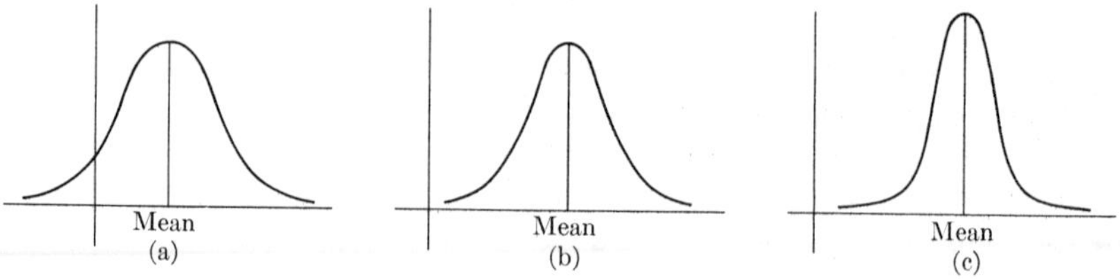

Fig. 22–3. Three different normal frequency curves.

curve but a class of curves which can be defined geometrically as the loci of points which are equidistant from a fixed point and a fixed line, and which are algebraically represented by the equation $y = (1/2a)x^2$ (for the proper choice of coordinate axes). The precise definition of the class of normal frequency curves will not be stated because the formula contains a function we have not studied. But we can characterize this class of curves well enough for our purposes. Figure 22–3 shows three different normal frequency curves. Their shapes resemble bells, and the curve is therefore often described as bell-shaped. Each curve is symmetric about a vertical line. The abscissa of the base point of this vertical line is the mean of the data which are plotted as abscissas, for example, lengths. The means of the three curves will generally be different values. In fact one may be a mean length; another, a mean height; and so on. It is almost apparent from the symmetry of these curves that the mode and median coincide with the mean of each distribution, and the different widths indicate that each has its own standard deviation. The left-hand curve (22–3a) evidently has the largest standard deviation of the three because the data are more widely dispersed.

Despite these differences, all normal frequency curves are characterized by their mean and standard deviation. Regardless of the value of the mean and of the standard deviation, 68.2% of the data lie within σ on either side of the mean (Fig. 22–4); 95.4% of the cases lie within 2σ of the mean; and 99.8% of the cases lie within 3σ of the mean.* Thus if the curve of Fig. 22–4 represented the heights of 100,000 men, if the mean were 67 inches, and if σ were 2, then 68,200 men would have heights falling within the range 65 to 69 inches; 95,400 men would have heights between 63 and 71 inches; and so forth. If the standard deviation were 1 instead of 2, the curve would have a sharper hump around the middle because the dispersion would be smaller. But it would still be true that 68.2% of the population would lie within σ of the mean; i.e., 68,200 men would have heights between 66 and 68 inches. Thus knowing that some frequency

* The normal frequency curve, as a mathematical curve, extends indefinitely far to the right and to the left of the Y-axis although its so-called tails come closer and closer to the X-axis. However, these tails play almost no role since only 0.1% of the measured events can occur beyond 3σ to the right or to the left of the mean.

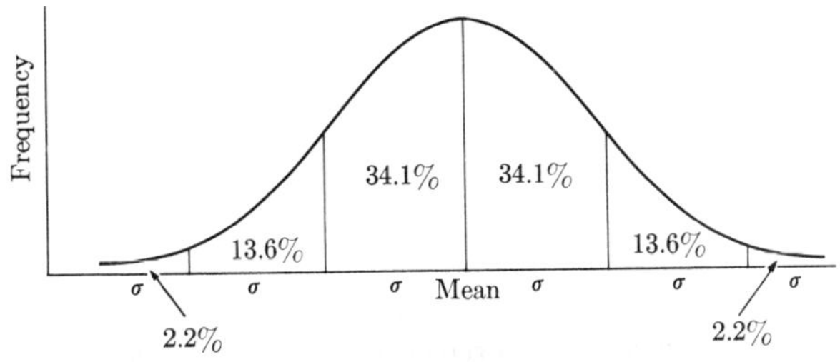

Fig. 22-4. The normal frequency curve.

distribution is normal and knowing the mean and standard deviation of this distribution, we can draw a great number of conclusions about the data.

About 1833 Quetelet decided to study the distribution of human traits and abilities in the light of the normal frequency curve. He took many of his data, incidentally, from the thousands of anatomical measurements made by the Renaissance artists, Alberti, Leonardo, Ghiberti, Dürer, Michelangelo, and others. He found what hundreds of successors have since confirmed. All mental and physical characteristics of human beings follow the normal frequency distribution. Height, the size of any one limb, head size, body weight, brain weight, intelligence (as measured by intelligence tests), the sensitivity of the eye to the various frequencies of the visible portion of the electromagnetic spectrum, all these properties are normally distributed within one genus, which may be race or nationality. The same is true of animals and plants. The sizes and weights of grapefruits of any one variety, the lengths of the ears of corn of any one species, the weights of dogs of any one breed, and so forth, are normally distributed.

Quetelet was struck by the fact that human traits and abilities follow the same distribution curve as do errors of measurement. He concluded that all human beings, like loaves of bread, are made in one mold and differ only because of accidental variations arising in the process of creation. Nature aims at the ideal man, but misses the mark and thus creates deviations on both sides of the ideal. The differences are fortuitous, and, for this reason, the law of error applies to these distributions of physical characteristics and mental abilities. On the other hand, if there were no general type to which men conform, measuring their characteristics, height, for example, would not reveal any particular significance in the graph or any definite numerical relationships in the data.

The typical man, according to Quetelet, emerges as the result of a great number of measurements. The mean of each of the characteristics, that is, the value having the largest ordinate, belongs to this typical, or "mean," man, who is, incidentally, the center of gravity around whom society revolves. The more measurements Quetelet made, the more he noted that individual variations are effaced and that the central characteristics of mankind tend to be sharply defined.

These central characteristics, he then declared, proceed from underlying forces or causes which fashion mankind. More than that, his results led him to believe that he had found decisive evidence for the existence of eternal laws of human society and of design and determinism in social phenomena.

Let us content ourselves, for the moment, with the observation that the applicability of the error curve to social and biological problems has led to knowledge in these fields and to laws. Indeed, the conviction that the distribution of any physical or mental ability must follow the normal curve is today so firmly entrenched that any measurements on a large number of people which do not lead to this result are suspect. If, for example, a new test given to a representative group does not lead to a normal distribution of grades, it is not the conclusion about the distribution of intelligence which is challenged; the test is declared invalid. Similarly, if measurements of velocity, force, or distance failed to follow a normal distribution, the scientist would blame his measuring instruments.

Another use of the normal frequency distribution occurs in manufacture. For example, manufactured wire is continually tested for quality. Suppose that 100 samples are taken each day from the day's production and tested for tensile strength. A graph can be drawn showing strength against the number of samples having that strength. Such graphs usually are normal distributions. Now if the distribution resembled the curve in Fig. 22–3(a), it would imply a wide variety in strength of samples and hence nonuniform production. On the other hand, a distribution such as the curve in Fig. 22–3(c) shows uniform production. The two distributions differ in dispersion, and therefore in their standard deviations. If uniformity is important—and often it is more important than superior quality because a defective piece of wire in an electric circuit may do a lot of damage—then a graph such as the left-hand one indicates the need for some change in the manufacturing process.

One must not presume, however, that all, or practically all, distributions follow the normal frequency curve. The distributions of incomes of families or individuals and the number of families owning 0, 1, 2, 3 or 4 cars are not normal. The failure of incomes to follow a normal curve raises an interesting point because physical and mental abilities are normally distributed and these qualities should determine income.

EXERCISES

1. What is a frequency distribution?
2. Describe the normal frequency curve.
3. Given a normal frequency distribution, what percentage of the data lie within a range of 2σ on both sides of the mean?
4. Suppose that the heights of 1000 college freshmen are measured and found to follow a normal frequency distribution, with a mean of 66 in. and a standard

deviation of 2 in. What percentage of the students has heights between 66 and 70 in.? between 60 and 72 in.?

5. The United States Army gives a well-designed intelligence test to all prospective soldiers and then rejects all those whose scores fall, say below one σ to the left of the mean. Draw a graph showing the frequency distribution of intelligence of the men accepted for service.
6. Where would the mean income of the distribution of incomes shown in Fig. 22–1 lie in relation to the modal income?
7. Suppose you measured the weights of 1000 grapefruits and made a frequency distribution of the various weights. If the resulting curve followed the solid line shown in Fig. 22–5, what would you conclude about the homogeneity of the species of grapefruit?

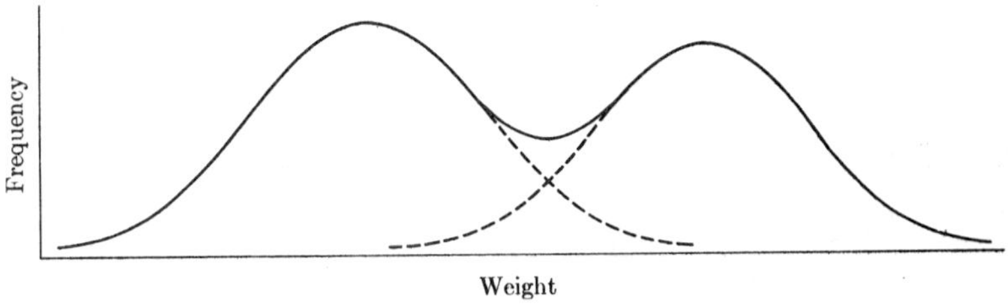

Fig. 22–5

22–6 FITTING A FORMULA TO DATA

We have seen that a great deal of information can be extracted from data by the application of averages, standard deviation, and graphs. When graphical methods are employed, we are particularly fortunate if the graph happens to be a normal frequency distribution. However, the major techniques of mathematics used to derive new information from given facts are designed to apply to formulas. If the data that we happen to be studying present a functional relationship, for example, the variation of population with time in some region, then it is extremely desirable to obtain a formula for this function.

Now the compression of data into formulas is usually possible, and the process is fraught with meaning which we shall examine later. For the present, we shall limit ourselves to illustrating the procedure and, for this purpose, we shall consider first a somewhat specialized and slightly oversimplified problem. By measuring the velocity of a falling body at various instants of time Galileo obtained the following data:

Time, in seconds	0	1	2	3	4
Velocity, in ft/sec	0	32	64	96	128

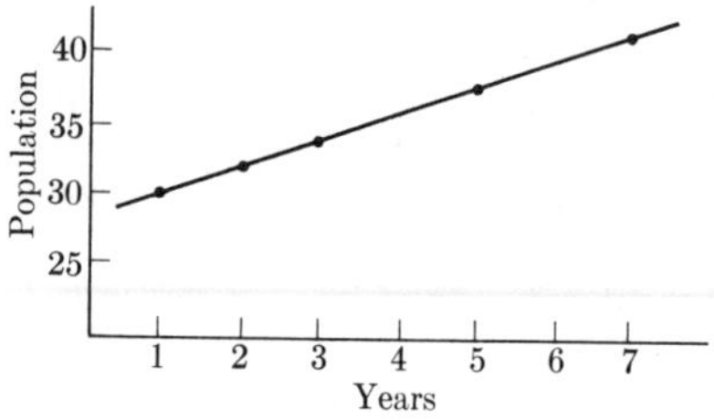

Fig. 22–6.
The graph of population versus years for a particular town.

By *inspecting* this table Galileo could see that the formula relating velocity and time is

$$v = 32t. \tag{1}$$

Let us treat next the problem of obtaining a formula for data when the result is not obvious by inspection. Most towns, cities, and even the country at large are concerned with studying population changes to predict the needs for housing, water, sewers, schools, and so on. Suppose a town has the following record of growth:

Year	1951	1952	1953	1955	1957
Population	3000	3200	3400	3800	4200

Can one find a formula to fit these data?

To simplify the graphing and calculation, let us count years from 1950 and population in hundreds. Thus 1951 will be regarded as the year 1 and the population for that year as 30. The graph of the data is shown in Fig. 22–6. A straightedge placed along the plotted points shows that they lie on a straight line. We know from our work on coordinate geometry (see Exercise 13, Section 12–3) that the formula or equation of a linear graph is of the form

$$y = mx + b. \tag{2}$$

Let y represent the population of the town and x the year. Since the straight line is the curve which fits our data, let us project the straight line backward to the point where it crosses the Y-axis. At this point, we see from the graph that the value of y is 28 and, of course, $x = 0$. Since formula (2) is to fit the graph, then for $x = 0$, y must be 28. If we substitute these values in (2), we have

$$28 = m \cdot 0 + b,$$

and we see that $b = 28$. So far, then, our formula is

$$y = mx + 28. \tag{3}$$

The quantity m is unknown. However, the graph tells us that when $x=5$, $y=38$. Let us therefore substitute these values in (3). We obtain

$$38 = m \cdot 5 + 28$$

or

$$5m = 10$$

or

$$m = 2.$$

Hence the final equation relating y and x is

$$y = 2x + 28, \tag{4}$$

and we have found a formula which fits the data of the above table.

With this formula we can now predict the population of the town, say for the year 1970. For 1970, $x=20$. Then, substituting 20 for x in (4), we obtain

$$y = 2 \cdot 20 + 28 = 68.$$

The formula predicts that the population in 1970 will be 68, that is, 6800 people. Of course, we are assuming that the factors which led to the increase in population from 1951 to 1957 will not only continue to operate, but will operate on the same level. Here we encounter one of the serious limitations in the use of statistics for social and economic phenomena. Since we do not know the fundamental forces which control such phenomena (although we may have some qualitative information), we cannot be at all certain that the population will continue to increase after 1957 in the same way as it did before. In fact, we can be fairly sure that it will not. Local, national, or global events slow down or accelerate the growth of a population. For example, during a financial depression young people cannot afford to marry and have children. By contrast, the study of physical phenomena reveals that the forces operating in nature are invariable.

As a matter of fact, there is a fundamental difference between the formulas obtained from data of the physical sciences and formulas derived from data of biology, psychology, the social sciences, and pedagogy. One may say that, in general, a formula developed from data of the first class *continues to hold* as added data are gathered. Three hundred years ago Kepler deduced his laws from observational data, and they are still correct. On the other hand, for problems of the second class, it does not happen very often that formulas continue to hold without corrections as additional data are gathered. We must constantly refit the formula to the enlarged collection of data. This inconstancy need not be interpreted to imply that lawlessness prevails in the social and biological sciences, for we have already encountered instances of what appear to be well-established laws in these areas.

The above example shows data which led to a linear function. But suppose that the graph were not a straight line or sufficiently close to one to be approximated by a straight line. An example will show that we can nonetheless fit a formula to nonlinear graphs. Suppose that, beginning with the year 1951, the profits of a business concern were as follows:

Year	1951	1952	1954	1956	1958
Profits, in million dollars	0.125	0.5	2	4.5	8

To find a formula fitting these data, we first plot the data. Let us record years after 1950 as abscissas and profits as ordinates. Figure 22–7 shows the graph.

The points plotted are connected by a smooth curve whose appearance suggests a parabola. We know from our work on coordinate geometry that parabolas placed with respect to the axes as shown in the figure have equations of the form [see Chapter 12, formula (9)]

$$y = \frac{1}{2a}x^2. \qquad (5)$$

Let x in (5) stand for time and let y denote the profits. To fit formula (5) to our data, we must first determine the value of a. We choose the coordinates of any one point on the curve. Thus in 1954 the profits were 2. The corresponding point on the curve has coordinates $x = 4$, $y = 2$. Substituting these values in (5), we obtain

$$2 = \frac{1}{2a} \cdot 16.$$

Then $a = 4$, and formula (5) becomes

$$y = \tfrac{1}{8}x^2. \qquad (6)$$

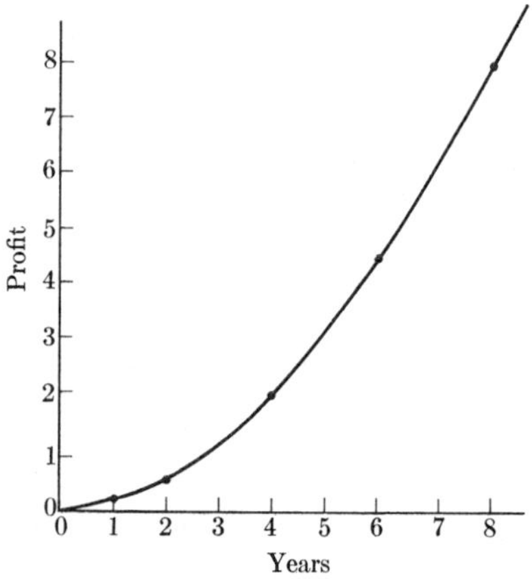

Fig. 22–7.
The graph of profits versus years for a given firm.

Of course, in classifying the curve as a parabola, we judged by appearance. Hence we should check whether (6) fits other data on the graph. Thus, for example, the coordinates of the point corresponding to 1956 are $x = 6$ and $y = 4.5$. If we substitute these values in (6), we see that the left side does indeed equal the right side, and so we have verified that the parabola does fit the data.

If, for example, the profits for the year 1956 had been 4.6 instead of 4.5, then the result of substituting 6 for x in (6) would not have yielded the exact figure.

However, one might still accept formula (6) as a good approximation to the data. If the substitution into (6) of one or more x-values of points on the graph had not yielded good approximations to the corresponding y-values, then our judgment about the parabolic nature of the graph would have been wrong, and we would have had to fit a different type of formula to the data. There are techniques which aid in deciding what formulas to try, but these are valuable only for the specialist.

These few examples, obviously chosen for their simplicity, show how formulas can be fitted to data. At the same time we see that the procedure presupposes a knowledge of coordinate geometry, that is, the relationship between equation and curve.

EXERCISES

1. The following data are given for two variables, x and y:

x	0	1	2	3	4
y	7	10	13	16	19

Find by inspection a formula of the form $y = mx + b$ relating x and y.

2. When a spring is stretched, it exerts a force which opposes the stretching force. To determine the relationship between F, the force exerted by the spring, and d, the amount of stretch or displacement, the following data are available:

d, in inches	1	2	3	4
F, in pounds	4	8	11.9	16.2

Graph these data and find a formula which relates F and d. The result is Hooke's law (Chapter 18) for the particular spring.

3. Suppose that experimentation has yielded the following data on the distance fallen by a body in various time intervals:

t, in seconds	1	$1\frac{1}{2}$	2	$2\frac{1}{2}$
d, in feet	16	36	64	100

Though you undoubtedly know the answer, graph the data, and fit a formula relating d and t to the graph.

4. People are constantly concerned with the rise of prices. These are measured by an index number which represents an average cost of vital commodities and services. Suppose that the index numbers for a number of years are as follows:

Year (Y)	1951	1953	1955	1957	1959
Index number (N)	7	8.9	11.1	13	15.2

Find a formula relating the index number N and the year Y. Check the formula obtained by trying it out on those data of the table that are not used to determine the constants.

5. The profits of a concern for the years following 1950 are:

Years after 1950 (t)	1	2	3	4	5
Profits, in thousands of dollars (P)	6	12	22	36	54

Fit a formula of the form $P = at^2 + b$ to the data.

6. The volume of a given quantity of water will vary with temperature because, as we have had occasion to learn in the past, water expands or contracts with temperature. Hence suppose you had gathered the following data on the volume in cu. in. of a fixed quantity of water at various temperatures (in degrees centigrade):

T	0	2	4	6	8	10	12	14
V	2.3	1.3	1	1.3	2.2	3.7	5.8	8.3

Determine the formula relating T and V. [*Suggestion:* Try the formula $V = a + bT + cT^2$, which, of course, represents a parabola. Then since you know V for $T = 0$, you can immediately find a. Next use two sets of data to determine b and c by solving two equations in two unknowns.]

7. What are the advantages of fitting a formula to data?

8. How would a knowledge of coordinate geometry be helpful to a scientist who is attempting to fit a formula to data?

22-7 CORRELATION

The process of fitting a formula to data is useful in that it summarizes data and may permit further mathematical work with the formula. However, it is not always possible to fit a formula to data. The very notion of a functional relation demands that there be a unique value of y for each value of x in the range studied. Many types of data do not meet this condition. Suppose, for example, one wished to study the relationship between height and weight of men. To any one height there correspond many weights, or if one starts with weight, to any one weight there correspond many heights. Hence one cannot ask for a formula which relates weight and height. Nevertheless there is some correspondence between these two variables, and one might wish to determine the extent and nature of this relationship.

Sir Francis Galton, a cousin of Charles Darwin and founder of the science of eugenics, faced the above problem in his study of human characteristics. Galton was a doctor who used statistics to study heredity. In particular, in his famous *Natural Inheritance*, Galton undertook to investigate the relationship between the heights of fathers and the heights of their sons. It is immediately apparent that to any given height of a father there correspond many heights of sons. Galton introduced a notion now known as correlation. This mathematical concept permits one to measure the closeness of the relationship between two sets of data which may not be functionally related.

Galton found a close relationship between the heights of fathers and their sons. Tall parents have tall children. He also discovered, incidentally, that the mean height of all the sons of tall fathers was closer to the mean of the entire population than was the mean of all the fathers. Thus, while the trait of tallness or shortness is on the average inheritable, succeeding generations regress toward a norm. He also found that the same conclusions apply to intelligence. Talent is, on the average, inherited, but the children are more mediocre than the parents. (Hence parents know better what is good for their children than do the children themselves!). After finding that the same law also holds for other human characteristics, Galton concluded, first, that human physiology is stable and, secondly, that all living organisms tend towards types. Leaving aside Galton's broad inferences, we find in his work examples of biological laws obtained solely by the use of statistics and the simplest of mathematics. Moreover, it was possible to establish these conclusions without any knowledge of the mechanism of heredity.

The precise mathematical measure of correlation which is widely used today was formulated by Karl Pearson. His formula yields a number which lies between -1 and 1. A correlation of 1 indicates that the given variables are directly related; when one variable increases or decreases, so does the other; when one of the variables assumes a high numerical value, so does the other. A correlation of -1 means that the behavior of one variable is directly opposite to that of the other; as the values of the first variable increase, those of the second decrease, and conversely. A correlation of zero means that the behavior of one variable has nothing to do with the behavior of the other; they proceed independently of each other. A correlation of three-fourths, say, indicates that the behavior of one variable is similar to, but not identical with, that of the other.

A knowledge of correlations can be extremely valuable. If stock prices correlate highly with industrial production, one can use a knowledge of the former to study and predict the behavior of the latter. This approach has a definite advantage since stock prices are much more easily compiled than data on industrial production. If general intelligence correlates highly with ability in mathematics, then a person with good intelligence can expect to do well in mathematics. If the total earnings of a nation's wage earners correlate highly with prosperity as measured by the total profits of business concerns, then industry in its own interest ought to consider the prudence of diverting a larger share of its earnings to its employees. If the correlation between the frequencies of occurrence of two diseases is high, the successful analysis of one may be expected to lead to an equally favorable result for the other. Knowledge of the correlation between success in high school and success in college or between success in college and financial success in later life can be extremely valuable in predicting the future of groups of individuals.

EXERCISES

Suppose that a study of 1000 students reveals a very high correlation between general intelligence and ability in mathematics. A particular student is known to be very intelligent. What would you expect his ability in mathematics to be?

22-8 CAUTIONS CONCERNING THE USES OF STATISTICS

Since statistics are now widely employed in our society to bolster arguments on both sides of controversial issues, it might be advisable on this account as well as for a general understanding of the nature of statistically established conclusions to become aware of the pitfalls in applying statistical methods and in interpreting the mathematical results. One of the first difficulties in applying statistics is to decide the meaning of the concepts involved. Suppose that one wished to make a statistical study of unemployment, say over a period of years. Who are the unemployed? Should the term include those people who do not have to work, but would like to? Or people who are employed two days a week and are looking for full-time employment? Or the well-trained engineer who cannot find a job corresponding to his qualifications and has to drive a cab? Or the man unfit for employment? Should a study of passenger cars include taxis, station wagons, and passenger cars used by salesman for business purposes?

After one has decided what objects or groups of people are to be included in a given term, the question of whether the data are reliable arises. For example, any study of crime rates must take into account that police departments occasionally change their practices of recording and classifying crimes. A study of the incidence of mental diseases among men and women must take into account that women are less frequently hospitalized than men.

The clear delineation of the problem to be investigated is also often a difficult matter. Suppose that one wishes to compare deaths due to automobiles in the United States and Great Britain. The number of deaths is certainly larger in the United States but so are the population and the number of automobiles. Should one compute the number of deaths per inhabitant, per automobile, or per mile of automobile travel?

The largest single problem which arises in the process of using statistics is the problem of sampling. To study the incidence of tuberculosis in the United States, for example, one does not examine every person; rather a group of people *believed to be typical* of the whole population is selected for study. This group is called a *random sample.* Similarly, all physiological and mental characteristics of human beings are studied by sampling. The level of retail food prices is gauged by selecting a few important food items which are considered to be representative of all foods. The study of wages in an industry is conducted by selecting a random sample of workers. The doctor studies a person's blood by sampling a small quantity which he believes to be typical because the blood is continually circulating throughout the body. A sociologist interested in the life of families with a given income will study a selected, typical group rather than the entire class. The Gallup poll studying the country's attitude towards public questions and the astronomer studying the number and sizes of stars in a region of the sky use sampling.

Since the conclusions of a statistical study are based on the sample, it is evident that the sample should be chosen with care. If one is studying the output of a machine by sampling its products, it is essential that the products be picked at various times during the day rather than all at one time. In the morning, before there is any chance of overheating, the machine may do better work than in the afternoon

Given a truly random sample, the next question concerns the extent to which the information derived from the sample can be trusted to be indicative of the entire population. This problem involves probabilities, and we shall discuss this subject in the next chapter.

The evaluation of statistical results presents problems of its own. Let us suppose that the meanings of terms are satisfactorily established and that representative samples have been chosen, or that the entire population has been covered so that the questions raised by sampling do not enter. Statistics do show that graduates of Harvard make more money later in life than graduates of any other university. What shall we conclude? Does Harvard's education ensure greater success for the average student there as opposed to the average student at some other university? Hardly. Many of the students at Harvard come from well-established families who take their sons into the family business or profession.

Statistics show that in 1954 among fatal accidents due to automobiles 25,930 occurred in clear weather, 370 in fog, 3640 in rain, and 860 in snow. Do these statistics show that it is safest to drive in fog? Obviously not. Fogs occur more often at night when fewer cars are on the road. When fogs occur, many people refrain from driving, and others drive more cautiously. Finally, fogs are rare.

To see more clearly the danger of drawing hasty conclusions from statistics, one might resort to some extreme and even ridiculous examples. It has been noted that among people who sit in the front rows of burlesque houses bald heads predominate. May one conclude that close observation of burlesque shows produces baldness?

These difficulties in the compilation and evaluation of statistics are real enough. They have led to false inferences and to derogatory remarks such as that there are romances, grand romances, and statistics, or to the definition of statisticians as men who draw precise lines from indefinite hypotheses to foregone conclusions. One must indeed be careful in the uses of statistics. Especially where sampling is involved, statistics do not prove anything; they tend to show. They give us guides to action. Almost always statistical results do not tell us anything certain about an individual, but they indicate what is very likely to hold among a class of individuals as, for example, the distribution of intelligence.

However, the difficulties are easily overshadowed by the effectiveness of the statistical approach in studies of population changes, stock market operations, unemployment, wage scales, cost of living, birth and death rates, extent of drunkenness and crime, distribution of physical characteristics and intelligence, and incidence of diseases. Statistics are the basis of life insurance, social security systems, medical treatments, governmental policies, educational studies, and the numbers racket. Modern business enterprises are using statistical methods to locate the best markets, test the effectiveness of advertising, gauge the interest in a new product, and so forth. Pure speculation, haphazard guesses, and the captiousness of individual judgments are being supplanted by statistical studies. Indeed statistical methods have been decisive in turning undeveloped and backward fields into sciences, and they have become a way of approaching problems and thinking in all fields.

EXERCISES

1. Compare the methodology of the deductive approach to a field of investigation with the statistical approach.
2. Why have economists not succeeded in finding a deductive approach to the economic system of a country?
3. Assuming a reasonable definition of an unemployed person, suppose that statistics show rising unemployment in the United States for a period of five years. Do these statistics imply that the economic condition of the country has been getting worse during those years?
4. Statistics show that every year more people die of cancer than in preceding years. Is cancer caused by factors which are becoming more common in our civilization?
5. Statistics show that cancer occurs much more frequently among men who smoke heavily than among men who smoke little or not at all. Is smoking a cause of cancer?
6. It has been shown statistically that older fathers produce more intelligent children. Do these statistics imply that men should have children later rather than earlier?

7. The average age at death of people with false teeth is higher than that of people possessing natural teeth. Do false teeth enable you to live longer?

Topics for Further Investigation

1. The work of Sir William Petty.
2. The work of John Graunt.
3. The work of Sir Francis Galton.
4. The concept of correlation.
5. The deductive approach to the social sciences.

Recommended Reading

ALDER, HENRY L. and EDWARD B. ROESSLER: *Introduction to Probability and Statistics,* Chaps. 1 to 4, W. H. Freeman & Co., San Francisco, 1960.

FREUND, JOHN E.: *Modern Elementary Statistics,* 2nd ed., Chaps. 1 through 6, Prentice-Hall, Inc., Englewood Cliffs, 1960.

HUFF, DARRELL: *How to Lie with Statistics,* W. W. Norton & Co., New York, 1954.

KLINE, MORRIS: *Mathematics in Western Culture,* Chap. 22, Oxford University Press, New York, 1953.

KLINE, MORRIS: *Mathematics: A Cultural Approach,* Chap. 28, Addison-Wesley Publishing Co., Reading, Mass., 1962.

NEWMAN, JAMES R.: *The World of Mathematics,* Vol. III, pp. 1416–1531 (selections on statistics), Simon and Schuster, Inc., New York, 1956.

REICHMANN, W. J.: *Use and Abuse of Statistics,* Chaps. 1 through 13, Oxford University Press, New York, 1962.

WOLF, ABRAHAM: *A History of Science, Technology and Philosophy in the Sixteenth and Seventeenth Centuries,* 2nd ed., Chap. 25, George Allen and Unwin Ltd., London, 1950. Also in paperback.

CHAPTER 23 *

THE THEORY OF PROBABILITY

Life is the art of drawing sufficient conclusions from insufficient premises.

SAMUEL BUTLER

23-1 INTRODUCTION

Mathematics has been created by man to help him understand the universe and utilize the resources of the physical world. But the physical world of civilized man also includes such activities as throwing dice, playing cards, betting on horse races, playing roulette, and other forms of gambling. To understand and master these very phenomena, a new branch of mathematics, the theory of probability, was created. However, the theory now has depth and significance far beyond the sphere for which it was originally intended.

The first look on the subject was written by the Renaissance roué Jerome Cardan. Cardan, being a mathematician as well as a gambler, decided that if he were to spend time on gambling, he might as well apply mathematics and make the game profitable. He thereupon proceeded to study the probabilities of winning in various games of chance and in a rare moment of altruism decided to let others profit by his thinking on the subject. He compiled his results in his *Liber De Ludo Aleae* (The Book on Games of Chance), which is a gambler's manual with advice on how to cheat and detect cheating.

In 1653 another gambler and amateur mathematician, the Chevalier de Méré, became equally interested in using mathematics to determine bets in games of chance. Since his own talent was limited, he sent some problems on dice to Pascal, and the latter in collaboration with Fermat decided to develop further the subject of probability. Whereas Cardan solved just a few problems of probability, Pascal envisioned a whole science. He aimed

> *to reduce to an exact art, with the rigor of mathematical demonstration, the incertitude of chance, thus creating a new science which could justly claim the stupefying title: the mathematics of chance.*

Cardan, Pascal, and Fermat were attracted to probability by problems of gambling. The subject was taken up by others, notably Laplace, whose interests, equally impractical, were in the heavens. In attempting to solve major astronomical problems, Laplace found himself compelled to consider the accuracy of astronomical observations. As we shall soon see, this problem leads to the theory of probability.

The theory might have remained a minor and largely amusing branch of mathematics were it not for the fact that the use of statistical methods made recourse to probability a necessity. Perhaps the most significant statistical problems which call for probabilistic thinking arise from the process of sampling. Statistical studies must, as a practical matter, proceed by sampling, and sampling unavoidably involves the possibility of error. If a survey of wages in the steel industry is based on data collected in two or three supposedly representative mills, we cannot be sure that we shall obtain *exact* facts about the entire industry from a study of this sample. If the output of a machine is tested by sampling, the conclusion derived from the sample may not hold for the entire output. To decide upon the efficacy of a new medical treatment doctors try it on a small group of patients. Now no treatment is perfect because its effect often depends upon other factors; a good therapeutic procedure for diabetes may be disastrous to a patient with an unusually weak heart. Suppose that the new treatment cures 80% of the people on whom it is tried, whereas some older therapy effected cures in 60% of the patients. Is the new treatment really better or is the difference in percentage an accident due to the particular sample on which it was tried?

All scientific work depends upon measurement. However, all measurements are approximate. Scientists attempt to eliminate this inaccuracy by making many measurements of a given quantity and then taking the mean of the values obtained. It is true that measurements of a quantity form a normal distribution, and we have good reason to believe that the mean of the entire distribution is the true value. But a scientist cannot obtain the entire distribution of measurements in order to find the mean. He can perform twenty or even fifty measurements of a quantity and determine their mean value, but this value is not the mean of the entire distribution. How reliable is the mean computed from the actual measurements?

Although the absence of certainty in some phases of scientific work is deplorable, it is not an insuperable obstacle. As a matter of fact, very little of what we look forward to in our futures is certain. How do we proceed in the face of uncertainty? Descartes stated the course which we all consciously or unconsciously follow: "When it is not in our power to determine what is true we ought to act in accordance with what is most probable." In our daily evaluation of probabilities, we are satisfied with rough estimates; i.e., we merely wish to know whether the probability is high or low. Crossing the street involves uncertainties, but we do cross because without calculation we know that the probability of doing so safely is high. In scientific and large business ventures, however, we must do better. We can no longer accept rough estimates, but must calculate probabilities exactly, and here the mathematical theory of probability serves.

23-2 PROBABILITY FOR EQUALLY LIKELY OUTCOMES

Suppose that we wished to calculate the probability of throwing a three on one throw of a die. One could resort to experience, as many people do anyway, and throw a die 100,000 times. He would find that threes show up on about one-sixth of the throws and conclude that the probability of throwing a three is $\frac{1}{6}$. However, resorting to experience as a means of determining a probability is burdensome and sometimes not even possible. Pascal and Fermat suggested the following approach. In the case of throwing a die, there are six possible outcomes (if we exclude the possibility of the die's coming to rest on an edge). Each of these possible outcomes is equally likely, and of these six, one is favorable to the throw of a three. Hence the probability that a three will show is $\frac{1}{6}$.

If we were interested in the probability that a three or a four will turn up in the throw of a die, we would still have six possible outcomes, but now two of the six would be favorable. In this instance, Pascal's and Fermat's approach would lead to the conclusion that the probability of obtaining a three or a four is $\frac{2}{6}$. If the problem were to calculate the probability of *not* throwing a three, the answer would be $\frac{5}{6}$, because in this problem there are five favorable outcomes out of the six possible ones.

In general, the definition of a quantitative measure of probability is this:

If, of n equally likely outcomes, m are favorable to the happening of a certain event, the probability of the event happening is m/n and the probability of the event failing is $(n - m)/n$.

From this general definition of probability it follows that if no possible outcomes were favorable, that is, if the event were impossible, the probability of the event would be $0/n$, or 0. If all n possible outcomes were favorable, that is, if the event were certain, the probability would be n/n, or 1. Hence the numerical measure of probability can range from 0 to 1, from impossibility to certainty.

As another illustration of this definition consider the probability of selecting an ace in a draw of one card from the usual deck of 52 cards. Here we have 52 equally likely outcomes, of which 4 would be favorable to the drawing of an ace. Hence the probability is $\frac{4}{52}$ or $\frac{1}{13}$.

There is often some question concerning the significance of the statement that the probability of drawing an ace from a deck of 52 cards is $\frac{1}{13}$. Does it mean that if one draws a card 13 times (each time replacing the card drawn), then one draw will be an ace? No, it does not. One can draw a card 30 or 40 times without obtaining an ace. However, the more times one draws, the better will the ratio of the number of aces drawn to the total number of draws approximate the ratio 1 to 13. This is a reasonable expectation because the fact that all outcomes are equally likely means that in the long run each outcome will occur its proportionate share of times.

Suppose that a coin has fallen heads five times in a row, and one now asks for the probability that the sixth throw will also be a head. Many people would argue that the probability of a head showing up on the sixth throw is no longer $\frac{1}{2}$, but is less. The argument generally given is that the number of heads and tails must be the same, and so a tail is more likely to show up after 5 throws of heads. But this is not so. Undoubtedly in a large number of throws, the number of heads will about equal the number of tails, but no matter how many heads have appeared already, the probability of a head on the next throw is still $\frac{1}{2}$. The goddess of fortune has no desire to atone for past misbehavior.

Let us consider another illustration of the definition of probability. Suppose that two coins are tossed up into the air. What are the probabilities of (a) two heads, (b) one head and one tail, and (c) two tails? To calculate these probabilities we must note first that there are *four* different, but equally likely, ways in which these coins can fall, namely: two heads, two tails, a head on the first coin and a tail on the second, and a tail on the first coin with a head on the second one. Of these four possible outcomes only one is favorable to obtaining two heads. Hence the probability of a throw of two heads is $\frac{1}{4}$. Likewise the probability of two tails is $\frac{1}{4}$. The probability that one head and one tail will show is $\frac{2}{4}$ because two of the four ways in which the coins can fall produce this result.

We consider next the probabilities of tossing heads and tails on a throw of three coins. The possible outcomes are:

HHH	*HTH*	*THH*	*TTH*
HHT	*HTT*	*THT*	*TTT*.

We see that there are eight possible outcomes. To calculate the probability of throwing three heads, we observe that only one possible outcome is favorable. Hence the probability of three heads is $\frac{1}{8}$. The probability of throwing two heads and one tail is, however, $\frac{3}{8}$, for of the eight possible outcomes three are favorable. Likewise, the probability of two tails and one head is $\frac{3}{8}$, and the probability of three tails is $\frac{1}{8}$.

Let us note that instead of considering the probability of throwing, say three heads on *one throw of three coins*, we could equally well consider the probability of throwing three heads on *three consecutive throws of one coin.* When three coins are tossed, each falls independently of the other two; hence the fact that they are thrown simultaneously is irrelevant; they could be thrown successively, and the result would be the same. Further, if we let the first coin take the place of the second and then of the third, the result would still be the same because one coin is just like another. Hence three throws of one coin should lead to the same probability as one throw of three coins. We know that the probability of three heads on one throw of three coins is $\frac{1}{8}$. On the other hand, the probability of a head on a throw of one coin is $\frac{1}{2}$. If we multiply the three probabilities of heads on the three throws of one coin, that is, if we form $\frac{1}{2} \cdot \frac{1}{2} \cdot \frac{1}{2}$, we can obtain the result of $\frac{1}{8}$. This example merely illustrates a general result: the probability of many separate events all happening, if the events are independent of one another, is the product of the separate probabilities.

The definition of probability we have been illustrating is remarkably simple and apparently readily applicable. Suppose that one were to argue, however, that the probability that a person will cross a street safely is $\frac{1}{2}$ because there are two possible outcomes, crossing safely and not crossing safely, and of these two, only one is favorable. If this argument were sound, people in large urban centers could not look forward to long lives. The fallacy in the argument is that the two possible outcomes, crossing and not crossing safely, are *not equally likely*. And this is the fly in the ointment. The definition given by Fermat and Pascal can be applied only if one can analyze the situation into equally likely possible outcomes.

One of the most impressive applications of the above concept of probability can now be made on the basis of some work done by Gregor Mendel (1822–1884), abbot of a monastery in Moravia, who in 1865 founded the science of heredity with his beautifully precise experiments on hybrid peas. Mendel started with two pure strains of peas, yellow and green. After cross-fertilization, the peas of the second generation, despite the mixture of green and yellow, proved to be all yellow. When the peas of this second generation were cross-fertilized, three-quarters of the resulting crop of peas were yellow and one-quarter, green. Such proportions had been observed before in the breeding of two pure species, but the explanation of the rather surprising results had eluded biologists.

Mendel supplied the interpretation. He argued that the gamete, or germ cell, of the pure yellow pea contained only a yellow particle, now called a gene,* and the germ cell of the pure green pea contained only a green gene. When the two germ cells were mated, seed developed which contained two genes, one from each parent. Thus each seed contained a yellow gene and a green gene. Why, then, were the peas of this second generation all yellow? Because, said Mendel, yellow was the dominant color. What happens when the hybrid peas are mated? The gamete of the hybrid pea contains only one gene of the pair determining color, and hence may contain a yellow gene or a green gene. Either of these mates with the gamete of another hybrid pea which may contain either a yellow or a green gene. The seed of the offspring contains two genes, one from each parent. Hence the seed may contain one of the following combinations: yellow-yellow, yellow-green, green-yellow, or green-green. All seeds which contain a yellow gene will give rise to yellow peas because yellow is the dominant color. Hence, if all combinations are equally likely, three-

* The genes are contained in chromosomes, but here we wish to concentrate, in particular, on the particles which determine color.

fourths of the third generation should be yellow; this is precisely the proportion Mendel obtained.

Let us now look at Mendel's results from the standpoint of probability. The gamete of a hybrid pea of the second generation can be yellow or green. This is analogous to head and tail of a coin. When two such gametes mate, the combinations are analogous to the combination of heads or tails on two coins. The probability of obtaining at least one head on the two coins is three-fourths because the outcomes of head-head, head-tail, tail-head, and tail-tail are equally likely. The probability of throwing at least one head is the same as that of breeding seed with at least one yellow gene. Hence the laws of probability predict the proportion of yellow peas that will appear in the third generation.

The theory of probability may now be used to predict the proportion of yellow peas in the fourth generation or the proportions of various strains which will result when several different pairs of characteristics, such as yellow and green, tall and short, smooth and wrinkled, are simultaneously interbred. Needless to say, the theory predicts precisely what happens.

This knowledge is now used with excellent practical results by specialists in horticulture and animal husbandry to create new fruits and flowers, breed more productive cows, improve strains of plants and animals, grow wheat free of rust, perfect the stringless string bean, and produce turkeys with plenty of white meat.

The use of the theory of probability in the study of human heredity is especially valuable. Scientists cannot control the mating of men and women, and even if they could do so, it would not be possible to obtain experimental results quickly and easily. Hence they must deduce the facts of heredity from just such considerations as were illustrated above. Moreover, because prejudices frequently enter into judgments of human characteristics, the objectiveness of the mathematical approach is far more essential in the study of human heredity than in studies of plants and animals.

EXERCISES

1. What is the largest value that the probability of an event can have? What does this probability mean?
2. What is the smallest value that the probability of an event can have? What does this probability mean?
3. What is the probability of throwing a two on a single throw of a die? What is the probability of throwing a three? What is the probability of throwing a three or a larger number?
4. The probability that Mr. X will live one more year is $\frac{1}{2}$ because there are two possible outcomes, namely, that he will be alive or dead at the end of the year, and only one of these possibilities is favorable. Do you accept this reasoning?

5. There are 4 caramels and 6 pure chocolate pieces in a box of candy. If a piece of candy is picked at random, what is the probability that it will be pure chocolate?
6. What is the probability of picking a diamond when one card is drawn from the usual deck of 52 cards?
7. What is the probability of throwing 3 heads and 1 tail on a throw of 4 coins?
8. The probability of throwing 4 heads and 1 tail on a throw of 5 coins is $\frac{5}{32}$. What is the probability of not throwing exactly 4 heads and 1 tail?
9. What is the probability of throwing a four or higher on a single throw of a die?
10. If the probability of an event is one in a million, is the event improbable? Is it impossible?
11. What is the number of possible outcomes in throwing two dice? [*Suggestion:* A three on one die and a five on the other is not the same as a five on the first die and a three on the second.]
12. How many of the outcomes in throwing two dice yield a total of 5 on the two faces?
13. What is the probability of throwing a five on a throw of two dice?
14. In Chapter 4 we found that the number of possible variations in the genetic make-up of any one child a husband and wife may have is 2^{48}. What is the probability that two children (not identical twins) will be exactly alike? (Identical twins come from the same fertilized ovum.)
15. Since it is correct to regard a single throw of three coins as equivalent to three successive throws of one coin, present an argument proving that the probability of tossing 2 heads and 1 tail in successive throws is also $\frac{3}{8}$.
16. A young man who dates two girls has to travel by a northbound train to see one, and by a southbound train to visit the other. He argues that since these trains run equally often in both directions, he can take the first one that comes along and, over many trips, will see both girls equally often. But suppose that at his station the train schedule is as follows.

Northbound:	8:00	8:05	8:10	8:15
Southbound:	8:04	8:09	8:14	8:19

Suppose, further, that the young man usually enters the station at any time during the 5-min intervals. Show that the probability that the "girl in the south" is favored (?) is $\frac{4}{5}$. The moral of this problem is not to let chance determine your dates.
17. A term frequently used in discussions of probability is "odds," which means the ratio of the probability in favor of an event to the probability against the event. What are the odds in favor of throwing a head on a single throw of a coin?
18. What are the odds in favor of throwing at least one head on a single throw of two coins?

23-3 PROBABILITY AS RELATIVE FREQUENCY

The concept of probability which we have examined presupposes that we can recognize equally likely outcomes and then take into account those that are favorable to a certain event. But what is the probability that Jones, now forty years old, will live to be sixty?

One cannot say that the two possible outcomes, life or death at age sixty, are equally likely. One might then try to determine the probability that a person of age forty will die of cancer by the age of sixty, perform similar calculations to determine the probability of death due to other causes, such as heart disease, diabetes, fatal accidents, etc., and, assuming that these probabilities could be determined, somehow manage to combine these partial findings to obtain the final probability that an individual of age forty will live 20 more years. This approach would hardly be successful. Yet the probabilities that, starting from a given age, a man will live any specified number of years are of vital importance to insurance companies. Hence these companies had to find ways of determining life expectancies and mortality rates and they went about it as follows. They collected the birth and death records of 100,000 people, and found, for example, that of 100,000 people alive at age ten, 78,106 people were still alive at age forty. They then took the ratio 78,106/100,000 or 0.78 as the probability that a person of age ten will live to be forty. Of the 78,106 alive at age forty, 57,917 were alive at age sixty. The probability of living from forty to sixty was then taken to be the ratio 57,917/78,106, or about 0.74.

This approach to probability is a basic one. It is, in essence, an appeal to experience to determine the favorable outcomes out of the total number of possible outcomes. Of course, the probabilities so obtained are not exact, but a sample of 100,000 people is large enough to ensure fairly reliable probabilities. The reliance upon experience may seem quite different from the calculation of probabilities based on equally likely outcomes, but the difference is not nearly so great as it may appear to be. Why do we decide that each face on a die is equally likely to show up? Actually it is experience with dice which makes us accept the intuitively appealing argument that all faces are equally likely to appear.

Even though probabilities of life expectancies, accidents of various kinds, and the incidence of diseases are obtained from data based on past experience, once they are obtained, mathematics can be employed to calculate with these probabilities. We saw earlier that the probability of throwing two heads on one throw of two coins or two successive throws of one coin is $\frac{1}{2} \cdot \frac{1}{2}$ or $\frac{1}{4}$. If an insurance company is asked to insure the lives of a husband and wife for a twenty-year period, it is important to know the probability that both will be alive twenty years from the date on which the policy is issued. Let us suppose that both are fifty years old. Now the probability that one person of age fifty will live to be seventy is about 0.55 because of about 70,000 people alive at age fifty, 38,500 are alive at age seventy. The probability that *both* will live to be seventy can be determined in the same way as the probability of throwing two heads on two throws of one coin, namely, $0.55 \cdot 0.55$, or about 0.30. Thus, once the probability that a person of age fifty will live to be seventy has been obtained, mathematics can be employed to calculate the probability that two fifty-year old people will *both* live to be seventy. The foregoing is a fairly simple, run-of-the-mill problem. As might be expected, mathematics is employed for the solution of far more complicated probability problems arising in insurance. Thus the theory of probability, which was first developed to solve problems of gambling, takes the gamble out of the insurance business.

EXERCISES

1. How would you determine the probability that a person of age 40 will live to be 60?
2. Of 100,000 ten-year old children, 85,000 reach the age of 30 and 58,000 reach the age of 60. What is the probability that a person of age 10 will live to be 30? What is the probability that a person of age 30 will live to be 60?
3. Suppose that the probability that any one person of age 40 will live to be 70 is 0.5. What is the probability that three particular people of age 40 will live to be 70?
4. Suppose that it is known from long experience that 50% of the people afflicted with a certain disease die of it, say within one year. A doctor who believes that he has developed a new treatment tries it on 4 people and all 4 recover (do not die within one year). How much reliance would you place on the treatment? [*Suggestion:* What is the probability that 4 people having the disease will recover without any treatment?]

23-4 PROBABILITY IN CONTINUOUS VARIATION

For the problems of probability considered so far, the number of possible outcomes was finite. Thus, for example, the number of possible outcomes in throwing three coins is 8, and the number of people who, out of a total sample of 100,000, may live 20 more years can vary from 0 to 100,000. Let us consider, however, the heights of human beings even if limited to values between 4 ft 6 in. and 8 ft 6 in. The number of possible heights in this range is infinite, since any two heights can differ by arbitrarily small amounts. Moreover we know from experience that all possible values in the range from 4 ft 6 in. to 8 ft 6 in. are not equally likely heights of human beings. What, then, is the probability that a man selected at random is between 70 and 71 in. tall?

The theory of probability can treat such problems. We shall consider the most important case, namely, where the frequencies of the various possibilities form a normal distribution (Chapter 22). Since we are now thinking in terms of probability rather than, as in the preceding chapter, of a frequency distribution of some quantity such as height or income, we should restate the normal frequency distribution in terms of probability. Let us consider the possible heights of men from the standpoint of the probability of various heights occurring. In our earlier discussion of normal frequency distributions we said that 34.1% of the cases lie within one σ, that is, one standard deviation, to the right of the mean. What this means is that if, for example, the frequency distribution of 1000 heights is normal, then 341 heights would fall within one σ to the right of the mean. If the mean should be 67 in. and σ should be 2 in., then 341 out of the 1000 men would have heights between 67 and 69 in. Then the probability that a man chosen at random out of the thousand has a height between 67 and 69 in. is 341/1000 or 0.341 because the probability is the ratio of the favorable possibilities or outcomes to the total number of possibilities (Fig. 23–1). Likewise, since in a normal frequency distribution 13.6% of all cases (or, one says, of the entire population) lie between σ and 2σ to the right of the mean, it follows that the probability of any individual's height being between σ and 2σ to the right of the mean is 0.136. In other words, every percentage which appears in the normal frequency distribution becomes a probability under the latter point of view. Thus the normal frequency curve can also be regarded as giving the probabilities of events whose possibilities are normally distributed, and it is therefore also known as the *normal probability curve*.

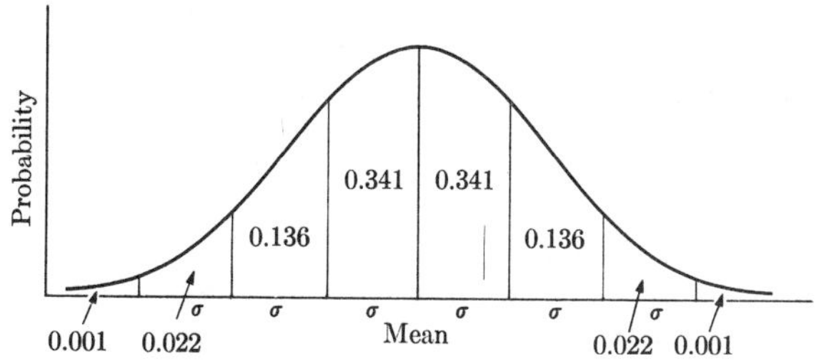

Fig. 23–1. The normal probability curve.

We can now answer questions about the probability of events when the frequencies of the various possibilities are normally distributed. For example, given that the heights of men are normally distributed with a mean of 67 in. and a standard deviation of 2 in., what is the probability of finding a man with height greater than 73 in.? Since 73 in. is 3σ to the right of the mean, a height greater than 73 in. is more than 3σ to the right and, as Fig. 23–1 shows, the probability that a height will fall more than 3σ to the right is 0.001; that is, about one man in a thousand is taller than 73 in.

Note that in dealing with infinite populations, we do not ask for the probability of a particular value, for example, the height 69 in. This probability is zero because it is one possibility in an infinite number of possibilities. But the

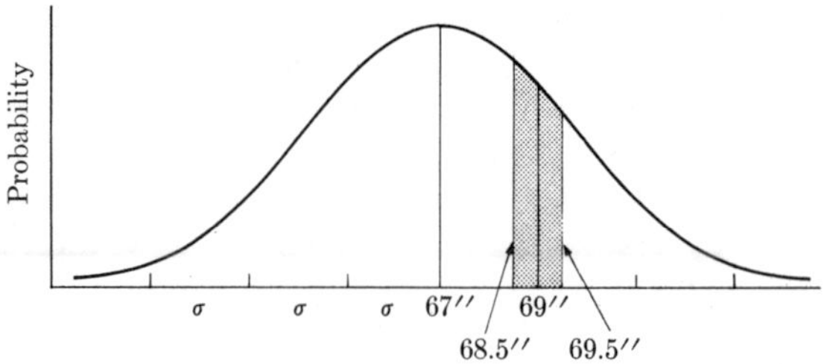

Fig. 23–2

question is not too significant. All measurements yield approximate values only. Let us suppose that the accuracy of measurement is 0.5 in. Then all we should be concerned with is a height between 68.5 in. and 69.5 in.; that is, if we are interested in men with a height of 69 in. and our accuracy of measurement is 0.5 in., then the practical problem we should set is finding the probability that a height falls between 68.5 and 69.5 in. To answer the question, What is the probability that a man chosen at random is between 68.5 and 69.5 in. tall, we would first note that this height lies between $3\sigma/4$ and $5\sigma/4$ to the right of the mean. We would next determine what percentage of the entire population in a normal frequency distribution lies between $3\sigma/4$ and $5\sigma/4$ (Fig. 23–2). We shall not bother to calculate the percentages or the corresponding probabilities which lie within fractional parts of a σ-interval because we wish to restrict our discussion to the simpler percentages or probabilities. However, the probability of a height falling within any given range can be calculated by a method of the calculus.

EXERCISES

1. Suppose that the heights of all Americans are normally distributed, that the mean height is 67 in., and that the standard deviation of this distribution is 2 in. What is the probability that any person chosen at random is between 67 and 73 in. tall? between 63 and 71 in. tall?
2. A manufacturer of electric light bulbs finds that the life of these bulbs is normally distributed and that the mean life is 1000 hr with a standard deviation of 50 hr. What is the probability that any bulb chosen at random will fail to burn at least 950 hr?
3. Given the data of the preceding problem, what is the probability that any bulb chosen at random will burn at least 1100 hr?
4. The grades obtained by a large group of students in an examination were normally distributed with a mean of 76 and a standard deviation of 3. What is the probability that a student had a grade between 76 and 79?
5. The weights of a large number of grapefruits were found to be normally distributed with a mean of 1 lb and a standard deviation of 3 oz. What is the probability that any one grapefruit has a weight between 1 lb 3 oz and 1 lb 6 oz?

23-5 BINOMIAL DISTRIBUTIONS

Let us return for a moment to the subject of tossing coins. When one coin is tossed, there are two possible outcomes: one head and one tail. When two coins are tossed (or one coin is tossed twice), there are four possible outcomes: one yielding two heads, two yielding one head and one tail, and one yielding two tails. When three coins are tossed (or one coin is tossed three times), there are eight possible outcomes: one yielding three heads, three yielding two heads and one tail, three yielding one head and two tails, and one yielding three tails. We could now calculate the total number and distribution of outcomes in tossing four coins, five coins, and so on.

Thinking about this problem of tossing many coins led Pascal to make use of the following "triangle" (now named after him):

$$
\begin{array}{ccccccccccccc}
 & & & & & & 1 & & & & & & \\
 & & & & & 1 & & 1 & & & & & \\
 & & & & 1 & & 2 & & 1 & & & & \\
 & & & 1 & & 3 & & 3 & & 1 & & & \\
 & & 1 & & 4 & & 6 & & 4 & & 1 & & \\
 & 1 & & 5 & & 10 & & 10 & & 5 & & 1 & \\
1 & & 6 & & 15 & & 20 & & 15 & & 6 & & 1 \\
\cdot & & \cdot & & \cdot & & \cdot & & \cdot & & \cdot & & \cdot
\end{array}
$$

Each number in this triangle is the sum of the two numbers immediately above it (zero must be supplied where one of these two numbers is missing). Thus 4 in the fifth row down is the sum of 1 and 3, and 6 is the sum of 3 and 3. Pascal discovered how well this triangle represents the probabilities of getting heads or tails in throwing coins. Take the case of *three* coins, for example. The number of possible outcomes is 8, and this is the sum of the numbers in the *fourth* row. The probabilities involved: $\frac{1}{8}$, $\frac{3}{8}$, $\frac{3}{8}$, $\frac{1}{8}$, are obtained from the individual numbers in the fourth row, namely 1, 3, 3, 1. Similarly, the probabilities of the various alternatives which can arise in the process of flipping five coins will be found in the sixth row of the triangle, and so on. (The number 1 in the first row tells us that the probability of winning on a throw of zero coins is 1. This is the only case in which one is certain to win.)

The numbers which appear in the second row, for example, are the coefficients of a and b in $a + b$, that is, 1 and 1. The numbers which appear in the third row are the coefficients of $(a + b)^2$; for since

$$(a + b)^2 = a^2 + 2ab + b^2,$$

we see that the coefficients are 1, 2, 1. The numbers which appear in the fourth row are the coefficients of $(a + b)^3$ for

$$(a + b)^3 = a^3 + 3a^2b + 3ab^2 + b^3.$$

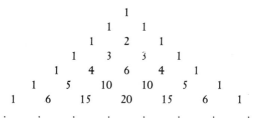

The relationship holds generally. The coefficients of $(a + b)^n$ are the numbers in the $(n + 1)$-row. The quantity $a + b$ is called a binomial because it consists of two terms. Hence distributions which follow any one row of Pascal's triangle are called *binomial distributions*.

If one wished to calculate the probabilities of the various outcomes in tossing 50 coins, he could employ some standard reasoning in the theory of probability or he could extend Pascal's triangle to the fifty-first row. It is, however, clear that this latter process is laborious and, as a matter of fact, calculating the various probabilities by means of the formulas of the theory would be equally laborious. There is an alternative. Let us observe the seventh row of Pascal's triangle. The numbers in this row refer to the throw of six coins, and their sum is 64. Hence they tell us, for example, that the probability of six heads is $\frac{1}{64}$; that of five heads and one tail, $\frac{6}{64}$; that of four heads and two tails, $\frac{15}{64}$; and so on. If we plot the number of possible heads as abscissas and the probabilities of these various numbers of heads as ordinates and draw a smooth curve through the points, we obtain the graph shown in Fig. 23–3. The shape of this graph suggests the normal probability curve. And, indeed, were we to calculate the tenth and the twentieth lines of Pascal's triangle and plot the corresponding graphs, we would find that as the number of coins increases, the graph of the probabilities of the various outcomes approaches the normal probability curve. For a large number of coins, 20 or more, the approximation is so good that we may as well use all the knowledge we have about the normal probability curve and abandon the calculation of the probabilities by special formulas or by extending Pascal's triangle.

To apply the normal probability curve, we must know the mean and the standard deviation. A glance at Pascal's triangle shows that the number of

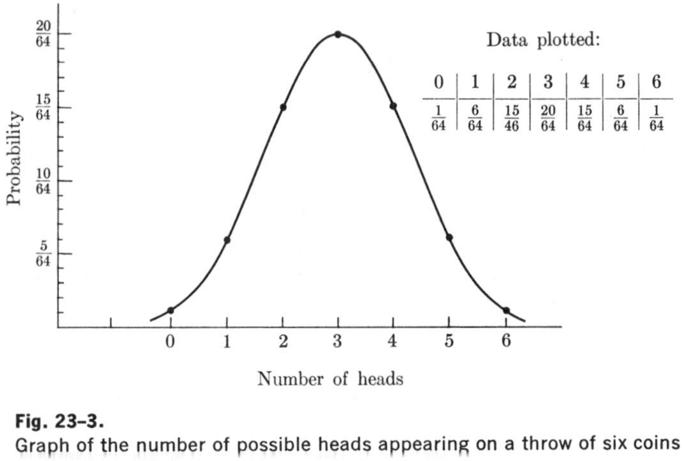

Fig. 23–3.
Graph of the number of possible heads appearing on a throw of six coins versus probability.

heads which has the highest frequency, and therefore the highest probability, is the middle number (or one of the two middle numbers if there is no one middle number). Thus in the seventh row the outcome of 3 heads has the greatest probability. Now this number 3 is the probability of throwing a head, namely, $\frac{1}{2}$, multiplied by the number of coins thrown, that is, 6. This example illustrates a general result which we shall state but not prove. *If n coins are tossed, the mean number of heads is n/2.**

The standard deviation could be determined by applying the procedure for finding the standard deviation of a frequency distribution to the frequencies involved in tossing coins. We would find that

$$\sigma = \sqrt{n \cdot \tfrac{1}{2} \cdot \tfrac{1}{2}} = \tfrac{1}{2}\sqrt{n}.$$

Both of these results hold for the throw of n coins and do not depend upon our approximating the frequencies of the various numbers of heads by the normal frequency distribution. However, if we do make the approximation, we can use the mean and standard deviation just indicated in connection with the normal frequency curve or with the normal probability curve.

The reader may have the impression that mathematicians have become overly absorbed in coin-tossing and accept this extensive interest as another quirk of queer minds. But, in fact, coin-tossing is merely a useful and concrete example which paves the way for more serious applications. Suppose that the probability of death from a certain disease is $\frac{1}{2}$; that is, half the people who contract the disease die within some definite period of time. A new medical treatment is given to 20 afflicted people and only 3 die within that crucial period. Is the treatment really effective or is it just chance that only 3 out of this group of 20 died? After all, to say that the probability of death from a disease is $\frac{1}{2}$ does not mean that in any one group of 20 people 10 will die. It means, as does any probability, that in the long run or in a large number half of those afflicted will die. How should we decide whether the medical treatment is effective?

The possible outcomes, that is, the possible numbers of those remaining alive without treatment among 20 afflicted people, are precisely the same as the possible outcomes in a throw of 20 coins. Just as the number of heads which may show up can vary from 0 to 20, so among the 20 people, none may live beyond the specified period, one may, and so on up to 20. Since the probability that any one person will remain alive is the same as that of tossing a head on a single throw, the probabilities of the various outcomes will be the

---

* In books on probability, this result is stated in more general form. One may, for example, be dealing with dice, and the probability of throwing ones on each die. This probability is $\frac{1}{6}$ and the mean number of ones is $n/6$ or, in general, if p is the probability of the single event, the mean number in n repetitions is np.

same in the two situations. Hence the mean number of those remaining alive is $(\frac{1}{2}) \cdot 20$, or 10. The standard deviation of this distribution is $\sigma = \frac{1}{2}\sqrt{n}$. In our case,

$$\sigma = \tfrac{1}{2}\sqrt{20} = \tfrac{1}{2}(4.47) = 2.24.$$

Then 3σ to the right of the mean is $10 + 6.72$, or 16.72. If we now use the normal probability approximation to our distribution of probabilities (Fig. 23–1), we can say that the probability that more than 16.72 out of 20 afflicted people remain alive without treatment is less than 0.001. But actually 17 people out of the 20 treated remained alive. The probability of this happening in any group of 20 afflicted people is so small that we must credit the treatment with the remarkable record in this group of 20.

Thus when the theory of probability developed for coin-tossing is applied to a serious medical problem, it produces a highly useful conclusion. As a matter of fact, it is precisely this theory which is used to determine the effectiveness of most medical treatments such as the Salk vaccine for poliomyelitis.

Let us consider another application. There is a common belief that boy and girl babies are equally likely or that the probability of a boy being born is $\frac{1}{2}$. Suppose that in 2500 births 1310 proved to be boys. Are these data consistent with the accepted probability of one boy in two births? We can answer this question. The possible numbers of boys in 2500 births are precisely the same as the possible numbers of heads in a throw of 2500 coins (or 2500 throws of one coin). Then the mean number of boys is $(\frac{1}{2})2500$, or 1250. The standard deviation of the distribution of frequencies of these various numbers of boys is $\sigma = \frac{1}{2}\sqrt{n}$. In our case,

$$\sigma = \tfrac{1}{2}\sqrt{2500} = \tfrac{1}{2} \cdot 50 = 25.$$

Now if we use the normal probability distribution as a good approximation to our binomial distribution, we may say that the normal probability curve has a mean of 1250 and a standard deviation of 25. Then 2σ to the right of the mean is 1300. An examination of Fig. 23–1 shows that the probability of 1300 or more boys is 0.023. This means that in only 23 cases out of 1000, on the average, will there be 1300 boys or more in 2500 births.

We are now faced with a problem rather than a conclusion, namely with the occurrence of a very improbable number of births. An event which is very improbable can, of course, occur. However, the entire reasoning is based on the assumption that boy and girl babies are equally likely or that the probability of a boy is $\frac{1}{2}$. It seems far more reasonable to question this hypothesis. As a matter of fact, more extensive records show that the ratio of boys to girls is 51 to 49 instead of 50 to 50. This slight difference in ratio makes a lot of difference in the result. Though our theory does not cover this case, we could show that substituting the probability $\frac{51}{100}$ for $\frac{1}{2}$ leads to the probability of about

$\frac{1}{6}$ that 1300 or more boys will occur in 2500 births. That an event should occur whose probability is $\frac{1}{6}$ is by no means surprising.

In concluding that the medical treatment was effective and in rejecting the hypothesis of equally likely boy and girl births, we relied upon a probability. In the first case we decided that the occurrence of an event whose probability is less than 0.001 implied that the treatment was effective. In the second case we decided that the occurrence of an event whose probability is 0.023 discredited the belief that boy and girl babies are equally likely. The question of what probability to accept as evidence for or against a hypothesis must be decided by the individual concerned, and his judgment will undoubtedly be influenced by the consequences that are likely to arise from his decision.

A recent and most interesting application of the above theory has been made to "prove" the possibility of extrasensory perception, that is, the ability of some people to discern undisclosed facts by extraordinary mental power, for example, to read hidden cards. In the actual tests made by Professor J. B. Rhine and others, subjects were asked to name certain cards held face down, and these people were able to give the correct answers a far greater number of times than the mathematical probabilities of sheer guesses would predict. Thus suppose that the 4 sixes of a deck were held face down on a table. If a subject were asked to select the six of diamonds, he should be able, on the basis of sheer guess, to do so about $\frac{1}{4}$ of the times. But suppose that the subject selects the right card $\frac{1}{3}$ of the times in a large number of trials. Such an unexpectedly large ratio of correct choices is interpreted by Rhine to mean an unusual mental faculty, i.e., extrasensory perception. Of course, the argument here centers about the interpretation of the results in the light of the theory of probability. Rhine's claim is that his experiments also point to telepathy, clairvoyance, prescience, and psychokinesis (the power of the mind to control material objects).

EXERCISES

1. Form the eighth row in Pascal's triangle and calculate from it the probability of throwing 4 heads and 3 tails in a throw of seven coins.
2. What is the probability of 2 heads in a toss of 6 coins?
3. Suppose that 6 coins are tossed 2000 times. Approximately how often will 2 heads appear?
4. What is the probability of getting at least 5100 heads in a toss of 10,000 coins?
5. Suppose that 50% of people afflicted with a certain disease die. A medical treatment is tried on 100 people, and 65 survive. What is the probability that the treatment is effective?
6. Suppose that the conditions in Exercise 5 are changed to 1000 people of whom 650 survive. Does the probability of the treatment's effectiveness change? Justify your finding by a qualitative argument.

7. Suppose 860 boys are born in 1600 births. Do these facts support or discredit the hypothesis that boys and girls are equally likely?
8. Suppose 860 heads turn up in a toss of 1600 coins. What conclusion would you draw?
9. The probability that a person of age 40 will live to be 70 is $\frac{1}{2}$. Out of 400 people in a certain industry who were forty years old, 150 were alive at the age of 70. What may you conclude about the death rate of workers in this industry?

23-6 THE PROBLEMS OF SAMPLING

When one knows the distribution of the probabilities of some variable, one can calculate the probability that a particular value or a range of values will occur and draw a conclusion from this probability. Thus if one knows the distribution of the probabilities of various heights, one can calculate the probability of finding a man with a height, say, between 69 and 70 in. Likewise, if one knows the distribution of the probabilities of the various numbers of heads on a throw of, say 1000 coins, one can calculate the probability of obtaining, for example, 600 or more heads on a particular throw.

In situations of this kind one knows the probabilities of the various alternatives and calculates the probability of a particular one. Now let us consider the following problem. A manufacturer turns out millions of units of the same product each year. He sets up a standard for his product which depends upon the use to which it is put, the price at which it is to sell, the kind of machine which makes it, and other factors. Thus he might decide what, for example, the standard or mean size of his product should be. It is desirable, of course, that all articles produced by one machine be exactly alike, but even under the remarkable accuracy of modern machine performances, complete uniformity cannot be achieved. Hence he also allows for variations, and fixes the standard deviation from the mean which his machinery should hold to. However, machines deteriorate and wear out, or some part may function poorly and affect the output, and the manufacturer therefore checks the output. Since it is too expensive to test each item, he resorts to sampling. Let us say that he examines a sample of 100 units each day. But a sample may have accidental variations which may not be representative of the entire product, just as the mean height of 100 people chosen at random is not necessarily the mean height of all people, even if these people were members of an ethnically homogeneous group. By applying the theory of probability he can decide whether, on the basis of the sample, his machinery is functioning properly, that is, is producing articles with the intended mean and standard deviation.

For the problem of quality control just described, the mean and standard deviation of the entire distribution or population are known, and one judges from a sample whether the mean, say, of the population is being adhered to. There are, however, many problems of sampling in which the mean and standard deviation of the original population are not known, and one is asked to determine them by sampling. For example, a manufacturer of electric light bulbs uses new materials or a new process to make the bulbs and wishes to know how long they will burn on the average. He could test 10,000 bulbs and obtain an answer, but this is not at all necessary. He can test 100 bulbs and find the mean life of this sample. By applying methods of probability theory, which we shall not present, to the mean of his sample, he can determine the mean life of the entire output. The estimate he obtains for the mean of the entire population will not be a precise figure but will lie between certain limits with a probability of almost 1. If he wishes to obtain a better estimate, he can use a larger sample, but the surprising fact is that rather small samples give good estimates of the mean of the entire output.

REVIEW EXERCISES

1. Let us consider a game in which there are 38 possible outcomes (this is true of roulette). Let us suppose that these are numbered 1 to 38. Find the probability
   a) that an even number will win,
   b) that an odd number will win,
   c) that a multiple of four will win.
2. Consider the usual deck of 52 cards. What is the probability in choosing a single card that it will be
   a) an ace? b) a diamond? c) an ace or a diamond?
3. What is the probability of throwing a 3 or a 5 on a single throw of a die?
4. There are 4 red balls and 3 black balls in an urn. If one ball is drawn, what is the probability that it will be
   a) black? b) red? c) red or black?
5. An urn contains 4 red balls, 3 black ones, and 2 white ones. Find the probability that a single ball drawn from the urn will be
   a) black, b) red, c) red or black.
6. a) What is the probability of throwing 2 ones on a single throw of 2 dice? Remember that there are 36 possible outcomes on a throw of 2 dice.
   b) What is the probability of throwing an ace twice in two successive throws of a single die?
7. Out of 100,000 people alive at the age of 10 about 70,000 reach the age of 50 and 65,000 reach the age of 55.
   a) What is the probability that a 10-year-old child will reach the age of 55?
   b) What is the probability that a 50-year-old person will reach the age of 55?
   c) Would you conclude from your answers to parts (a) and (b) that the probability of living to the age of 55 is much better at the age of 50 than at the age of 10?
8. Suppose that the heights of American women are normally distributed and that the mean height is 64 in. and that the standard deviation is $1\frac{1}{2}$ in. What is the probability that the height of a woman chosen at random is less than 61 in.?

Topics for Further Investigation

1. Probability applied to games of chance.
2. The evidence for extrasensory perception.
3. The tests of hypotheses by sampling.
4. Probability applied to the study of heredity.
5. The statistical view of nature.

Recommended Reading

ALDER, HENRY L. and EDWARD B. ROESSLER: *Introduction to Probability and Statistics*, Chaps. 5 through 9, W. H. Freeman & Co., San Franciso, 1960.

BOHM, DAVID: *Causality and Chance in Modern Physics*, Routledge & Kegan Paul Ltd., London, 1957.

BORN, MAX: *Natural Philosophy of Cause and Chance*, Oxford University Press, New York, 1949.

COHEN, MORRIS R. and ERNEST E. NAGEL: *Introduction to Logic and Scientific Method*, Chaps. 15 and 16, Harcourt Brace and Co., New York, 1934.

FREUND, JOHN E.: *Modern Elementary Statistics*, 2nd ed., Chaps. 7 through 11, Prentice-Hall, Inc., Englewood Cliffs, 1960.

GAMOW, GEORGE: *One Two Three . . . Infinity*, Chaps. 8 and 9, The New American Library Mentor Books, New York, 1947.

KASNER, EDWARD and JAMES R. NEWMAN: *Mathematics and the Imagination*, Chap. 7, Simon and Schuster, Inc., New York, 1940.

KLINE, MORRIS: *Mathematics in Western Culture*, Chap. 24, Oxford University Press, New York, 1953.

LAPLACE, P. S.: *A Philosophical Essay in Probabilities*, Dover Publications, Inc., New York, 1951.

LEVINSON, HORACE C.: *The Science of Chance*, Holt, Rinehart and Winston, Inc., New York, 1950.

MORONEY, M. J.: *Facts from Figures*, Chaps. 1 through 14, Penguin Books Ltd., Harmondsworth, England, 1951.

ORE, OYSTEIN: *Cardano, The Gambling Scholar*, pp. 143–241, Princeton University Press, Princeton, 1953.

REICHMANN, W. J.: *Use and Abuse of Statistics*, Chaps. 14 through 17, Oxford University Press, New York, 1962.

RHINE, J. B.: *Parapsychology, Frontier Science of the Mind*, Thomas and Co., Springfield, 1957.

SCHRÖDINGER, ERWIN: *Science and the Human Temperament*, W. W. Norton & Co., New York, 1935. Reprinted under the title *Science, Theory and Man*, Dover Publications, Inc., New York, 1957.

WEAVER, WARREN: *Lady Luck*, Doubleday and Co., Inc., Anchor Books, New York, 1963.

CHAPTER 24

THE NATURE AND VALUES OF MATHEMATICS

This, therefore, is Mathematics: she reminds you of the invisible forms of the soul; she gives life to her own discoveries; she awakens the mind and purifies the intellect; she brings to light our intrinsic ideas; she abolishes oblivion and ignorance which are ours by birth . . .

PROCLUS DIADOCHUS

24-1 INTRODUCTION

We have been studying the ideas, technical content, scientific applications, and cultural influences of mathematics. In our concentration on details we may have failed to note a few broader features. Also, much of what we said at the outset about the nature of mathematics was deliberately limited to ideas which would be helpful in undertaking the study of the subject. But, as we now know, both the nature and the content changed radically during the centuries. It may profit us therefore to survey the subject as mathematicians see it today, in order to gain some insights which could not be attempted at the beginning of our study.

24-2 THE STRUCTURE OF MATHEMATICS

Mathematics, viewed as a whole, is a collection of branches. The largest branch is that which builds on the ordinary whole numbers, fractions, and irrational numbers, or what, collectively, is called the real number system. Arithmetic, algebra, the study of functions, the calculus, differential equations, and various other subjects which follow the calculus in logical order are all developments of the real number system. We shall refer to this branch as the mathematics of number. A second branch is Euclidean geometry. Projective geometry and each of the several non-Euclidean geometries are branches as are various other arithmetics and their algebras. Were we to pursue mathematics still further, we would find that it contains many more divisions.

Each branch has the same logical structure. It begins with certain concepts, such as the whole numbers in the mathematics of number, and such as point, line, and triangle in Euclidean geometry. These concepts must obey explicitly stated axioms. Some of the axioms of the mathematics of number are the associative, commutative, and distributive properties and the axioms about equalities. Some of the axioms of Euclidean geometry are that two points determine a line, all right angles are equal, and the axiom on parallel lines. The non-Euclidean geometry of Gauss, Lobatchevsky, and Bolyai contains the same axioms as Euclidean geometry does, except for the parallel axiom. From the concepts and axioms, theorems are deduced. Hence, from the standpoint of structure, the concepts, axioms, and theorems are the essential components. We shall discuss these in turn.

The basic concepts of the elementary branches of mathematics are abstractions from experience. Whole numbers and fractions were certainly suggested by obvious physical counterparts. But it is noteworthy that many more concepts are introduced which are, in essence, creations of the human mind with or without partial help from experience. Irrational numbers, such as $\sqrt{2}$, were forced upon mathematicians to represent all the lengths occurring in Euclidean geometry, for example, the length of the hypotenuse of a right triangle whose arms are both one unit long. The notion of a negative number, though perhaps suggested by the need to distinguish debits from credits, is nevertheless not wholly derived from experience, for the mind had to conceive of the notion of an entirely new type of number to which operations such as addition, multiplication, and the like can be applied. The notion of a variable to represent the quantitative values of some changing physical phenomenon, such as temperature or time, is also at least one mental step beyond the mere observation of change. The concept of a function, or a relationship between variables, is almost entirely a mental creation. The farther one proceeds with the mathematics of number, the more remote from experience are the concepts introduced and the larger is the creative role played by the mind. The derivative of a function, for example, the instantaneous rate of change of distance compared with time, is a wholly man-made and, one might also say, an ingenious construction. We do not experience instants or instantaneous velocities but, rather, small intervals of time and speeds over small intervals of time.

The gradual introduction of new concepts which more and more depart from forms of experience finds its parallel in geometry. Though point, line, triangle, circle, and a few other elementary concepts are no more than abstractions from experience, this is not true of most of the curves which geometry considers. The approach to the conic sections as sections of a cone made by a plane or by the locus definitions (Chapter 6) subsequently used to define parabola, ellipse, and hyperbola was conceived by the mind. The notions of projection and section of projective geometry and many of the specific concepts of projective geometry such as cross ratio are entirely mental creations.

This brief review of the origin of mathematical concepts may serve to emphasize several major facts. The first of these is growth. As mathematicians continue to work in any given branch, they discover new concepts which are worth introducing and developing. Secondly, as their work advances, the new concepts are less and less drawn from experience and more and more from the recesses of human minds. Moreover, the development of concepts is progressive, later concepts being built on earlier ones. These facts have unfortunate consequences. Because the more advanced ideas are purely mental creations rather than abstractions from immediate experience and because they are defined in terms of prior concepts, it is more difficult to comprehend them. At least, one cannot usually find simple and familiar physical pictures or experiences to illustrate their meanings.

Axioms constitute the second major component of any branch of mathematics. The axioms of the mathematics of number and of Euclidean geometry were suggested by experience. Up to the introduction of non-Euclidean geometry, the idea that man chose axioms suggested by experience would have been regarded as absurd. Axioms were understood to be basic, self-evident truths about the concepts involved. While minor variations in the choice and wording were tolerated, the universal belief was that man had to accept what was clearly true. These truths were supposed to be written into the universe and were considered to be inescapable. Some philosophers, such as Plato and Descartes, believed that these truths were already planted in our minds by God. We know now that this view must be discarded. The seemingly self-evident nature of the axioms of the mathematics of number and of Euclidean geometry is really the consequence of limited experience and relatively superficial observation. In the simpler uses of numbers and in the limited regions of the universe accessible to man until recent times, the well-known axioms of number and Euclidean geometry seemed inescapably true. But we can choose other axioms and produce other geometries and even other number systems. Though we can make arbitrary choices, we do not do so because we wish our new axioms to yield systems of mathematics as significant and as useful as the older systems. Were we able to continue our study of mathematics proper, we would find that there are many other systems of axioms which lead to important branches of mathematics.

The fruit of mathematical activity consists of the theorems deduced from a set of axioms. The theorems offer new knowledge by no means immediately discernible in the axioms. Whether or not the reader may prize the knowledge, he learns by pursuing the implications of the Euclidean axioms that the sum of the angles of a triangle is 180° and that the area of a circle is π times the square of the radius. The amount of information that can be deduced from some sets of axioms is almost incredible. Euclid deduced about 500 theorems from his set. The axioms of number give rise to the results of algebra, properties of functions, the theorems of the calculus, the solutions of various types of differential equations, and many other results we have not surveyed.

Mathematical theorems, as we know, must be deductively established. Since observation, measurement, induction, and a variety of other methods of obtaining knowledge are available and are used in all other scientific pursuits, the requirement that theorems be deductively established is very stringent. We have pointed out in Chapter 3 why the Greeks insisted on deductive arguments in mathematics. We are now in a position to see how much has been gained by exploring the implications of axioms. The development of precise and reliable methods of doing arithmetic, the solution of equations for unknowns, and the results of Euclidean geometry were but a first step. Simple deductive arguments about right triangles and admittedly some physical data enabled man to determine the sizes and distances of the heavenly bodies and thus obtain the first real knowledge of the solar system. By adding to mathematical axioms such physical axioms as Newton's laws of motion and gravitation, man was able to calculate the motions of projectiles, planets, the moon, and even artificial satellites. Equally valuable results were deduced about the behavior of light and sound from mathematical and physical principles. The decision to explore deductively the axioms of a non-Euclidean geometry led to the construction of totally new geometries, one of which has actually been applied to the study of physical phenomena. A brief review of the contents of the preceding chapters will recall many other achievements of the deductive process. The Greeks themselves would have been astonished by the rich results which deduction produced and which the more common methods of obtaining knowledge could not have yielded. Thus much of our deepest knowledge, which we would otherwise not have had, was obtained by deductive reasoning.

The restriction of mathematics to results deduced from explicit axioms has had several other major values. It obliged human beings to apply their reasoning faculty. Though, as we pointed out in Chapter 3, imagination and invention play key roles in suggesting what to prove and how to prove it, the mathematician's goal is not any new knowledge whatever but knowledge which he has some reason to believe can be deduced from axioms. Hence it is the deductive requirement which has caused man to explore and utilize his own mental capacities to an extent which no other subject has demanded.

The third major value of the deductive process is that it enables man to predict. In a sense all reasoning enables man to predict. If one measures the angles of a dozen triangles and finds in each case that the sum of the angles is 180°, he can then predict that the sum of the angles in any other triangle is also 180°. But, as we have often emphasized, conclusions obtained by induction or analogy are not certain, whereas mathematical predictions are. Equally important is the recognition that predictions which amount to merely applying a general result to a special case are logically shallow. The predictions of mathematics are themselves general results; they are the outcome of dozens of hard-won and by no means obvious deductive arguments and are therefore profound facts which most likely could not be otherwise obtained.

The fourth value of the deductive process is its power to organize knowledge. If all the results now available in Euclidean geometry had been obtained by a multitude of observations, inductions, or measurements, they would make an unwieldy mass that could not be assimilated. The value of the information would hardly be realized. But deductive organization permits the mind to survey the whole readily, grasp what is fundamental and what is subordinate, and see the interrelationships of the many conclusions. Comprehension is vastly aided.

We have been discussing the components of mathematical branches and the values of the deductive structure. There is another feature of mathematics which a backward glance reveals, namely growth. There is no doubt of the growth of mathematics, but there is some disagreement among mathematicians as to just what it is that they are adding to their structures. Succinctly phrased, the issue might be called one of discovery versus creation. Do the concepts, axioms, and theorems exist in some objective world and are merely detected by man or are they entirely human creations?

In Greek times the axioms of mathematics were regarded as necessary truths. Accordingly, mathematical theorems were also believed to be truths about the universe already incorporated in the design of the world. Hence each new theorem was regarded as a discovery, a disclosure of what already existed. That the alternate interior angles of parallel lines are equal was preordained. Mathematicians were merely uncovering theorems, but because human minds were limited, they had to labor hard and long to recognize what really lay open before them. To the mind of God all this knowledge was immediate. One might say that from this standpoint mathematics was like a mine whose riches were all there from the beginning, but had to be brought to the surface by patient digging. The existence of these riches was as independent of man as the stars and planets appear to be. This view of mathematics was undoubtedly the dominant one until well into the eighteenth century and is held by some even today.

The contrary view holds that mathematics, its concepts, axioms, and theorems, are created by man. Man distinguishes objects in the physical world and invents numbers, for example, as a way of representing one aspect which he has singled out from experience. Axioms, too, are man's generalization of how physical lines and figures seem to behave, with no guarantee given that figures actually behave this way or that the axioms really incorporate fundamental facts. Theorems may very logically follow from the axioms, but one could hardly claim more reality for them than for the axioms. Mathematics, according to this view, is a human creation in every respect. It is a consequence of what human beings are and how they think rather than what the physical world or some objective ideal world really contains.

Is then mathematics a collection of diamonds hidden in the depths of the universe and gradually unearthed one by one or is it a collection of synthetic stones manufactured by man but nevertheless so brilliant that it bedazzles those mathematicians who are already partially blinded by pride in their own creations? Several considerations incline us to the latter point of view. Historically, mathematics has not had an invariable character. To the pre-Greeks it was a practical tool. To the Greeks it was a body of pre-existing truths. It was practical and mystical knowledge in medieval Europe. The seventeenth and eighteenth centuries identified it in subject matter and method with science. Non-Euclidean geometry not only forced a separation but revealed the arbitrariness inherent in the axioms. It thus seemed to establish that mathematics is not an idealized account of the physical world but has only a correspondence with the physical world. As scientific knowledge increases, new mathematical creations are suggested and employed. It would appear as though mathematics is the creation of human, fallible minds rather than a fixed, eternally existing body of knowledge. The subject seems very much dependent upon the creator. As Alfred North Whitehead put it, "The science of pure mathematics may claim to be the most original creation of the human spirit." Only the *relatively* universal acceptance of mathematics (as opposed to the acceptance of religious, political, and ethical doctrines) may lure us into granting that subject an objective existence.

EXERCISES

1. What are the fundamental components of a branch of mathematics?
2. Would you distinguish between whole numbers and irrational numbers in respect to derivation from experience?
3. Why are axioms necessary in a deductive system?
4. Criticize the statement: Mathematics is a fixed body of thought created in Greek times.
5. Name some branches of mathematics.
6. What are the factors that make possible the growth of mathematics?
7. What can research in mathematics mean?
8. What advantages has the requirement that man reason deductively yielded to the mathematical sciences?

24-3 THE VALUES OF MATHEMATICS FOR THE STUDY OF NATURE

Mathematics proper, as we have often emphasized, deals with numbers, geometrical figures, and generalizations or extensions of ideas involving numbers and geometrical figures. Mathematics proper does not deal with forces, weight, velocity, light, or the planets. The task of the so-called pure mathematicians is to find and establish the implications of the axioms about mathematical concepts, that is, to prove theorems. There is much to be said for the study of mathematics itself, and we shall discuss it in a later section. However, the primary value of mathematics is not so much what the subject itself offers but what it helps man to achieve in the study of the physical world.

The greatest mathematicians from Greek times onward were interested in the physical world and in the use of mathematics to study the physical world. Although many of these men surely liked mathematics and tackled questions of mathematics proper, without thinking of immediate or even potential application, it is fairly certain that they were willing to devote time to such problems only because they were already convinced of the value of the subject for science. Actually many of the greatest mathematicians were also the greatest physicists and astronomers of their ages, and, until very recent times when the increase in knowledge forced specialization, almost all mathematicians contributed to science. One finds among the supreme mathematicians men, such as Newton, Lagrange, and Laplace, who even cared little or nothing for mathematics proper, but felt compelled to take up mathematical problems in order to solve physical problems.

We shall not review the facts of history to substantiate the above assertions. Rather we wish to summarize the ways in which mathematics works with science and the values which science derives from this collaboration.

Of course, the study of numbers and geometrical figures is to some extent physical knowledge. Quantity is an important physical fact as are the properties of geometrical figures since these are but forms of physical objects. Geometry is, moreover, a study of space. Although the belief that Euclidean geometry expresses the laws of space has been proved wrong, the various geometries that men have constructed are at least possible and, in some cases, useful descriptions of physical space.

However, the greater importance of mathematics for science lies in the fact that the physical universe is explored so effectively with mathematics. To the Greeks, who first proclaimed that the universe is mathematically designed, and to scientists up to 1600, applying mathematics to nature meant searching for geometrical patterns in nature. This quest in itself yielded Ptolemaic theory, some laws of light and mechanics, and the heliocentric theory. A more powerful mathematical approach to nature was forged by Galileo when he decided that science must seek to establish quantitative laws. Such laws as Newton's second law of motion, the law of gravitation, and Hooke's law, though they belong to science, are quantitative relationships among variables, or mathematical formulas. As we saw in earlier chapters, mathematical processes can be applied to these formulas to deduce new ones. When these new formulas are interpreted physically, new physical information is revealed.

Mathematics serves, then, to express physical laws, and the processes of mathematics are used to derive new physical information from the basic physical laws. But mathematics does far more for science. The goal of scientific efforts is not a collection of facts, whether obtained experimentally or deduced from other already established facts. The over-all goal is the formulation of theories, such as the theory of motion, the theories of light, sound, and electromagnetic waves, the theory of relativity, and quantum theory. These theories are mathematical structures. When fully developed, they are entirely analogous to the mathematics of number or to Euclidean geometry. The foundations on which any scientific theory rests are concepts and axioms, though the latter are usually called physical principles. The heart of any scientific theory is a series of results mathematically deduced from basic principles. Thus Kepler's laws are deduced from the laws of motion and gravitation and are, since Newton's work, an integral part of the theory of motion. In other words, mathematical deduction provides the structure of any scientific theory; it is the bond between one law and another. A scientific theory is a comprehensible and consistent collection of facts, and it is comprehensible and consistent because the facts are arranged in the form of a series of mathematical deductions.

In discussing earlier the value of deduction, we pointed out that, among other advantages, it permits prediction. Since science utilizes mathematical deduction, it too can predict. The predictions of the height or range of projectiles, of eclipses, and of radio waves are remarkable just because they are beyond the capabilities of any other method. The value of prediction cannot be overemphasized. On the practical side, it underlies all large-scale engineering ventures. The waste that would be entailed if inferences had to be drawn solely from models or from experiments would be enormous in even such relatively simple projects as building a bridge or a skyscraper. From the standpoint of pure science, prediction is valuable in that it confirms the scientific principles on which the predictions are based. The principles stand or fall on their predictive value, and mathematical arguments are the necessary link between the basic principles and the predictions.

The abstractness of mathematically formulated scientific principles has great value for science. Because the same abstract mathematical laws may govern two entirely different physical situations, the scientist may discover some unsuspected relationship between the two situations. Thus, for example, the trigonometric functions apply to all wave motions, sound, radio, light, water waves, waves in gases, and many other types of wave motions. The person who understands trigonometric functions and their properties understands in one swoop all the phenomena governed by these functions. He has but to interpret the variables to suit the physical situation, and he can immediately comprehend a host of facts about the phenomenon because he knows the mathematical properties of the functions. So, too, can knowledge of the normal law of distribution be applied to heights, intelligence, the sensitivity of the ear to frequencies, and to other biological phenomena. We see, then, that the mathematics of such related areas provides an integrative value; it features the common contents. Only the names, so to speak, of the different phenomena are different. Herein lies one great value of mathematics. Abstract mathematical relationships, seemingly outside the realm of physical reality, are the key to large classes of physical phenomena.

Another value of mathematics for science has already been treated in our presentation of gravitation. Quantitative physical principles, such as the law of gravitation, described purportedly physical concepts such as the force of gravitation. But closer examination revealed that these concepts were physical mysteries and that all we knew about them were certain quantitative laws and their mathematical consequences. The force of gravitation could be and undoubtedly is a fiction. Hence our only precise knowledge about these supposedly real phenomena consists of a number of mathematical formulas. Mathematics, then, is the essence of our best scientific theories. Those who, admitting the paradox, deplore the fact that to achieve success, the physical sciences have to pay the price of mathematical abstractness must reconsider what it is they would look for in the ultimate scientific exposition of the nature of the physical world.

Science is indebted to mathematics in many ways and not the least of these is that mathematics provides concepts to represent physical notions. A function is a mathematical concept, but it provides the very tool for the representation of physical laws. The derivative and integral of the calculus are immensely effective in studying physical processes. The conic sections are the curves which proved to be just right for projectile motion and astronomy. Over and above concepts, mathematics provides entire theories with which to systematize and express scientific results. A non-Euclidean geometry is a complete theoretical system into which one can fit facts about space and the behavior of figures in space; such a geometry enabled Einstein to formulate his theory of relativity. This major function of mathematics is sometimes described by the statement that mathematics provides models for the scientific description of reality. The concepts and models are what really determine the thinking and theories of science, for scientists seeking to represent their ideas in precise language and to organize their findings readily adopt convenient mathematical ideas.

It is highly important that many of these models, developed on behalf of some physical problems, turned out to be just the right ones for totally new applications. The conic sections were probably created on behalf of investigations of light and to answer basic questions in the already significant Euclidean geometry. Given the conic sections and their properties, Kepler saw the right use for them. It seems unlikely that Kepler would have had the strength and inspiration to create the mathematics of conic sections *and* determine that they were the correct paths of the planets. The same applies to Galileo and the parabola. In recent times Einstein, in developing his theory, took full advantage of the already existing non-Euclidean geometry. The conception and development of such a geometry climaxed, as we saw, the work of hundreds of men and required, in addition, the genius of Gauss to perceive the true significance of these efforts.

EXERCISES

1. Describe some of the values which mathematics offers to science.
2. Why is reasoning about concepts, such as numbers, formulas, and geometric figures, likely to be more fruitful than reasoning about concrete physical phenomena?

24-4 THE AESTHETIC AND INTELLECTUAL VALUES

We have asserted that the primary value of mathematics is the assistance it renders in the study of nature, and we have summarized the ways in which mathematics supports and even molds science. This is not to deny that many of these mathematical concepts, methods, and results were suggested by physical thought. But mathematicians often carry the development of a theme or a whole branch of their subject far beyond the needs of science. The conic sections and non-Euclidean geometry are examples of such activity. If there was little or no scientific use for these extensions, why were they explored?

A partial answer is that mathematicians, already convinced of the extraordinary usefulness of some ideas, for example, the whole numbers and simple geometrical figures, were satisfied that almost any results concerning these concepts would be worth having for their potential applicability. Many men who earn enough money to satisfy their needs strive to earn more because they know that money is helpful and that uses for the surplus earnings will arise. When non-Euclidean geometry proved to be useful despite its strange, seemingly inapplicable nature, mathematicians were all the more reinforced in their proclivity to pursue their own themes. Although nature is the womb from which most basic mathematical ideas are born, mathematicians have always felt free to amplify and extend these ideas without regard for applicability to nature, confident that at some time the extensions will prove their worth.

However, immediate need in science and potential usefulness have not been the only motivations for mathematical creations. The individual may well recognize that mathematical activity is guided by physical needs, but he himself may study mathematics simply because he likes it. He may be content to know that the subject he has chosen for investigation is important for the understanding or mastery of nature and yet care little himself about the scientific bearing of the mathematics. He may even choose some problem of his own fabrication and pursue it (see Chapter 3).

In other words there are men who pursue mathematics for its own sake; they are attracted by the fact that mathematics is an art. Among the branches of mathematics that we have examined, the subject of projective geometry and the subject of congruences (Chapter 21), a topic in the branch of mathematics called the theory of numbers, were motivated largely by aesthetic interests.

What artistic qualities do some men find in mathematics? Though experience, observation, measurement, and even guesswork suggest some results, imagination, intuition, and insight are required for major creations, and the exercise of such talents is one of the attractions offered by an art. Indeed the mathematician must be able to discern possible conclusions and methods of proof where the average person would see no hint at all. Though reason enters, on the one hand, as a guard to ward off far-fetched hypotheses and, on the other, as a guide and prompter, creation is hardly a matter of logic. If we may borrow the words of Rheticus, the man who prepared for publication the major work of Copernicus:

> *The mathematician . . . is surely like a blind man who, with only a staff to guide him, must make a great, endless, hazardous journey that winds through innumerable desolate places. What will be the result? Proceeding anxiously for a while and groping his way with his staff, he will at some time, leaning upon it, cry out in despair to heaven, earth, and all the gods to aid him in his misery. God will permit him to try his strength for a period of years, that he may in the end learn that he cannot be rescued from threatening danger by his staff. Then God compassionately stretches forth His hand to the despairing man, and with His hand conducts him to the desired goal.*

Since persistence is required for all creative work, the mathematician, too, must have the stamina to wrestle with a problem until he has succeeded in solving it. He must have confidence in his powers. He may be driven to creative activity, as is the poet or painter, by pride in his reasoning faculty, the spirit of exploration, and the desire to express himself, but he must persist. The greatest mathematicians have stressed the concentration and time they have devoted to problems. Gauss said, perhaps over-modestly but sincerely, "If others would but reflect on mathematical truths as deeply and as continuously as I have, they would make my discoveries."

Perhaps the best reason for regarding mathematics as an art is not so much that it affords an outlet for creative activity as that it provides spiritual values. It puts man in touch with the highest aspirations and loftiest goals. It offers intellectual delight and the exaltation of resolving the mysteries of the universe.

Many who accept the above described values of mathematics nevertheless insist that an art form provide emotional satisfactions. Actually a true art appeals primarily to the mind; in fact, some arts—witness modern abstract painting—can hardly have emotional import. However, mathematics meets this criterion too. There are positive and negative emotional responses to mathematics. On the negative side, there are the intense feelings of dislike which many have for the subject. On the positive side, there are pleasures ranging from the quiet satisfaction felt by many laymen who read in the subject, to the thrill of success which even young students experience when they have solved a problem, and to the real delight which mathematicians who do original research derive from their work. There are satisfactions, much like those offered by great paintings, to be obtained from surveying orderly chains of reasoning. This order and harmony can be found in the development of mathematical themes, and there is the harmony which mathematics imposes on nature and which minds such as Ptolemy, Copernicus, Kepler, Newton, and Einstein fashion. Coleridge has said that the essence of beauty is the discovery of unity within and beyond obvious variety. The mathematicians are artists who use nature as their model and provide their own orderly and unifying interpretations.

Somewhat distinct from the aesthetic value of mathematics is the intellectual challenge. Mathematicians respond to this challenge much as business men respond to the excitement of making money. Mathematicians enjoy the excitement of the quest, the thrill of discovery, the sense of adventure, the satisfaction of mastering difficulties, the pride of achievement or, if one wishes, the exaltation of the ego, and the intoxication of success. Mathematics is particularly attractive to people who enjoy such challenges because it offers sharp, clear problems. The fields of political theory, economics, and ethics are far more complex, and it is more difficult not only to isolate and formulate the problems but to be at all sure that one has the information which can lead to a decisive solution of these problems. By contrast an exercise in geometry, though it may contain no impressive results, offers a clear and circumscribed problem.

EXERCISES

1. What are the motivations for mathematical activity?
2. Defend or attack the thesis that mathematics is an art.
3. What is commonly meant by pure and applied mathematics?
4. Are mathematicians free to create what they wish to?

24-5 MATHEMATICS AND RATIONALISM

Among the values which mathematics offers is one which transcends the subject and its relationship to the physical sciences. Mathematics has been the advocate, essence, and embodiment of rationalism. Rationalism is not a whim. It is a spirit which stimulates, invigorates, challenges, and drives human minds to function at the highest mental level in exploring and establishing the deepest implications of knowledge already at hand. It calls for the courage to discard one's dearest and most cherished beliefs if these do not satisfy rational criteria. Mathematics also holds forth the ideal of detachment and of objective judgment. It has fostered independence of thought, adherence to the dictates of reason, careful scrutiny of arguments, and a spirit of criticism.

Mathematicians have set the highest standards of reasoning by their persistence and indomitableness. We have seen that they have worked for centuries to *prove* what intuition or experience would have us believe is unquestionable. For two thousand years mathematicians sought an exact method of trisecting an angle, squaring the circle, and doubling the cube, though practical constructions of as great an accuracy as desired were theirs for the asking. In this case the long-sought methods proved to be nonexistent; that is, it is not possible to perform these constructions with straightedge and compass. But the important point at the moment is the search. Likewise for two thousand years mathematicians sought to replace the parallel axiom of Euclid by a more reliable statement. Though in this instance, too, the outcome of their search was a surprise, the important point again is the magnitude and persistence of the effort and the ultimate rejection of many substitute statements which re-examination showed would not do. To reason mathematically is to seek perfection in reasoning. The commonly used phrase "mathematical exactness" pays homage to this ideal of mathematicians.

As man's greatest and most successful intellectual experiment, mathematics demonstrates manifestly how powerful our rational faculty is. It is the finest expression of man's intellectual strength. His reason has, for example, far outstripped his imagination. He can think about stars so distant that only numbers convey any meaning, about spaces which cannot be pictured, and about electrons too small to be seen with the most powerful microscopes.

Rationalism and exact thought patterned after mathematics can be applied to many fields. In this broader sense at least and often more concretely mathematics has penetrated almost all domains of inquiry and has served as the model of all intellectual enterprises. In Plato's time the word mathematics meant rational systematic knowledge, the modern equivalent of which is the German term *Wissenschaft*. Later with Aristotle it came to mean the specific subjects we study in mathematics courses. It is Plato's meaning which the subject of mathematics really encompasses today, even though the word itself still retains the more limited meaning.

24-6 THE LIMITATIONS OF MATHEMATICS

Yet there are some limitations on what mathematics and mathematical methods seem to be able to achieve. Some aspects of the physical world and of human behavior have not yielded to either. Thus, touch, taste, and smell are sense perceptions which, unlike seeing and hearing, have defied mathematical analysis or even measurement, though the sense organs involved are physiologically simpler than the eye and the ear. Human character, desires, motivations, and emotions are more successfully studied by advertising men than by mathematicians. Neither axioms nor theorems have been found which would furnish a model of man's behavior, of the operation of his mind, and of the economic and political systems most beneficial to human society. Numbers and geometrical forms do not seem to be the applicable concepts. Only inanimate nature and, in fact, merely portions of that have yielded to mathematical analysis. It is true that statistical methods have given us some ability to predict human characteristics and behavior, but statistical conclusions wipe out all individual nuances and yield only gross effects. There is, of course, still hope that current research on mathematical models may serve other domains of inquiry and in particular solve the problems of society.

One should also question the extent to which mathematics really represents the physical world. The discipline has been effective in treating some abstractions: space, time, mass, velocity, weight, force, the frequency of light and sound, and other such concepts. It treats those physical concepts which can be represented by numbers or geometrical figures. But physical objects possess other properties as well. We do not usually think of human beings as chunks of matter moving in space and time. Nor would a poet or an artist be content to say that the mathematical laws of planetary motion represent the essence of the planets. We have become so accustomed to the analysis of the physical world in terms of space, time, form, mass, and the like, that we tend to overlook the fact that these concepts represent just some properties and narrow ones at that. They cause us to look at the world with blinders. The mathematical approach may not be the deepest possible or the most illuminating; it certainly does not answer the question of whether the solar system is designed for any special ends. Scientists may say that this question does not fall within the province of science, but it is nevertheless a question which human beings would like to see answered. The refusal of scientists to consider it does not wipe out the question, but only reveals a limitation of the mathematical approach.

The plight of man is pitiable. We are wanderers in a vast universe, helpless before the havocs of nature, dependent upon nature for food and other necessities, and uninformed as to why we were born and what to strive for. Man is alone in a cold and alien universe. He gazes upon the mysterious, rapidly changing, and endless world about him and is confused, baffled, and even frightened by his own insignificance. As Pascal put it,

> *For after all what is man in nature? A nothing in relation to infinity, all in relation to nothing, a central point between nothing and all and infinitely far from understanding either. The end of things and their beginnings are impregnably concealed from him in an impenetrable secret. He is equally incapable of seeing the nothingness out of which he was drawn and the infinite in which he is engulfed.*

Montaigne and Hobbes said the same thing in other words. The life of man is solitary, poor, nasty, brutish, and short. He is the prey of trivial happenings.

Endowed with a few limited senses and a brain, man began to pierce the mystery about him. By utilizing what the senses reveal immediately or what can be inferred from experiments man adopted axioms and applied his reasoning powers. His quest was the quest for order, his goal, to build sound knowledge as opposed to transient sensations. Amid the chaos of life and his environment, he has sought patterns of explanation and systems of knowledge that might help him to attain some mastery over his environment. The chief tool proved to be the product of man's own reason, and its accomplishments were described by Fourier.

> *It brings together the most diverse phenomena and discovers hidden conformities which unite them. If matter evades us, such as the air and light, because of its extreme thinness, if objects are located far from us in the immensity of space, if man wishes to understand the performance of the heavens for the successive periods which separate a large number of centuries, if the forces of gravity and of heat be at work in the interior of a solid globe at depths which will be forever inaccessible, mathematical analysis can still grasp the laws of these phenomena. It renders them present and measurable and seems to be a faculty of the human reason destined to make up for the brevity of life and for the imperfection of the senses; and what is more remarkable still, it follows the same method in the study of all phenomena; it interprets them in the same language, as if to affirm the unity and simplicity of the plan of the universe, and to make still more manifest the immutable order which presides over all natural events.*

Over the centuries man has created such grand structures as Euclidean geometry, Ptolemaic theory, the heliocentric theory, Newtonian mechanics, electromagnetic theory, and in recent times the theory of relativity and quantum theory. In all of these and in other significant and powerful bodies of science, mathematics, as we now know, is the method of construction, the framework, and indeed the essence. Mathematical theories have enabled us to know something of nature, to embrace in comprehensive intelligible accounts varieties of seemingly diverse phenomena. Mathematical theories have revealed whatever order or plan man has found in nature and have given us mastery or partial mastery over vast domains.

It may be that man has introduced some limited and even artificial concepts and only in this way has managed to institute some order in nature. Man's mathematics may be no more than a workable scheme. Nature itself may be far more complex or have no inherent design. Nevertheless, mathematics remains the method *par excellence* for the investigation, representation, and mastery of nature. In those domains where it is effective it is all we have; if it is not reality itself, it is the closest to reality we can get.

Mathematics then is a formidable and bold bridge between ourselves and the external world. Though it is a purely human creation, the access it has given us to some domains of nature enables us to progress far beyond all expectations. Indeed it is paradoxical that abstractions so remote from reality should achieve so much. Artificial the mathematical account may be, a fairy tale perhaps, but one with a moral.

In the last analysis it is the picture which an age forms of its world which is its most valuable possession, for man seeks primarily to know himself, and this understanding is inseparable from his understanding of the cosmos. The knowledge so gleaned filters through philosophy, literature, religion, the arts, and social thought. It thereby fashions the whole culture and provides whatever answers man has to the major questions he raises about his own life.

Recommended Reading

BRONOWSKI, J. and B. MAZLISH: *The Western Intellectual Tradition*, Harper and Row, New York, 1960.

BURY, J. B.: *The Idea of Progress*, Dover Publications, Inc., New York, 1955.

ELLIS, HAVELOCK: *The Dance of Life*, Chap. 3, The Modern Library, New York, 1929.

HARDY, G. H.: *A Mathematician's Apology*, Cambridge University Press, London, 1940. (Keep a copious quantity of salt on hand while reading this book.)

KLINE, MORRIS: *Mathematics in Western Culture*, Chap. 28, Oxford University Press, New York, 1953.

NEWMAN, JAMES R.: *The World of Mathematics*, Vol. III, pp. 1756–1795, Vol. IV, pp. 2051–2063, Simon and Schuster, Inc., New York, 1956.

POINCARÉ, HENRI: *The Value of Science*, Chaps. 1 to 3, Dover Publications, Inc., New York, 1958.

POINCARÉ, HENRI: *Science and Method*, Chaps. 1 to 3, Dover Publications, Inc., New York, 1952.

RANDALL, JOHN HERMAN, JR.: *The Making of the Modern Mind*, rev. ed., Houghton Mifflin Co., Boston, 1940.

RUSSELL, BERTRAND: *Our Knowledge of the External World*, George Allen & Unwin Ltd., London, 1926 (also in paperback).

RUSSELL, BERTRAND: *Mysticism and Logic*, Longmans, Green and Co., New York, 1925.

SAWYER, W. W.: *Prelude to Mathematics*, Chaps. 1 to 3, Penguin Books Ltd., Harmondsworth, England, 1955.

SPENGLER, OSWALD: *Decline of the West*, Vol. I, Chap. 2, A. A. Knopf, Inc., New York, 1926.

SULLIVAN, J. W. N.: *Aspects of Science*, Second Series, pp. 80–105, A. A. Knopf, Inc., New York, 1926.

APPENDIX

Table of Trigonometric Ratios

Angle	Sine	Tangent	Cotangent	Cosine	
0°	0.0000	0.0000	.....	1.0000	90°
1	0.0175	0.0175	57.290	0.9998	89
2	0.0349	0.0349	28.636	0.9994	88
3	0.0523	0.0524	19.081	0.9986	87
4	0.0698	0.0699	14.300	0.9976	86
5	0.0872	0.0875	11.430	0.9962	85
6	0.1045	0.1051	9.5144	0.9945	84
7	0.1219	0.1228	8.1443	0.9925	83
8	0.1392	0.1405	7.1154	0.9903	82
9	0.1564	0.1584	6.3138	0.9877	81
10	0.1736	0.1763	5.6713	0.9848	80
11	0.1908	0.1944	5.1446	0.9816	79
12	0.2079	0.2126	4.7046	0.9781	78
13	0.2250	0.2309	4.3315	0.9744	77
14	0.2419	0.2493	4.0108	0.9703	76
15	0.2588	0.2679	3.7321	0.9659	75
16	0.2756	0.2867	3.4874	0.9613	74
17	0.2924	0.3057	3.2709	0.9563	73
18	0.3090	0.3249	3.0777	0.9511	72
19	0.3256	0.3443	2.9042	0.9455	71
20	0.3420	0.3640	2.7475	0.9397	70
21	0.3584	0.3839	2.6051	0.9336	69
22	0.3746	0.4040	2.4751	0.9272	68
23	0.3907	0.4245	2.3559	0.9205	67
24	0.4067	0.4452	2.2460	0.9135	66
25	0.4226	0.4663	2.1445	0.9063	65
26	0.4384	0.4877	2.0503	0.8988	64
27	0.4540	0.5095	1.9626	0.8910	63
28	0.4695	0.5317	1.8807	0.8829	62
29	0.4848	0.5543	1.8040	0.8746	61
30	0.5000	0.5774	1.7321	0.8660	60
31	0.5150	0.6009	1.6643	0.8572	59
32	0.5299	0.6249	1.6003	0.8480	58
33	0.5446	0.6494	1.5399	0.8387	57
34	0.5592	0.6745	1.4826	0.8290	56
35	0.5736	0.7002	1.4281	0.8192	55
36	0.5878	0.7265	1.3764	0.8090	54
37	0.6018	0.7536	1.3270	0.7986	53
38	0.6157	0.7813	1.2799	0.7880	52
39	0.6293	0.8098	1.2349	0.7771	51
40	0.6428	0.8391	1.1918	0.7660	50
41	0.6561	0.8693	1.1504	0.7547	49
42	0.6691	0.9004	1.1106	0.7431	48
43	0.6820	0.9325	1.0724	0.7314	47
44	0.6947	0.9657	1.0355	0.7193	46
45°	0.7071	1.0000	1.0000	0.7071	45°
	Cosine	Cotangent	Tangent	Sine	Angle

ANSWERS TO SELECTED EXERCISES AND REVIEW EXERCISES

Chapter 3

Section 3–4

7. (a) No (b) Yes (c) No
   (d) No (e) No (f) Yes
   (g) Yes (h) No (i) No

REVIEW EXERCISES

5. (a), (b) 6. None 7. No 9. Nothing

Chapter 4

Section 4–3

Second set:

1. (a) $\sqrt{3} + \sqrt{5}$ (b) $\sqrt[3]{2} + \sqrt[3]{7}$ (c) $\sqrt[3]{2} + \sqrt{7}$
   (d) $2\sqrt{7}$ (e) $\sqrt{21}$ (f) $\sqrt[3]{10}$
   (g) 2 (h) 6 (i) $\sqrt{\frac{5}{2}}$
   (j) 2 (k) $\sqrt[3]{5}$
2. (a) $5\sqrt{2}$ (b) $10\sqrt{2}$ (c) $5\sqrt{3}$

Section 4–4

1. —2 2. —13 3. +3
4. —5 5. 500, 500

Section 4–5

3. (a) $12a$ (b) $12a$ (c) $5a + \sqrt{2}a$
   (d) $-2a$ (e) $6a + 12b$ (f) $28a + 35b$
   (g) $a^2 + ab$ (h) $a^2 - ab$ (i) $16a$
   (j) a^2b
4. $a^2 + 5a + 6$ 5. $n^2 + 2n + 1$ 6. Yes
7. Yes 8. Yes 9. No

Section 4–6

First set:

1. No, $4\frac{1}{2}$ hr 2. No
3. $\dfrac{15a + 10b}{6(a + b)}$ 4. No 5. $\frac{5}{12}$ of a ditch per day

Second set:

1. 16 2. 12 3. 12 4. 36

Third set:

4. 13, 14, 20, 100, 120, 244 5. 5, 6, 8, 12, 36
6. 0.3 7. $\frac{1}{3}$ 8. 3 9. 6

REVIEW EXERCISES

1. (a) $\frac{41}{35}$ (b) $\frac{1}{35}$ (c) $-\frac{1}{35}$ (f) $\frac{23}{36}$ (g) $-\frac{23}{36}$ (h) $\dfrac{ad + bc}{bd}$
2. (b) $-\frac{4}{15}$ (c) $\frac{4}{15}$ (e) $\dfrac{ac}{bd}$ (i) 2 (k) 1 (n) $\frac{21}{44}$
3. (a) 140 (b) $4ab$ (d) $6\,xy$
4. (a) $\frac{5}{14}$ (b) $1 + 2a$ (e) $b + c$
5. (c) $\frac{3}{2}$ (e) 3 (f) 4 (h) $\sqrt{\frac{10}{3}}$
6. (a) $4\sqrt{2}$ (c) $6\sqrt{2}$ (f) $\frac{3}{2}\sqrt{2}$ (h) $\frac{3}{2}\sqrt{\frac{3}{2}}$
9. (b) False (d) False (e) False (f) False (h) False
10. (b) 11 (c) 101 (e) 1000 (g) 10,011
11. (b) 5 (d) 13 (e) 9

Chapter 5

Section 5–2

5. (a) $3x + 4$ (b) $3x^2 + 4$

Section 5–3

First set:

1. (a) 5^{10} (d) x^5 (e) 5^3 (g) x^2 (j) $\dfrac{1}{5^3}$ (k) 1 (l) 5
3. (a) False (c) False (e) False

Second set:

1. (a) 3^{12} (c) 5^8 (d) 10^6 (f) 10^4
2. (a) 100,000 (b) 6 (c) $\frac{1}{8}$ (d) ab^3
3. (c) False (e) False (f) False

Section 5–4

1. (a) $15x^2$ (b) $x^2 + 9x + 20$ (c) $3x^2 + 19x + 20$
   (d) $x^2 - 9$ (e) $x^2 + 5x + \frac{25}{4}$ (f) $x^2 - \frac{25}{4}$
2. (a) $(x + 5)(x + 4)$ (b) $(x + 2)(x + 3)$ (c) $(x - 2)(x - 3)$
   (d) $(x + 3)(x - 3)$ (e) $(x + 4)(x - 4)$ (f) $(x + 9)(x - 2)$

Section 5–5

1. 1177 ft 2. 22 mi/hr 3. 73
4. 50 5. 80 6. 1250

Section 5–6

First set:

1. (a) 6, 2 (b) −9, 2 3. (a) $-6 \pm 3\sqrt{3}$ (b) $6 \pm 3\sqrt{3}$

Second set:

1. (a) $4 + \sqrt{6}, 4 - \sqrt{6}$ (b) $-4 + \sqrt{6}, -4 - \sqrt{6}$
   (c) $3 + 3\sqrt{2}, 3 - 3\sqrt{2}$ (d) −3, −1
   (e) 4, 4

REVIEW EXERCISES

1. (b) $x^2 + 5x + 6$ (c) $x^2 + 5x - 14$
   (e) $x^2 + 5x + \frac{21}{4}$ (f) $2x^2 + 5x + 2$
2. (a) $(x + 3)(x - 3)$ (e) $(x + 3)^2$ (f) $(x + 6)(x + 1)$
   (i) $(x - 6)(x - 1)$ (l) $(x + 8)(x - 2)$ (m) $(x + 9)(x - 3)$
4. (c) 1 (d) 2 (e) $\frac{18}{7}$ (g) $\frac{54}{25}$ (h) $\dfrac{b - 2}{a}$
5. $12\frac{1}{2}$
6. 95
7. (a) 5 and 1 (b) 7 and -1 (f) 7 and -2
8. (b) 9 and 1 (c) $-5 \pm \sqrt{\frac{76}{4}}$ (e) $6 \pm \sqrt{21}$
9. (a) $-6 \pm \sqrt{27}$ (c) $-6 \pm \sqrt{42}$ (e) $-3 \pm \sqrt{6}$

Chapter 6

Section 6–2

8. 103 ft
10. 2π ft

Section 6–3

2. 418,500 mi
3. 100 by 100
4. $\dfrac{p}{4}$ by $\dfrac{p}{4}$
5. 25 by 50
7. $\sqrt{2Rh + h^2}$
8. $\sqrt{2Rh}$
9. 63 mi

Section 6–5

2. 2
4. 5

Chapter 7

Section 7–2

3. 1 to 0
4. 0 to unlimitedly larger and larger values

Section 7–3

1. 445 ft
2. 14,265 ft
3. 11,500 ft
4. Yes, by 19 ft
5. 3944 mi

Section 7–4

1. 30°
5. 139 mi
6. 263 mi
7. 18,960 mi

Section 7–5

1. 93,000,000 mi
2. 428,000 mi
3. 1065 mi
4. 36,000,000 mi

Section 7–6

1. 32°
2. 0° to 42°
4. 172,000 mi/sec

REVIEW EXERCISES

3. (a) $\frac{5}{13}, \frac{12}{13}, \frac{5}{12}$ (c) $\sqrt{\frac{3}{7}}, \dfrac{2}{\sqrt{7}}, \dfrac{\sqrt{3}}{2}$ (d) $\dfrac{\sqrt{3}}{3}, \dfrac{\sqrt{6}}{3}, \sqrt{\frac{1}{2}}$
5. $\dfrac{\sqrt{3}}{2}, \sqrt{3}$
7. 83.91 ft
9. 48°

11. 2384 ft
12. 3682 mi, $2\pi \cdot 3682$ mi
13. 107π mi, approximately
15. 32°, approximately

Chapter 11

Section 11–2

4. $\frac{5}{4}$

Chapter 12

Section 12–3

3. (a) $y = 2x$ (b) $y = \frac{1}{\sqrt{3}}x$ (c) $y = -4x$
(d) $y = 4x$ (e) $y = -4x$
5. Yes
8. Yes
10. m
11. (0, 7)
12. $(0, b)$
13. m; $(0, b)$

Section 12–4

3. $x = -\frac{1}{12}y^2$
4. (0, 2)
5. (a) $y = \frac{1}{16}x^2$
(b) $y = \frac{1}{24}x^2$ (c) $y = -\frac{1}{20}x^2$ (d) $x = \frac{1}{16}y^2$
6. 1, 4, 9, 16, 25
7. $1\frac{1}{10}$, $4\frac{2}{5}$, $9\frac{9}{10}$, $17\frac{3}{5}$, $27\frac{1}{2}$

Section 12–6

3. 10
5. 2, 8

Section 12–7

2. A surface
3. A sphere
4. A plane

REVIEW EXERCISES

1. 0
4. (a) $y = 4x$ (b) $y = 4x + 2$
5. (a) $y = -4x$ (b) $y = -4x + 2$ (c) $y = -4x - 2$

Chapter 13

Section 13–4

4. 0, 96, 224, etc.
6. 100, 256, 784
7. 2, $2\frac{1}{2}$, 4
8. +3, −3
9. 32, 212

Section 13–5

3. $10\frac{2}{3}$ mi/hr
4. 128 ft/sec; 64 ft/sec; $t = 2$
7. 400 ft; 676 ft, 1600 ft
11. 256 ft/sec, approximately
15. 121 ft
16. 121 ft
17. $d = v^2/22$
18. 352 ft, 88 ft
19. 576 ft

Section 13–6

3. 192 ft/sec, 432 ft; 240 ft/sec, 756 ft

Section 13–7

2. (a) 384 ft (b) 32 ft/sec (c) 400 ft
3. 0
5. 15,625 ft
7. 870 ft, 18.1 sec
8. 960 ft/sec
11. 7 sec

REVIEW EXERCISES

2. (a) 4 (b) 5 (c) $12\frac{1}{2}$ (d) $\frac{1}{2}$
3. (a) 80 (b) 128
5. $v = gt$
6. (a) $gt/2$ (b) $gt^2/2$
8. (a) 2 (b) $\sqrt{6}$ (c) 3 (d) $\sqrt{\frac{25}{2}}$
   (e) $\frac{13}{4}$
10. (a) 1 (b) $\frac{9}{8}$ (c) $7\frac{3}{8}$ (d) $2\frac{11}{16}$
11. (a) $v = 100 + gt$ (b) $d = 100t + gt^2/2$
12. (a) 2 (b) 4
13. (a) $v = 128 - 32t$ (b) $d = 128t - 16t^2$
15. -112; the result is physically meaningless.
16. -112; the object is 112 ft below the roof.

Chapter 14

Section 14–2

1. $y = 3x/2$ 2. $y = 5x^2/16$ 3. $y = 5x + x^2$

Section 14–3

3. 5.5 sec, approximately; 550 ft, approximately
4. 3000 ft/sec 5. 1200 ft/sec
6. 5450 ft, approximately

Section 14–4

1. 230 ft/sec, 193 ft/sec 2. $y = -x^2/25 + 3x/2$
4. 37.5 ft 5. 14.1 ft, approximately
6. 1000 ft/sec

Section 14–5

1. $y = -16x^2/V^2\cos^2 A + x\sin A/\cos A$
2. $V\sin A/32$ 3. $V^2\sin^2 A/64$
5. 123,000 ft, approximately 6. 20,000 ft
7. $(62.5)\sqrt{2}$ sec

REVIEW EXERCISES

2. (a) $y = 7x/3$ (b) $y = 5x^2/9$
   (c) $x = 3y^2/25$ (e) $x^2 + y^2 = 25$
   (f) $y = (5x^2/4) + 3x/2$
3. (a) $x = 240 \cdot 5280t$, $y = 16t^2$ (b) $\sqrt{330}$ sec
   (c) $\sqrt{330}$ sec (d) $\sqrt{660}$ sec
   (e) $240 \cdot 5280 \quad \sqrt{660}$ ft
4. (a) $1000\sqrt{2}$, $1000\sqrt{2}$ (b) $x = 1000\sqrt{2}\,t$, $y = 1000\sqrt{2}\,t - 16t^2$
   (c) 125,000 ft (d) 31,250 ft

Chapter 15

Section 15–3

11. 165 years

Section 15–7

2. 848 poundals or $26\frac{1}{2}$ lb
3. 4320 lb

Section 15–8

2. 20.5 ft/sec^2, approximately
4. 150 lb
5. 1125 lb

Section 15–9

1. 37.5 lb
4. 15 lb
7. 2400 lb
10. Yes

Section 15–10

3. 500 ft/sec^2
4. 0.00897 ft/sec^2, approximately

Section 15–11

2. 26,000 ft/sec, approximately
3. 3300 ft/sec, approximately
4. 27.8 days, approximately
5. 24,000 ft/sec, approximately

REVIEW EXERCISES

1. (a) 0.001 (b) 0.000001 (c) 0.00000001
   (d) 0.00002 (e) 0.0001
2. (a) $5.8 \cdot 10^4$ (b) $5.879 \cdot 10^4$ (c) $6.34 \cdot 10^4$
   (e) $5/10^2$
3. $1.833 \ldots \cdot 10^6$ (b) $4/10^4$ (c) 7.2
   (d) $\dfrac{2.18}{10}$, approximately
4. $9.1 \cdot 10^7$
5. $5.97 \cdot 10^{27}$ gm
6. $1.98 \cdot 10^{33}$ gm
7. (a) 14.2 ft/sec^2 (b) 2.6 ft/sec^2
8. (a) 2840 poundals (b) 520 poundals
9. 1060 poundals

Chapter 16

Section 16–5

4. 80 ft/sec, 144 ft/sec
5. 160 ft/sec
6. 32 ft/sec

Section 16–6

3. (a) $4x$ (b) $4t$ (c) x
   (d) $12x^2$ (e) $-4x$ (f) $-32t$
   (g) $-32t + 128$ (h) $128 - 32t$
4. 228 ft/sec
5. (b) $\dot{V} = 4\pi r^2$
6. (a) $\dot{y} = 4ax^3$
7. $\dot{A} = 2x$
8. $\dot{A} = l$

Section 16–7

1. (a) $\frac{3}{50}$ (c) 0
2. (a) 2 (b) 2

Section 16–8

1. 0
3. 25 by 50
4. 25 by $\frac{50}{3}$
5. $r = \sqrt{50/3\pi}$, $h = 2r$

REVIEW EXERCISES

1. (a) 96 ft/sec (b) 176 ft/sec (c) 192 ft/sec

3. (a) 40 (b) 60 (c) $20a$
4. 32 ft/sec^2
5. (a) 14 (b) -6 (c) 6 (d) -14
6. 6
7. -1
8. (a) -25, minimum (b) -25, minimum (c) 25, maximum
9. 0, no, no
10. 324, $\frac{9}{2}$ sec
11. $20/\sqrt{3}$, $20/\sqrt{3}$, $10/\sqrt{3}$

Chapter 17

1. (a) x^3 (b) $5x$ (c) $x^2/2$
   (d) $3x^2/2$ (e) t^2 (f) $16t^2$
   (g) $32t$ (h) $t^2 + 10t$ (i) $-16t^2 + 128t$
   (j) $-32t$ (k) $16t^2$

Section 17–3

1. $150 - 32t$, $150t - 16t^2$
2. $16t^2 + 50$
3. $-32t$, $75 - 16t^2$
4. $100 - 32t$, $100t - 16t^2 + 50$
5. $-100 - 32t$, $50 - 100t - 16t^2$

Section 17–4

1. $69\frac{1}{3}$
2. $50\frac{2}{3}$
3. 10
4. 90

Section 17–5

1. 257,561,000 ft-lb
2. 264,000,000 ft-lb
3. 40,000 ft-lb

Section 17–6

1. 36,500 ft/sec, approximately

REVIEW EXERCISES

1. (a) $d = 200t - 16t^2$ (b) 625 ft (c) -200 ft/sec
2. (a) $d = 200t - 16t^2$ (b) $d = 100 + 200t - 16t^2$
3. (a) $d = 200t - \dfrac{5.3}{2}t^2$ (b) 3585 ft, approximately
6. $3297 \cdot 10^6$ poundals
7. 5700 ft/sec, approximately

Chapter 18

Section 18–3

First set:

1. (a) $\sin 60°$ (b) $\sin 30°$ (c) $-\sin 30°$
   (d) $-\sin 80°$ (e) $-\sin 90°$ (f) $-\sin 60°$
   (g) $-\sin 10°$ (h) $\sin 10°$ (i) $-\sin 50°$
   (j) $\sin 30°$
2. 1, -1
3. 90°
8. 2

Second set:

1. $\pi/2$ or 1.57, approximately; $\pi/6$ or 0.52, approximately; π, $3\pi/2$, 2π, $8\pi/3$
2. 90°, 120°, 450°, 540°, $-90°$, 57°, approximately
3. (a) 0 (b) 1 (c) $\sqrt{3}/2$
   (d) -1 (e) 0 (f) 1

Third set:

1. 1, 2, 2, $\sqrt{3}$ 2. 3, -3 3. 4

4. (a) 1, 0, -1, 0 (b) 1, 0, -1, 0

8. 10

9. (a) 1, 0, $-\sqrt{3}/2$ (b) $-\sqrt{3}/2$, 0, 0 (c) 0, -2, 2

Section 18–5

1. 128 2. $5/2\pi$ 3. 0.01 sec

4. $y = 3 \sin 2\pi \cdot 50t$ 5. $y = 3 \sin 5t$ 6. $50/4\pi^2$

Chapter 19

Section 19–2

2. $y = 0.0005 \sin 2\pi \cdot 300t$

3. 540, 0.002 4. 20

Section 19–3

2. (a) 1 (b) 1 (c) 1 (d) 100

Section 19–4

2. 240, 0.01

3. $y = 0.01 \sin (2\pi\, 500t) + 0.002 \sin (2\pi\, 1000t) + 0.005 \sin (2\pi\, 1500t)$

4. 2160

Chapter 21

Section 21–3

3. $\dfrac{-1}{-2}$ 4. Yes

Section 21–4

6. 9 + any positive or negative multiple of 12

14. (a) 5 (b) 3 (c) 6

15. (a) 1, 4 (b) 0, 2, 4

Section 21–5

4. *A* 5. *A*

Chapter 22

Section 22–3

2. 8113, 1200 3. 6.1, 5, 10 4. 31, 35, 40

Section 22–4

1. 3.24 2. 3 3. 12.8

Section 22–5

3. 95.4 4. 47.7, 99.8

Section 22–6

1. $y = 3x + 7$ 2. $F = 4d$

3. $d = 16t^2$ 4. $N = Y - 1944$

5. $P = 2t^2 + 4$ 6. $V = 2.3 - 0.675T + 0.0075T^2$

Chapter 23

Section 23–2

3. $\frac{1}{6}, \frac{1}{6}, \frac{4}{6}$
5. $\frac{3}{5}$
6. $\frac{1}{4}$
7. $\frac{1}{4}$
8. $\frac{27}{32}$
9. $\frac{1}{2}$
11. 36
12. 4
13. $\frac{4}{36}$
17. 1 to 1
18. 3 to 1

Section 23–3

2. 0.85, 0.68
3. $\frac{1}{8}$

Section 23–4

1. 0.499, 0.954
2. 0.159
3. 0.023
4. 0.341
5. 0.136

Section 23–5

1. $\frac{35}{128}$
2. $\frac{15}{64}$
3. $2000(\frac{15}{64})$
4. 0.977
5. 0.999
7. Discredit

REVIEW EXERCISES

1. (a) $\frac{1}{19}$ (b) $\frac{1}{19}$ (c) $\frac{9}{38}$
2. (a) $\frac{1}{13}$ (b) $\frac{1}{4}$ (c) $\frac{16}{52}$
3. $\frac{1}{3}$
4. (a) $\frac{3}{7}$ (b) $\frac{4}{7}$ (c) 1
5. (a) $\frac{3}{9}$ (b) $\frac{4}{9}$ (c) $\frac{7}{9}$
6. (a) $\frac{1}{36}$ (b) $\frac{1}{36}$
7. (a) $\frac{65}{100}$ (b) $\frac{65}{70}$ (c) Yes
8. 0.023

Additional Answers and Solutions for

MATHEMATICS FOR THE NONMATHEMATICIAN

The following chapter contains the complete text of the Instructor's Manual originally published for use with this work. It contains additional solutions and answers to the problems in the text.

CONTENTS

PART I

PLANS OF COURSES BASED ON THIS TEXT

This text is addressed to liberal arts students and to other students who do not intend to major in mathematics or science. The audience then may consist of liberal arts students who take mathematics as part of their cultural preparation for work in the social sciences, the arts and the humanities; prospective elementary school teachers who should have a broader background on which to base their teaching of arithmetic and simpler facts of geometry; and prospective high school teachers in nonscientific subjects.

For the general liberal arts student and the prospective teacher a course based on the material as arranged in this text would seem to the author to be most suitable, for the book presents mathematics as an integral component of western culture. From a purely pedagogical standpoint, the presentation of mathematics as a development intimately involved with other branches of our culture may awaken interest in a subject which has failed notoriously to attract the nonscientist.

Even for a broad liberal arts mathematics course this text may contain more material than can be covered in the time allotted. However, many of the chapters as well as sections in chapters are not essential to the logical continuity. These chapters and sections have been starred. Thus Chapter 10 on painting shows historically how mathematicians were led to projective geometry (Chapter 11), but from a logical standpoint, Chapter 10 is not needed in order to understand the succeeding chapter. Chapter 19 on musical sounds is an application of the material on the trigonometric functions in Chapter 18 but is not essential to the continuity. The two chapters on the calculus are not used in the succeeding chapters. Desirable as it may be to give students some idea of what the calculus is about, it may still be necessary in some classes to omit these chapters. The same can be said of the chapters on statistics (Chapter 22) and probability (Chapter 23).

There is also the possibility of covering most or all of the chapters by omitting certain sections within chapters. Chapter 6 on Euclidean geometry may well serve as an illustration. The mathematical material of this chapter is intended as a review of some basic ideas and theorems of Euclidean geometry and as an introduction to the conic sections. Some of the familiar applications are given in Section 6-3 (see the Table of Contents) and probably should be taken up. However, the applications to light in Sections 6-4 and 6-6 and the discussion of cultural influences in Section 6-7 can be omitted.

Some of the material, whether or not included in the recommendations for particular groups, can be left to student reading. In fact, the first two chapters were deliberately fashioned so that they could be read by students. The objective here, in addition to presenting intrinsically important ideas, was to induce students to read a mathematics book, to give them the confidence to do so, and to get them into the habit of doing so. It seems necessary to counter the students' impression, resulting no doubt from elementary and high school instruction in mathematics, that whereas history texts are to be read, mathematics texts are essentially reference books for formulas and homework exercises.

For a one-semester liberal arts course, the basic content can be as follows:

Chapter 2	on a historical orientation,
Chapter 3	on logic and mathematics,
Chapters 4 and 5	on the number system and elementary algebra,
Chapter 6	through Section 6-5, on Euclidean geometry,
Chapter 7	through Section 7-3, on trigonometry,
Chapter 12	on coordinate geometry,
Chapter 13	on functions and their uses,
Chapter 14	through Section 14-4, on parametric equations,
Chapter 15	through Section 15-10, on the further use of functions in science,
Chapter 20	on non-Euclidean geometry,
Chapter 21	on different algebras.

Any additional material would enrich the course but would not be needed for continuity.

Colleges concerned primarily with the training of prospective elementary school teachers may feel that they must give preference to topics that these students will soon be obliged to teach. It is possible to choose material that will meet such objectives.

For courses emphasizing the number concept and its extension to algebra, one can take advantage of the logical independence of numerous chapters and use Chapters 3 through 5 on reasoning, arithmetic, and algebra and Chapter 21 on different algebras. To pursue the development of this theme into the area of functions one can include Chapters 13 and 15.

Courses emphasizing geometry can utilize Chapters 6, 7, 11, 12, and 20 on Euclidean geometry, trigonometry, projective geometry, coordinate geometry, and non-Euclidean geometry respectively. Some algebra, that reviewed in Chapter 5, is involved in Chapters 7 and 12. If knowledge of the material of Chapter 5 cannot be presupposed, this chapter must precede the treatment of geometry.

The essence of the two preceding suggestions may be diagramed thus:

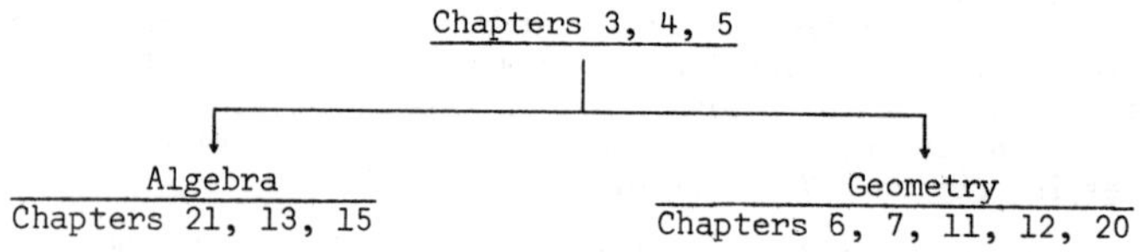

Of course, starred sections in these chapters are optional. If the essential material is covered, other chapters can, of course, be included.

PART II

SUGGESTIONS FOR PRESENTING INDIVIDUAL CHAPTERS

Chapter 1

There is rather little that one should teach, in the formal sense of teaching, of the contents of this chapter. The purpose of the chapter (and to a large extent of the next one) is to combat and hopefully eradicate dislike and fear of mathematics which many students have acquired in their elementary and high school experiences. By openly debating the question of whether mathematics is worth studying and by offering material that the student can read easily there is the possibility that he will find the approach at least disarming and perhaps refreshing and inviting.

Moreover, because he can read and understand the material he may be encouraged to tackle the later chapters with equal success and so overcome the psychological handicap which hinders many college students.

There is some effort to at least name the values of mathematics, the material benefits of mathematics to engineering, the value in aiding man to understand natural phenomena, the uses in the social sciences, the importance for philosophy as, for example, in finding truths, the uses in the arts of painting and music, and the values of mathematics proper. The latter are mathematics as an art and as an intellectual challenge. Puzzles are a good example of an intellectual challenge on a low level.

Chapter 2

Though, as already noted, this chapter too may serve to encourage the student that he can master mathematics because he finds that he can read the material, there is more positive content to be advanced than in Chapter 1. The student should learn that mathematics was not created once and for all in some ancient time but that there have been successive periods of development and these have been tied to and stimulated by problems and movements of various civilizations. The civilizations which should be discussed are the Egyptian and Babylonian, the classical Greek, the Alexandrian Greek, the Roman, the Hindu and Arabic, the medieval, and Western European since about 1400. The Roman and medieval civilizations are singular because despite achievements in other areas, they were nonproductive in mathematics. These exceptional cases could provoke a discussion of what factors in a civilization promote mathematical activity. The question cannot be answered at this stage but students might be asked to keep this in mind and to note that the answer is important because it would tell us what we must do in our civilization to foster mathematics.

Chapter 3

Though this chapter does emphasize that mathematics is distinguished from other fields by its insistence on deductive proof, the teacher should make sure that the students see that the creation of mathematics is not deductive. The important implication for the learning of mathematics is that students should be encouraged to guess, conjecture, experiment and pursue ideas even though they see no immediate prospect of success. They must be assured that even the greatest mathematicians were baffled for years in their efforts to solve problems.

Chapter 4

This chapter is largely a review of arithmetic. It is the author's contention that students should learn to operate with numbers on the basis of the physical meanings of numbers. Thus -2+(-5) is -7 because two debts of 2 dollars and 5 dollars amount to a debt of 7 dollars. Moreover these operations must ultimately become automatic and the mind freed for thinking about more advanced and more complicated mathematics. The axiomatic approach is presented in order to assure students that mathematics does live up to the standard of ultimately proving all its results deductively. But in the case of numbers we know so surely what is correct on intuitive grounds that we do not need to cite axioms to argue that 3+5=5+3.

There are three distinct applications of numbers in Section 4-6. These are not essential to the continuity but they may be intriguing and they may convince students that even a knowledge of numbers is significant for applications. One of these applications deals with the subject of other bases for the number system, a topic receiving much emphasis today.

Chapter 5

This chapter is largely a review of algebra. Most students have forgotten (or never really learned) their high school algebra and for them it is important to review. The topics covered suffice for all uses of algebra in later chapters. If the course were to be devoted to elementary algebra the author would use a different organization and presentation. But to help students recall earlier work it seems necessary to use the presentation to which they are accustomed. The method of obtaining the quadratic formula is novel but the reason for using it is that it is possible to motivate its discovery. To keep the notion of proof in algebra before the students, they should be asked to cite axioms on equalities, such as equals added to equals.

Chapter 6

This chapter is also largely review work. However, there is a decided effort to show that Euclidean geometry can be useful in solving practical as well as scientific problems. The high school geometry courses are usually purely mathematical, and students do not see why the many theorems should be learned. The starred sections can of course be omitted but some applications should be taken up. The emptiness and pointlessness of mathematics, as far as the student is concerned, must be combated and one can do so here with rather simple applications. Moreover, the major thesis that mathematics is vital to the study of nature can already be supported substantially.

Chapter 7

Though many students now take trigonometry in high school, this chapter undertakes to treat the subject as new to the student and the presentation is more detailed. That is, the review goals of Chapters 4, 5, and 6 are now dropped. Here too some applications should be included. Many of these, such as finding the distance to the moon and the radius of the moon, demonstrate the remarkable power of even simple mathematics.

Chapter 8

This chapter supplies historical continuity. It also puts together the evidence which the Greeks acquired for concluding that nature is mathematically designed. In addition the chapter serves as a breather. Students who fear mathematics and are scared of symbolism may find once again that there is something here they can grasp. The chapter can be left to student reading.

Chapter 9

This chapter sets the stage for the appearance of mathematics in Western Europe. It is remarkable that though civilization in Europe (apart from Greece and Rome) can be said to date from Roman times and certainly from the fifth century of our era, no mathematics of consequence was developed in Europe until about 1400 A.D. What trifles were known prior to that time were taken from the poor Roman texts. The Europeans became interested in mathematics when they learned, through a series of historical events which are related in the chapter, of the Greek works. Other events in Europe, for example, the geographical explorations and the breakdown of the totally religious outlook, encouraged the study and development of mathematics in Europe. This chapter can be read by the students and need not be discussed in class if the time does not permit it.

Chapter 10

This chapter on painting, though of course intrinsically important as a major example of how mathematics has been applied to and determines painting styles, is also intended to stimulate interest in mathematics by students who like the arts and fail to see values in mathematics. The use of mathematical perspective is basic in the Renaissance paintings and one who would appreciate these creations must know something about the mathematical structure. The relationship then of the two fields is not incidental but fundamental. The chapter can, of course, be recommended as just interesting reading and not made part of the course.

Chapter 11

Since the entire text cannot be covered in the usual 6-point course, this chapter on projective geometry could be omitted for prospective elementary school teachers. It can be of interest to liberal arts students in that it shows how some mathematical problems arise from ideas that were first explored in behalf of realistic painting. The chapter is descriptive and so does not call upon any detailed prior knowledge of geometry.

Chapter 12

Coordinate geometry is fundamental in mathematics and should be included in almost any course. The graphing of simple equations and the relationship between curve and equation is used in later chapters. The application of this chapter's material is made in later chapters.

Chapter 13

Here is an introduction to functions and a treatment of simple linear and quadratic functions. The study of functions as a topic of pure mathematics is rather pointless. Hence the attempt has been made to introduce these functions in connection with definite and concrete physical situations. Historically the physical phenomena considered were precisely the ones which led to the introduction of functions.

Chapter 14

This chapter on parametric equations makes further use of functions and also introduces another idea of coordinate geometry, namely, the representation of a curve by two parametric equations instead of one direct relationship between x and y. The motion of projectiles was studied intensively in the sixteenth and seventeenth centuries (it still is today) and, as the next chapter will show, suggested to Newton how to unify in one theory the motions that take place on earth and the motions of the heavenly bodies.

Chapter 15

This chapter, on the mathematical side, is still concerned with functions, notably the inverse square function. However, as noted in connection with Chapter 13, the study of mathematical functions appears empty to young people and so the chapter shows how functions can be used to determine the mass of the earth and the mass of the sun, to establish Kepler's laws, and in other applications. These few examples given in the chapter are necessarily simple but they are taken from the greatest mathematical and scientific work of our era - the work of Newton in establishing the laws of our solar system. This work gave enormous prestige to mathematics and stimulated the most important mathematical developments of the eighteenth century. A few physical concepts are utilized in the applications but these are fully discussed in the chapter. No prior knowledge of science is presupposed.

Chapters 16 and 17

The two chapters on the calculus can do no more than give students an inkling of what the calculus is about. To some the mere thought that they can study ideas of the famous calculus is thrilling. Others will not get the ideas in this brief treatment and the attempt to present them will fail.

Chapters 18 and 19

These two chapters present trigonometric functions. Again to relieve the boredom of studying functions as an end in themselves, Chapter 18 tries to motivate the topic by seeking to learn about the motion of a bob on a spring. This physical motivation is, from the student's standpoint, not an exciting one, though at least it is concrete. The application that is scientifically important and that appeals to many students is the analysis of musical sounds. The latter can be approached on the basis of what is done with the bob on the spring. The introduction of the subject of musical sounds is another effort to show students that seemingly bare and uninviting mathematical functions enable man to understand another extensive area of our physical world, the area of musical sounds.

Chapter 20

Non-Euclidean geometry does attract students because it challenges ideas they understand and are convinced of. In the intellectual development of Western man it is, next to the very founding of mathematics itself, the most significant and portentous development. This creation forced the realization that mathematics is not portraying laws inherent in the design of the universe but is merely providing man-made schemes or models which we can use to deduce conclusions about our world only to the extent that the model is a good idealization.

Chapter 21

The realization that geometry does not offer truths forced the realization that arithmetic likewise fails to do so. Mathematicians were obliged to recognize that there are many arithmetics and the corresponding algebras. This chapter, too, arouses interest because it challenges ideas the students had thought unquestionable.

Chapters 22 and 23

Though a few applications to social phenomena appear earlier in the text, there is little opportunity to consider such applications until one takes up the mathematical topics of statistics and probability. These topics are important in the modern world and, though the chapters have been starred because they are not essential for the mathematical continuity, they are valuable in giving students some idea of how social and biological phenomena are approached.

Chapter 24

This is a summarizing chapter. It is possible at the end of the course to characterize mathematics, its values for science, and the limitations of mathematics meaningfully because the students now have some experience with the subject on which to draw. One can refer to earlier material to illustrate what one means by a deductive structure, mathematical models, and so on.

PART III

SOLUTIONS AND ANSWERS

Chapter 1, page 10

1. The oarsman takes the goat across (the wolf does not eat cabbage). He comes back and takes the wolf across and takes the goat back. He leaves the goat on the original shore and takes the cabbage across. Then he goes back and brings the goat over.

2. The man fills the 5-pt jar and then fills the 3-pt jar from the 5-pt one. He empties the 3-pt jar, and then pours the 2 pt remaining in the 5-pt jar into the 3-pt jar. He now fills the 5-pt jar and pours enough into the 3-pt jar to fill it. Since there were 2 pt in the 3-pt jar, one pint is drawn out of the 5-pt jar, leaving 4 pt in it.

3. Refer to the two couples as Mr. and Mrs. A and Mr. and Mrs. B. Mr. A takes Mrs. A across. He comes back and takes Mr. B across. Mrs. A now goes back and brings Mrs. B across.

Chapter 2, page 28

1. Egyptian, Babylonian, classical Greek, Alexandrian Greek, Hindu, Arab and Western European (from 1500 on)

2. The methods and formulas gave results which worked; that is, the results were correct as far as experience showed. Thus if they calculated the area of a rectangular field by using length times width, a larger rectangular area would grow more crops.

3. The pre-Greek Egyptian and Babylonian civilizations thought in physical terms. A rectangle was a piece of land. Even in the case of whole numbers and fractions, though these civilizations did work with pure numbers, they did not consciously think of numbers as entities in themselves and seek to understand them as such. The Greeks thought of numbers and geometrical forms as ideas existing as such and even superior to the concrete interpretations of numbers and geometrical figures.

4. To start with assured premises and to establish to conclusions deductively

5. They preserved the Greek works and took over ideas from the Hindus. The Europeans learned Greek and Hindu mathematics from Arabic works (though later the Europeans also acquired Greek manuscripts).

6. The Greeks were the first to think of mathematics as an independent discipline, and to formulate standards for obtaining mathematical conclusions. They also built up a large body of geometry.

7. We see from the survey in the chapter that far more was created after Greek times than during that period.

Chapter 3, page 38

1. Arithmetic tells us only that 4 · 5 = 20. Whether the answer is to be interpreted as 20 men or 20 trucks depends entirely on the physical situation to which the numbers are applied.

2. One cannot multiply a number of pennies by a number of pennies whether in the form of 25¢ or 0.25 dollar. One can multiply 25 by 25 or 0.25 by 0.25. Whether the answer makes physical sense depends on whether there is any physical sense in the original problem. There is no physical point to 25¢ by 25¢. There is physical point in asking, for example, how much money will 25 times 25¢ amount to. This problem cannot be reformulated as 0.25 times 0.25 because the first 25 is a number of times and not 0.25 dollar.

3. Justice, liberty, democracy, good and evil

4. One cannot divide 30 books by 5 people. One can divide 30 by 5. Whether the answer is 6 books can be decided only on physical grounds.

5. As in Exercise 2, one cannot multiply dollars by dollars or cents by cents. One can multiply 6 by 1 or 600 by 100. The correct mathematics in this situation is 6 times 1 where physically the 6 means 6 repetitions and the 1 means 1 dollar. For this physical interpretation the 6 cannot be replaced by 600.

6. Mathematics deals with the concepts themselves and not the physical interpretations or realizations. Thus mathematics can deal with 3 · 5 but not with 3 people times 5 books. Likewise mathematics deals with the concept of a triangle but not with a triangularly shaped piece of land.

7. The abstract concepts are general and the reasoning about them can be applied to hundreds of situations. Furthermore, it is easier to work with abstractions, at least to the extent that irrelevant details are ignored. The Greeks also thought that the abstract concepts were the essence of knowledge.

Chapter 3, page 39

1. Abstraction refers to selecting the key or basic properties of some physical object, as the rectangular shape of a piece of land. The fact that the boundary and surface are formed by particles of earth is ignored. In idealization we ignore some significant fact for the sake of a simpler model. Thus regarding the earth as a point in astronomical work is contrary to the fact that the earth is an extended body but the idealization is good enough for some scientific work.

2. For some purposes it is correct to regard the lines to the sun as parallel. Thus two shadows of two nearby objects may be regarded as formed (outlined) by parallel lines of light from the sun. However, if A and B are far apart on the earth's surface then the lines to the sun may usually not be taken as parallel. The direction of the sun will differ and this difference may be important (as in calculating the distance of the sun from the earth).

3. Yes. The thickness would not usually matter. If it did one might have to regard the pole as a cylinder or as a tapering cylinder (cone).

Chapter 3, page 44

1. That the coin will always fall heads

2. Deductive reasoning consists of those ways or modes of combining premises so that the conclusion is necessary or indubitable.

3. It can often be done more readily, less expensively; and in some cases it is the only method available. Above all the conclusions are certain, whereas they are not in induction and analogy.

4. No. One may not have the premises with which to reason.

5. One could do this by adopting suitable premises. (It is interesting to get students to suggest premises.) However, not all people would agree to the premises.

6. The answers are in the answer section of the textbook. In each case, drawing circles to represent the classes of objects or people involved will show whether or not the conclusion must hold. Only if it must is the reasoning valid.

7. At least one of the premises must be false.

8. The reasoning may be correct but the conclusion untrue because at least one of the premises may be untrue.

Chapter 3, page 49

1. The Egyptians and Babylonians relied on the usefulness of the result to accept the mathematical procedure or formula. The Greeks insisted on deductive proof.

2. The scientist uses induction from observations or experiments and reasoning by analogy as well as deductive reasoning. The mathematician uses only deductive reasoning to prove his conclusions.

3. The Egyptian and Babylonian reliance upon adopting formulas and procedures if they were good enough for daily use is an empirical basis.

The Greeks started with axioms and reasoned deductively. Hence they were sure of their conclusions without having to test them at all.

4. (a) The argument is valid. However, the first premise is not true.
(b) Valid and in this case also true

5. No. From the information given, the square of an even number could also be odd.

6. The argument relies on induction, which is not acceptable in mathematics. It is not obvious that the square of any larger number is even.

7. Yes. If the square of a number is even and the number itself were odd, then by the first premise the square of the number would be odd. But we are given that the square is even.

8. Primarily because they wished to be sure that the conclusions were correct

9. If the sides are a, b, c and the opposite angles are A, B, C then because a = b, A = B. Because b = c, B = C. Then A = B = C because things equal to the same things are equal to each other.

10. By reasoning deductively from the known truths. Since conclusions established deductively are indubitable, there would be new truths.

Chapter 3, page 51

1. The Greeks believed the axioms were truths.

2. They adopted abstract mathematical concepts; they insisted on deductive proof; and they chose axioms which they believed were truths.

3. Yes. The insistence on obtaining truths, as opposed to useful or approximate results, is the position of philosophers. Historically the early Greek mathematicians were philosophers and these were the men who set the standards for mathematics. One could argue that mathematics began with the Egyptians and Babylonians, who were not philosophers. But the distinguishing features of mathematics were imposed by philosophers.

Chapter 3, page 54

1. It seems as though the triangles should be congruent.

2. It seems as though EFGH should be a parallelogram. (It is. EF joins the midpoints of two sides of triangle ABC. Hence EF is parallel to and equal to one half of AC. Likewise GH is parallel to and equal to one half of AC as a side of triangle ADC. Hence EF is parallel to GH and equal to GH.)

3. No. The proof is inductive and as a matter of fact the conclusion is not true. When n = 41, the formula yields 41^2 and this is not prime.

4. If corresponding sides are equal and if one angle of one equals the corresponding angle of the other, the quadrilaterals are congruent.

5. $1^3 + 2^3 + \cdots + n^3 = (1 + 2 + \cdots + n)^2$.

Chapter 3, page 55

1. The results were useful.

2. On the whole the Egyptians and the Babylonians thought in terms of the physical meanings of the mathematical concepts. The Greeks thought of the abstract concepts as objects in themselves.

3. To start with truths (the axioms) and to deduce further truths

4. The distinguishing features of mathematics as we understand the subject today were imposed or set by the Greeks.

5. (a) and (b) follow necessarily from the premises.

6. No one of the conclusions stated necessarily follows.

7. No. All quadrilaterals need not be parallelograms.

8. John will not be successful.

9. Nothing. Smith might go to the movies even when it does not rain.

10. The statement "I go to the movies only if it rains" means that "if I go to the movies, it must rain." Since Smith went to the movies, it must have rained.

Chapter 4, page 65

1. The position of a digit in a number determines the quantity represented by the digit.

2. Because otherwise we would not be able to distinguish the meaning of 5, say, in 507 and 57

3. We can operate with it as we operate with all other numbers. There is one exceptional fact: We cannot divide by 0.

4. As a quotient of integers or as decimals

5. The definitions must give results which agree with experience.

Chapter 4, page 66

1. Any even number can be written as 2m, where m is an integer. Then $(2m)^2 = 4m^2 = 2 \cdot 2m^2$. Since the $(2m)^2$ contains 2 as a factor, $(2m)^2$ is even.

2. If we multiply any number ending in 1, 3, 5, 7, or 9 by itself, the product ends in 1, 3, 5, 7, or 9 and so must be odd.

3. If a were odd, then by Exercise 2, a^2 would be odd. But a^2 is given as even. Hence a cannot be odd.

4. The assertion is false because, for example, $2^2 + 3^2 = 13$ and 13 is not a square.

Chapter 4, page 72

1. The answers are given in the answer section of the textbook.

2. The answers are given in the answer section of the textbook.

3. The argument is false. It is true that no irrational can be expressed with a finite number of decimal places. But there may be and there are other numbers which cannot be expressed with a finite number of decimal places.

Chapter 4, page 75

1 to 5. The answers are given in the answer section of the textbook.

Chapter 4, page 81

1. The only basis for belief other than actual calculation in this case is that the principle holds for small numbers where actual calculation shows it is correct.

2. We can prove that the principle is correct. $a(b - c) = a[b + (-c)] = ab + a(-c)$ by the distributive axiom (Axiom 9). Then $ab + a(-c) = ab + (-ac)$ because the product of a positive and a negative quantity is negative. Then $ab + (-ac) = ab - ac$ because to add a negative quantity is equivalent to subtracting the corresponding positive quantity.

3. The answers are given in the answer section of the textbook.

4. $(a + 3)(a + 2) = (a + 3)a + (a + 3)2 = a^2 + 3a + 2a + 6 = a^2 + 5a + 6$.

5. $(n + 1)(n + 1) = (n + 1)n + (n + 1)1 = n^2 + n + n + 1 = n^2 + 2n + 1$.

6. Equals divided by equals give equals.

7. Subtract 2 from both sides; equals subtracted from equals give equals.

8. Yes, by the distributive axiom

9. No. To show that it is false for all a, b and c we have but to find one case where it is false. Let a = 2, b = 3, c = 4.

Chapter 4, page 83

1. The average speed is the speed that would enable him to do the entire trip in the same time as under the actual speeds. One tends to think that the speed upstream and the speed downstream can be averaged. But these speeds are used for different periods of time, 3 hr on the way up and 1 1/2 hr on the way down. Hence they cannot be averaged to obtain the average speed.

2. The correct average price should be that which produces the same money as if the apples and oranges were sold separately. The correct average price will depend upon how many of each are sold. Thus if he sells 12 apples and 12 oranges at 5 pieces for 10¢, he receives (24/5)(10) or 48¢. But if he sells 12 apples at 2 for 5¢ he receives 30¢ and for 12 oranges at 3 for 5¢ he receives 20¢ or a total of 50¢. Hence the average of 5 pieces for 10¢ is wrong.

3. As noted in Exercise 2, the correct average price depends on how many of each are sold. a apples should bring a(5/2). b oranges should bring in b(5/3). Hence the total income should be (5a/2) + (5b/3). Then the average price should be [(5a/2) + (5b/3)]/(a + b).

4. As the answer to Exercise 3 shows, the average price depends on a and b. It is not possible to simplify the average price so as to eliminate a and b.

5. The first man digs 1/2 of the ditch per day and the second 1/3 of the ditch per day. It is correct to say that together they dig 5/6 of the ditch in one day. Then the average per man is 5/12 because the 2 men would dig twice this much in one day.

Chapter 4, page 84

1. Each ace can be paired with any one of 4 kings. Hence 16 pairs in all.

2. Each choice of color can be made with each choice on the heater and with each choice on the radio. Thus 3 • 2 • 2.

3. The reasoning is the same as in Exercise 2.

4. Each of 6 numbers on one die can be paired with 6 on the other. Hence 6 • 6.

Chapter 4, page 88

1. When the sums are 6 or more one must convert them to base six. Thus 3 + 2 = 5 in base 6. 3 + 4 = 11 in base 6. 5 + 5 = 14 in base six; etc.

2. When the products are 6 or more one must convert them to base 6. Thus 3 • 4 = 20 in base six.

3. 0 + 0 = 0, 0 + 1 = 1, 1 + 1 = 10. 0 · 0 = 0, 0 · 1 = 0, 1 · 1 = 1.

4-7. These are answered in the answer section of the textbook.

8. Let the base be b. Then $1 \cdot b^2 + 0 \cdot b + 1 = 10$. Then b = 3.

9. To express any number from 0 to 63, we can use base 2, base 3, etc. If we use base 2 and have separate weights for 1, 2, 2^2, 2^3, 2^4, and 2^5 we shall need six weights and we know that we can handle all weights in question because all numbers from 0 to 63 can be expressed with 0 or 1 in the six places a, b, c, d, e, f. If we use base 3 or some higher base we need multiples of the separate powers. Thus to represent 53 in base 3 we write $1 \cdot 3^3 + 2 \cdot 3^2 + 2 \cdot 3 + 2$. Thus we would need a weight for 3^3, 2 weights of 3^2, etc. This in itself requires 7 weights. Thus to use base 3 or a higher base we need more weights. If we seek to build up a set of weights which do not correspond to writing numbers in any base, we find that we must have 1, 2, 4 or 2 two's, 8, etc. and end up with more weights.

Chapter 4, page 91. Review Exercises

The answers to follow complete the set of answers given in the back of the textbook.

1. (d) $\frac{23}{36}$ (e) $-\frac{7}{36}$ (i) $\frac{ad - bc}{bd}$ (j) $\frac{ad + bc}{bd}$ (k) $\frac{2 + x}{2x}$

2. (a) $\frac{4}{15}$ (d) $\frac{4}{15}$ (f) $\frac{c}{b}$ (g) 1 (h) $-\frac{ac}{bd}$

(j) $\frac{14}{9}$ (l) 2 (m) $\frac{ad}{bc}$ (o) 4

3. (c) 6ab (e) 24xyz

4. (c) a + 2b (d) 2x + 4y

5. (a) 7 (b) 11 (d) $\frac{9}{4}$ (g) $2\sqrt{2}$

6. (b) $4\sqrt{3}$ (d) $2\sqrt{2}$ (e) $\frac{3}{2}$ (g) $\frac{3\sqrt{3}}{2}$

7. (a) $\frac{294}{1000}$ (b) $\frac{3742}{10000}$ (c) $\frac{8}{100}$ (d) $\frac{3}{1000}$

8. (a) 1.7 (b) 2.2 (c) 2.5

9. (a) Correct (c) Correct (g) Correct (i) False

10. (a) 1 (d) 111 (f) 10,000

11. (a) 1 (c) 6 (e) 9

Chapter 5, page 97

1. Symbols are used to secure brevity and generality as in the case of ax + b and often to avoid ambiguity by using different symbols for different meanings of a word as in the use of = and ≡ for the word equal. The symbol = is to be used in $x + 4 = 7$ whereas the symbol ≡ is to be used in $(x + 1)(x - 1) \equiv x^2 - 1$.

2. The statement is vague. Equal in what respects? In intelligence? In physical characteristics? In political rights?

3. As an example of the use of words, consider (c). We must say: The product of some number times the sum of that number multiplied by itself and the product of that number and another number.

4. Yes

5. (a) $3x + 4$ (b) $3x^2 + 4$

Chapter 5, page 100

1. The answers to those parts not in the back of the text are:
(b) 6^{10} (c) 10^9 (f) 10^3 (h) $1/10^3$
(i) 10^5

2. Yes. We have but to remember the meaning of positive integral exponents to see that the statement is correct.

3. (a) False (b) Correct (c) False (d) Correct
(e) False

Chapter 5, page 101

1. (a) 3^{12} (b) 3^{12} (c) 5^8 (d) 10^b
(e) 10^{12} (f) 10^4 (g) 3^{10} (h) 30^4

2. (a) 10^5 or 100,000 (b) $6^4/6^3$ or 6 (c) $8^5/8^6$ or 1/8
(d) a^3b^3/a^2 or ab^3 (e) a^3b^3/a^2b^2 or ab.

3. (a) Correct (b) Correct (c) False (d) Correct
(e) False (f) False

Chapter 5, page 104

1. The answers are given in the answer section of the textbook.

2. The answers are given in the answer section of the textbook.

3. Use the distributive axiom and the theorems on exponents.

4. Try numbers for x and y. If for these numbers the left side does not equal the right side, the equality is false. But if for specific numbers the left side does equal the right, the statement is merely verified but not proved.

6. The answer is correct but the reasoning is not.

7. The flaw lies in the step of dividing both sides by a-b. Since a = b, the divisor is 0.

Chapter 5, page 111

1. Let x be the length of the rod. Then $(x/1100) - (x/16,850) = 1$. Solving for x gives 1177 ft.

2. By the Pythagorean theorem $x^2 + (2640)^2 = (2641)^2$. Thus $x^2 = (2641)^2 - (2640)^2 = (2641 - 2640)(2641 + 2640) = 5281$. $x = \sqrt{5281} = 73$ ft

3. $800/(200 + x) = 640/(200 - x)$. Multiply both sides by $(200 + x)(200 - x)$. Then x = 22 mi/hr.

4. Let x be the number of years. Then $10,000 + 600x = 20,000 + 400x$. Solving for x gives x = 50 years.

5. Let x be the height of the staff. Then $(x + 2)^2 = x^2 + (18)^2$. Solving for x gives x = 80 ft.

6. Let x be the number of copies. Then $5000 + x = 5x$. Solving for x gives x = 1250 copies.

7. The mathematical equation must relate numbers only. Hence as a mathematical equation 1/4 = 25 is incorrect.

8. The same reason as in Exercise 7

Chapter 5, page 115

1. (a) $(x - 6)(x - 2) = 0$ (b) $(x + 9)(x - 2) = 0$

2. (a) Since the sum of the roots is -8 we form a new equation in y where $y = x - 4$. Then $x = y + 4$. Substitute in the original equation. $(y + 4)^2 - 8(y + 4) + 12 = 0$ or $y^2 - 4 = 0$. Hence $y = +2$ and -2 and $x = 6,2$.

(b) Let $y = x + (7/2)$. Then $x = y - (7/2)$. Substituting in the given equation yields $y^2 - (121/4) = 0$. Then $y = 11/2, -11/2$ and $x = 2,-9$.

3. (a) Let $y = x + 6$. Then $x = y - 6$. Substituting in the original equation gives $y^2 - 27 = 0$. Then $y = \sqrt{27}, -\sqrt{27}$. Then $x = \sqrt{27} - 6, -\sqrt{27} - 6$. Of course $\sqrt{27} = 3\sqrt{3}$.

(b) Let $y = x - 6$. Then $x = y + 6$. Substituting in the original equation gives $y^2 - 27 = 0$. Then $y = \sqrt{27}, -\sqrt{27}$. $x = 6 + \sqrt{27}, 6 - \sqrt{27}$.

Chapter 5, page 118

1. In each case we first make sure that the coefficient of x^2 is 1. If it is not, as in (a), we divide through by that coefficient. We then substitute the values of p and q in (30). The answers are in the back of the text.

Chapter 5, page 121. Review Exercises

1. The answers not given in the answer section of the textbook are: (a) $6x + 18$, (d) $x^2 - 9$, (g) $x^2 - y^2$.

2. The answers not given in the answer section of the textbook are: (b) $(x - 4)(x + 4)$, (c) $(x - a)(x + a)$, (d) $(a - b)(a + b)$, (g) $(x + 1)(x + 4)$, (h) $(x - 3)^2$, (k) $(x - 3)(x - 4)$.

3. $2x = -2$; $x = -1$.

4. The answers not given in the answer section of the textbook are: (a) $x = 3/2$, (b) $x = -3/2$, (f) $x = 1/6$, (i) $x = (c + b)/a$.

5. Let x = number of grams of acid to be added. The number of grams of water in the 50 grams is $0.75(50) = 37.5$ gr. Then $37.5/(50 + x) = 0.60$. Clear of fractions and solve for x. $x = 12\ 1/2$ gr

6. Let x be the unknown grade. Then $(60 + 70 + x)/3 = 75$. Then $x = 95$.

7. (a) $(x - 5)(x - 1) = 0$ (b) $(x - 7)(x + 1) = 0$
(c) $(x - 6)(x - 1) = 0$ (d) $(x + 9)(x - 3) = 0$
(e) $(x - 4)(x - 3) = 0$ (f) $(x - 7)(x + 2) = 0$

8. (a) Let $y = x + 5$. Answer: -9 and -1.
(b) Let $y = x - 5$. Answer: 9 and 1.
(c) Let $y = x + 5$. Answer: $-5 \pm \sqrt{76/4}$
(d) Let $y = x - 5$. Answer: $5 \pm \sqrt{76/4}$.

(e) Let $y = x - 6$. Answer: $6 \pm \sqrt{21}$
(f) Let $y = x + 6$. Answer: $-6 \pm \sqrt{21}$

9. The answers not given in the answer section of the textbook are:

(a) $-6 \pm \sqrt{30}$, (b) $6 \pm \sqrt{30}$, (d) $6 \pm \sqrt{42}$, (f) $-\frac{9}{2} \pm \sqrt{\frac{61}{4}}$
(g) $-5 \pm \sqrt{33}$.

10. Let t be the number of hours both ships travel until they meet. Then the ship at B travels 2t miles in t hours. The ship at A travels 5t miles in t hours. Then by the Pythagorean theorem $25t^2 = 10^2 + 4t^2$ or $21t^2 = 100$ and $t = \sqrt{100/21}$. This is the time it takes the two ships to meet. In the text, x was the distance BC. But BC = 2t. Hence $x = 2\sqrt{100/21} = \sqrt{400/21}$. Thus we get the same answer.

Chapter 6, page 130

1. Axioms are statements whose truth we accept as evident. Theorems are proven on the basis of the axioms.

2. Because they seemed to be self-evident truths

3. Using the construction suggested in the text we have $\angle A = AC'B$, because they are base angles of the isosceles triangle $AC'B$. Now $\angle AC'B$ is an exterior angle of triangle ACC'. Hence $\angle AC'B > \angle C$. Then $\angle A > \angle C$. But angle A is given equal to $\angle C$. Hence BC is not greater than BA. Likewise if we assume BC less than BA we can obtain the same contradiction by laying off BC on BA.

4. Use the figure and the suggestion in the text. If we draw GH so that $\angle 1' = \angle 2$ then GH is parallel to CD by Theorem 2. But AB is given parallel to CD. By Axiom 5 there can be just one line through the point where AB and GH intersect which is parallel to CD. Hence $\angle 1$ cannot be greater than $\angle 2$. If we assume $\angle 1$ is less than $\angle 2$ we can repeat the argument by drawing a line through the intersection of CD and EF so as to introduce an angle $2'$ equal to $\angle 1$.

5. Referring to Fig. 3-7, we have that $\angle 1 = \angle 2$ because they are alternate interior angles of parallel lines. Likewise $\angle 3 = \angle 4$. But $\angle 1 + \angle A + \angle 3 = 180^\circ$ because they form a straight angle. Then $\angle 2 + \angle A + \angle 4 = 180^\circ$ because we have replaced $\angle 1$ and $\angle 3$ by equal angles.

6. If two adjacent sides and the included angle of one equal the corresponding sides in the other

7. If the ratio of two adjacent sides in one equals the ratio of the corresponding sides in the other

8. Let x be the other arm. Then $x^2 + (5280)^2 = (5281)^2$ or $x^2 = (5281)^2 - (5280)^2 = 1(5281 + 5280)$. Then $x = 103$ ft.

9. The areas of two similar triangles are to each other as the square of the ratio of any two corresponding sides. In the present case the

ratio of corresponding sides is 3. Hence the ratio of the areas is 9 or one area is 9 times the other. But the price of the larger plot is only 5 times that of the smaller one. Hence the larger plot is the better buy.

10. The circumference of the roadway is $2\pi(21{,}120{,}000 + 1) = 2\pi(21{,}120{,}000) + 2\pi$. Hence the circumference of the roadway exceeds the circumference of the earth by 2π ft.

11. Parallel lines do not meet by the very definition of parallel lines. Euclid's assumption is Axiom 5 which is a quite different statement.

Chapter 6, page 137

1. If we draw the lines AE and BE we have two triangles ACE and BCE. These are congruent by s • a • s = s • a • s. Hence AE = BE.

2. We have similar triangles and $AD/OD = A'D'/OD'$. We know that OD = 93,000,000 mi, $OD' = 1$ ft, and $D'D' = 0.0045$ ft. Then we can calculate AD. One must convert the miles to feet. AD = 418,500 mi.

3. By the argument in the text the rectangle of maximum area should be a square. Hence the dimensions are 100 by 100.

4. The rectangle must be a square. Hence $p/4$ by $p/4$

5. Use the idea that is discussed on p. 134. Suppose the farmer had to enclose a rectangular area on both sides of the lake edge with 200 ft. of fencing. This rectangle must be a square with dimensions of 50 ft by 50 ft. Then the half which lies on one side of the edge has dimensions 25 by 50.

6. Think of a rectangle with dimensions a and b which add up to 12. The perimeter is 24. This rectangle is largest when $a = b = 6$.

7. By the Pythagorean theorem, $(R + h)^2 = R^2 + x^2$. Hence $x = \sqrt{2Rh + h^2}$.

8. Since h is small compared to R, we can drop h^2 compared to 2Rh.

9. We can use the result of Ex. 8. $x = \sqrt{2Rh} = \sqrt{2 \cdot 4000 \cdot (1/2)} =$ 63 miles, approx.

10. Suppose A and B are two rectangles with the same area and A is a square. If the perimeter of A is not less than B, replace A by a square A′ which has the same perimeter as B does. Then the area of A′ will be greater than that of B. Since A and B have the same area, A′ has more area than A. But A has more perimeter than B, and A′ has the same perimeter as B. Hence A has more perimeter than A′ but less area. This is impossible because if p is the perimeter of a square the area is $p^2/16$.

Chapter 6, page 141

1. The mirror image of A is on a perpendicular from A to the mirror and as far behind the mirror as A is in front.

2. The physical problem can be translated into light rays. If a light ray goes from A to point P on m to A', the shortest path is the one for which AP and $A'P$ make equal angles with m. As pointed out on page 141, this is a geometrical fact. Hence it applies to the location of the pier also.

3. He should aim at the point P on m for which AP and $A'P$ make equal angles with m.

4. Since billiard balls behave like lights rays, the final direction will be parallel to the original one.

5. As the text suggests, if $\measuredangle 1 \neq \measuredangle 2$, then pick P' so that there $\measuredangle 1' = \measuredangle 2'$. Then $AP' + P'A'$ is the shortest path from A to the mirror to A'. But APA is given as the shortest path. Hence $\measuredangle 1$ must equal $\measuredangle 2$.

Chapter 6, page 144

1. The circle is an ellipse for which the two foci coincide. Then $PF = PF'$ and this is the radius.

2. Since $PF + PF' = 10$ and $PF' = PF + FF' = 6$, $PF + PF + 6 = 10$. Hence $PF = 2$.

3. Because $PF + PF'$ is the sum of two sides of the triangle PFF'. This sum must always be greater than the third side.

4. The point on the axis must be equidistant from focus and directrix. Hence it is 5 units from the focus (and directrix).

Chapter 6, page 148

1. $QP + QF_1 > PF_1$ because the sum of 2 sides of a triangle is greater than the third side. Hence $QP + PF_2 + QF_1 > PF_2 + PF_1$. Now $QF_2 = QP + PF_2$. Hence $QF_2 + QF_1 > PF_2 + PF_1 = a$.

2. If Q is any point on t other than P, then by Exercise 1, $QF_2 + QF_1 > PF_2 + PF_1$. Hence the shortest path from F_2 to t to F_1 is $F_2P + PF_1$. By Exercise 5 of Section 6-4, F_2P and F_1P must make equal angles with t.

3. Each light ray will strike some point P on the ellipse, such as P in Fig. 6-32 and be reflected to F_1, because, as proven in Exercise 2,

F_2P and F_1P make equal angles with t, and light is reflected in accordance with just this equality of angles.

4. The lengths F_1P and F_2P become coincident radii. Then since both F_1P and F_2P make equal angles with t and the total angle is 180°, the radii are perpendicular to the tangent.

Chapter 6, page 150. Review Exercises

1. Consider Δ ACD and Δ ABE. AC = AB; $\measuredangle$ A is common; and AD = AE. Hence the triangles are congruent and so BE = DC. Then triangles CBD and BCE are congruent, so that $\measuredangle$ DBC = $\measuredangle$ BCE. From this, the base angles of ABC are equal because they are supplements of equal angles.

2. (a) $\measuredangle$ 3 = $\measuredangle$ 1 because they are vertical angles. Then because $\measuredangle$ 3 = $\measuredangle$ 2, $\measuredangle$ 1 = $\measuredangle$ 2 and so we have alternate interior angles equal.

(b) Since $\measuredangle$ 1 is also supplementary to $\measuredangle$ 4, $\measuredangle$ 1 = $\measuredangle$ 2 because supplements of the same angle are equal. But then we have equal alternate interior angles.

3. The theorem on p. 141 applies here.

4. It should take the path of a parabola whose focus is at A and directrix is m. Then the ship is as far as it can get from any gun at A or on the shore (simultaneously).

5. Let P be the center of any circle T and let us use C′ and D′ to denote the centers of circles C and D. Then if we draw a radius from C to the common point E of tangency of circles C and T, this line must pass through P because CE and PE must both be perpendicular to the common tangent at E. Then if r is the radius of T, C′P = c - r. Similarly the lines from P and D′ to the common point F of tangency of T and D must form one straight line because PF and D′F are both perpendicular to the common tangent. Then PD′ = r + d. Now PC′ + PD′ = c - r + r + d = c + d. This quantity is the same no matter where T is. Hence the points P satisfy the definition of an ellipse.

Chapter 7, page 162

1. $\sin 45^\circ = \sqrt{2}/2$, $\cos 45^\circ = \sqrt{2}/2$, $\tan 45^\circ = 1$

2. (b) $\sin 70^\circ = 0.9397$ (d) $\cos 55^\circ = 0.5736$ (f) $\tan 80^\circ = 5.6713$

3. cos A varies from 1 to 0.

4. tan A varies from 0 to unlimitedly large values.

5. sin A = BC/AB, cos B = BC/AB, etc.

6. 90°-A is the other acute angle B. Hence we are back to Exercise 5.

7. $\sin^2 A = BC^2/AB^2$; $\cos^2 A = AC^2/AB^2$. Then $\sin^2 A + \cos^2 A = (BC^2 + AC^2)/AB^2$. But by Pythagorean theorem the numerator equals the denominator. If we know sin A, we can compute cos A or vice versa.

8. sin D = FE/FD, cos D = ED/FD, tan D = FE/ED

Chapter 7, page 164

1. Use tan A = BC/AC. Then A = 56° and AC = 300. Answer: 445 ft

2. Let x = distance from the observer to the bldg. Then tan 5° = 1248/x or x = 1248/tan 5°. Answer: 14,265 ft

3. Let x = the length. Then sin 5° = 1000/x or x = 1000/sin 5°. x = 11,500 ft.

4. tan 50° = 380/x. Then x = 380/tan 50°. x = 319 ft. Hence the ship is 19 feet beyond the rocks.

5. tan A = 75/100 and tan B = 100/75.

6. From the given equation we have (r + 3) sin 87°46' = r or (r + 3) (0.99924) = r. Then 0.99924r + 2.99772 = r. Hence 0.00076r = 2.99972. Then r = 3944 mi.

Chapter 7, page 170

1. The angle POV equals the measured angle of 30°. To see this, drop a perpendicular from P to OV. Hence the latitude is 30°.

2. From 90° south latitude to 0° at the equator and then from 0° to 90° north latitude at the North Pole

3. The longitude increases from 0° to 180° west longitude. Then he crosses the 180° longitude line and the longitude becomes east longitude and decreases from 180° to 0°.

4. The 30° circle

5. He travels 2/360 of the earth's circumference or (2/360)(25,000) or 139 miles, approximately.

6. The radius of the 41° latitude circle is computed on p. 168 and is 3019 miles. Then the man travels 5/360 of the circumference or (5/360) $2\pi(3019)$ which is 263 miles, approximately.

7. He travels the full circumference of the 41° latitude circle. Since (Exercise 6) the radius is 3019, the circumference is 18,960 mi.

Chapter 7, page 175

1. cos E = 4000/ES. Then ES = 4000/cos E = 4000/0.000043 = 93,000,000 mi.

2. sin 16° = RS/ES. RS = ES sin 16' ≐ 93,000,000 (0.0046) = 428,000 mi

3. From the given equation we have (241,000 + r)sin 15' = r or (241,000 + r)(0.0044) = r. Solve for r. r = 1065 mi

4. Using Fig. 7-23(b) with M replacing V, we have sin E = MS/ES. Then MS = ES sin E = 93,000,000 sin 23° = 36,000,000 mi.

Chapter 7, page 184

1. sin 45°/sin r = 4/3. Then sin r = (3/4) sin 45°. r = 32°

2. sin i/sin r = 2/3. Then sin i = (2/3)sin r. The largest value possible for r is 90°. Then sin r = 1 and the largest value for sin i = 2/3 = 0.6666. Then the largest i value is about 42°.

3. Let i and r be the angles of incidence and refraction at the air-to-glass boundary. Then sin i/sin r = v_1/v_2 where v_1 and v_2 are the velocities of light in air and glass. Now the angle of incidence at the glass-to-air boundary is r. Call the angle of refraction r´. Then sin r/sin r´ = v_2/v_1. If we multiply this equation by the preceding we have sin i/sin r´ = 1 or sin i = sin r´ or i = r´. Then the initial and final rays are parallel.

4. sin 50/sin 45 = v_1/v_2. v_1 is 186,000. Solve for v_2 which proves to be 172,000 mi/sec.

Chapter 7, page 184. Review Exercises

1. (a) 19° (b) 28° (c) 71° (d) 62°

2. The hypotenuse is 2. Then sin 45° = $\sqrt{2}/2$, cos 45° = $\sqrt{2}/2$, tan 45° = 1.

3. The answers not given in the back of the textbook are:
(b) 12/13; 5/13; 12/5 (e) $1/\sqrt{10}$; $9/\sqrt{10}$; 1/9

4. 4/5; 3/4

5. $\sqrt{3}/2$; $\sqrt{3}$

6 $3/\sqrt{34}$; $5/\sqrt{34}$

7. tan 40° = AB/100; AB = 100 tan 40° = 83.91 ft

8. Let x = height of the pole. Then $\tan 20^\circ = x/15$ or $x = 15 \tan 20^\circ = 5.46$ ft.

9. Let the angle be denoted by A. Then $\cos A = 40/60$. Then $A = 48^\circ$.

10. Let x be the distance. Then $\tan 35^\circ = x/60$ or $x = 60 \tan 35^\circ = 42$ ft.

11. Let x be the distance from target to gun. Then $\tan 50^\circ = x/2000$. $x = 2000 \cdot \tan 50^\circ = 2384$ ft.

12. Following the method on p. 168 we have $\cos 23^\circ = O'P/OP$. Then $O'P = OP \cos 23^\circ = 4000(0.9205) = 3682$ mi. This is the radius. The circumference is $2(3682)\pi$.

13. In Exercise 12 we have the circumference of the 23° latitude circle. Then the man travels $(5/360)\ 2\pi \cdot 3682 = 107$ miles, approximately.

14. Again following the method of p. 168 we have $\cos 67^\circ = O'P/OP$. Then $O'P = OP \cos 67^\circ = 4000(0.3907) = 1{,}560$ miles.

15. $\sin i/\sin r = 4/3$. Angle $i = 45^\circ$. Then $r = 32^\circ$ approximately.

16. $\sin i/\sin r = 3/4$. Then $\sin r = (4/3) \sin i$. Here $i = 30^\circ$. Then $\sin r = 4/6$ and $r = 42^\circ$ approximately.

Chapter 8, page 195

1. The pre-Greek views were unscientific. The planets were associated with gods who ruled the affairs of men. The Greek views offer mathematical schemes which show regular patterns of planetary motion.

2. All reality reduces to numbers and relationships among numbers.

3. All motions take place about a fixed earth.

4. Deferent and epicycle (see p. 191)

5. If the planet moves clockwise on the epicycle while the epicycle moves anticlockwise on the deferent, the path is practically an ellipse.

6. Nature follows a rational pattern, a pattern obtained by human reasoning. With the Greeks each of the several domains of nature which they studied had a mathematical pattern.

7. The planets follow a plan or pattern which is described by mathematics. Presumably the universe was designed so that the planets would follow this pattern.

8. The properties of space and of objects in space were found to be described by Euclidean geometry. Again the presumption is that space and the objects were designed to possess the properties which follow from the axioms of geometry.

Chapter 9, page 207

1. Mathematics seems to flourish in civilizations that are interested in the physical world.

2. The rediscovery of the Greek works; a revived interest in the physical world stimulated by the Greek works; the geographical explorations; the employers' and laborers' interest in materials; a freer intellectual atmosphere brought about by the Reformation; new inventions such as gunpowder and lenses which led to mathematical problems.

3. God designed the universe mathematically.

Chapter 10, page 230

1. The artists were impelled to portray nature as human beings saw it. To do so they had to create a mathematical system of perspective.

2. A conceptual system uses principles and conventions which may have very definite meaning but which are not what one actually sees in the real world. Thus the use of a gold background on which angels are portrayed is intended to suggest that the angels live in heaven. An optical system portrays what the eye actually sees in real scenes.

3. Alberti, Uccello, Piero della Francesca, and Leonardo da Vinci

4. The eye sees what a section of the projection actually contains.

5. See the italicized statements on pp. 220 and 221. A fourth theorem (p. 222) is that horizontal lines making a 45° angle with the canvas must be drawn so as to pass through a diagonal vanishing point.

Chapter 11, page 238

1. (a) It will be a triangle. (b) It will be a quadrilateral.

2. Because the figure and the section look alike to the eye

3. (a) The two triangles lie in one plane.
(c) Desargues' theorem holds for any two triangles whether they lie in different planes or the same plane.

4. 5/4

5. The geometrical property which holds for some figure will also hold for the figure obtained by a section of a projection of that figure from some point.

6. DA and DB become infinite and so their ratio approaches 1. Then C becomes the midpoint of AB.

Chapter 11, page 242

2. The three points of intersection lie on one line.

Chapter 11, page 246

1. Four lines, no three of which pass through the same point

2. We can interchange the words point and line in a theorem about figures lying in one plane and obtain a new theorem.

3. Four lines all passing through one point.

4. Interchange point and line and draw the figure.

5. If one dualizes a theorem then, according to the principle of duality, he automatically obtains a new theorem.

Chapter 11, page 248

1. To discover geometrical properties common to a figure and a section of a projection of that figure, or common to two different sections of the same projection of that figure, or two sections of two different projections

2. Projective geometry is concerned with properties that are invariant or remain the same under any projection and section. These properties usually deal with intersections of points and lines, points lying on a line, and the property of being a conic section. Euclidean geometry is concerned with congruence and similarity of figures. These properties hold only under special projections and sections.

3. The theme is that projective geometry arose from the work on realistic painting.

Chapter 12, page 256

1. He took over the axiomatic, deductive method of mathematics.

2. See the middle paragraph of p. 252.

3. The Europeans needed some efficient and more effective methods, especially for new and more complicated curves coming into mathematics.

4. The new astronomy, the design of a clock, paths on maps, the design of lenses, projectile motion

Chapter 12, page 263

1. Two numbers which locate the position of the point with respect to two perpendicular lines

2. It is an equation in x and y which is satisfied by the coordinates of any points on the curve and only these points.

3. See answer section of textbook.

4. (a) For example, $x = 51$, $y = 10$. (b) For example, $x = 0$, $y = 6$.

5. Yes, because $(-3)^2 + 2(5)^2 = 59$.

6. Since $(3)^2 + (-2)^2 = 4 \cdot 3 + 1$, the answer is yes.

7. (a) A straight line with a slope of 3, cutting the y-axis at (0,7)
(b) A circle with center at the origin and radius of 7
(c) A circle with center at the origin and radius of $\sqrt{20}$
(d) Write the equation as $y = -(x/2) + 3$. Hence a straight line with slope of $-1/2$, cutting the y-axis at (0,3)
(e) Same as (c)

8. Yes. These coordinates have a different geometrical meaning but serve to locate points with respect to the equator and the 0° -longitude (half) circle.

9. Yes. If its position with respect to the coordinate axes is different, the same curve will have a different equation.

10. Yes, the slope is m.

11. (0,7)

12. (0,b)

13. Slope of m; (0,b)

14. There are no values of x and y which belong to both equations. Hence it is meaningless to subtract one equation from the other. The x's and y's of one are not the same as the x's and y's of the other.

Chapter 12, page 267

Exercises 3 to 7 are answered in the answer section of the textbook.

Chapter 12, page 270

2. Write the given equation as $y + 25 = (x - 5)^2$ and introduce x' and y' such that $x' = x - 5$ and $y' = y + 25$. Follow the text procedure on pp. 269-270.

3. The curve of $y = -x^2 + 6x$ is the same as the curve of $-y = x^2 - 6x$. The latter curve differs from $y = x^2 - 6x$ in that for the same x-value, each y-value is the negative of the other y-value. Hence the curve is reversed with respect to the x-axis, or one can say the curve is the reflection in the x-axis of the curve of $y = x^2 - 6x$ which is shown in Fig. 12-16.

5. Each point on $y = x^2 - 6x + 9$ is 9 units higher than the point on $y = x^2 - 6x$ with the same x-value.

6. One can always find the pairs of x- and y-values which satisfy the equation, and the points having these pairs of values form a curve. (One could point out that in odd cases the curve may consist of one point, e.g., $x^2 + y^2 = 0$, or there may be no curve, e.g., $x^2 + y^2 = -9$.

7. (c) This is one form of the hyperbola.

Chapter 12, page 273

3. The length in question is determined by letting $y = 0$. Then $x = \pm 5$. Hence the total length is 10.

4. Equation (18) becomes $\sqrt{x^2 + y^2} + \sqrt{x^2 + y^2} = 10$ or $x^2 + y^2 = 25$. This represents a circle.

5. We calculated in Exercise 3 the coordinates of the point where the path crosses the positive x-axis. These are (5,0). Hence the distance from (3,0) is 2. Likewise the point where the path crosses the negative x-axis has coordinates (-5,0). The distance of this point from (3,0) is 8.

Chapter 12, page 275

The answers are given in the answer section of the textbook.

Chapter 12, p. 278

1. We can write equations in four letters and speak about them as though they represented figures in a four-dimensional space. But there is no implication that a real four-dimensional space exists.

2. A hyperplane

3. Yes. They introduced the idea of representing a curve by an equation in x and y and applying algebra to the equation to deduce facts about the curve.

4. To an extent. We still need the older, purely geometric Euclidean geometry for basic facts about simple figures such as lines, triangles, and circles. Beyond that we can use either geometric proofs or coordinate geometry to establish other geometrical facts. Sometimes one approach is simpler than the other.

Chapter 12, page 278. Review Exercises

1. 0 2. 0 3. The x-coordinates are the same.

4 and 5. See answers in back of text.

6. Because $(2)^2 + (\sqrt{21})^2 = 25$, the point lies on the circle.

7. These curves are all parabolas. See Section 12-4.

8. These curves are all parabolas but displaced from the standard position. See Section 12-5.

9. See Eq. (19) and Fig. 12-17.

10. The curve is a hyperbola.

11. Given $ax^2 + bx + c = 0$, graph $y = ax^2 + bx + c$ and find where the graph cuts the x-axis.

Chapter 13, page 290

1. Shapes and sizes of objects and motion

2. A qualitative explanation is one which deals with physical concepts such as heaviness, rising, falling, motion, heat, cold, natural place, force, attraction, and the like. A quantitative explanation relates the measures of varying quantities. It is really a description rather than an explanation.

3. To isolate the primary qualities and not try to treat both primary and secondary ones; to idealize a physical phenomenon, as in neglecting the friction of air in studying motion, rather than to tackle the full physical problem with all factors included; to obtain basic physical principles from observation and experiment rather than to accept those which the human mind thinks must hold; and to seek quantitative description rather than physical explanation.

4. The Greeks sought to understand the workings of nature. Bacon and Descartes sought to use scientific knowledge to help men secure the necessities of life, maintain health, and even secure comforts.

5. Galileo sought quantitative description. The description would use mathematical formulas. Moreover, he planned to deduce other formulas from known ones by mathematical means.

Chapter 13, page 292

1. A relationship between variables

2. A formula is a symbolic representation of a function by an equation involving the variables.

4. 0, 96, 224, etc.

5. r is the independent variable and A, the dependent one.

6 and 7. See answer section of textbook.

8. $t = \pm 3$. The negative value does not have physical significance if the formula represents the distance a dropped body falls in t seconds.

9. When $C = 0$, $F = 32$. This is the freezing temperature of water. When $C = 100$, $F = 212$. This is the temperature at which water boils.

Chapter 13, page 298

1. No. A force is needed to overcome the friction of air.

2. With no gravity, the man would stay in the same place. With gravity, he is accelerated downward.

3. In 59 min it travels (59/60)10. In 1 min it travels (1/60)50. The total distance is 64/6 mi in 1 hr and so the average speed is 10 2/3 mi/hr.

4. From $v = 32t$ we have $v = 128$. The average speed is $(0 + 128)/2$ or 64 ft/sec. This is the actual speed at $t = 2$.

5. Acceleration is the rate of change of speed.

6. The graph is a straight line through the origin with a slope of 32.

7. 400; 676; 1600 ft

8. The graph is a parabola.

9. The parabola opens downward.

10. He falls according to the formula $d = 16t^2$.

11. $v = 8\sqrt{1000} = 256$ ft/sec approximately

12. They would attain larger velocities as they fell (when d is greater than 1).

13. $v = 5.3t$; $d = (5.3/2)t^2$

14. $v = at$; $d = (1/2)at^2$

15. Square both sides in (6). Then $v^2 = 64d$. Hence $d = v^2/64$. When $v = 88$, $d = 121$ ft.

16. We can, following the suggestion, use $v = 32t$ and ask what is t when $v = 88$. Answer: 2 3/4 sec

17. 32 is replaced by 11. Hence $d = v^2/22$.

18. We have $d = v^2/22$. This is the distance traveled when the velocity starts from 0 and the acceleration is 11 ft/sec^2. When $v = 88$, $d = 352$. If the object starts with 88 ft/sec and decelerates at 11 ft/sec^2, it will travel 352 ft before reaching 0 velocity. In one second the automobile travels 88 ft.

19. Let d be the depth of the well (to the water). Then $d = 16t^2$ relates the time and distance of fall. $t = \sqrt{d/16}$ is the time for the stone to fall. $d/1152$ is the time it takes the sound to reach the man. Then $\sqrt{d/16} + (d/1152) = 6\ 1/2$. To solve this for d, write $\sqrt{d}/4 = (13/2) - (d/1152)$. Now square both sides and we obtain a quadratic equation in d. However, the coefficients are large. A simpler solution is obtained by letting t_1 be the time it takes the stone to reach the water. Then $16t_1^2$ is the distance to the water and $16t_1^2/1152$ is the time it takes the sound to come back. Consequently $t_1 + (16t_1^2/1152) = 13/2$. This quadratic in t_1 is easy to solve if we first replace 16/1152 by 1/72. Then $t_1 = 6$.

Chapter 13, page 300

1. The speed and distance will be greater because the ball has the speed of 128 ft/sec in addition to the speed acquired by the action of gravity.

2. $v = 128 + 32t$; $d = 128t + 16t^2$.

3. Substitute the t values in (7) and v = 192 and 240 ft/sec. Substitute the t-values in (8) and d = 432 and 756 ft.

4. The curve is a parabola opening upward.

5. Write $d = 16(t^2 + 6t)$. Now complete the square. Then $d + 144 = 16(t^2 + 6t + 9) = 16(t + 3)^2$. Let $t' = t + 3$ and $d' = d + 144$.

Chapter 13, page 303

1. Cf. (10); $d = 128t - 16t^2$.

2. (a) When t = 4, d = 384 ft; (b) When t = 4, v = 32 ft/sec.
(c) We must first find t when v = 0. This t-value is 5. When t = 5, d = 400 ft.

3. d = 0. This means the ball has just returned to the ground.

4. The height of 512 ft is reached on the way up and 4 sec later on the way down.

5. $v = 1000 - 32t$; $d = 1000t - 16t^2$. v = 0 when t = 1000/32. Substitute this value of t in the formula for d. d = 15,625 ft

6. $d = 100 - 16t^2$

7. $v = 96 - 5.3t$, $d = 96t - (5.3/2)t^2$, $v = 0$ when $t = 96/5.3 =$ 18.1 sec, approximately. Substitute this value of t in the formula for d. d = 870 ft, approximately.

8. $d = Vt - 16t^2$, where V is the initial speed. We know that d = 0 when t = 60. Hence $0 = 60V - 16 \cdot 60^2$. Then V = 960 ft/sec.

9. The initial velocity at the suggested origin is 300 mi/hr or $300 \cdot 5280$ ft/hr. Then $d = 300 \cdot 5280t - 16t^2$.

10. $v = 96 - 32t$; $d = 96t - 16t^2$.

11. Since the origin in Exercise 10 is at the roof, when the ball returns to the ground, d = -112. Substitute this value of d in $d = 96t - 16t^2$ and solve for t. t = 7 sec

Chapter 13, page 304. Review Exercises

1.	(a) 224	(b) 80	(c) 112	(d) 155	(e) 288
2.	(a) 4	(b) 5	(c) 12 1/2	(d) 1/2	

3. (a) (0 + 160)/2 or 80   (b) (0 + 256)/2 = 128

4. Yes. At t = 7, v = 224; at t = 8, v = 256. The average speed during the eighth second is (224 + 256)/2 = 240 ft/sec.

5. $v = gt$; $d = gt^2/2$.

6. (a) Average $v = gt/2$ (b) $d = gt^2/2$

7.	(a) 256	(b) 784	(c) 196	(d) 225	(e) 441
8.	(a) 2	(b) $\sqrt{6}$	(c) 3	(d) $\sqrt{25/2}$	(e) 13/4
9.	(a) 160	(b) 176	(c) 224	(d) 332	(e) 288
10.	(a) 1	(b) 9/8	(c) 7 3/8	(d) 2 11/16	

11. $v = 100 + gt$; $d = 100t + gt^2/2$.

12. In each case we let d be the given distance and solve the resulting quadratic equation in t.

(a) 2   (b) 4   (c) 1.9 sec, approximately
(d) 1.1 approximately

13. (a) $v = 128 - gt$ (b) $d = 128t - (gt^2/2)$

14. (a) 144   (b) 140   (c) 80   (d) 44

15 and 16. See answers in back of text.

17. Yes. The only change would be that the acceleration due to gravity, which is 32 ft/sec^2 on the earth, would have to be replaced be the value that fits the particular body. Thus on the moon, 32 is replaced by 5.3.

Chapter 14, page 309

1. Since $t = x/2$, $y = 3x/2$. This represents a straight line.

2. Since $t = x/4$, $y = 5x^2/16$. This represents a parabola.

3. Since $t = x/2$, $y = 5x + x^2$. This represents a parabola.

4. Give values to t and calculate the corresponding x and y. Plot the x and y. The graph is a straight line.

Chapter 14, page 313

1. The object would have the horizontal speed of the airplane in any case. Without gravity the object would move horizontally at the velocity (speed) of 100 mi/hr and stay alongside the plane. (This is similar to the astronaut's walk in space except that the motion there is not in a straight line.)

2. The time required will be the same because the vertical motion is given by $y = 16t^2$ in both cases.

3. The equations $x = 100t$ and $y = 16t^2$ describe the motion. When $y = 500$, $t = 5\sqrt{5}/2$. This is the time to reach the ground. When $t = 5\sqrt{5}/2$, $x = 100(5\sqrt{5}/2) = 550$ ft, approximately.

4. The bullet has the speed of the plane plus the speed with which it is fired or 3000 ft/sec.

5. The equations describing the motion are $x = Vt$, and $y = 16t^2$ when x and y are measured from the point of fire and y is positive downward. We know that $x = 300$ and $y = 1$ are correct values. Hence $300 = Vt$ and $1 = 16t^2$. Then $t = 1/4$ and $V = 1200$ ft/sec.

6. The bomb falls 5280 ft. The vertical motion is given by $y = 16t^2$. When $y = 5280$, $t = \sqrt{330}$. In this time the bomb travels $300\sqrt{330}$ ft. Hence it should be released $300\sqrt{330}$ ft before the plane reaches the target. $300\sqrt{330} = 5450$ ft, approximately.

7. The plane is directly over the place where the bomb strikes the ground because both have the same horizontal motion.

Chapter 14, page 317

1. Horizontal velocity = 300 cos 40° = 230 ft/sec; vertical velocity = 300 sin 40° = 193 ft/sec.

2. Since $t = x/20$, $y = (-x^2/25) + (3x/2)$. It is a parabola opening downward.

3. Make a table of values by letting t = 0, 1, 2, etc.

4. When the projectile strikes the ground y = 0. The t = 30/16. Hence x = 20(30/16) = 37.5 ft.

5. At the maximum height the vertical velocity is 0. The vertical velocity is given by v = 32t + 30. (The 30 which appears in $y = -16t^2 + 30t$ is the vertical velocity. Compare (7) and (8).) v = 0 when t = 30/32. At this value of t, y = 14.1 ft, approximately.

6. When the shell strikes the ground, y in (8) is 0. Then t = 500/16. The vertical velocity at this value of t is given by (7). At t = 500/16, v = -500. The horizontal velocity is always the same and is 866 ft/sec. We see that the terminal velocity is the same as the initial velocity except that the vertical component is directed downward. The magnitude of the terminal velocity is $\sqrt{(866)^2 + (500)^2}$ = 1000 ft/sec. See Fig. 14-4, where OR = $\sqrt{(OP)^2 + (OQ)^2}$.

Chapter 14, page 323

1. From (20) we have t = x/V cos A. If we substitute this in (22), we obtain $y = -(16x^2/V^2 \cos^2 A) + x(\sin A/\cos A)$.

2. From (21), when $v_y = 0$, we have time = V sin A/32.

3. We substitute the value of t obtained in Exercise 2 in (22). Then max $y = V^2 \sin^2 A/64$.

4. When the projectile reaches the ground, y = 0. From (22), the time equals V sin A/16. If we look at the time to reach maximum height (see Exercise 2) we see that it takes twice as long to reach the ground again.

5. We must use (25), where now A = 40° and V = 2000. Then x_1 = 123,000 ft.

6. We can use the result on p. 322. Max $x_1 = V^2/32 = (800)^2/32$ = 20,000 ft

7. The maximum range is given by max $x_1 = V^2/32 = (2000)^2/32$. The angle of fire must be 45°. Now for any x_1 and t, x = Vt cos A [see Eq. (20)]. Hence let $x = (2000)^2/32$, A = 45°, V = 2000 and solve for t. $t = (62.5)\sqrt{2}$ sec

8. In Exercise 3 we found that the max y for any angle of fire is $V^2 \sin^2 A/64$. For fixed V, this quantity is a maximum when $A = 90^\circ$.

Chapter 14, page 324. Review Exercises

1. The sketches are made by making a table of values. One can find the direct equation and check.

2. See answers at the back of the text for (a), (b), and (c). As for (d), $t = (x - 7)/3$ and $y = 5[(x - 7)/3] + 9 = (5x/3) - (8/3)$.

3. (a) $x = 240 \cdot 5280t$, $y = 16t^2$
   (b) Let $y = 5280$ and solve for t. $t = \sqrt{330}$ sec
   (c) Same time as in (b) because the vertical motion is the same
   (d) Let $y = 2 \cdot 5280$ in $y = 16t^2$ and solve for t. $t = \sqrt{660}$ sec
   (e) Since it takes $\sqrt{660}$ sec to fall, the bomb will travel $x = 240 \cdot 5280 \cdot \sqrt{660}$ ft.

4. (a) $2000\sqrt{2}/2$ and $2000\sqrt{2}/2$
   (b) $x = 1000\sqrt{2}t$; $y = -16t^2 + 1000\sqrt{2}t$
   (c) Use (25) or since the 45° angle of fire gives maximum range, use $V^2/32$. Answer: 125,000 ft
   (d) In Exercise 3 of the preceding list we found that the maximum height is given by $V^2 \sin^2 A/64$. Here $V = 2000$ and $A = 45^\circ$. Answer: 31,250 ft

Chapter 15, page 333

1. In a geocentric system the earth is fixed and the motions of the other bodies are described as they appear from the earth. In a heliocentric system the sun is regarded as fixed and the motions are described in relation to the sun.

2. No. He continued to use the deferent and epicycle scheme.

3. The sun is at one focus of each ellipse.

4. His main innovation was to replace each system of deferent and epicycles for any one planet by an ellipse. He also gave three new laws of motion.

5. If the planet moves counterclockwise while the epicycle moves clockwise, the path is a crude ellipse. The main point is that a single curve, if reasonably simple, say an ellipse, is easier to comprehend and work with than a combination of curves.

6. See pp. 330-331.

7. Because the scheme was mathematically simpler

8. The student would have to admit that he accepts it because he was taught to do so.

9 and 10. See pp. 328-9.

11. Let D be the given distance and calculate T from $T^2 = D^3$. Then T = 165 yr.

Chapter 15, page 341

1. Because one is a constant times the other so that a large mass feels heavy

2. The man's mass is also 160 pounds (of mass). His weight on the moon is 5.3 times his mass or 5.3(160) = 848 poundals. This is also 848/32 pounds of weight.

3. The man's mass is 160 lb. His weight on the sun is 27 • 32 • 160 poundals or 27 • 160 pounds of weight.

Chapter 15, page 343

2. The acceleration is $a = GM/(5000 \cdot 5280)^2$. We do know that $32 = GM/(4000 \cdot 5280)^2$. If we divide the first equation by the second, we have $a/32 = (4000 \cdot 5280)^2/(5000 \cdot 5280)^2 = 16/25$. Than a = (16/25(32) = 20.5 ft/sec^2 approximately.

3. Strictly speaking, no, because the formula $d = 16t^2$ was derived on the assumption that the acceleration is 32 ft/sec^2 all along the path.

4. 150 lb (of mass)

5. The mass of the automobile is 3000 lb. Then, by (3), F = 3000 • 12 = 36,000 poundals = 36000/32 lb (of force).

Chapter 15, page 345

1. The weight or force of gravity is given by (6). Then $F = GM \cdot 150/(8000 \cdot 5280)^2$. We know that $150 \cdot 32 = GM \cdot 150/(4000 \cdot 5280)^2$. Then $F/150 \cdot 32 = (4000 \cdot 5280)^2/(8000 \cdot 5280)^2 = 16/64 = 1/4$. Hence F = 150 • 32(1/4) poundals or 150(1/4) pounds of weight.

2. The law shows how the same mass m has a weight which varies with distance from the center of the earth.

3. Let the two masses be m and 2m. Then the two forces are GMm/r^2 and $GM \cdot 2m/r^2$. The latter is twice the former.

4. The man's weight is given by $F = GMm/r^2$ where M is the mass of the earth. If M were one tenth as much, then F would be one tenth as much or 15 lb.

5. By (5) the acceleration would be twice as much. The body would acquire more velocity each second and reach the ground sooner.

6. By (5) the acceleration of the body higher up would be less.

7. In place of $F = GMm/(4000 \cdot 5280)^2$ we would have $F_2 = GMm/(1000 \cdot 5280)^2$. Since $F = 150$, we have $F_2/150 = (4000 \cdot 5280)^2/(1000 \cdot 5280)^2 = 16$. Hence $F_2 = 16 \cdot 150$ lb.

8. In the case of $F = GMmr/R^3$, the force increases linearly with r. In the case of $F = GMm/r^2$, the force decreases as the square of r.

9. The weight would still decrease as the distance from the center of the earth increased but it would not decrease so much.

10. Yes. By (6), $F/m = GM/r^2$ and for a fixed r, this ratio is the same for any mass m and its weight F.

Chapter 15, page 351

1. Yes. According to Newton's first law of motion, if no force acts on an object, the object remains at rest if it is at rest or it moves at a constant speed in a straight line if it is in motion. If an object does not move in a straight line, a force must act on it and this force, by Newton's second law, causes an acceleration.

2. No. We have but to review the derivation to see that it does not.

3. By (24), $a = (50)^2/5 = 500$ ft/sec^2. The force is exerted by the hand.

4. The moon's velocity is given by $2\pi \cdot 240000 \cdot 5280/(26/3) \cdot 24 \cdot 60 \cdot 60$ in ft/sec. We now compute v^2/r where v is the value just given and $r = 240{,}000 \cdot 5280$. The result is $a = 0.00897$ ft/sec^2 approximately.

5. The mass of the sun is $4.40 \cdot 10^{30}$ lb. The volume of the sun is $(4/3)(432{,}000 \cdot 5280)^3$. The ratio of mass to volume is about 90.

Chapter 15, page 355

1. We cannot consider the formula for the variation of mg with latitude. All we can say is that as the latitude increases, more centripetal force is required to keep an object rotating with the earth. Hence less gravitational force is available to supply the weight, and weight decreases as one moves from either pole to the equator.

2. By (30), $v^2 = GM/r$. If the satellite is close to the earth $r = R$. Also $GM = 14 \cdot 10^{15}$. Then $v^2 = 14 \cdot 10^{15}/4000 \cdot 5280$. Then $v = 26000$ ft/sec, approximately.

3. By (30), $v^2 = GM/240{,}000 \cdot 5280 = 14 \cdot 10^{15}/240{,}000 \cdot 5280$. Then $v = 3300$ ft/sec, approximately.

4. The circumference of the moon's path is $2\pi \cdot 240{,}000 \cdot 5280$. The velocity, by Exercise 3, is 3300 ft/sec. Then the time is the distance divided by the velocity or approximately 27.8 days.

5. We use (30), in which $r = 4500 \cdot 5280$ ft and $GM = 14 \cdot 10^5$. Then $v = 24{,}000$ ft/sec, approximately.

Chapter 15, page 359

1. It applies to all masses in the universe insofar as all uses of it show. The calculations based on it agree with experience.

2. He derived them mathematically from his laws of motion and gravitation.

3. Since Kepler's laws proved to be mathematical consequences of the laws of motion and gravitation -- laws that applied to all other motions -- it was more likely that the heliocentric view, which gives rise to Kepler's laws, is correct.

4. A check on the correctness of physical principles is that one can derive from them conclusions which agree with experience. Since Newton's principles lead to Kepler's laws and the latter are known to be observationally correct, Newton's principles are likely to be correct

Chapter 15, page 361. Review Exercises

1. See answers in answer section of textbook.

2. See answers in answer section of textbook for (a), (b), (c), and (e). (d) $4.675 \cdot 10$ (f) $7.4/10^3$

3. See answers in answer section of textbook for (a), (b), (c), and (d). (e) $2.8/10^3$ (f) 6

4, 5, and 6. Answers are given in the answer section of the textbook.

7. (a) By (5), $a = GM/(6000 \cdot 5280)^2$. We know that $32 = GM/(4000 \cdot 5280)^2$. Hence $a/32 = (4000 \cdot 5280)^2/(6000 \cdot 5280)^2 = 16/36$. Then $a = 32(16/36) = 14.2$ ft/sec^2.

(b) The reasoning is the same except that 14,000 replaces 6000. Then $a = 2.6$ ft/sec^2.

8. (a) A man who weighs 200 lb at the surface has 200 lb of mass. When he is 2000 miles above the surface of the earth, then $GM/(6000 \cdot 5280)^2 = 14.2$. Hence the pull of gravity F is m times this latter amount or $F = 14.2(200) = 2840$ poundals or 2840/32 lb.

(b) Use 7(b) and argue as in 8(a). Answer: 520 poundals

9. The man's mass is 200 lb. On the moon $a = GM/r^2$, where M is the mass of the moon and r its radius. Now a = 5.3. Hence F = 200(5.3) = 1060 poundals = 1060/32 lb.

Chapter 16, page 373

1. The rate of change of distance compared to time is the change in the distance divided by the interval of time in which that change takes place.

2. Average speed is the average over some interval of time or the distance traveled in that interval divided by the interval of time. Instantaneous speed is the speed at an instant of time as opposed to the average over an interval of time.

3. The notion of limit

4. The distance fallen in 5 seconds is $16 \cdot 5^2$ or 400 ft. The average speed is 400/5 or 80 ft/sec. After 4 sec the body falls $16 \cdot 4^2$ or 256 ft. Hence during the fifth second the body falls 400 - 256 or 144 ft. The average speed during this second is 144/1 ft/sec.

5. Repeat the process on pp. 371-372 with t = 5 instead of t = 4. In step (6) we obtain k/h = 160 + 16h. The limit of k/h as h approaches 0 is 160.

6. At t = 3, $d_3 = 128 \cdot 3 - 16 \cdot 3^2$. At t = 3 + h, $d_3 + k = 128(3 + h) - 16(3 + h)^2 = 128 \cdot 3 + 128h - 16 \cdot 9 - 16 \cdot 6h - 16h^2$. Then $k = 128h - 96h - 16h^2$. $k/h = 32 - 16h$. As h approaches 0, k/h approaches 32.

Chapter 16, page 378

1. (a) $y_1 = bx_1$; $y_1 + k = b(x_1 + h)$. Then k = bh and k/h = b. Since b is constant, the limit of k/h or $\dot{y} = b$.

(b) $y_1 = ax_1^3$; $y_1 + k = a(x_1 + h)^3 = ax_1^3 + 3ax_1^2h + 3ax_1h^2 + ah^3$. Then $k = 3ax_1^2h + 3ax_1h^2 + ah^3$ and $k/h = 3ax_1^2 + 3ax_1h + ah^2$. As h approaches 0, $3ax_1h$ and ah^2 approach 0 and k/h approaches $3ax_1^2$. Hence $\dot{y} = 3ax_1^2$ and since this holds for any x, $\dot{y} = 3ax^2$.

(c) $y_1 = c$, $y_1 + k = c$; k = 0; hence k/h = 0 and the limit of k/h = 0. That is, $\dot{y} = 0$.

2. $y_1 = x_1^2 + 5$; $y_1 + k = (x_1 + h)^2 + 5 = x_1^2 + 2x_1h + h^2 + 5$. Then $k = 2x_1h + h^2$ and $k/h = 2x_1 + h$. The limit of k/h as h approaches 0 is $2x_1$ or $\dot{y} = 2x_1$. The example suggests that the constant in

the original function makes no contribution to the derivative.

3. The answers are given in the answer section of the textbook.

4. If we apply (25), we get $\dot{d} = 100 + 32t$. At $t = 4$, $\dot{d} = 228$ ft/sec.

5. (a) The instantaneous rate of change of the volume with respect to the radius is the surface of the sphere.
   (b) If $V = (4/3)\pi r^3$, then by (23), $\dot{V} = 4\pi r^2$ and this is the expression for the area of the surface of the sphere.

6. $\dot{y}$ should be $4ax^3$.

7. If $A = x^2$, then $\dot{A} = 2x$. The result is intuitively reasonable because if we think of a square and think of a slightly larger one (see figure alongside) then the rate at which the area increases depends on the length of 2 adjacent sides AB and BC or we could say that the actual increase for an increase h in x is $2xh + h^2$ and the rate for the increase h in x is $2x + h$. For very small h the rate of increase in area is $2x$.

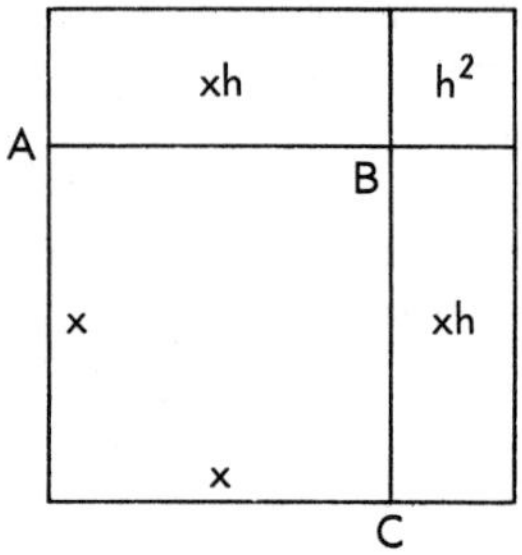

8. $\dot{A} = \ell$. Geometrically, for a small change h in w, the increase in area is ℓ/h. Then the average rate is ℓ and in this case ℓ is also the instantaneous rate.

Chapter 16, page 382

1. (a) $\dot{y} = (1/100)2x$. At $x = 3$, $\dot{y} = 6/100$. This is the slope.
   (b) At $x = 5$, $\dot{y} = 10/100$. The slope is steeper at $x = 5$.
   (c) At $x = 0$, $\dot{y} = 0$. The tangent to the curve is horizontal at $x = 0$.

2. (a) $\dot{y} = 4 - 2x$. At $x = 1$, $\dot{y} = 2$. The direction is that of the tangent which has a slope of 2.
   (b) We must find when $\dot{y} = 0$. Since $\dot{y} = 4 - 2x$, $\dot{y} = 0$ when $x = 2$.

3. At **A** the slope and therefore the derivative is some positive number. The slope and therefore the derivative decreases to 0 at the peak then the derivative becomes more and more negative. About halfway toward the lowest point the derivative becomes less negative and then becomes 0 at the lowest point. After that it becomes more and more positive.

4. $y = x^2$ and $y = x^2 + 5$ have the same graph except that the latter is 5 units higher. Then at any given value of x, the slopes are the same. Hence the derivatives must be the same.

Chapter 16, page 385

1. $\dot{d} = 128 - 32t$. At $t = 4$, $\dot{d}$ or v is 0. This means physically that the body has reached the highest point of its motion. Geometrically it means that the graph of $d = 128t - 16t^2$ has the slope 0 at $t = 4$ or the tangent is horizontal there.

2. $A = xy$. $y = (p/2) - x$. Then $A = x[(p/2) - x] = (px/2) - x^2$. Then $\dot{A} = (p/2) - 2x$. When $\dot{A} = 0$, $x = p/4$. Then $y = p/4$. The rectangle is a square. The calculus method is direct. We know how to proceed. It would take more fumbling to find the geometrical proof.

3. $A = xy$. $y + 2x = 100$ or $y = 100 - 2x$. Then $A = x(100 - 2x) = 100x - 2x^2$. Hence $\dot{A} = 100 - 4x$. When $\dot{A} = 0$, $x = 25$. Then $y = 50$. Here too, the calculus method is more direct.

4. Let y be the vertical dimension and x, the horizontal one. Then $A = xy$. But $3y + 2x = 100$ or $y = (100 - 2x)/3$. Then $A = x(100 - 2x)/3 = (100/3)x - (2/3)x^2$. $\dot{A} = (100/3) - (4/3)x$. When $\dot{A} = 0$, $x = 25$. Then $y = 50/3$.

5. As the text shows, $V = 50r - \pi r^3$. Then $\dot{V} = 50 - 3\pi r^2$. When $\dot{V} = 0$, $r = \sqrt{50/3\pi}$. If we substitute this value of r in $h = (50 - \pi r^2)/\pi r$ we get $2(50/3\pi)/r = 2\sqrt{50/3\pi} = 2r$.

Chapter 16, page 386. Review Exercises

1. (a) When $t = 6$, $d = 576$. Hence the average speed is 576/6 or 96 ft/sec.
   (b) When $t = 5$, $d = 400$. Hence during the sixth second the average speed is $(576 - 400)/1 = 176$ ft/sec.
   (c) $\dot{d} = 32t$. At $t = 6$, $\dot{d} = v = 192$ ft/sec.

2. Calculate the instantaneous speed at $t = 6$. This, by Exercise 1 (c), is 192 ft/sec. This value must be close to the average speed in the interval from 5.9 to 6.

3. $\dot{y} = 20x$. Hence: (a) 40 (b) 60 (c) 20a

4. $v = \dot{d} = 32t$. Hence $a = \dot{v} = 32$.

5. The answers are given in the answer section of the textbook.

6. $\dot{y} = 2x + 2$ and at $x = 2$, $\dot{y} = 6$.

7. $\dot{y} = 2x + 2$. When $\dot{y} = 0$, $x = -1$.

8. (a) $\dot{y} = 2x + 10$. When $\dot{y} = 0$, $x = -5$ and $y = -25$. A minimum
   (b) $\dot{y} = 2x - 10$. When $\dot{y} = 0$, $x = 5$ and $y = -25$. A minimum
   (c) $\dot{y} = -2x + 10$. When $\dot{y} = 0$, $x = 5$ and $y = 25$. A maximum
   (d) $\dot{y} = -2x + 6$. When $\dot{y} = 0$, $x = 3$ and $y = 9$. A maximum
   (e) $\dot{y} = -2x + 6$. When $\dot{y} = 0$, $x = 3$ and $y = 11$. A maximum

9. $\dot{y} = 3x^2$. At $x = 0$, $\dot{y} = 0$. The function does not have a minimum or a maximum at $x = 0$.

10. $\dot{h} = 144 - 32t$. When $\dot{h} = 0$, $t = 9/2$ sec. and $h = 324$ ft.

11. Let x = a side of the base and y, the height. Then $V = x^2y$. But $x^2 + 4xy = 400$ or $y = (400 - x^2)/4x$. Then $V = 100x - (x^3/4)$. Hence $\dot{V} = 100 - (3x^2/4)$. When $\dot{V} = 0$, $x = \sqrt{400/3} = 20/\sqrt{3}$. Then $y = 10/\sqrt{3}$.

12. Using x and y as in Exercise 11, we have $V = x^2y$ and $2x^2 + 4xy = 400$ or $x^2 + 2xy = 200$. Then $y = (200 - x^2)/2x$. $V = 100x - (x^3/2)$. Then $\dot{V} = 100 - (3x^2/2)$. When $\dot{V} = 0$, $x = \sqrt{200/3}$. Then $y = \sqrt{200/3}$. Hence the box is a cube.

Chapter 17, page 390

1. See answer section of textbook.

Chapter 17, page 394

1. $\dot{v} = -32$. Thus $v = -32t + C$. When $t = 0$, $v = 150$. Hence $v = -32t + 150$. Now $d = -16t^2 + 150t + C$. But $d = 0$ when $t = 0$. Then $C = 0$ and $d = -16t^2 + 150t$.

2. Let distance be measured downward from the point 50 ft above where the object is dropped. $\dot{v} = 32$ and so $v = 32t + C$. When $t = 0$, $v = 0$. Then $C = 0$ and $v = 32t$. Hence $d = 16t^2 + C$. When $t = 0$, $d = 50$. Then $C = 50$ and $d = 16t^2 + 50$.

3. Since the distance is measured upward from the ground, $\dot{v} = -32$ and $v = -32t + C$. Now $v = 0$ when $t = 0$. Then $C = 0$ and $v = -32t$. Hence $d = -16t^2 + C$. When $t = 0$, $d = 75$. Then $C = 75$ and $d = -16t^2 + 75$.

4. Since distance is measured upward from the ground, $\dot{v} = -32$; $v = -32t + C$. When $t = 0$, $v = 100$. Then $C = 100$ and $v = -32t + 100$. Next, $d = -16t^2 + 100t + C$. When $t = 0$, $d = 50$. Hence $C = 50$ and $d = -16t^2 + 100t + 50$.

5. Distance is measured upward from the ground. Then $\dot{v} = -32$ and $v = -32t + C$. When $t = 0$, $v = -100$. Hence $C = -100$ and $v = -32t - 100$. Then $d = -16t^2 - 100t + C$. When $t = 0$, $d = 50$. Then $C = 50$ and $d = -16t^2 - 100t + 50$.

Chapter 17, page 397

1. $\dot{A} = x^2$. Then $A = (x^3/3) + C$. When $x = 2$, $A = 0$. Then $C = -8/3$. Hence $A = (x^3/3) - 8/3$. When $x = 6$, $A = (6^3/3) - (8/3) = 69\ 1/3$.

2. As in Exercise 1, $A = (x^3/3) + C$. When $x = 4$, $A = 0$. Then $A = (x^3/3) - (64/3)$. When $x = 6$, $A = (216/3) - (64/3) = 50\ 2/3$.

3. $\dot{A} = x$. Then $A = (x^2/2) + C$. When $x = 4$, $A = 0$. Then $C = -8$ and $A = (x^2/2) - 8$. When $x = 6$, $A = 10$.

4. $\dot{A} = x^2 + 9$. Then $A = (x^3/3) + 9x + C$. When $x = 3$, $A = 0$. Then $C = -36$. Hence $A = (x^3/3) + 9x - 36$. When $x = 6$, $A = 90$.

Chapter 17, page 401

1. We may use (29). The 500-lb weight has a mass of 500 lb. Hence $W = 32(4000 \cdot 5280)500\left(1 - \dfrac{4000 \cdot 5280}{4100 \cdot 5280}\right) = 32(4000 \cdot 5280)500(1/41)$. If we neglect the factor 32, the answer in ft-lb is about 257,000,000.

2. If the gravitational force were always what it is at the surface of the earth it would be GMm/R^2 [see Eq. (23)]. However (p.400), $GM = 32R^2$. Hence the force would be $32m$. Hence the work would be $32 \cdot 500 \cdot 100 \cdot 5280$. If we neglect the 32 to keep the answer in ft-lb, the result is 264,000,000.

3. From $k = [300 + 2(100 - \bar{x})]h$ we have that $k/h = 300 + 2(100 - \bar{x})$. When h approaches 0, k/h approaches $\dot{W}$ and $\bar{x}$ approaches x. Hence $\dot{W} = 300 + 2(100 - x) = 500 - 2x$. Then $W = 500x - x^2 + C$. When $x = 0$, $W = 0$. Hence $W = 500x - x^2$. To raise the tool completely $x = 100$. Then $W = 40{,}000$ ft-lb.

4. $W_1 = c/r_1$. $W_1 + k = c/(r_1 + h)$. Then $k = [c/(r_1 + h)] - [c/r_1] = -ch/(r_1 + h)r_1 = -ch/(r_1^2 + r_1h)$. Hence $k/h = -c/(r_1^2 + r_1h)$. When h approaches 0, $\dot{W} = -c/r_1^2$ or for any r, $\dot{W} = -c/r^2$.

Chapter 17, page 403

1. Use (33) with $R = 4000 \cdot 5280$ and $d = 240{,}000 \cdot 5280$. Then $V =$ 36,500 ft/sec, approximately.

2. The derivation of (34) applies to escape velocity from the earth. However, the only place in which we used a value which applies specifically to the earth was in deriving (29) where we used the fact that $GM = 32R^2$. For M the mass of the moon, $GM = 5.3R^2$ where R is the radius of the moon. Then (34) becomes, for the moon, $V = \sqrt{2(5.3)R}$.

Chapter 17, page 412

1. The basic new idea is the limit concept.

2. The history of the calculus shows that mathematicians go through a period of thinking and fumbling before they arrive at the correct logical approach.

Chapter 17, page 413. Review Exercises

1. (a) Starting with $\dot{v} = -32$, we have $v = -32t + C$ and since $v = 200$ when $t = 0$, $v = -32t + 200$. Then $d = -16t^2 + 200 + C$. But $d = 0$ when $t = 0$. Hence $d = -16t^2 + 200t$.
   (b) At the maximum point $\dot{d}$ or $v = 0$. Hence $t = 200/32$ and $d = -16(200/32)^2 + 200(200/32) = 625$ ft.
   (c) When the object hits the ground, $d = 0$. Then $t = 200/16$. Substitute this value of t in $v = -32t + 200$ and we obtain -200 ft/sec.

2. (a) The height above the roof is obtained exactly as in (a) of Exercise 1. Then $d = -16t^2 + 200t$.
   (b) The height above the ground is $d = -16t^2 + 200t + 100$.

3. (a) $\dot{v} = -5.3$. Hence $v = -5.3t + C$. When $t = 0$, $v = 200$. Hence $C = 200$ and $v = -5.3t + 200$. Then $d = -(5.3/2)t^2 + 200t + C$. When $t = 0$, $d = 0$. Then $d = -(5.3/2)t^2 + 200t$.
   (b) The maximum height is attained when $\dot{d}$ or v is 0. Then $t = 200/5.3$. Substitute this value of t in the formula for d and we obtain $d = 3585$ ft.
   (c) The object loses less velocity per second as it rises because the downward acceleration is only 5.3 ft/sec^2 on the moon and 32 ft/sec^2 on the earth. Hence it takes longer for the velocity to become 0.

4. (a) $\dot{A} = 3x$. Then $A = (3x^2/2) + C$. When $x = 0$, $A = 0$. Hence $C = 0$ and $A = 3x^2/2$. When $x = 4$, $A = 24$.
   (b) The triangle is a right triangle with arms of 4 and 12. Hence the area is $(1/2)(4)(12) = 24$.

5. (a) $\dot{A} = 2x + 7$. Then $A = x^2 + 7x + C$. When $x = 4$, $A = 0$. Then $C = -44$. Hence $A = x^2 + 7x - 44$. When $x = 6$, $A = 34$.
   (b) The area of the trapezoid is $(1/2)(2)(15 + 19) = 34$.

6. Use (29) with $m = 200$, $R = 4000 \cdot 5280$ and $r = 4100 \cdot 5280$. Then $W = 3297 \cdot 10^6$ poundals.

7. Use (33) with $R = 4000 \cdot 5280$ and $d = 100 \cdot 5280$. Then $V = 5700$ ft/sec, approximately.

Chapter 18, page 422

1. Answers are given in the answer section of the textbook.

2 and 3. Answers are given in the answer section of the textbook.

4. Because every 360° the function repeats the values it takes on in the preceding 360° interval

5. It describes the height of Q above or below 0.

6. The function $y = \sin A$ first of all emphasizes the relation between two variables, A and y. Moreover, the function is defined for all values of A.

7. It starts from 0, rises to 1 at 90°, falls to 0 at 180°, falls farther to -1 at 270° and then rises to 0 at 360°. Thereafter it repeats this behavior every 360°.

8. Two values.

9. The period is 360°. The cycle refers to the set of y-values going from 0 to 1 to -1 to 0.

Chapter 18, page 424

1, 2, and 3. The answers are given in the answer section of the textbook.

4. The variation of A from 0 to 2π produces the same variation in sin A as when A goes from 0 to 360°. Hence see Exercise 7 of the preceding list.

Chapter 18, page 426

1, 2, 3, and 4. See the answer section of the textbook.

5. The shape is the same as when t varies from 0 to 1. See Fig. 18-12.

6. The shape is similar to that of Fig. 18-12 except that there are 3 full cycles of y-values in the interval from $t = 0$ to $t = 1$.

7. The shape is that of Fig. 18-12 except that all y-values are twice as large (or as small when y is negative).

8. Comparing with (7), we see that $f = 10$.

9. See the back of the textbook.

Chapter 18, page 433

1. From (15) we have $32 \cdot 2 = (1/2)k$ or $k = 128$.

2. Use (22). $f = (1/2\pi)\sqrt{50/2} = 5/2\pi$

3. $T = 1/f = 0.01$ sec

4. By (21), $y = \sin 2\pi ft$ provided the amplitude or maximum displacement is 1. If it is D, then $y = D \sin 2\pi ft$. [Cf. Eq. (24)]. In our case $f = 50$ and $D = 3$ in.

5. Use (24). $D = 3$ in. $= 1/4$ ft. $k = 75$ and $m = 3$. Hence $y = 3 \sin 5t$ or, in feet, $y = (1/4) \sin 5t$.

6. In (25), $T = 1$ and $k = 50$. Hence $m = 50/4\pi^2$ lb.

7. (a) The mass makes 5 oscillations or has a frequency f of 5 and it moves 4 units above and below its equilibrium position.
   (b) $2\pi f = 10$. Hence $f = 5/\pi$. The amplitude is again 4.

8. By (22), $f = (1/2\pi)\sqrt{k/m}$. Make m larger to decrease f.

9. Compare $a = -(GM/R^3)r$ with (18) where $a = -(k/m)y$. In each case the acceleration is a constant times the displacement. Now (18) leads to (24). Hence by the very same steps, except that r replaces y and GM/R^3 replaces k/m, we get $r = D \sin \sqrt{GM/R^3}\, t$. Thus the man oscillates back and forth in the tunnel. Since D is the initial displacement, the man will go from one end of the tunnel to the other and back again repeatedly.

Chapter 19, page 442

1. $y = D \sin 2\pi ft$. Here the D is the amplitude or the maximum displacement which a typical air molecule undergoes. f is the number of times per second that the typical molecule makes a complete oscillation or a complete back-and-forth motion about its mean or rest position.

2. $y = 0.0005 \sin 2\pi \cdot 300t$.

3. The frequency is 540 and the amplitude is 0.002 in.

4. 400/20 or 20 cycles/sec

Chapter 19, page 445

1. (a) $y = \sin 2\pi t$ has a frequency of 1 in one second along the t-axis and $y = \sin 6\pi t$ has a frequency of 3. Both have an amplitude of 1. Sketch both on the same axes and add ordinates as is done in Fig. 19-5.
   (b) $y = \sin 2\pi t$ has a frequency of 1 in one second along the t-axis and an amplitude of 1. $y = (1/2)\sin 4\pi t$ has a frequency of 2 and an amplitude of 1/2. Sketch both on the same set of axes and add ordinates as is done in Fig. 19-5.
   (c) $y = \sin 2\pi t$ has a frequency of 1 in one second. $y = \sin 3\pi t$ has a frequency of 1 1/2. This means that the graph will go through 1 1/2 cycles in one second. The amplitude is 1. Graph both on the same set of axes and add ordinates as is done in Figure 19-5.

2. (2) $y = \sin 8\pi t$ repeats its behavior 4 times in one second. Hence the entire function will repeat when $y = \sin 2\pi t$ repeats, that is, after one second. The frequency of the entire function is 1.
   (b) The frequency of $y = 2 \sin 2\pi t$ is 1 and that of $y = \sin 4\pi t$ is 2. Hence the two together will repeat when the first one does; that is, the frequency of the entire function is 1.

(c) As in (b), all three functions will repeat when the first one does. Hence the frequency of the entire function is 1.
(d) The respective frequencies of the three terms are 100, 200, and 300. The entire function repeats when the first one does, that is, after 1/100 of a second. The frequency of the entire function is 100.

Chapter 19, page 448

1. See p. 446, lines 11 to 13.

2. The frequency of the entire sound is 240. The amplitude of the third harmonic is 0.01.

3. See answer section of textbook.

4. The third harmonic has a frequency three times that of the fundamental or $3 \cdot 720$.

5. The harmonics beyond the first one (the fundamental tone) repeat their behavior many times in each cycle of the first harmonic. Hence the entire sound repeats when the first harmonic does. If this one repeats itself, say 100 times a second or has a frequency of 100, then the others certainly repeat themselves each time the first one does; so the entire sound repeats when the first one does.

Chapter 19, page 451

1. The first has a larger amplitude and so is louder. The second one has a larger frequency and so is higher pitched.

2. A simple sound is one in which the motion of a typical air molecule can be represented by $y = D \sin 2\pi ft$.

3. A musical sound repeats its behavior many times a second or is periodic. Noise is aperiodic or irregular. Hence the musical sound can be represented by a sum of sine functions according to Fourier's theorem.

4. The pitch of a complex sound is set by the frequency of the first harmonic. The quality depends on which harmonics are present and their amplitudes.

5. Musical sounds can be represented mathematically but music appeals to the ear, and the physiological processes and individual reactions are not contained in the mathematics.

Chapter 20, page 458

1. See p. 453, line 8.

2. Euclid's parallel axiom is a somewhat complicated statement, and therefore was not as self-evident a truth as axioms were supposed to be.

3. The alternative axioms such as Playfair's made assertions about what happens far out in space beyond man's experience. As mathematicians began to appreciate this fact, they became dissatisfied with such assertions as axioms.

4. Saccheri proposed to take the two alternative possibilities -- no parallel lines to a given line and more than one parallel line to a given line -- and show that when each of these is used as an axiom in conjunction with Euclid's other axioms (of course, aside from Euclid's parallel axiom), then a contradiction can be deduced. Thus the only remaining possible state of affairs, the existence of one and only one line parallel to a given line, would have to be true.

5. No. Saccheri did not arrive at a contradiction when he assumed that there could be more than one parallel line to a given line, but he did deduce such strange theorems that he, nevertheless, decided that such a geometry was impossible. He did arrive at a contradiction by assuming that no parallel lines exist. His conclusion was that Euclid's geometry is the only possible geometry.

6. They realized that an alternative geometry was possible (there was no contradiction in it) and that such an alternative geometry could apply to physical space.

Chapter 20, page 460

1. See p. 458, middle.

2. In Euclid's geometry, parallel lines are lines (in the same plane) which have no point in common. In Gauss's geometry, the term parallel lines is reserved for the two lines (m and n on p. 458) which separate those lines through point P which do meet a given line ℓ from those which do not.

3. See p. 459, the next-to-the-last paragraph.

4. See p. 459, the last paragraph.

5. The triangle with angle sum of 170°.

6. Both triangles have an angle sum of 170°. Their areas must be equal because, if one area were smaller than the other, the triangle of smaller area would have to have an angle sum closer to 180°.

7. The triangles are similar.

Chapter 20, page 462

1. Riemann pointed out that experience tells us only that a line is endless, not that it is infinite.

2. There are no parallel lines.

3. Theorems which can be derived from those axioms which are alike in Gauss's, Euclid's and Riemann's geometries will be the same.

4. Gauss's geometry is more like Euclid's because only one axiom is changed -- the axiom on parallel lines.

5. In Euclid's geometry the sum is 180°; in Gauss's, the sum is always less than 180°; and in Riemann's geometry the sum is always more than 180°.

6. See p. 461, last paragraph.

7. A geometry whose axioms differ from Euclid's and, consequently, whose theorems differ from Euclid's

8. It is obliterated. Similar triangles must be congruent.

Chapter 20, page 464

1. Not if one uses figures of ordinary or small size.

2. Yes, because all agree to within possible measurements

3. Because the sum of 179° 59' 58" was subject to experimental error so that the true sum could be more, equal to, or less than 180°

Chapter 20, page 471

1. The concept of line in Euclidean geometry is undefined and must merely possess the properties stated in the axioms. This is true of the "lines" on a cylindrical surface.

2. Definitions presuppose other concepts in terms of which the definitions are made. Hence one must start with undefined concepts.

3. If we interpret line to be great circle on the sphere.

4. Yes. The geometries created by mathematicians are intended to fit or describe physical space and do so rather well, but no one of them may be the precise description of the geometrical properties of physical space.

5. In some instances, yes. Builders and carpenters often do, but surveyors and astronomers use light rays.

6. The character of the geometry would almost certainly be some kind of non-Euclidean geometry.

7. Same answer as Exercise 6.

Chapter 20, page 474

1. Since several geometries which differ from one another all fit physical space, we can no longer decide which is the true one. We are forced to conclude that mathematics is not necessarily a body of truths.

2. The mathematical space merely has some correspondence with physical space. The former is constructed by human beings; the latter is fixed in the real world and independent of human beings.

3. Yes, in the sense that the applicability of its results must be checked. Yet mathematics differs primarily in the concepts with which it deals and in its insistence on deductive proof.

Chapter 20, page 477

1. That mathematics is not a body of truths about the physical world

2. It caused scientists to face the fact that scientific theories are not truths and that science cannot seek truths because these may be unattainable. Scientists can construct useful theories.

3. In both cases, law and mathematics, fundamental axioms or principles are assumed and consequences are deduced. But the principles of two different legal systems will differ, and so will the consequences. This is the case in geometry also.

4. Yes. If one recognizes that all science can do is provide a theory which agrees with experience, the existence of non-Euclidean geometries gives more latitude in the formation of useful theories.

Chapter 21, page 481

1. No. From the standpoint of what happens physically, 10 + 10 does not give 20.

2. No. We may to the best of our ability to measure decide that the smaller bags each contain 10 lb. But when we put them together, we may be able to measure the error and find that the total weight is more or less than 20 lb.

3. If we assume that the 5-lb weights are very accurate, the scale will still balance. The axiom involved is that equals added to equals give equals.

4. If we assume that the 100-ft/sec velocity and the 32t-ft/sec velocity are precise figures, then the two velocities add.

5. No. The frequencies do not add. The readers who covered Chapter 18 would recognize that the combination of the two sound waves would have a frequency of 50 cycles/sec.

6. The mixture has a temperature of 45°.

Chapter 21, page 484

1. No. We have seen (p. 483, bottom) that in adding batting averages, we cannot replace 2/3 by 4/6 or 8/12.

2. We must show that

$$\frac{a}{b}\left(\frac{c}{d}+\frac{e}{f}\right) = \frac{ac}{bd}+\frac{ae}{bf}$$

or

$$\frac{a}{b}\left(\frac{c+e}{d+f}\right) = \frac{ac+ae}{bd+bf}$$

or

$$\frac{a(c+e)}{b(d+f)} = \frac{ac+ae}{bd+bf}$$

or

$$\frac{ac+ae}{bd+bf} = \frac{ac+ae}{bd+bf} \quad .$$

3. Any fraction whose numerator is -1. Thus 1/2 + -1/2 = 0/4 = 0.

4. Yes.

Chapter 21, page 490

1. As examples of entries in the addition table, we have $5 + 4 \equiv 2$, $6 + 1 \equiv 0$, $6 + 2 \equiv 1$, $6 + 3 \equiv 2$, etc.

2. As examples of entries in the multiplication table we have $5 \cdot 4 \equiv 6$, $5 \cdot 5 \equiv 4$, $6 \cdot 1 \equiv 6$, $6 \cdot 2 \equiv 5$, $6 \cdot 3 \equiv 4$, etc.

3. Yes.

4. To divide 4 by 2 we need some number x such that $4 \equiv 2x$, modulo 6. The possible values of x are 0,1,2,3,4, and 5. For x = 2, we have $4 \equiv 4$, modulo 6 and this is correct. Hence one answer is 2.

To divide 3 by 2 we need an x such that $3 \equiv 2x$, modulo 6. Again the possible values are 0,1,2,3,4, and 5. No one of these numbers serves as an answer.

5. To find an x such that $4 \equiv 2x$, modulo 7, we have x = 2. To find an x such that $3 \equiv 2x$, modulo 7, we have x = 5. If the modulus is a prime number, we can always divide one number by another in that modular arithmetic.

6. 9 or 9 ± any multiple of 12

7. $a \equiv b$, modulo m, means $a = b + pm$ and $c \equiv d$, modulo m, means $c = d + qm$. Then $a - c = b - d + (p - q)m$. Since $(p - q)m$ is a multiple of m, $a - c \equiv b - d$, modulo m.

8. $a = b + pm$; $c = d + qm$. Then $ac = bd + bqm + dpm + pqm^2 = bd + (bq + dp + pqm)m$. We see that $ac - bd$ is a multiple of m. Hence $ac \equiv bd$, modulo m.

9. We cannot always divide one congruence by another.

10. Two numbers each congruent to a third, modulo m, are congruent to each other, modulo m.

11. The sum is 1220. The sum of the digits in this sum is 5. The sum of the digits in 578 is 20 or 2, modulo 9. The sum of the digits in 642 is 12 or 3, modulo 9. We see that the sum of the sums of the digits, that is, 2 + 3, is congruent to the sum of the digits in 1220.

12. The product is 371,076. The sum of the digits in this product is 24 or 6, modulo 9. The sum of the digits in 578 is 2, modulo 9, and the sum of the digits in 642 is 3, modulo 9. The product of these sums, that is, $2 \cdot 3$, equals the sum of the digits in the original product.

13. (a) Yes. $(a + b) - (b + a) = 0$ and 0 is a multiple of m.
(b) Yes. $a + (b + c) \equiv (a + b) + c$, modulo m, because $a + (b + c) = (a + b) + c$.
(c) Yes. $a(b + c) \equiv ab + ac$, modulo m, because $a(b + c) = ab + ac$.

14. (a) 5 (b) 3 (c) 6 (d) 2 (e) 4 (f) 3

15. (a) 1, 4 (b) 0, 2, 4 (c) 1, 3, 5

<u>Chapter 21, page 497</u>

1. Since the union is B, all the objects in A are in B.

2. All the objects in A must also be in B.

3. Because many of the properties of addition, e.g., the commutative and associative properties, hold for union as well as ordinary addition

4. $A \cap A = A$ and $A \cup A = A$.

5. The objects common to A and $B \cup A$ are just those in A. Hence the answer is A.

6. Let A be the set of all professors, B the set of all intelligent persons, and C the set of all students. The premise that all professors are intelligent becomes $A \cap B = A$. The premise that no students are

intelligent becomes $B \cap C = 0$. Then by "multiplying" the two equations (see p. 496, Step 14) we have $(A \cap B) \cap (B \cap C) = A \cap 0$ or $A \cap B \cap C = 0$ or, since $A \cap B = A$, $A \cap C = 0$. This conclusion means that no professors are students.

7. An object is in $A \cap (B \cup C)$ if it is in A and in B or C. Such an object is then in $A \cap B$ or in $A \cap C$. Hence it is in $(A \cap B) \cup (A \cap C)$. Conversely, if an object is in $(A \cap B) \cup (A \cap C)$, it must be in $A \cap B$ or $A \cap C$ or both. If it is in $A \cap B$, it is in A and in B. Hence it will be in $B \cup C$ and in $A \cap (B \cup C)$. Thus, any object in one side of the given equation is in the other and the two sides must be co-extensive.

8. (a) $A \cap B$ can contain no more than is in A. Then $A \cup (A \cap B)$ is A.
   (b) $A \cup B$ surely contains all the objects in A. Then $A \cap (A \cup B)$ is A.
   (c) $A - B$ contains all the objects in A that are not in B. Then $(A - B) \cup B$ contains what is in A but not B and what is in B, the latter also containing the rest of A. Then $(A - B) \cup B = A \cup B$.

9. We have to show that for at least some A's and B's there is no X such that $A = B \cap X$. Suppose B is contained in A. No matter what X is, $B \cap X$ cannot be greater than B. Hence $B \cap X$ cannot equal A because A contains objects not in B.

Chapter 22, page 503

1. See p. 502.

2. 8113; 1200.

3. 6.1; 5; 10. The mean is the best representative because it takes into account that 6 students got a grade of 10. If only 1 has a grade of 10 and 6 a grade of 8, the median would be the same but the mean would change.

4. 31; 35; 40. Here it is more difficult to say which average best represents the data. The mean or median would be a reasonable choice.

5. If the word average means median than the statement is automatically correct by the definition of median and so affirms nothing. If the word average means mean or mode (assuming there is some measure of intelligence), then it need not be true.

6. No, because the depth may be 10 ft in one area and still there may be a mean depth of 4 ft.

Chapter 22, page 505

1. 3.24 2. 3 3. 12.8

4. The variations from the mean height are greater in the city where the standard deviation is 3 in.

5. A grade of 75 where the standard deviation is 5 means that the student did much better than most of the students because most grades were grouped more closely about 65 in this case.

6. The standard deviation is a measure of how closely the data are grouped about the mean or how much they depart from the mean.

Chapter 22, page 510

1. It is a graph or table which shows a set of data (scores, grades, wages) and how many people, for example, are associated with each one of the data.

2. See p. 508. 3. 95.4% 4. 47.7%; 99.8%

5. The graph is a normal frequency curve with the portion to the left of one σ to the left of the mean chopped off.

6. The modal income is 95. The mean would lie to the left because the number of people earning salaries that are smaller than 95 is much larger than the number of people earning larger salaries.

7. There are two different species, each with its own mean and variations from that mean.

Chapter 22, page 515

1 and 6. See answer section of textbook.

7. The formula is a concise representation of the data. Moreover, we can use the formula to calculate values of the dependent variable for given values of the independent variable (these values may not be in the given data). Of course, to do so assumes the formula has a validity beyond that in the given data.

8. He would know a good deal about what types of equations fit curves of given shapes and so be able to select the form of the equation to fit a given curve.

Chapter 22, page 518

1. One would expect high ability in mathematics.

Chapter 22, page 520

1. The deductive approach starts with axioms that seem to apply to the field and then deduces conclusions. The statistical approach starts with data and uses techniques to obtain more or less reliable information from the data.

2. The phenomena of economics are so complex that one cannot find basic axioms.

3. No. The population may be increasing and the percentage of unemployed may actually be decreasing. Or, if women lose jobs, their husbands may be earning more and so the families can be better off.

4. No. The records of what deaths are due to cancer may be more accurate or the identification of cancer as the cause of death may be surer.

5. No. The cause may be some physiological or neurological factor which prompts men possessing it to smoke.

6. No. The intelligence of the children may be due to the greater intelligence of the parents. More intelligent parents may marry late because they enter professions requiring a lot of schooling. Hence it may not be the age of the parents but their intelligence.

7. No. The people who live longer may have better physical constitutions, but because they live so long get to need false teeth.

Chapter 23, page 527

1. One. It means the event is certain to happen.

2. Zero. It means the event is impossible.

3. The probability of a 2 is 2/6. The probability of a 3 is 3/6. The probability of 3 or more is 4/6.

4. No. The two possible outcomes, living or dying, are not equally likely.

5. 6/10

6. 13/52

7. Distinguish the coins as, say, penny, nickel, dime, quarter. There are 16 possible ways in which they can fall. Four of these would be 3 heads and a tail. Hence 4/16

8. 1 - (5/32) or 27/32

9. 3/6

10. It is improbable but not impossible.

11. 6 • 6 or 36 in all

12. 4 and 1; 1 and 4; 2 and 3; and 3 and 2. Hence 4

13. 4/36 in view of Exercises 11 and 12.

14. $1/2^{48}$.

15. On p. 525 we find that the probability of 2 heads and a tail on a single throw of three coins is 3/8. But, as pointed out on p. 525, to throw one coin 3 times is the same as throwing three coins simultaneously.

16. During the five-minute interval from 8 to 8:05, say, there are four minutes for which the first train will be southbound and only one minute for which the first train will be northbound. If the young man enters the station at any time during the five-minute interval he will be 4 times as likely to get there between 8 and 8:04 as between 8:04 and 8:05. Hence the probability of getting a southbound train is 4/5.

17. $\frac{1/2}{1/2}$ or 1.

18. The probability of two tails is 1/4; the probability of at least one head is 3/4. Hence the odds are $\frac{3/4}{1/4} = \frac{3}{1}$.

Chapter 23, page 530

1. Take the number of people alive at age 40 and divide it into the number of people alive at age 60.

2. 85/100; 58/85.

3. The problem is the same as the probability of 3 heads in a throw of three coins. Hence 1/8.

4. The probability that all four will recover without any treatment is the same as the probability of 4 heads in one throw of four coins. This is 1/16. This is not very unlikely. Hence it is not clear that the treatment is really the reason that all four recovered.

Chapter 23, page 532

1. The probability of a height falling within 3σ to the right of the mean is 0.499. The probability of a height falling within 2σ on either side of the mean is 0.954.

2. The probability of a bulb falling beyond 1σ to the left of the mean is 0.159.

3. The probability of falling beyond 2σ to the right of the mean is 0.023.

4. The probability of falling within 1σ to the right of the mean is 0.341.

5. The probability of falling between 1σ and 2σ to the right of the mean is 0.136.

Chapter 23, page 537

1. The eighth row is 1, 7, 21, 35, 35, 21, 7, 1. The sum is 128. The probability of 4 heads and 3 tails is 35/128.

2. From the seventh row of Pascal's triangle we obtain 15/64.

3. The probability of two heads (Exercise 2) is 15/64. Hence (15/64)2000.

4. The mean number of heads is 10,000/2 or 5000. The standard deviation of the frequency distribution of heads is $(1/2)\sqrt{n}$ or 50. To get 5100 heads or more, one must get a number of heads which lies more than 2σ to the right of the mean. This probability is 0.023.

5. The mean number of those who normally survive out of 100 is 50. The standard deviation of the probability curve which shows the probability of 1, 2, ..., 100 surviving is $(1/2)\sqrt{100} = 5$. The chance of 65 or more surviving is the probability of a number more than 3σ to the right of the mean, and this is 0.001. The likelihood of 65 or more surviving without treatment is very small. Hence the likelihood that the treatment is effective is high.

6. Here the mean is 500 and the standard deviation of the probability curve of 1, 2, ..., 1000 people surviving is $(1/2)\sqrt{1000} = 15.5$, approximately. The probability of 650 people surviving is the probability of something which lies about 10σ to the right of the mean. This probability is very small. Since under treatment 650 did survive, the likelihood that the treatment was effective is very great.

7. The mean number of boys, on the hypothesis that boys and girls are equally likely, is 800, and the standard deviation of the probability curve of the various possible number of boys is $(1/2)\sqrt{1600} = 20$. The number 860 lies 3σ to the right of the mean, and the probability of at least 860 boys is 0.001. Hence the hypothesis is to be discredited.

8. If the coins are unbiased, the mean number of heads is 800 and the standard deviation of the number of heads is $(1/2)\sqrt{1600} = 20$. The probability of at least 860 heads is 0.001. Hence the hypothesis that the coins are unbiased is very unlikely. However, one must be more cautious in this case because there is, after all, the possibility of 860 heads in a toss of 1600 coins.

9. The mean number of those who should be alive at the age of 70 if the probability of living is 1/2, is 200. The standard deviation of the probability distribution that 1, 2, 3, ... people will be alive at age 70 is $(1/2)\sqrt{400} = 10$. Since only 150 were alive, the number of

those alive is 5σ below the mean. The probability of an event 5σ or more below the mean is very small. Hence under the 50% death rate, the probability that only 150 should be alive at age 70 is very small. One must conclude that the death rate in that industry is not 1/2 but much greater.

Chapter 23, page 539. Review Exercises

1. (a) 1/19 (b) 1/19 (c) 9/38

2. (a) 4/52 (b) 13/52 (c) 16/52

3. 2/6

4. (a) 3/7 (b) 4/7 (c) 1

5. (a) 3/9 (b) 4/9 (c) 7/9

6. (a) 1/36 (b) 1/36

7. (a) 65/100 (b) 65/70 (c) Yes

8. Since 61 inches is 2σ below the mean, the probability of a height 2σ or more below the mean is 0.023.

Chapter 24, page 546

1. Concepts, axioms, and theorems.

2. Yes. The whole numbers are direct abstractions from experience with individual objects. The irrational numbers were created to provide ideally exact lengths. Such numbers are not arrived at by the experience of measurement.

3. We must start with some premises in order to make deductive proofs.

4. The growth of mathematics since Greek times has been enormous. Moreover, our understanding of what the Greeks produced differs radically from what the Greeks took mathematics to be.

5. Algebra, Euclidean geometry, projective geometry, trigonometry, non-Euclidean geometries, etc.

6. A willingness to tackle problems suggested by phenomena of nature. As more phenomena are explored, more problems are suggested. There must be support for men who spend their time in solving such problems, and communication among mathematicians so that all can take advantage of what has been created.

7. The formulation and solution of new problems whether suggested by scientific work or devised by minds speculating about numbers or geometrical figures

8. Reliability of the conclusions; the discovery by reasoning alone of implications of the axioms; prediction of physical events

Chapter 24, page 550

1. Concepts and rational schemes or theories to represent physical phenomena; comprehension to the extent that such schemes explain; an ordering of physical knowledge; prediction of physical events.

2. Because the concepts embody the essential features and are not encumbered by irrelevant physical properties. Moreover, the same concepts and results apply to many different physical phenomena or situations.

Chapter 24, page 552

1. The study of nature, the aesthetic values, and intellectual activity

2. The arguments for mathematics as an art are given on pp. 550-552. One could attack the thesis on the ground that an art must appeal to the senses or at least through the senses.

3. Pure mathematics is concerned with ideas and problems that are investigated primarily because they are aesthetically or intellectually attractive. Applied mathematics is concerned with ideas and problems directed to the investigation of nature.

4. Yes. However, to produce worthwhile mathematics they must offer either usefulness to science and engineering or aestheticallly valuable material.

INDEX

A CATALOG OF SELECTED

DOVER BOOKS

IN ALL FIELDS OF INTEREST

A CATALOG OF SELECTED DOVER BOOKS IN ALL FIELDS OF INTEREST

DRAWINGS OF REMBRANDT, edited by Seymour Slive. Updated Lippmann, Hofstede de Groot edition, with definitive scholarly apparatus. All portraits, biblical sketches, landscapes, nudes. Oriental figures, classical studies, together with selection of work by followers. 550 illustrations. Total of 630pp. 9⅝ × 12¼.
21485-0, 21486-9 Pa., Two-vol. set $25.00

GHOST AND HORROR STORIES OF AMBROSE BIERCE, Ambrose Bierce. 24 tales vividly imagined, strangely prophetic, and decades ahead of their time in technical skill: "The Damned Thing," "An Inhabitant of Carcosa," "The Eyes of the Panther," "Moxon's Master," and 20 more. 199pp. 5⅜ × 8½. 20767-6 Pa. $3.95

ETHICAL WRITINGS OF MAIMONIDES, Maimonides. Most significant ethical works of great medieval sage, newly translated for utmost precision, readability. Laws Concerning Character Traits, Eight Chapters, more. 192pp. 5⅜ × 8½.
24522-5 Pa. $4.50

THE EXPLORATION OF THE COLORADO RIVER AND ITS CANYONS, J. W. Powell. Full text of Powell's 1,000-mile expedition down the fabled Colorado in 1869. Superb account of terrain, geology, vegetation, Indians, famine, mutiny, treacherous rapids, mighty canyons, during exploration of last unknown part of continental U.S. 400pp. 5⅜ × 8½. 20094-9 Pa. $6.95

HISTORY OF PHILOSOPHY, Julián Marías. Clearest one-volume history on the market. Every major philosopher and dozens of others, to Existentialism and later. 505pp. 5⅜ × 8½. 21739-6 Pa. $8.50

ALL ABOUT LIGHTNING, Martin A. Uman. Highly readable non-technical survey of nature and causes of lightning, thunderstorms, ball lightning, St. Elmo's Fire, much more. Illustrated. 192pp. 5⅜ × 8½. 25237-X Pa. $5.95

SAILING ALONE AROUND THE WORLD, Captain Joshua Slocum. First man to sail around the world, alone, in small boat. One of great feats of seamanship told in delightful manner. 67 illustrations. 294pp. 5⅜ × 8½. 20326-3 Pa. $4.95

LETTERS AND NOTES ON THE MANNERS, CUSTOMS AND CONDITIONS OF THE NORTH AMERICAN INDIANS, George Catlin. Classic account of life among Plains Indians: ceremonies, hunt, warfare, etc. 312 plates. 572pp. of text. 6⅛ × 9¼. 22118-0, 22119-9 Pa. Two-vol. set $15.90

ALASKA: The Harriman Expedition, 1899, John Burroughs, John Muir, et al. Informative, engrossing accounts of two-month, 9,000-mile expedition. Native peoples, wildlife, forests, geography, salmon industry, glaciers, more. Profusely illustrated. 240 black-and-white line drawings. 124 black-and-white photographs. 3 maps. Index. 576pp. 5⅜ × 8½. 25109-8 Pa. $11.95

SUNDIALS, Albert Waugh. Far and away the best, most thorough coverage of ideas, mathematics concerned, types, construction, adjusting anywhere. Over 100 illustrations. 230pp. 5⅜ × 8½. 22947-5 Pa. $4.50

PICTURE HISTORY OF THE NORMANDIE: With 190 Illustrations, Frank O. Braynard. Full story of legendary French ocean liner: Art Deco interiors, design innovations, furnishings, celebrities, maiden voyage, tragic fire, much more. Extensive text. 144pp. 8⅜ × 11¾. 25257-4 Pa. $9.95

THE FIRST AMERICAN COOKBOOK: A Facsimile of "American Cookery," 1796, Amelia Simmons. Facsimile of the first American-written cookbook published in the United States contains authentic recipes for colonial favorites—pumpkin pudding, winter squash pudding, spruce beer, Indian slapjacks, and more. Introductory Essay and Glossary of colonial cooking terms. 80pp. 5⅜ × 8½. 24710-4 Pa. $3.50

101 PUZZLES IN THOUGHT AND LOGIC, C. R. Wylie, Jr. Solve murders and robberies, find out which fishermen are liars, how a blind man could possibly identify a color—purely by your own reasoning! 107pp. 5⅜ × 8½. 20367-0 Pa. $2.50

THE BOOK OF WORLD-FAMOUS MUSIC—CLASSICAL, POPULAR AND FOLK, James J. Fuld. Revised and enlarged republication of landmark work in musico-bibliography. Full information about nearly 1,000 songs and compositions including first lines of music and lyrics. New supplement. Index. 800pp. 5⅜ × 8¼. 24857-7 Pa. $14.95

ANTHROPOLOGY AND MODERN LIFE, Franz Boas. Great anthropologist's classic treatise on race and culture. Introduction by Ruth Bunzel. Only inexpensive paperback edition. 255pp. 5⅜ × 8½. 25245-0 Pa. $5.95

THE TALE OF PETER RABBIT, Beatrix Potter. The inimitable Peter's terrifying adventure in Mr. McGregor's garden, with all 27 wonderful, full-color Potter illustrations. 55pp. 4¼ × 5½. (Available in U.S. only) 22827-4 Pa. $1.75

THREE PROPHETIC SCIENCE FICTION NOVELS, H. G. Wells. *When the Sleeper Wakes, A Story of the Days to Come* and *The Time Machine* (full version). 335pp. 5⅜ × 8½. (Available in U.S. only) 20605-X Pa. $5.95

APICIUS COOKERY AND DINING IN IMPERIAL ROME, edited and translated by Joseph Dommers Vehling. Oldest known cookbook in existence offers readers a clear picture of what foods Romans ate, how they prepared them, etc. 49 illustrations. 301pp. 6⅛ × 9¼. 23563-7 Pa. $6.50

SHAKESPEARE LEXICON AND QUOTATION DICTIONARY, Alexander Schmidt. Full definitions, locations, shades of meaning of every word in plays and poems. More than 50,000 exact quotations. 1,485pp. 6½ × 9¼. 22726-X, 22727-8 Pa., Two-vol. set $27.90

THE WORLD'S GREAT SPEECHES, edited by Lewis Copeland and Lawrence W. Lamm. Vast collection of 278 speeches from Greeks to 1970. Powerful and effective models; unique look at history. 842pp. 5⅜ × 8½. 20468-5 Pa. $11.95

THE BLUE FAIRY BOOK, Andrew Lang. The first, most famous collection, with many familiar tales: Little Red Riding Hood, Aladdin and the Wonderful Lamp, Puss in Boots, Sleeping Beauty, Hansel and Gretel, Rumpelstiltskin; 37 in all. 138 illustrations. 390pp. 5⅜ × 8½. 21437-0 Pa. $5.95

THE STORY OF THE CHAMPIONS OF THE ROUND TABLE, Howard Pyle. Sir Launcelot, Sir Tristram and Sir Percival in spirited adventures of love and triumph retold in Pyle's inimitable style. 50 drawings, 31 full-page. xviii + 329pp. 6½ × 9¼. 21883-X Pa. $6.95

AUDUBON AND HIS JOURNALS, Maria Audubon. Unmatched two-volume portrait of the great artist, naturalist and author contains his journals, an excellent biography by his granddaughter, expert annotations by the noted ornithologist, Dr. Elliott Coues, and 37 superb illustrations. Total of 1,200pp. 5⅜ × 8.
Vol. I 25143-8 Pa. $8.95
Vol. II 25144-6 Pa. $8.95

GREAT DINOSAUR HUNTERS AND THEIR DISCOVERIES, Edwin H. Colbert. Fascinating, lavishly illustrated chronicle of dinosaur research, 1820's to 1960. Achievements of Cope, Marsh, Brown, Buckland, Mantell, Huxley, many others. 384pp. 5¼ × 8¼. 24701-5 Pa. $6.95

THE TASTEMAKERS, Russell Lynes. Informal, illustrated social history of American taste 1850's–1950's. First popularized categories Highbrow, Lowbrow, Middlebrow. 129 illustrations. New (1979) afterword. 384pp. 6 × 9.
23993-4 Pa. $6.95

DOUBLE CROSS PURPOSES, Ronald A. Knox. A treasure hunt in the Scottish Highlands, an old map, unidentified corpse, surprise discoveries keep reader guessing in this cleverly intricate tale of financial skullduggery. 2 black-and-white maps. 320pp. 5⅜ × 8½. (Available in U.S. only) 25032-6 Pa. $5.95

AUTHENTIC VICTORIAN DECORATION AND ORNAMENTATION IN FULL COLOR: 46 Plates from "Studies in Design," Christopher Dresser. Superb full-color lithographs reproduced from rare original portfolio of a major Victorian designer. 48pp. 9¼ × 12¼. 25083-0 Pa. $7.95

PRIMITIVE ART, Franz Boas. Remains the best text ever prepared on subject, thoroughly discussing Indian, African, Asian, Australian, and, especially, Northern American primitive art. Over 950 illustrations show ceramics, masks, totem poles, weapons, textiles, paintings, much more. 376pp. 5⅜ × 8. 20025-6 Pa. $6.95

SIDELIGHTS ON RELATIVITY, Albert Einstein. Unabridged republication of two lectures delivered by the great physicist in 1920–21. *Ether and Relativity* and *Geometry and Experience*. Elegant ideas in non-mathematical form, accessible to intelligent layman. vi + 56pp. 5⅜ × 8½. 24511-X Pa. $2.95

THE WIT AND HUMOR OF OSCAR WILDE, edited by Alvin Redman. More than 1,000 ripostes, paradoxes, wisecracks: Work is the curse of the drinking classes, I can resist everything except temptation, etc. 258pp. 5⅜ × 8½. 20602-5 Pa. $4.50

ADVENTURES WITH A MICROSCOPE, Richard Headstrom. 59 adventures with clothing fibers, protozoa, ferns and lichens, roots and leaves, much more. 142 illustrations. 232pp. 5⅜ × 8½. 23471-1 Pa. $3.95

PLANTS OF THE BIBLE, Harold N. Moldenke and Alma L. Moldenke. Standard reference to all 230 plants mentioned in Scriptures. Latin name, biblical reference, uses, modern identity, much more. Unsurpassed encyclopedic resource for scholars, botanists, nature lovers, students of Bible. Bibliography. Indexes. 123 black-and-white illustrations. 384pp. 6 × 9. 25069-5 Pa. $8.95

FAMOUS AMERICAN WOMEN: A Biographical Dictionary from Colonial Times to the Present, Robert McHenry, ed. From Pocahontas to Rosa Parks, 1,035 distinguished American women documented in separate biographical entries. Accurate, up-to-date data, numerous categories, spans 400 years. Indices. 493pp. 6½ × 9¼. 24523-3 Pa. $9.95

THE FABULOUS INTERIORS OF THE GREAT OCEAN LINERS IN HISTORIC PHOTOGRAPHS, William H. Miller, Jr. Some 200 superb photographs capture exquisite interiors of world's great "floating palaces"—1890's to 1980's: *Titanic, Ile de France, Queen Elizabeth, United States, Europa,* more. Approx. 200 black-and-white photographs. Captions. Text. Introduction. 160pp. 8⅞ × 11¾. 24756-2 Pa. $9.95

THE GREAT LUXURY LINERS, 1927–1954: A Photographic Record, William H. Miller, Jr. Nostalgic tribute to heyday of ocean liners. 186 photos of Ile de France, Normandie, Leviathan, Queen Elizabeth, United States, many others. Interior and exterior views. Introduction. Captions. 160pp. 9 × 12. 24056-8 Pa. $9.95

A NATURAL HISTORY OF THE DUCKS, John Charles Phillips. Great landmark of ornithology offers complete detailed coverage of nearly 200 species and subspecies of ducks: gadwall, sheldrake, merganser, pintail, many more. 74 full-color plates, 102 black-and-white. Bibliography. Total of 1,920pp. 8⅝ × 11¼. 25141-1, 25142-X Cloth. Two-vol. set $100.00

THE SEAWEED HANDBOOK: An Illustrated Guide to Seaweeds from North Carolina to Canada, Thomas F. Lee. Concise reference covers 78 species. Scientific and common names, habitat, distribution, more. Finding keys for easy identification. 224pp. 5⅜ × 8½. 25215-9 Pa. $5.95

THE TEN BOOKS OF ARCHITECTURE: The 1755 Leoni Edition, Leon Battista Alberti. Rare classic helped introduce the glories of ancient architecture to the Renaissance. 68 black-and-white plates. 336pp. 8⅜ × 11¼. 25239-6 Pa. $14.95

MISS MACKENZIE, Anthony Trollope. Minor masterpieces by Victorian master unmasks many truths about life in 19th-century England. First inexpensive edition in years. 392pp. 5⅜ × 8½. 25201-9 Pa. $7.95

THE RIME OF THE ANCIENT MARINER, Gustave Doré, Samuel Taylor Coleridge. Dramatic engravings considered by many to be his greatest work. The terrifying space of the open sea, the storms and whirlpools of an unknown ocean, the ice of Antarctica, more—all rendered in a powerful, chilling manner. Full text. 38 plates. 77pp. 9¼ × 12. 22305-1 Pa. $4.95

THE EXPEDITIONS OF ZEBULON MONTGOMERY PIKE, Zebulon Montgomery Pike. Fascinating first-hand accounts (1805–6) of exploration of Mississippi River, Indian wars, capture by Spanish dragoons, much more. 1,088pp. 5⅜ × 8½. 25254-X, 25255-8 Pa. Two-vol. set $23.90

A CONCISE HISTORY OF PHOTOGRAPHY: Third Revised Edition, Helmut Gernsheim. Best one-volume history—camera obscura, photochemistry, daguerreotypes, evolution of cameras, film, more. Also artistic aspects—landscape, portraits, fine art, etc. 281 black-and-white photographs. 26 in color. 176pp. 8⅜ × 11¼. 25128-4 Pa. $12.95

THE DORÉ BIBLE ILLUSTRATIONS, Gustave Doré. 241 detailed plates from the Bible: the Creation scenes, Adam and Eve, Flood, Babylon, battle sequences, life of Jesus, etc. Each plate is accompanied by the verses from the King James version of the Bible. 241pp. 9 × 12. 23004-X Pa. $8.95

HUGGER-MUGGER IN THE LOUVRE, Elliot Paul. Second Homer Evans mystery-comedy. Theft at the Louvre involves sleuth in hilarious, madcap caper. "A knockout."—Books. 336pp. 5⅜ × 8½. 25185-3 Pa. $5.95

FLATLAND, E. A. Abbott. Intriguing and enormously popular science-fiction classic explores the complexities of trying to survive as a two-dimensional being in a three-dimensional world. Amusingly illustrated by the author. 16 illustrations. 103pp. 5⅜ × 8½. 20001-9 Pa. $2.25

THE HISTORY OF THE LEWIS AND CLARK EXPEDITION, Meriwether Lewis and William Clark, edited by Elliott Coues. Classic edition of Lewis and Clark's day-by-day journals that later became the basis for U.S. claims to Oregon and the West. Accurate and invaluable geographical, botanical, biological, meteorological and anthropological material. Total of 1,508pp. 5⅜ × 8½.
21268-8, 21269-6, 21270-X Pa. Three-vol. set $25.50

LANGUAGE, TRUTH AND LOGIC, Alfred J. Ayer. Famous, clear introduction to Vienna, Cambridge schools of Logical Positivism. Role of philosophy, elimination of metaphysics, nature of analysis, etc. 160pp. 5⅜ × 8½. (Available in U.S. and Canada only) 20010-8 Pa. $2.95

MATHEMATICS FOR THE NONMATHEMATICIAN, Morris Kline. Detailed, college-level treatment of mathematics in cultural and historical context, with numerous exercises. For liberal arts students. Preface. Recommended Reading Lists. Tables. Index. Numerous black-and-white figures. xvi + 641pp. 5⅜ × 8½.
24823-2 Pa. $11.95

28 SCIENCE FICTION STORIES, H. G. Wells. Novels, *Star Begotten* and *Men Like Gods,* plus 26 short stories: "Empire of the Ants," "A Story of the Stone Age," "The Stolen Bacillus," "In the Abyss," etc. 915pp. 5⅜ × 8½. (Available in U.S. only)
20265-8 Cloth. $10.95

HANDBOOK OF PICTORIAL SYMBOLS, Rudolph Modley. 3,250 signs and symbols, many systems in full; official or heavy commercial use. Arranged by subject. Most in Pictorial Archive series. 143pp. 8⅛ × 11. 23357-X Pa. $5.95

INCIDENTS OF TRAVEL IN YUCATAN, John L. Stephens. Classic (1843) exploration of jungles of Yucatan, looking for evidences of Maya civilization. Travel adventures, Mexican and Indian culture, etc. Total of 669pp. 5⅜ × 8½.
20926-1, 20927-X Pa., Two-vol. set $9.90

DEGAS: An Intimate Portrait, Ambroise Vollard. Charming, anecdotal memoir by famous art dealer of one of the greatest 19th-century French painters. 14 black-and-white illustrations. Introduction by Harold L. Van Doren. 96pp. 5⅜ × 8½.
25131-4 Pa. $3.95

PERSONAL NARRATIVE OF A PILGRIMAGE TO ALMANDINAH AND MECCAH, Richard Burton. Great travel classic by remarkably colorful personality. Burton, disguised as a Moroccan, visited sacred shrines of Islam, narrowly escaping death. 47 illustrations. 959pp. 5⅜ × 8½. 21217-3, 21218-1 Pa., Two-vol. set $17.90

PHRASE AND WORD ORIGINS, A. H. Holt. Entertaining, reliable, modern study of more than 1,200 colorful words, phrases, origins and histories. Much unexpected information. 254pp. 5⅜ × 8½. 20758-7 Pa. $5.95

THE RED THUMB MARK, R. Austin Freeman. In this first Dr. Thorndyke case, the great scientific detective draws fascinating conclusions from the nature of a single fingerprint. Exciting story, authentic science. 320pp. 5⅜ × 8½. (Available in U.S. only) 25210-8 Pa. $5.95

AN EGYPTIAN HIEROGLYPHIC DICTIONARY, E. A. Wallis Budge. Monumental work containing about 25,000 words or terms that occur in texts ranging from 3000 B.C. to 600 A.D. Each entry consists of a transliteration of the word, the word in hieroglyphs, and the meaning in English. 1,314pp. 6⅞ × 10.
23615-3, 23616-1 Pa., Two-vol. set $27.90

THE COMPLEAT STRATEGYST: Being a Primer on the Theory of Games of Strategy, J. D. Williams. Highly entertaining classic describes, with many illustrated examples, how to select best strategies in conflict situations. Prefaces. Appendices. xvi + 268pp. 5⅜ × 8½. 25101-2 Pa. $5.95

THE ROAD TO OZ, L. Frank Baum. Dorothy meets the Shaggy Man, little Button-Bright and the Rainbow's beautiful daughter in this delightful trip to the magical Land of Oz. 272pp. 5⅜ × 8. 25208-6 Pa. $4.95

POINT AND LINE TO PLANE, Wassily Kandinsky. Seminal exposition of role of point, line, other elements in non-objective painting. Essential to understanding 20th-century art. 127 illustrations. 192pp. 6½ × 9¼. 23808-3 Pa. $4.50

LADY ANNA, Anthony Trollope. Moving chronicle of Countess Lovel's bitter struggle to win for herself and daughter Anna their rightful rank and fortune—perhaps at cost of sanity itself. 384pp. 5⅜ × 8½. 24669-8 Pa. $6.95

EGYPTIAN MAGIC, E. A. Wallis Budge. Sums up all that is known about magic in Ancient Egypt: the role of magic in controlling the gods, powerful amulets that warded off evil spirits, scarabs of immortality, use of wax images, formulas and spells, the secret name, much more. 253pp. 5⅜ × 8½. 22681-6 Pa. $4.50

THE DANCE OF SIVA, Ananda Coomaraswamy. Preeminent authority unfolds the vast metaphysic of India: the revelation of her art, conception of the universe, social organization, etc. 27 reproductions of art masterpieces. 192pp. 5⅜ × 8½.
24817-8 Pa. $5.95

CHRISTMAS CUSTOMS AND TRADITIONS, Clement A. Miles. Origin, evolution, significance of religious, secular practices. Caroling, gifts, yule logs, much more. Full, scholarly yet fascinating; non-sectarian. 400pp. 5⅜ × 8½.
23354-5 Pa. $6.50

THE HUMAN FIGURE IN MOTION, Eadweard Muybridge. More than 4,500 stopped-action photos, in action series, showing undraped men, women, children jumping, lying down, throwing, sitting, wrestling, carrying, etc. 390pp. 7⅞ × 10⅝.
20204-6 Cloth. $19.95

THE MAN WHO WAS THURSDAY, Gilbert Keith Chesterton. Witty, fast-paced novel about a club of anarchists in turn-of-the-century London. Brilliant social, religious, philosophical speculations. 128pp. 5⅜ × 8½. 25121-7 Pa. $3.95

A CEZANNE SKETCHBOOK: Figures, Portraits, Landscapes and Still Lifes, Paul Cezanne. Great artist experiments with tonal effects, light, mass, other qualities in over 100 drawings. A revealing view of developing master painter, precursor of Cubism. 102 black-and-white illustrations. 144pp. 8¾ × 6⅜. 24790-2 Pa. $5.95

AN ENCYCLOPEDIA OF BATTLES: Accounts of Over 1,560 Battles from 1479 B.C. to the Present, David Eggenberger. Presents essential details of every major battle in recorded history, from the first battle of Megiddo in 1479 B.C. to Grenada in 1984. List of Battle Maps. New Appendix covering the years 1967–1984. Index. 99 illustrations. 544pp. 6½ × 9¼. 24913-1 Pa. $14.95

AN ETYMOLOGICAL DICTIONARY OF MODERN ENGLISH, Ernest Weekley. Richest, fullest work, by foremost British lexicographer. Detailed word histories. Inexhaustible. Total of 856pp. 6½ × 9¼.
21873-2, 21874-0 Pa., Two-vol. set $17.00

WEBSTER'S AMERICAN MILITARY BIOGRAPHIES, edited by Robert McHenry. Over 1,000 figures who shaped 3 centuries of American military history. Detailed biographies of Nathan Hale, Douglas MacArthur, Mary Hallaren, others. Chronologies of engagements, more. Introduction. Addenda. 1,033 entries in alphabetical order. xi + 548pp. 6½ × 9¼. (Available in U.S. only)
24758-9 Pa. $11.95

LIFE IN ANCIENT EGYPT, Adolf Erman. Detailed older account, with much not in more recent books: domestic life, religion, magic, medicine, commerce, and whatever else needed for complete picture. Many illustrations. 597pp. 5⅜ × 8½.
22632-8 Pa. $8.95

HISTORIC COSTUME IN PICTURES, Braun & Schneider. Over 1,450 costumed figures shown, covering a wide variety of peoples: kings, emperors, nobles, priests, servants, soldiers, scholars, townsfolk, peasants, merchants, courtiers, cavaliers, and more. 256pp. 8⅜ × 11¼. 23150-X Pa. $7.95

THE NOTEBOOKS OF LEONARDO DA VINCI, edited by J. P. Richter. Extracts from manuscripts reveal great genius; on painting, sculpture, anatomy, sciences, geography, etc. Both Italian and English. 186 ms. pages reproduced, plus 500 additional drawings, including studies for *Last Supper, Sforza* monument, etc. 860pp. 7⅞ × 10¾. (Available in U.S. only) 22572-0, 22573-9 Pa., Two-vol. set $25.90